HYDROINFORMATICS

DATA INTEGRATIVE APPROACHES IN COMPUTATION, ANALYSIS, AND MODELING

HYDROINFORMATICS

DATA INTEGRATIVE APPROACHES IN COMPUTATION, ANALYSIS, AND MODELING

Praveen Kumar
University of Illinois
Urbana, Illinois

Jay Alameda
National Center for SuperComputing Applications (NCSA)
University of Illinois
Urbana, Illinois

Peter Bajcsy
National Center for SuperComputing Applications (NCSA)
University of Illinois
Urbana, Illinois

Mike Folk
National Center for SuperComputing Applications (NCSA)
University of Illinois
Urbana, Illinois

Momcilo Markus
Illinois State Water Survey
Champaign, Illinois

CRC Press
Taylor & Francis Group
Boca Raton London New York

CRC Press is an imprint of the
Taylor & Francis Group, an **informa** business
A TAYLOR & FRANCIS BOOK

First published 2006 by CRC Press

Published 2019 by CRC Press
Taylor & Francis Group
6000 Broken Sound Parkway NW, Suite 300
Boca Raton, FL 33487-2742

First issued in paperback 2019

ISBN-13: 978-0-367-45397-8 (pbk)
ISBN-13: 978-0-8493-2894-7 (hbk)

Visit the Taylor & Francis Web site at
http://www.taylorandfrancis.com

and the CRC Press Web site at
http://www.crcpress.com

Library of Congress Card Number 2005050561

Library of Congress Cataloging-in-Publication Data

Hydroinformatics : data integrative approaches in computation, analysis, and modeling / Praveen Kumar ... [et al.].
p. cm.
Includes bibliographical references and index.
ISBN 0-8493-2894-2 (alk. paper)
1. Hydrology--Data processing. I. Kumar, Praveen.

GB656.2.E43H92 2005
551.48'0285--dc22 2005050561

Dedication

To Charu and Ilina
Praveen

To Susie, Emily and Josie
Jay

To Zuzana
Peter

To Karen
Michael

To Aleksandra, Simon, and Danilo
Momcilo

Authors and Contributors

Praveen Kumar
Department of Civil and Environmental Engineering
University of Illinois
Urbana, Illinois

Jay C. Alameda
National Center for Supercomputing Applications
University of Illinois
Urbana, Illinois

Peter Bajcsy
National Center for Supercomputing Applications
University of Illinois
Urbana, Illinois

Mike Folk
National Center for Supercomputing Applications
University of Illinois
Urbana, Illinois

Momcilo Markus
Center for Watershed Science
Illinois State Water Survey
Champaign, Illinois

John J. Helly
San Diego Supercomputer Center
University of California, San Diego
San Diego, California

Seongeun Jeong
University of California, Berkeley
Berkeley, California

Xu Liang
Department of Civil and Environmental Engineering
University of California, Berkeley
Berkeley, California

Yao Liang
Bradley Department of Electrical and Computer Engineering
Virginia Polytechnic Institute and State University
Blacksburg, Virginia

Barbara Minsker
Department of Civil and Environmental Engineering
University of Illinois
Urbana, Illinois

Michael Piasecki
Department of Civil, Architectural, and Environmental Engineering
Drexel University
Philadelphia, Pennsylvania

Benjamin L. Ruddell
University of Illinois
Urbana, Illinois

Dragan Savic
Centre for Water Systems
University of Exeter
Exeter, Devon
United Kingdom

Lydia Vamvakeridou-Lyroudia
National Technical University
Athens, Greece

Amanda White
University of Illinois
Urbana, Illinois

Authors and Contributors

Praveen Kumar, Ph.D. has been a faculty member in the Department of Civil and Environmental Engineering at the University of Illinois since 1995. Prior to joining the University of Illinois, he was a research scientist at the Universities Space Research Association (USRA) and Hydrologic Sciences Branch, NASA — Goddard Space Flight Center, Greenbelt, Massachusetts. His expertise is in large scale hydrologic processes with an emphasis on hydroclimatology, hydroecology, estimation and data assimilation, geomorphology, and hydroinformatics. He obtained his Bachelor of Technology from the Indian Institute of Technology, Bombay (1987), Master of Science from the Iowa State University (1989), and Ph.D. from the University of Minnesota (1993), all in civil engineering.

Jay C. Alameda has been with the National Center for Supercomputing Applications (NCSA) at the University of Illinois at Urbana-Champaign since 1990. Currently, Alameda is the lead for the NCSA effort in middleware, which encompasses grid computing environments for applications as well as grid development. In this role, he has worked to develop cyberinfrastructure in service of advanced environments for atmospheric discovery and advanced multiscale chemical engineering, in the form of configurable, reuseable workflow engines and client-side tools and supporting services. In developing these advanced environments, Alameda has worked with partners in the Alliance Expeditions (MEAD and Science Portals), and is now working with the Linked Environments for Atmospheric Discovery (LEAD) and Open Grid Computing Environments Consortium (OGCE) efforts. Through the course of developing application grid computing infrastructure, Alameda also works to ensure that NCSA and TeraGrid production resources are capable of supporting these new application use modalities. Alameda continues to cultivate collaborations with new discipline areas, such as the Consortium of Universities for the Advancement of Hydrologic Science, Inc. (CUAHSI), in order to help develop advanced capabilities to meet their scientific needs, and better inform the development of cyberinfrastructure, and is working to connect developments within NCSA through service architectures. He earned his Master of Science in nuclear engineering from the University of Illinois at Urbana-Champaign (1991), and Bachelor of Science in chemical engineering from the University of Notre Dame (1986).

Peter Bajcsy, Ph.D. has been with the Automated Learning Group at the National Center for Supercomputing Applications at the University of Illinois at Urbana-Champaign (UIUC), since 2001 working as a research scientist on problems related to automatic transfer of image content to knowledge. He has been an adjunct assistant professor of the Department of Electrical and Computer Engineering and the Department of Computer Science at UIUC since 2002 teaching and advising graduate students. Before joining NCSA, he had worked on real-time machine vision problems for the semiconductor industry and synthetic aperture radar (SAR) technology for the government contracting industry. He has developed several software systems for automatic feature extraction, feature selection, segmentation, classification, tracking, and statistical modeling from electro-optical, SAR, laser and hyperspectral datasets. Dr. Bajcsy's scientific interests include image and signal processing, statistical data analysis, data mining, pattern recognition, novel sensor technology, and computer and machine vision. He earned his Ph.D. from the Electrical and Computer Engineering Department, University of Illinois at Urbana-Champaign, 1997, M.S. degree from the Electrical Engineering Department, University of Pennsylvania, Philadelphia, 1994, and Diploma Engineer degree from the Electrical Engineering Department, Slovak Technical University, Bratislava, Slovakia, 1987.

Mike Folk, Ph.D. has been with the Hierarchical Data Format (HDF) Group at the National Center for Supercomputing Applications at the University of Illinois at Urbana-Champaign since 1988, leading the HDF group during that period. Dr. Folk's professional interests are primarily in the area of scientific data management. Through his work with HDF, Dr. Folk is heavily involved with data management issues in NASA and the earth science community. He has also helped to lead the effort to provide a standard format to address data management needs for the Department of Energy's ASCI project, which involves data I/O, storage, and sharing among tera-scale computing platforms. Before joining NCSA, Dr. Folk taught computer science at the university level for 18 years. Among Dr. Folk's publications is the book *File Structures, A Conceptual Toolkit*. Dr. Folk earned his Ph.D. in computer science from Syracuse University in 1974, an M.A.T. in mathematics teaching from the University of Chicago in 1966, and B.S. in mathematics from the University of North Carolina, Chapel Hill, in 1964.

Momcilo Markus, Ph.D. has extensive research, industry, and teaching experience dealing with water issues in Europe and in the United States. He is currently a research hydrologist at the Illinois State Water Survey evaluating hydrologic changes in watersheds, nutrient load estimation methods, and nutrient-based watershed classification. Formerly Dr. Markus worked with the National Weather Service's river forecasting system, FEMA's flood insurance mapping program, and various other activities. His specialties include statistical/stochastic hydrology, hydrologic modeling, water resources management, data mining, pattern recognition, and water quality statistics. He earned his B.S. and M.S. at the University of Belgrade, and Ph.D. from the Colorado State University.

John J. Helly, Ph.D. is a scientist at the San Diego Supercomputer Center at the University of California, San Diego where he is the Director of the Laboratory for Earth and Environmental Science. Dr. Helly has a Ph.D. in computer science and M.S. in biostatistics from University of California, Los Angeles (UCLA), and an M.A. and B.A. from Occidental College in biology. He has research interests in many elements related to the Earth including hydrology, oceanography, atmosphere, and climate as well as computation, data analysis, modeling, and visualization.

Seongeun Jeong is a Ph.D. student under the supervision of Dr. Xu Liang at the University of California, Berkeley. His research interest is hydroinformatics.

Xu Liang, Ph.D. is an assistant professor in the Department of Civil and Environmental Engineering at the University of California, Berkeley. Her research interests include hydroinformatics; investigations of subgrid scale representations or parameterizations of physical and hydrological processes for land surface modeling as applied to weather, climate, and environmental studies; surface water and groundwater interactions and their impacts on land–atmosphere systems, and on environmental and ecological systems; applications of remote sensing in data assimilations; and hydrologic forecasts for ungauged watersheds.

Yao Liang, Ph.D. is an assistant professor in the Bradley Department of Electrical and Computer Engineering, Virginia Polytechnic Institute and State University. Prior to joining

Virginia Tech, he was a technical staff member in Alcatel, Raleigh, North Carolina. His research interests include network quality of service and adaptive resource allocation, data management and data mining, sensor networks, machine learning and applications, and distributed information systems.

Barbara Minsker, Ph.D. is an associate professor in the Department of Civil and Environmental Engineering at the University of Illinois, Urbana-Champaign. Dr. Minsker also has a joint appointment with National Center for Supercomputing Applications. She has research interests in investigating improved methods for modeling complex environmental systems so that informed management-level decisions can be made under conditions of uncertainty. Dr. Minsker is using machine learning approaches such as genetic algorithms, decision trees, support vector machines, and artificial neural networks, innovative and cost-effective solutions to investigate complex environmental problems.

Michael Piasecki, Ph.D. is an associate professor in the Department of Civil, Architectural, and Environmental Engineering at Drexel University. He obtained his Ph.D. in civil engineering in 1994 from the University of Michigan. His research interest includes the development of information systems in hydraulic and hydrologic engineering and environmental decision support systems using web-based technology. Particular focus is on development of metadata profiles in the GeoSciences and their implementation using domain ontologies using XML, RDF, and OWL.

Benjamin L. Ruddell is currently affiliated with the University of Illinois at Urbana-Champaign as an academic professional. He completed his Master's degree in civil and environmental engineering with Dr. Praveen Kumar at UIUC and his Bachelor's degree in engineering at Calvin College in Grand Rapids, Michigan. His research interests include hydrology, ecology, natural resources, spirituality in environmental issues, and finding better ways to synthesize and analyze geospatial data.

Dragan Savic is professor of hydroinformatics at the University of Exeter and a chartered (professional) engineer with more than 20 years of research, teaching, and consulting experience in various water-engineering disciplines. His interests include hydroinformatics, computer modeling, artificial intelligence techniques, optimization techniques, and risk/uncertainty modeling, with particular application to the operation and design of water supply, distribution, and wastewater systems. He jointly heads the University of Exeter's Centre for Water Systems with over twenty research staff.

Lydia Vamvakeridou-Lyroudia is a lecturer at the National Technical University of Athens, Greece, currently on leave at the University of Exeter and a professional engineer, with more than 20 years of experience. Her interests include computer modeling and artificial intelligence techniques, especially fuzzy logic and genetic algorithms, with particular application to water supply/distribution and irrigation systems for design optimization, expert systems, and real time models. She has published several papers and reports, while her computer models have been professionally applied to more than 100 irrigation and water supply networks (case studies) in Greece.

Amanda White is a graduate student at the University of Illinois at Urbana-Champaign, where she obtained her Master's degree in geomorphology and is currently working on her Ph.D. in applying statistical data mining techniques to hydrologic and ecologic remote-sensing data in order to extract scientific information. Her research interests include: inter-annual climate and vegetation variability, global climate change, water, carbon, energy cycling, geomorphology, and information science.

Preface

The concept of "hydroinformatics" was pioneered by Michael B. Abbott in 1989, as the application of advanced computation and communication technologies to problems of the aquatic environment. A biannual conference on this topic was initiated in 1994 and the *Journal of Hydroinformatics* was launched in 1999. This demonstrates the tremendous progress and expansion in the field. One of the rapidly expanding areas is the use of data driven techniques for understanding the water cycle, and the increasing use of this knowledge in related fields of science and engineering. Impetus for this interest is fueled by the availability of increasing volume of data related to the water cycle, from a variety of global, regional, and local data gathering activities. These include observations from satellites, state as well as national level hydrological and meteorological networks, experimental field campaigns, programs to establish long-term hydrologic observatories, etc. The challenges associated with the management, analyses, and applications of such large volumes of data are unique and daunting. Scientific innovations are being pursued to address them in a systematic way, largely through the collaboration of people with multidisciplinary backgrounds.

This book is an outcome of such a collaborative effort among the authors. All the authors cotaught a graduate level experimental course on "Hydroinformatics" in the Fall 2003 semester here at the University of Illinois. This provided a unique opportunity, not only for the students enrolled in the course, but also for us to learn from each other. We realized that a pedagogical compilation of the contents of the course under a single cover would be very valuable for students, particularly because it cuts across a variety of traditional disciplines. We also felt that the expertise represented by us left out several critical gaps in a book aimed at a comprehensive introduction. We are indebted to the several experts who have contributed chapters in an attempt to close this gap. We greatly benefited from the involvement of the students enrolled in the course and a few of them are coauthors of the chapters. The class exercises have occasionally been used as examples in the book.

Additional material is available from the CRC Web site: www.crcpress.com. Under the menu Electronic Products (located on the left side of the screen), click on Downloads & Updates. A list of books in alphabetical order with Web downloads will appear. Locate this book by a search, or scroll down to it. After clicking on the book title, a brief summary of the book will appear. Go to the bottom of this screen and click on the hyperlinked "Download" that is in a zip file.

The goal of this book is to provide a basic introduction to several emerging concepts to enable a reader to get a panoramic view of the domain. Armed with these fundamentals a reader may proceed with more advanced discovery in any of the several areas that are covered. As is the case with any book on an emerging topic, this book runs the risk of being obsolete relatively quickly. It is our hope that this book will contribute to its own obsolescence by educating a new generation of students who will take these concepts beyond what we can now imagine.

Praveen Kumar, Jay C. Alameda, Peter Bajcsy,
Mike Folk, and Momcilo Markus

Acknowledgments

Praveen Kumar
This work derives from the research conducted over the past several years with funding support from the National Science Foundation (NSF), National Aeronautics and Space Administration (NASA), National Oceanographic and Atmospheric Administration (NOAA), and the University of Illinois. Special thanks are due to National Center for Supercomputing Applications (NCSA), particularly the faculty fellows program, which provided me the opportunity to interact with the outstanding people at NCSA that made this book possible. I also thank several of my students, Amanda White, Francina Dominguez, Amenu Geremew, Srinivas Chintalapati, Kyungrock Paik, Hyunil Choi, Vikas Mehra, and Pratyush Sinha for reviewing the initial drafts of the chapters to ensure that they were student friendly. Special thanks to Benjamin Ruddell for helping pull all the material together. We would also like to thank the staff at Taylor & Francis, particularly Cindy Renee Carelli, for her support throughout this project.

Jay C. Alameda
I would like to thank my family, Susie, Emily, and Josie, and their patience through all the weekends and vacations lost in preparing this text. I would like to particularly thank the members of my group; Greg Daues, Shawn Hampton, and Albert Rossi, for their advice and contributions, and my research collaborators, especially Richard Alkire, Kelvin Droegemeier, Dennis Gannon, Praveen Kumar, and many more with whom I have been privileged to work with for the past eight years. Special thanks, too, to Praveen Kumar, not only for his patience with me in both the Hydroinformatics course and text preparation, but also for having the inspiration to pull together our group to develop the course and text. Finally, I would like to thank the students in the class, who astounded me with their projects catalyzed by the material we developed for the course; in particular, I would like to thank Francina Dominguez who provided the material for the hydrology example in my chapters. I would also like to acknowledge support from the National Science Foundation, in the form of grants for the National Computational Science Alliance, Linked Environments for Atmospheric Discovery, the Open Grid Computing Environments Consortium, and the Consortium of Universities for the Advancement of Hydrologic Science, Inc.

Peter Bajcsy
I would like to thank my wife for her continuous support during the book preparation that included the very many long evenings and nights spent on pulling together diverse materials for the educational benefits of the future readers. I would like to gratefully acknowledge all my collaborators. Special thanks to all students and researchers who have worked with me on multiple projects related to the presented material, namely, Rob Kooper, David Clutter, Peter Groves, Sunayana Saha, Tyler Jeffrey Alumbaugh, Sang-Chul Lee, Martin Urban, Yi-Ting Chou, Young-Jin Lee, and David Scherba. I would also like to acknowledge the contribution of the students who took the hydroinformatics class in Fall 2003. Their final reports were used to create the Appendix with homework assignment. Namely, Amanda White's report served as a prototype solution for the given homework. This work

was partially supported by research grants from the National Science Foundation (NSF), the National Institute of Health (NIH), the Office of Naval Research (ONR), the Air Force Research Laboratory (AFRL), the National Archives and Records Administration (NARA), the National Aeronautics and Space Administration (NASA), and the National Center for Supercomputing Applications (NCSA).

Michael J. Folk
I am especially grateful to Pedro Nunes Vicente for teaching me about hydroinformatics, developing the HDF assignments, and working tirelessly to help our students understand and complete the material. Thanks to Praveen Kumar, who cajoled me into the teaching part of his course, then turning the course material into text. The students from the hydroinformatics class helped hone the presentations with their good comments and questions. Others to whom I owe special gratitude for significant contributions include Robert E. McGrath and Elena Pourmal at NCSA, Norm Jones and his EMRL team at BYU, and Ramiro Neves of Instituto Superior Tecnico in Lisbon. And none of this would have happened without the support of the entire HDF Group and my countless colleagues from NASA, the DOE, and elsewhere who have helped me to understand the concepts of large scale scientific data management that are covered here. This work was supported by funding from the National Aeronautics and Space Administration (NASA), the Lawrence Livermore National Laboratory, the Department of Defense SciDAC project, the National Science Foundation (NSF), the National Archives and Records Administration (NARA), the National Center for Supercomputing Applications (NCSA), and the State of Illinois.

Momcilo Markus
I would like to thank Praveen Kumar, University of Illinois, for the invitation to participate in both teaching Hydroinformatics and writing this book. His excellent organizational skills and expertise greatly contributed to the success of both. I would also like to acknowledge Dr. Misganaw Demissie, Head, Center for Watershed Science, Illinois State Water Survey, for his support, knowledge, and for creating an inspiring work environment. In addition, I would like to thank Professors N. Sundararajan and P. Saratchandran, Nanyang Technological University, Singapore, authors of the MRAN method, for their valuable advice and for providing a lot of insight on the method. Finally, I would like to acknowledge the contributions of Geremew Amenu, Graduate Student, University of Illinois, who assisted in applying the MRAN method to hydrologic forecasting, and for reviewing the draft of the chapter on MRAN.

Any opinions, findings, and conclusions or recommendations expressed in this material are those of the authors and do not necessarily reflect the views of the funding agencies and should not be interpreted as representing their official policies, either expressed or implied.

Contents

1

Data Integrative Studies in Hydroinformatics

Praveen Kumar

Contents

1.1 What Is Hydroinformatics?

The global water cycle is a very complex process that involves physical, biogeochemical, ecological, and human systems (Figure 1.1). The physical system consists of components that provide stores and flow paths for water through the physical environment of the earth, including oceans, atmosphere, lakes, rivers, soils, aquifers, etc. Numerous processes are associated with the movement of water through these stores and flow paths such as evaporation, transpiration, precipitation, streamflow, groundwater flow, and sediment transport. The biogeochemical system consists of components that provide the chemical constituents of water and transformations therein. These include the nutrient cycles (e.g., nitrogen and phosphorous), carbon cycle, contaminants, and other chemicals such as pesticides and hormones, etc. The ecological system consists of life cycles of organisms that are dependent on their habitats, vegetations, and the impact of hydrologic variability and disturbance regimes on the functioning of the terrestrial and aquatic ecosystems. The human system consisting of components such as engineering works, water use, and water resources management, add to this complexity through their influence on the flow paths and residence time of water, their chemical composition and sediment load, as well their variability.

The complexity of understanding and predicting these systems is enormous. It requires identifying key linkages and feedback between them. A significant step in this direction has been a migration from a traditional laboratory approach to a natural laboratory approach. The former consists of studying processes in the controlled environment of a laboratory, while the latter consists of deploying observational systems in the natural environment. The challenges of each of these approaches are different. In the traditional laboratory setting it is very difficult, if not impossible, to fully reproduce the complexity of the processes. In the natural laboratory setting it is very difficult to identify and observe all the variables that give rise to the complexity of the natural system.

The natural laboratory approach is becoming increasingly prevalent and many new observational systems are being deployed across the globe. These systems either

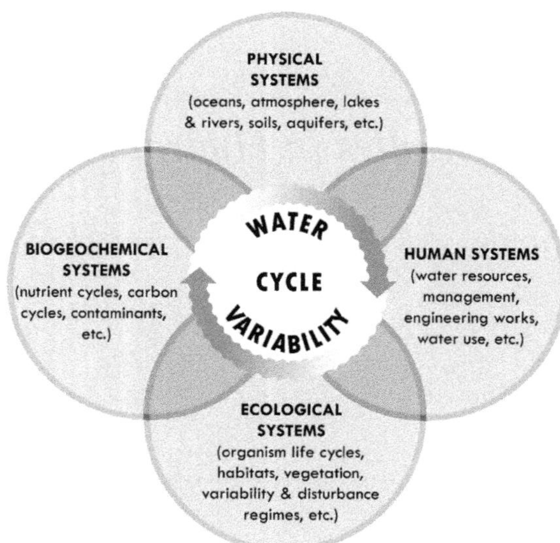

FIGURE 1.1
The variability of the water cycle affects several systems that are linked together.

provide long-term monitoring or support intensive field campaigns for short periods of time. Remote sensing is becoming an increasingly valuable tool, be it from satellite, airplane, or ground-based sensors, and the volume of data thus acquired is very high. For example, every day the Terra and Aqua satellites deployed by NASA (National Aeronautics and Space Administration) gather several terabytes of data related to a multitude of variables of the earth's atmosphere and land surface. These data are being ingested into models which then provide a dynamically consistent set of values for both the observed and unobserved variables (or states). This process of data acquisition and numerical model-based predictions are providing immense opportunities for formulating and testing new hypotheses regarding the dynamic behavior of the water cycle and the linkages between the different systems influencing the water cycle (Figure 1.1).

However, as the volume of data grows, we are faced with the new challenge of utilizing the data effectively in analysis, modeling, and prediction studies. Encompassed in this challenge are the problems related to storage, representation, search, communication, visualization, and knowledge discovery. Limitations in any one of these phases results in only a subset of the entire volume of data being used for a study thus limiting our ability to formulate and test hypotheses or make informed predictions. In addition, our scientific vision is stymied due to the use of fragmented and limited datasets. The value of the data is typically predicated on our ability to extract higher level of information from it: information useful for a better understanding of the phenomena, as well as for socioeconomic benefits. As the volume and the dimensionality (number of variables) of the data grows, it is important to develop and use technologies that support their effective utilization; otherwise, the value of the data itself is reduced and the investment squandered. In the recent years technologies have been developed that provide new ways of handling data, and have the potential to change the way we undertake scientific investigation, hence, providing better modeling and prediction support for societal benefit.

Hydroinformatics encompasses the development of theories, methodologies, algorithms and tools; methods for the testing of concepts, analysis and verification; and knowledge representation and their communication, as they relate to the effective use of data for the characterization of the water cycle through the various systems (Figure 1.1). This entails issues related to data acquisition, archiving, representation, search, communication, visualization, and their use in modeling and knowledge discovery. In this context, hydroinformatics is an applied science. It sits at the cross roads of hydrology, computer science and engineering, information science, geographic information systems, and high-end computing. Its application domain is broad and includes urban hydrology, river basin hydrology, and hydroclimatology, to name a few.

1.2 Scope of the Book

Hydroinformatics draws upon and builds on developments in several related disciplines such as computer science, computer networking, database engineering, and hydrologic science. As such, the basic principles of hydroinformatics are interspersed among a variety of fields making it difficult for a novice to obtain a comprehensive introduction to the basic concepts. The intent of this book is to bridge this void and provide a pedagogical introduction to a variety of related topics. While the contents of the book reflect the current state-of-the-art thinking, it is not meant to be an introduction to the state of the current research. Rather, the book is designed to introduce the basic concepts underlying lot of research developments in hydroinformatics. It is hoped that the reader will gain sufficient insight to pursue independent reading in any one or a combination of these areas.

The book is divided into five sections. Section I on "Data Driven Investigations in Hydrology" deals with several emerging issues in data representation. As much as a building's architecture is represented by a blueprint, a software's architecture is represented through a "Unified Modeling Language" or UML. UML is a graphical language for visualizing, specifying, constructing, and documenting the artifacts of a software intensive system. It provides a standard way to represent both conceptual entities such as processes and functions, as well as concrete things such as classes, schemas, and software components. An introduction to UML is provided in Chapter 2. This introduction will enable the reader to understand and create UML diagrams that are increasingly being used in areas such as geographic information systems (GIS). This understanding is also necessary to fully comprehend some of the latter chapters.

Chapter 3 deals with the concepts in the development of digital libraries for the publication of data. While the results of research are well archived in journal publications, the data on which these results are based are often not accessible to the community at large. This limits the ability for reanalyzing and reinterpreting the data, results, and their conclusions. The digital library concept is aimed at bridging this gap by providing a better avenue for the publication of data and the associated results.

Information about the data, often referred to as the metadata, is very important for making the data useful. It contains information regarding as to where the data was collected, what instruments were used for their characteristics, what errors are associated with the measurements, what variables were measured, etc. Metadata may also be associated with results of numerical simulations in order to describe the problem, domain, variables, time frame, etc. When looking for a certain set of data, we often search the metadata to identify the relevant data. Searching a metadata catalog is an important component of the digital library technology. In these regard, metadata is an extremely important component of the data management process, particularly as the number of records stored in a database grows significantly large. However, for metadata to be effective certain standards and protocols must be followed during its creation. Recently, there has been significant effort in establishing such a standard. Chapter 4 describes these issues in detail.

The way we represent data has an important bearing on the types of operations that can be performed efficiently. For example, the Roman representation of numbers is not as amenable to numerical operations as the Hindu–Arabic representation. As new data is available (often in different format than the old data) data fusion and the ensuing analysis becomes a challenging task. To facilitate such operations, new models for the representation of hydrologic data have recently been developed using the paradigm of object-oriented geodatabase design. The basic ideas behind contemporary data model design are described in Chapter 5, with a brief introduction to the Arc Hydro data model [1]. This data model is becoming popular among the surface water hydrology community, as it integrates several types of data into a common framework. Chapter 6 describes a new data model, Modelshed, which is built on top of Arc Hydro and enables integration of remote sensing and three-dimensional model data with that of surface water features.

The water cycle integrates a wide array of processes, thus the data associated with these processes come in a variety of formats. An effective conceptual framework for integrating and managing such data is needed for archiving and transferring from one system to another. The Hierarchical Data Format (HDF) has become a popular and an effective representation for such purposes. In Section II on "Managing and Accessing Large Datasets," Chapters 7 through 9 are devoted to describing the conceptual and practical elements of the HDF format, including the data model for storage and retrieval, different file formats, and an "under the hood" look at the new HDF version 5 protocol.

In recent years, the XML (Extended Markup Language) has become a machine independent way of sharing data across heterogeneous computing platforms. This protocol improves our ability to manage a variety of data formats distributed over a heterogeneous collection of systems (e.g., Windows, Linux, etc.), hence enabling us to use computing and data resources residing over a distributed network. The ability to transparently manage and communicate between these distributed resources has given rise to the development of "grid computing." This emerging technology enables a new and a powerful way of performing high-end computing using data sizes and computational speeds previously unimaginable. It brings to life the concept of "network is the computer." These developments are described in the first three chapters (Chapters 10 through 12) of Section III on "Data Communication." An example of an integrated data management system that brings together the several elements of metadata, data storage, and user interfaces is described in Chapter 13.

Careful processing is required to enable effective use of data for the study and modeling of the systems associated with the water cycle. Many steps are involved: (1) analysis of data sources (sensing, data acquisition, data transmission); (2) data representation (raster, boundary, or point data); (3) registration of spatial coordinate systems (spatial registration, georeferencing) and temporal synchronization; (4) data integration (integration and fusion of related or unrelated features over common spatial and temporal regions); (5) feature extraction; (6) feature selection; and (7) feature analysis and decision support. All eight chapters in Section IV on "Data Processing and Analysis" describe the basic principles of each of these steps in detail.

In Section V, under the broad concept of "Soft Computing," we present introductions to several analytical concepts useful in hydroinformatics. In Chapter 22, we first describe Statistical Data Mining concepts that are useful for automatically extracting information from large multivariate datasets. Data mining technologies are becoming increasingly more accessible through various software packages. In Chapter 23, we discuss the Neural Network approach to predictive modeling. This technique is particularly useful when there is a large uncertainty with regards to the knowledge about the physical system, but there is a need for reliable prediction. Chapter 24 discusses Genetic Algorithms, an optimization tool used in a wide variety of applications. In Chapter 25, we provide an introduction to Fuzzy Logic which moves us out of the probabilistic domain to enable us to include uncertain and subjective information in our analyses.

Reference

1. Maidment, D. R., *Arc Hydro: GIS for Water Resources*, ESRI Press, p. 220, 2002.

Section I

Data Driven Investigations in Hydrology

2

Unified Modeling Language

Benjamin L. Ruddell and Praveen Kumar

Contents

2.1 What Is UML?

The *Unified Modeling Language (UML)* is the product of two decades of thinking and synthesis on the analysis and the design of object-oriented software systems. The UML effort officially began in 1994 when James Rumbaugh and Ivar Jacobson joined Grady Booch at the Rational Software Corporation for unifying the different object-modeling languages developed through the 1970s and 1980s, including the Booch, Object-Oriented Software Engineering, and Object-Modeling Technique methods developed independently by these authors. With the support of the broader software community, the UML was developed and released in the late 1990s as a streamlined, flexible, and expressive method for planning and documenting software. The goals of this effort, as mentioned by the authors, are [1]

1. To model systems, from concept to executable artifact, using object-oriented techniques
2. To address the issues of scale inherent in complex, mission-critical systems
3. To create a modeling language usable by both humans and machines

By all means, the UML effort has succeeded at establishing an industry-standard method for software engineering. In fact, this method is useful outside the software industry,

Figure 2.1
Otto is satisfied with the results of this doghouse-modeling project. (Photography used by permission of Benjamin L. Ruddell.)

in applications where flowcharts, planning, structure diagrams, and process management are critical tasks. The UML is now an indispensable tool for software engineers, including those concerned with database systems.

So what is the UML? The name says it all. It is a rigorous, technical, graphical language used for the modeling of object-based systems. It is a standard method of blueprinting software systems. It is used to visualize, specify, construct, and document a software system. Graphical symbols are used to plot the structure and the function of a system in a visual manner, similar to a blueprint. These symbols have detailed attributes that work together to specify the exact arrangement of elements in a system. When used in a development environment, the UML may be read directly by software tools and used to program and construct the software. The detailed UML software *blueprint* makes a great reference because it documents the exact working of a system, simplifying future work and modification of the software.

Why use the UML? The UML is used to model. Why is *modeling* important? Software development is a lot like architecture, and this comparison is useful for understanding the UML. Take, for example, the architectural analogy of the doghouse and the skyscraper. If we are planning to build a doghouse, we probably do not need much modeling. A little bit of experience and practical knowledge should enable us to complete this project on time and on budget. It is just a matter of putting a few boards together, and adding a little paint and some shingles. Our user will be satisfied with the product (see Figure 2.1).

On the other hand, if we aim to build a skyscraper, the doghouse approach is not adequate. Some important differences exist. Our work is a part of a huge coordinated process that involves hundreds of people, several organizations, and companies. We need to meet the complex requirements of building codes, safety standards, and multiple clients. Because a failure or delay of the project will have additional cost, a lot of thinking needs to be done before the construction begins. This thinking process must be documented and communicated to the other architects, engineers, and contractors using blueprints that everyone understands. If we assume that the plans are followed, then the quality of the planning process determines whether the skyscraper will be completed on time and on budget, and whether the design satisfies the users (see Figure 2.2).

Today's software and information systems are more like skyscrapers than doghouses. They function within advanced operating system architectures, link many external libraries

FIGURE 2.2
A modern skyscraper is the product of a rigorous modeling effort because it is far more complex than a doghouse. (Photography used by permission of Benjamin L. Ruddell.)

and data sources, and serve the critical needs of science and business. Software is rapidly expanded and revised to meet new needs, and this expansion must be anticipated in a system's design. A team of programmers work together to complete the project, and individual programmers do a lot of work updating someone else's code. In this environment, development cannot proceed in a haphazard manner. To develop software of lasting quality, usefulness, and maintainability we need to design and document the system using a model that is understood by everyone.

The UML is a rich and an expressive language for many applications. This chapter presents an abbreviated high-level introduction to the framework of the UML, and a more focused treatment of elements of the UML used in data modeling and database applications. Specific attention will be given to Microsoft's *Component Object Model* and the *Object Model Diagrams* used in the design and deployment of many modern database systems.

2.2 The Framework of the UML

The UML is composed of three basic categories, which make up the "grammar" of the UML: things, relationships between things, and diagrams of the things and their relationships.

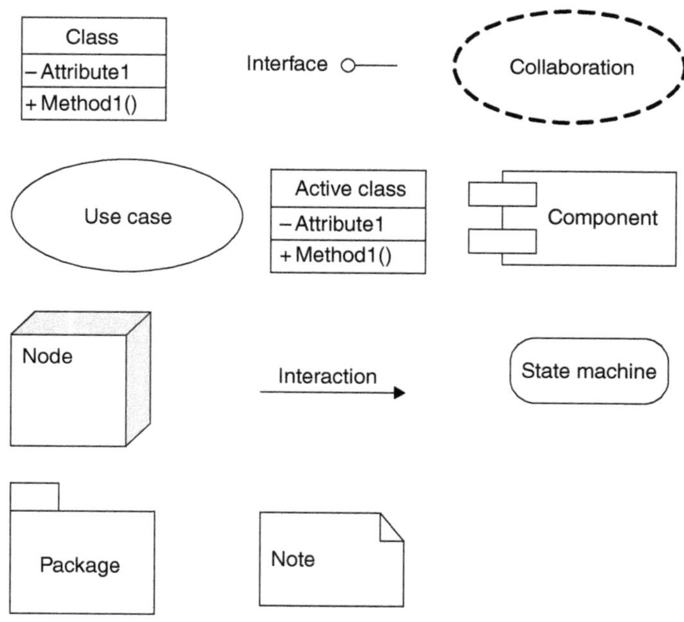

FIGURE 2.3
The graphical representation of UML "things."

2.2.1 UML Things

Things include structural software objects, and three other types: behavioral, grouping, and annotational things. The UML is a language and structural things are like nouns. They are concrete objects or collections of objects in a system. *Behavioral* things are like verbs, and represent actions and states of being. *Grouping* things are categories and collections of other things. Groupings are purely conceptual, so unlike structural and behavioral things, they are not part of a functional implementation. Like groupings, annotational things do not translate into the implementation of a software system; they are simply descriptive notes used by developers. Listed here are the specific subtypes of things within each category. All these things are graphically represented in the UML with specific shapes, shown in Figure 2.3. The most important things for data modeling are *classes, interfaces, packages,* and *notes*. Here is a list of the different UML "Things":

1. Structural things
   - *Class*: object class with methods and attributes
   - *Interface*: standardized access point for the service of classes
   - *Collaboration*: collection of interacting, interdependent things
   - *Use case*: description of the particular combination of interactions of objects that result in a collaboration
   - *Active class*: class whose objects control processes and initiate activities on their own authority
   - *Component*: modular physical package that contains classes and interfaces, such as a source code file or COM library
   - *Node*: point in a network of computational resources

2. Behavioral things
   - *Interaction*: a message between objects for a specific purpose
   - *State machine*: the behavior of an object as it responds to a series of events and interactions and initiates new activities (examples of states: waiting, calling object1)
3. Grouping things
   - *Package*: "folder" to conceptually organize other things for the purpose of explanation of a system
4. Annotational things
   - *Note*: "post-it" to document development notes on a diagram

2.2.2 UML Relationships

There are four types of relationships that exist in the UML: dependency, association, generalization, and realization (see Figure 2.4). A *dependency* between two things exists when change to one thing affects the other. An association describes a connection between the attributes of two classes. It includes *roles* describing the purpose or the point of connection at each end of the association, and "multiplicities" describing the number of objects that participate in the association at each end of the association. A special sort of association, the *aggregation*, exists when one class is built out of an assembly of other classes. Another special association, the *N-ary* association, connects three or more things together. *N*-ary is defined in contrast to the binary nature of other relationships. A *generalization* exists between a subclass and its parent class. Subclasses inherit the interfaces and the attributes of their parent classes, and are generalized as special types of the parent class. Association and

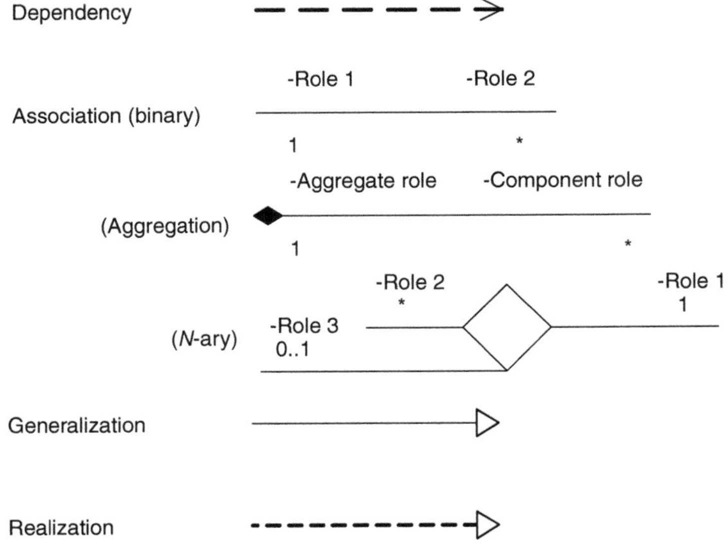

Figure 2.4
The graphical representation of UML "relationships." The shapes share similarities based on their conceptual likeness. Role names are found at each end, and the 0, 1, and * symbols represent the zero, one, and many multiplicities of the objects attached at each end of the association. Dependency, generalization, and realization relationships are simpler, needing only a distinctive arrow to convey meaning.

generalization relationships are essential to object-oriented database systems, and will be illustrated in this chapter. A *realization* is a cross between a generalization and a dependency, and exists where one thing implements and executes the purposes of the other. Realizations exist between two sets of things: interfaces and classes that implement them, and use cases and collaborations that implement them. Graphical UML relationships are drawn between two or more related things, and include some descriptive information about the relationship. Associations, aggregations, and generalizations are the most important relationships for data modeling.

2.2.3 UML Diagrams

If the things and relationships of the UML are like the lines and words on a blueprint, then UML *diagrams* are the blueprint itself. In the building of a skyscraper, the construction documents are composed of many blueprints. One blueprint has information about the electrical systems, another describes the mechanical and ventilation systems, and yet another the physical layout of the structure's beams and columns. Likewise, a UML diagram provides a specific functional view of a part of a system. One diagram contains the logical elements necessary to describe a particular system in sufficient detail, and omits other elements that would clutter the diagram. Diagrams are scalable in their detail; on a master diagram, each package would be represented by one symbol, but other specialized diagrams would describe the intricate behavioral function of the classes within each package. The UML's nine specific types of diagrams are listed here. Each focuses on a specific part of the architecture of a software system. The class and the object diagrams are most important for data modeling. Class and object diagrams are similar because they show classes, interfaces, and relationships. They are different because class diagrams show this structure from an abstract design perspective, whereas object diagrams show a specific "real-world" implementation of that structure. The nine types of UML diagrams are:

- Class
- Object
- Use Case
- Sequence
- Collaboration
- Statechart
- Activity
- Component
- Deployment

2.3 Object Model Diagrams

2.3.1 Object-Oriented Software Designs: Databases, HDF5, and COM

The UML is often used to express the design of object-oriented relational databases such as Oracle™, IBM's DB2™, Microsoft's SQL Server™, and Access™ applications, and other popular open-source systems such as MySQL and PostgreSQL. These technologies are based on different platforms (Unix/Windows, Java/COM), but are good candidates for the UML because they are object based. Object-oriented relational databases use the structured

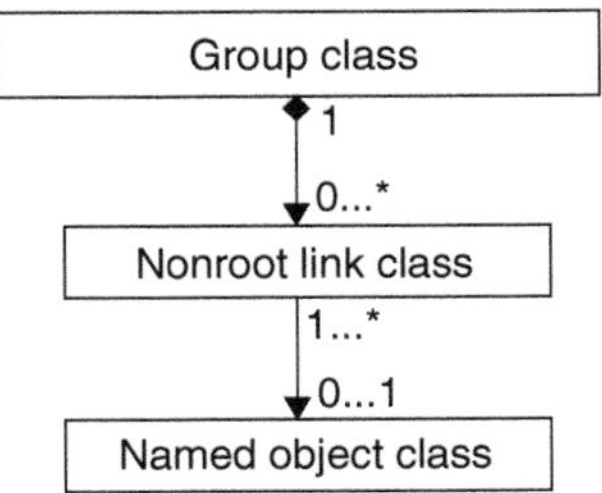

FIGURE 2.5
Part of the HDF5 data model. The HDF structure contains named object classes, where each (0 . . . 1) is linked to between one and many (1 . . .*) of nonroot link classes. Any number (0 . . .*) of link classes are aggregated to form one group class.

relationships of object classes to create a rigorous, scalable, and flexible platform to store the data. Database design is unique to each specific application, and can be expressed using a data model (see Chapter 5 for more on data models).

Another good application for the UML is in object-oriented file formats, such as the Hierarchical Data Format 5 (HDF5) developed by the National Center for Supercomputing Applications (NCSA). See Chapter 9 for more details on the HDF5. The HDF5 provides an efficient structure for data objects to be arranged in a hierarchy, with associated metadata, within a single file. That data and structure may be accessed and viewed piece-by-piece, and individual objects may be read and written within a much larger file. This allows a user to locate the needed data and access it without dealing with the entire file's contents, as a user would have to do with a typical binary file format. These capabilities are essential because HDF5 is used by the National Aeronautics and Space Administration (NASA) to store massive remote sensing datasets. A "data model" is used to conceptualize the structure of a HDF5 file, and this data model is encoded in an XML document known as a schema. Part of the HDF5 data model is illustrated in Figure 2.5 [2]. More on data models is explained in Chapter 5.

The Component Object Model (COM) framework is the foundation of Microsoft Windows™-based applications and database systems [3]. The COM is a protocol that defines the structure of objects and interaction between different software components. COM standard components are compiled in a binary format and are fully modular. This means that regardless of the language they were developed in, or what application they were developed for, COM components may be accessed and reused by other applications. COM separates interface from implementation, meaning that specific object classes implement interfaces. Client applications do not interact directly with objects. They interact with standard interfaces, which are implemented in a unique fashion by each object that supports the interface. This will be clarified by an illustration later in this section. Because the ESRI ArcObjects and incorporated geodatabase technologies are built on the COM standard, the UML Object Model Diagrams (OMDs) presented in this chapter illustrate COM-compatible class structure design. A few minor additions to the strict UML standards are made for COM class notation. First, CoClasses are represented by three-dimensional shaded boxes, while Standard Classes are represented by an unshaded three-dimensional box. The three-dimensional graphic symbolizes that these class types are able to be instantiated as real objects. The abstract class is shown as a two-dimensional shaded rectangle. In addition, a new relationship, the "instantiation," symbolizes that one class has been created by another. The instantiation relationship is similar to the "realization" type in standard UML. Finally, the ArcObjects do not support multiple inheritance. In other words, each class may only inherit from one parent class.

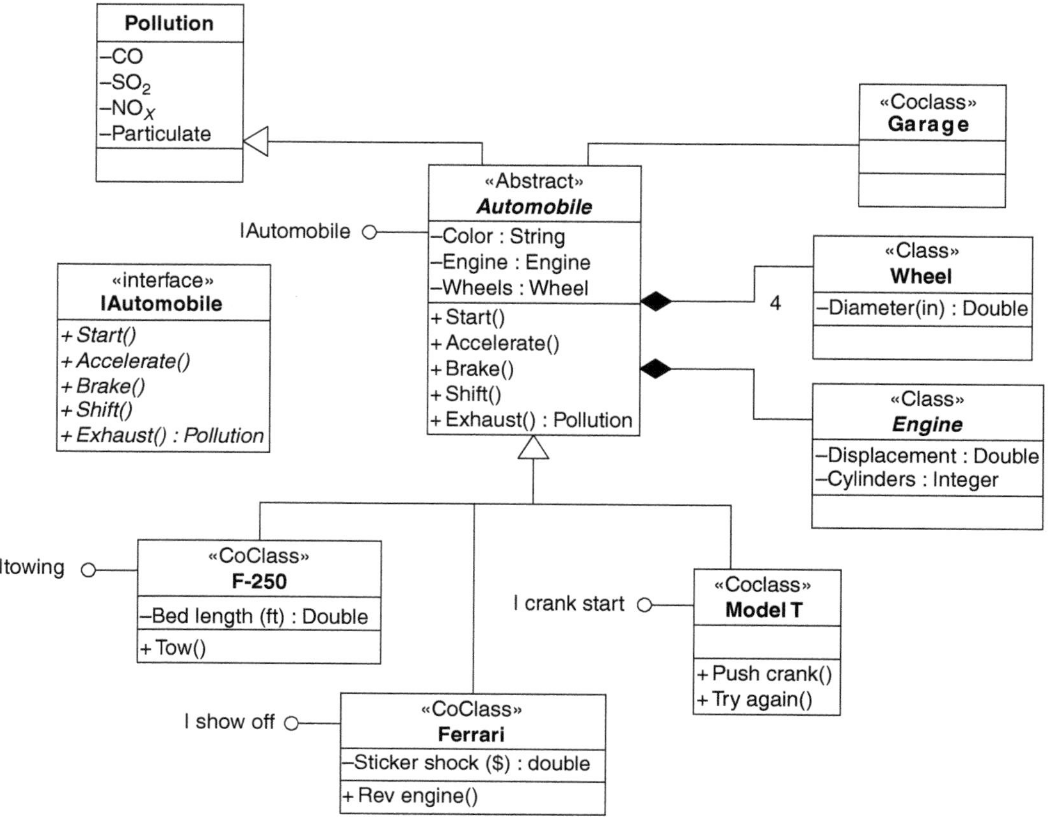

Figure 2.6
An automobile object model diagram. An automobile is composed of an aggregate of four wheels and an engine, and is associated with exactly one garage. Three subtypes are specified, each inheriting the attributes and interface of an automobile, but each having its own unique attributes and methods. An automobile realizes/creates pollution with its exhaust method. The IAutomobile interface provides access to the start, accelerate, brake, shift, and exhaust methods. All automobiles will implement these methods.

2.3.2 Classes and Relationships

We will use an OMD of automobiles to help explain COM and the structure of relational object-oriented systems. In a software system modeling automobiles, it would be useful to create objects that represent the different types of cars, their capabilities and attributes, their components, and other related data. Figure 2.6 is a static structure of a UML class diagram, meaning that it describes the architecture of the classes in a system, rather than a specific implementation of that system or the internal processes of the classes' methods and interactions. Look at the Automobile class in the center of the diagram. Note that it has some attributes that all automobiles possess, such as color and engine, and an interface, IAutomobile. The "I" in IAutomobile tells us that this is an interface. The IAutomobile interface specifies methods that all automobiles can use, such as start and accelerate. In fact, any object class that supports the IAutomobile interface is able to use its methods.

Now note that the Automobile class is "Abstract." In the real world, there is no such thing as an "automobile"; what we see driving around are specific types of automobiles, such as F-250's and Ferraris. In this example, "Automobile" is a type of category for four-wheeled vehicles; it is an abstract idea. In this object model concept there are three types of classes: abstract, concrete, and standard. Abstract classes, also known as superclasses or parent classes, are generalizations, and cannot be instantiated. Concrete classes, also known as CoClasses or implementation classes, may be instantiated as new objects, and can stand-alone. *Instantiation* occurs when a program takes a class definition, which describes an object, and creates a real functioning object in memory based on that description. Standard classes are like CoClasses, except that they cannot be created as stand-alone objects. Standard classes must be created by a method of an instantiated object. In this example, engines and wheels serve no purpose on their own, they only exist as parts of an automobile. Pollution is "realized," or created, by an Automobile when it executes its "exhaust" method.

Observe the relationships that connect the diagram's classes, represented by lines. An association exists between the garage and the automobile classes. This association has a one-to-one multiplicity because in this model, each automobile has one garage. The one-to-one multiplicity and "automobile"/"garage" role names are not displayed here because they are implicit. Aggregations exist between the engine, the wheel, and the automobile classes. Four wheels and an engine are necessary to form an automobile. The automobile has wheel and engine attributes where references to the associated wheel and engine standard object classes will be stored once the automobile is implemented as a specific object. The wheels and the engine are a part of the automobile, and must be created by a method of the Automobile. Pollution is "realized" by the exhaust method of the Automobile, which returns a Pollution object. Generalizations establish the F-250, Ferrari, and Model T as subtypes of the Automobile class.

2.3.3 Inheritance and Polymorphism

The CoClasses in this example, the F-250, Ferrari, and Model T, are specific types of Automobiles. The generalization relationship implies that these CoClasses inherit the attributes and the interfaces of their abstract parent class. This is the definition of "Inheritance": subclasses inherit from parent classes. If there is a lengthy hierarchy of inheritance above a CoClass in the OMD, the CoClass will inherit the attributes and methods of all classes above it, all the way up to "IUnknown," the base COM class interface from which all COM classes inherit. The F-250 class has a bed in the back, described by the Bed Length attribute, and has towing capabilities, accessed through the ITowing interface. Most subtypes will have their own unique attributes and interfaces, in addition to those inherited from the parent class.

Just like in the real world, interaction with objects is accomplished through their interfaces. If we know how to use one automobile, we know how to use them all, even if things work differently under the hood; the interface remains the same. Our Ferrari and Model T objects use different code mechanisms (fuel injection vs. carburetor, if you like) to implement the "accelerate" method, but they are both able to accelerate. This is the meaning of "polymorphism": because the interface and the implementation are separate, we can use the same interface on different types of objects that support the interface. To support an interface, a class must simply provide implementation for all its methods.

2.4 Database Design and Deployment

It is possible to graphically design and automatically deploy a well-formed data model to a database using today's advanced computer-aided software engineering tools. The UML makes this possible by providing a rigorous structured modeling language that is readable by both humans and computers. Once the design is properly laid out in the UML within a development environment, it can be checked for integrity, then compiled or formatted for export to a specific application. Many UML-based development environments and applications are in use today (such as Microsoft Visio, Visual Paradigm™, and SmartDraw™), exploiting the full spectrum of the UML's expressiveness. However, in keeping with the theme of hydroinformatics and geospatial data, the focus of this section is on the windows-based geodatabase technology and the Microsoft Visio™ software used to develop it.

2.4.1 The ESRI Geodatabase

The ESRI geodatabase technology is a rigorous object-oriented relational data model and is well suited to the UML. The geodatabase is a subset of the COM-compliant ArcObjects technology, which is the foundation of the ArcGIS™ software. The structural design of ArcGIS, geodatabase, and ArcObjects is a hierarchy of classes, interfaces, and relationships. At the top of the hierarchy, the ArcObjects inherit from the Windows™ COM class named "Class," and the interface "IUnknown." At the bottom of the geodatabase hierarchy we find the customizations and specialized class structures known as "data models."

The ESRI geodatabase object model has more than one hundred interconnected classes that support geospatial data and metadata storage by their attributes and methods. The

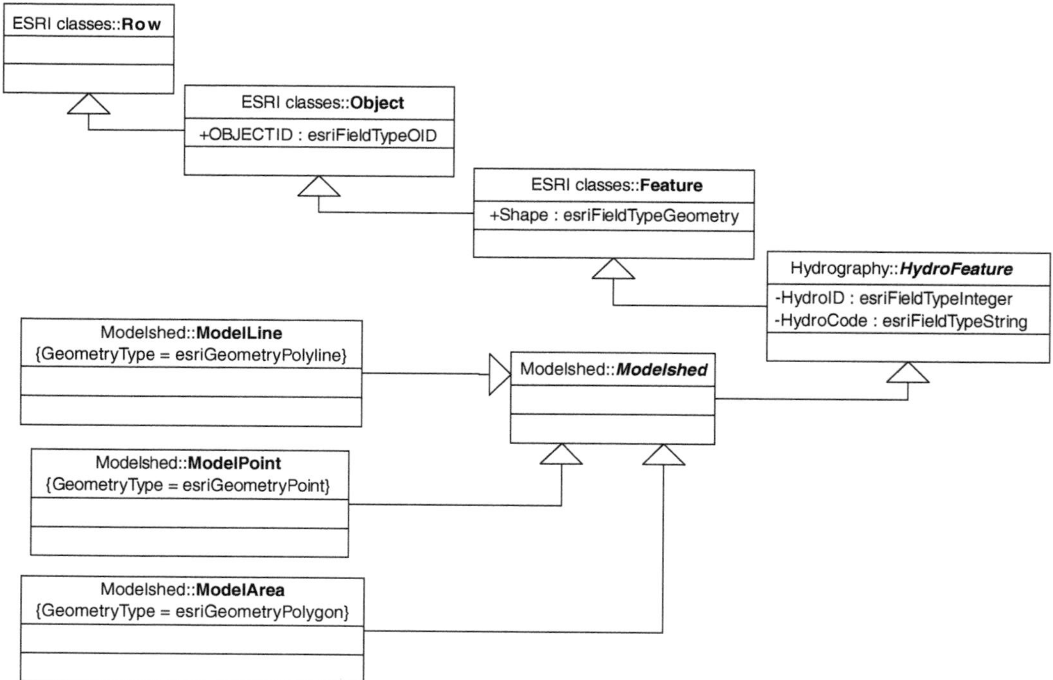

FIGURE 2.7
The Modelshed geodata model inherits from the geodatabase classes. Row, object, and feature are geodatabase classes. HydroFeature is a subtype of feature, with two additional attributes, specified by ArcHydro. ModelArea inherits from this structure, and is a type of row, object, feature, HydroFeature, and Modelshed. ModelArea will inherit the attributes and methods (HydroID, HydroCode, Shape, ObjectID, etc.) of all its parents. ModelArea is distinguished from ModelLine and ModelPoint by its GeometryType tag.

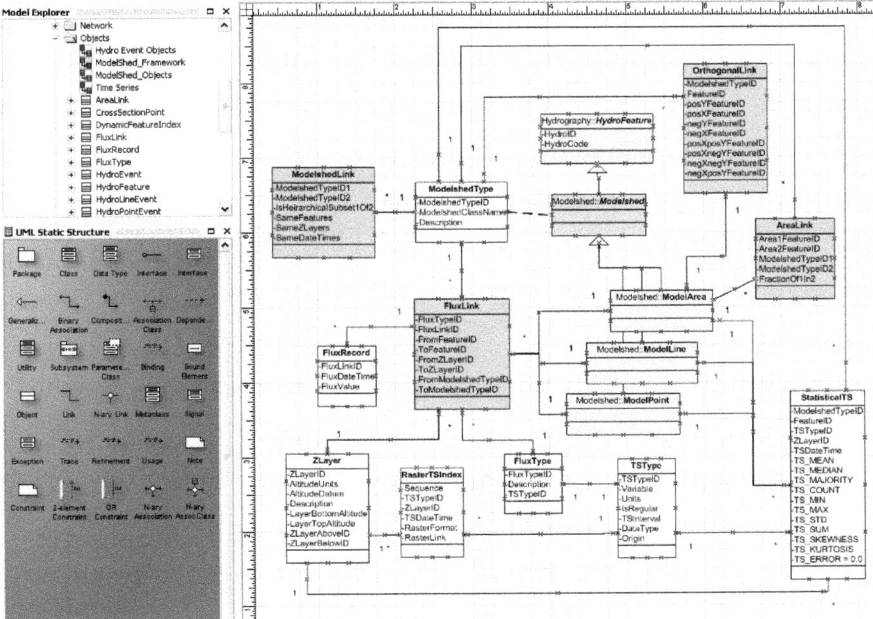

FIGURE 2.8
The Microsoft Visio UML object modeling environment. The data model is constructed by connecting UML shapes into structure diagrams. Visio helps organize these objects to form a coherent data model design, and manages different views of the data model. This design diagram may be modified to update and expand the model.

ArcHydro and Modelshed geodata models inherit from this structure (see Figure 2.7). The "Modelshed" is a type of HydroFeature (an ArcHydro Class), which is itself a subtype of feature, a geodatabase object class. Each inherits interfaces and attributes from the classes above. One advantage of this arrangement is that any software designed to use an interface on the "row" class can work with all of the objects, which inherit from "row." Customizations and expansions of the class structure are therefore automatically supported (in most cases) by existing software tools and methods. If you add something to the end of the hierarchy, there is no need to reengineer the whole system.

Elsewhere in the geodatabase structure, rows are organized into a "table," which is a type of "dataset." A number of datasets combine within a "workspace," which is the object class representing the program's read/write access point to a particular database that exists on a hard disk in the real world. Geodatabase workspaces may be formed from many different types of files. A geodatabase is not a file type; it is a class structure which may be fully or partially expressed within many different geospatial data storage formats, such as shape files, text files, Computer Aided Drafting (CAD) files, local Microsoft Access™ databases, and database servers such as Oracle. The whole geodatabase OMD is too complex to include here, but is available at the ESRI online support Web site.

2.4.2 Microsoft Visio and the Geodatabase

Microsoft Visio is to windows-based COM database programming what CAD is to engineering. Using CAD, engineers are able to quickly draw and document their designs, automatically generate printouts and plots from different views, and even send those plans to robotic manufacturing devices to be automatically turned into reality. Visio provides the same capabilities to a software engineer developing object-based applications and data structures. Using the UML, a programmer can graphically design the things and the relationships that make up a program, and use a variety of tools to validate and export those designs. Visio can construct and reverse-engineer the data models that define the structure of databases.

Construction of a data model is accomplished through a visual interface where the various shapes in the UML are available on a toolbar. They are arranged on a diagram and connected with relationships (see Figure 2.8). The properties, attributes, and methods of classes may be edited and programmed directly within the diagram. The various diagrams are automatically crosschecked and updated with changes to maintain the integrity of the object structure. External class libraries may be referenced, and their classes and data types included in the diagrams along with custom classes. For example, Java, C++, and Visual Basic have their own native data types, and all these types may be included together automatically in a Visio diagram. Once the development of a data system is completed, the diagrams and the documentations are preserved within the project file. If reference or update is required later, this project file contains all the information needed to find out exactly how the system was put together.

In the specific context of geodatabase design, Visio has been the tool of choice for ESRI and other developers working to store geospatial data in database repositories. The ArcHydro data model is available as a Visio project file, and it may be easily modified and expanded. ESRI provides a "semantics checker" with its Visio geodatabase template. This checker will scan a static class structure diagram to ensure that the diagram is well formed and compatible with the ArcObjects. Then Visio can export the structure of a data model using the XML Metadata Interchange (XMI) format. XMI is a format for the communication of an XML schema (metadata) between UML-based modeling tools and object-based metadata repositories. This XMI file contains the schema of the data model, and may be applied to a data/metadata repository by other tools.

References

1. Booch, G., Rumbaugh, J., and Jacobson, I. *The Unified Modeling Language User Guide*. Boston, MA: Addison-Wesley, 1999, pp. 3–14, 435–440.
2. Pourmal, E. HDF5 Abstract Data Model: Classes, Objects, and their Relationships. 9 September 1999. The National Center for Supercomputing Applications at the University of Illinois at Urbana-Champaign, Champaign, IL. http://hdfeos.gsfc.nasa.gov/hdfeos/workshops/WSthree/presentations/19.
3. Microsoft COM Web site. July 2004. http://www.microsoft.com/Com/.

3

Digital Library Technology for Hydrology

John J. Helly

CONTENTS

3.1 Introduction

The notion of digital libraries for scientific data has been evolving for decades; however, has really advanced rapidly since the late 1990s and will probably continue to do so and we hope that this document will contribute to that speed-up. The recent acceleration of progress is due to a number of factors including the ubiquitous distribution of computers and general access to the Internet-based World Wide Web (WWW) as well as the huge proliferation of WWW browsers and supporting tools due to commercialization. In addition to the proliferation of web sites and browsers is the effect of "CGI" techniques; the common gateway interface method that has provided customizable program interfaces to the WWW using typically the Perl and Java languages.

This rapid change has significantly altered our perceptions and practices about how to handle digital scientific resources, it is helpful to consider how some of this progress has been seen and accomplished by those of us who have been developing systems for the distribution of scientific data; a first-hand perspective as it were. So we will begin by

Table 3.1

Chronology of scientific data publication efforts at SDSC.

Project	Date	Purpose
San Diego Bay Environmental Data Repository	1994	Environmental monitoring
CEED (Caveat Emptor Ecological Data) Repository	1994	Ecological research
Ecological archives	1996	Ecological publication
SIOExplorer	1998	Oceanographic research
EarthRef	2000	Geochemical research
ERESE: Enduring Resources for Earth Science Education	2003	Earth science education
CUAHSI Hydrologic Information System	2004	Hydrologic research

examining a brief chronology of information technology advances from the point-of-view of a developer at the San Diego Supercomputer Center (SDSC). Opinions, like mileage, may vary.

As shown in Table 3.1, in about 1994, our team of computer scientists and students at SDSC began working on something called an environmental data repository for San Diego Bay. At that time, we did not call this a digital library, however, the idea behind this effort was to provide a locus for the collection of digital resources that had been produced by various federal, state, and local agencies, and individual academic investigators pertaining to the environmental condition of San Diego Bay. This was done to apply information technology to the problem of collaborative resource management in San Diego Bay. Our goal was to enable members of the San Diego Bay Interagency Water Quality Panel to focus on the problems at hand instead of "talking about what they were talking about" by having common access to a common set of information. At that time, if one went into a room of 100 people and asked, "Who has used the World Wide Web?," the number of hands raised might have been five. The interested reader can refer Helly (2001) for greater details on the nature and effectiveness of this project.

This collaboration between the SDSC at the University of California, San Diego, and San Diego Bay Interagency Water Quality Panel (Bay Panel) was innovative. The project was funded by members of the Bay Panel for (1) developing an environmental data repository to facilitate the acquisition and sharing of data and (2) the development of a visual model of the bay. The project was recognized among the most innovative applications of IT of its kind (National Science and Technology Council 1996).

Around the same time, a proposal was formulated by a team of ecologists around the United States and computer scientists at SDSC for submission to the National Science Foundation to develop a repository for ecological data in response to a report by the Ecological Society of America's Future-of-Long-Term-Data (FLED) Committee. This activity by the ESA was among the first in which an entire disciplinary community recognized the need for a broad strategy to ensure the integrity and longevity of scientific research data.

This decade-long series of projects that has led us to a coherent and consistent design strategy for digital libraries supporting various Earth science domains and most recent incarnation of this, the CUAHSI HIS, are described here in detail.

3.1.1 A Scholarly Model for Data Publication Using Digital Libraries

Most data and metadata in Earth sciences are published in the context of traditional, peer-reviewed publications in paper journals. Especially, the practice of publishing data electronically is extremely poor since electronic journals continue to be functionally similar

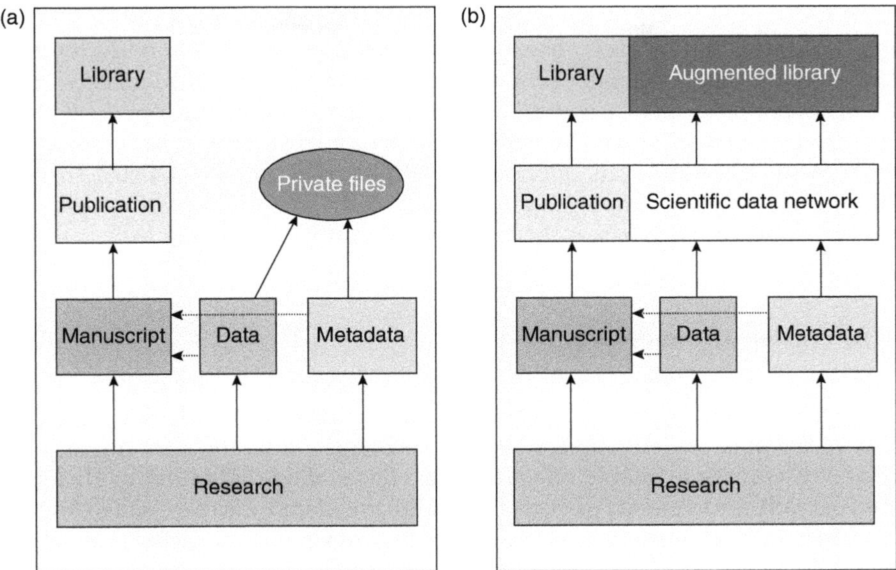

Figure 3.1
Difference between approaches to scholarly research before (a) and after the use of digital library technology (b) for publishing scientific data. (From Helly 2003.)

to paper journals. The confinement of most data publication to paper journals has the result that authors rarely publish data and even less frequently their metadata (Helly 1998).

Research effectively produces three types of data products: (1) data, (2) metadata, and (3) the interpretations of these data in the form of various illustrations included in manuscripts. Our current paper publishing protocol (Figure 3.1[a]) is mostly biased toward the final writing and illustrations. Digital data and metadata tend to be rarely published on paper copy or in its electronic equivalents (dashed lines). Page limitations in high profile journals actively work to eliminate or greatly reduce the actual publication of data and metadata. As a result, interpretations and figures based on data are widely published and archived in libraries, while most of the primary data are confined to research files of investigators. These private archives, however, do not provide sufficient access for future research that might result in a reinterpretation of the data. There exists no return flow of data into community-wide research activities and, therefore, these data are effectively lost. In the worst case of such data loss, a new investigator will have to reanalyze the original samples, if they are still available, with the end result of another cycle of data generation and loss. This puts the Earth sciences, as well as other sciences, in a position where the interpretations of data are published in reliable and accessible media, for a price, while the empirical observations and measurements have an uncertain fate. This effective loss of data or metadata due to nonpublication cannot be reconciled with our principles of scholarly science. Without them, scientific work cannot be duplicated or falsified.

Figure 3.1(b), is an alternative data publication method in which the generation of scientific products and its publication remains the same, but data and metadata are published in a scientific data network. This network takes on the role of libraries for the archiving and serving of data to the scientific community in order to support future research. In this new protocol the traditional publication of scientific results is accompanied by publication of its data and metadata making the "complete" scientific product available in a consistent and coherent manner (Helly et al. 1998, 2002). Such a data network may include a range of options, including the attachment of electronic data supplements to publications (GERM Steering Committee 2001). This is already the practice in a number of highly-regarded journals. However, unlike homogeneous transaction-oriented databases a data network, such as the one considered here, must accommodate a wide range of data file types (e.g., *.txt, *.jpg, *.xls) and content formatting and semantics. By that we mean the internal organization of the data file with considerations of encoding of data, computer word structure, floating-point and integer data type representations, and all the attendant platform-dependencies that might be entailed. We address this problem of multisource data next.

3.1.2 The Problem of Multisource Data

The main technical problem associated with the acquisition of data from a diverse set of sources or authors, assuming that individuals and organizations will release the data, is the variability of the data content and format. This is compounded by the difficulty of obtaining sufficient metadata to enable effective reuse of the data (Michener et al. 1997). These problems are well understood, documented, and translated into a problem for obtaining sufficient labor and expert resources to convert these data into a usable form. At this point there are no general solutions to this problem, and the best current approach is to encourage the conversion of any given data file into a common format by those most expert in the original data. This approach, a common data format, has limited practical application; however, since the range of data and related analysis software applications is very great and unlikely to diminish. This will be discussed in greater detail later in the chapter. As a result of this diversity, it makes sense that we have to build data networks that can accommodate different types of data both in the (1) file-system aspects such as size, naming conventions and bit or byte level data representations, and (2) in the internal organization, or semantics, of the data itself.

It is worth mentioning that the common format approach has been used successfully in at least some settings and is exemplified by the CIDS (C4 Integrated Data System) at Scripps Institution of Oceanography (http: //www-c4.ucsd.edu/~cids/) based on the NetCDF (Network Common Data Format) data file format (Rew et al. 1997). The CIDS typifies many of the scientific mission approaches in the Earth systems science community. This type of application is, however, relatively homogeneous in both its empirical data and user population. The application domains do exist, and when possible, it is a great simplification to encourage this type of data network architecture.

3.1.2.1 Types of Data

We have previously described the fact that there are different types of data. In computer science, data types mean specific things such as characters or strings or literals, integers or floating point numbers relating to the physical representation of data in encoded form. In an attempt to be more clear, in this context, by physical representation we mean the electronic, magnetic, optical, biological, or quantum mechanical means used to store the data within some form of persistent memory such as on a hard-disk within a file system or in volatile memory such as a computer system's primary storage devices (i.e., RAM). So this is an additional distinction to consider and account for in data system interoperability. Interoperability spans a distressingly wide range of issues at virtually every level of computer system architecture.

To further confuse the issue, in the broader realms of disciplinary science the term "data types" tend to mean something much more ambiguous. It can refer to streams of data from sensors to file-type formats (e.g., *.cdf, *.txt) to encoded data formats (e.g., GRIB).

It is therefore important to understand what is meant by those using this term in the specific context of the discussion. Here we use it in the term "types of data" to be careful to distinguish what we refer to from the computer science meaning of the term "data types."

Two broad types of data and, consequently as will be shown, data systems have been identified by a recent set of workshops, one at the San Diego Supercomputer Center and one at the Lamont–Doherty Earth Observatory, on data interoperability sponsored by the National Science Foundation. The report has not yet been published but the reader is encouraged to seek it out as it will have been published by this time. These data systems are:

1. Transaction-oriented systems including streaming, real-time or near-real-time data that may be buffered in primary storage media (i.e., random access memory) or record-oriented systems (i.e., database management systems or DBMS).
2. Nonreal-time data stored primarily in secondary (i.e., spinning disk) or tertiary (i.e., removable) storage media as files.

This is an important distinction that has considerable architectural significance and has led to the development of very different data systems that are designed to operate principally on data streams or records vs. principally on data files. There are many examples of each of these data system approaches and we will discuss examples later. For the time being it is sufficient to recognize that the difference exists and that it will likely continue to be an important design difference. However, because each type has unique strengths and weaknesses, we expect that future systems to be built to accommodate both types of data and therefore hybrid systems will evolve. In fact, the hydrologic information system that we will describe later may be the first of this type.

Another important dimension to the "types of data" considerations are those that have been variously described as structured vs. unstructured or gridded vs. ungridded, and raster vs. vector vs. scalar data. These descriptions actually apply to the logical data structures used to represent the data and not the manner in which they are stored physically. Some refer to this type of consideration as relating to the structure or semantics of the data but there is no universal agreement on this classification since it is highly subjective and context dependent.

In order to simplify our subsequent discussion and refrain from having to provide too many caveats and asides, we will focus on the file-oriented aspects of the problem and generally assume "data" to mean data files of all types. Later we will introduce the term "Arbitrary Digital Object (ADO)" as the preferred term but that is a little ahead of the game at the moment. Now we will turn to the large-scale issues of reproducibility of results and the implications pertaining to data integration; a very common purpose and application of multisource data systems. We will also go into considerable detail on a scheme for supporting data integration including quality control processing as they affect the reproducibility of scientific results.

3.1.2.2 Importance of and Considerations for Reproducibility of Results

One of the key considerations of a scholarly model of data publication that is often overlooked by computer scientists is the need to ensure scientific integrity of data, as opposed to physical integrity, its unique naming and persistence in unmodified, pristine form once published, and the ability to establish an audit trail from derived data products back to the source data. This traceability is needed to determine and evaluate the origin, history, and processing of the data used to support analytical interpretations and decision making and to enable computed results to be regenerated on-demand.

This type of traceability is also essential in (1) working backward from an error found in a derived data product to its source, (2) making the corrections, and (3) reprocessing the data from the point at which the error was detected. In other words, an effective method of traceability must provide enough information to debug a data anomaly (Helly et al. 2000). This, in addition to enabling independent analysis, is a key reason it is important to publish data as both derived products and source data. Consequently, it is important that data distribution systems provide the means for the:

1. Backward traceability (i.e., decomposition) of a composite or derived data object to enable the unambiguous identification of its constituents and the related processing.
2. Forward traceability from constituent data objects to derived data objects to support the proper attribution of intellectual property rights in authorship or ownership of source data.

A methodology to support these requirements should ensure a minimal level of self-consistency of fundamental digital objects, such as files, according to domain-specific standards as well as enabling the declaration of the dependencies of an integrated (i.e., composite) or derived digital object on its components. This idea of dependency is similar to the notion of functional dependency found in the relational data model as described by Basir and Shen (1996) but pertains to data files rather than the attributes (or parameters) within a file or database. For example, attributes may span multiple data files. Such a method must also define the means by which self-consistency is achieved, the way it is verified and how these features are provided for the integrated (i.e., composite) data object. What is described later can provide a basis for new methods of quality control that preserve the information needed to meet these requirements.

Methodology for Data Integration and Coupling to Applications

One of the key features and motivations for digital libraries is to provide access to a wide range of data. The desire for this type of access is motivated by a desire to integrate data from these sources for the purpose of correlative and synthetic analysis. This requires data integration processing and the data transformations necessary to enable this integration (Figure 3.2). However, every step in this processing chain of transformation and integration has great potential for introducing errors into the resultant integrated data products so we shall describe here the approach we are following in the design of the Hydrologic Information System (HIS).

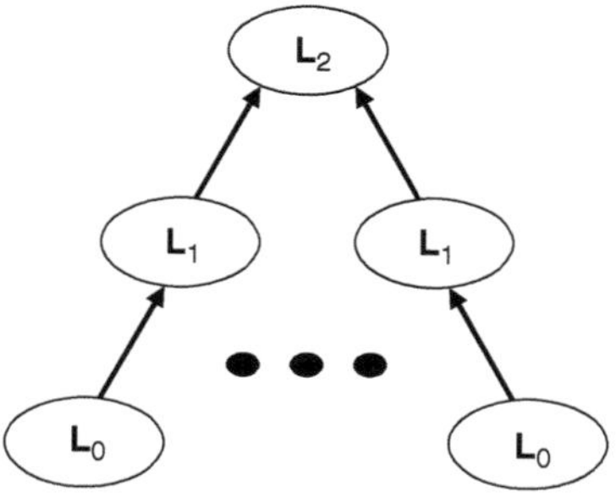

FIGURE 3.2
Hierarchy of data levels where L_0 is the raw or source data, L_1 is any modification to an L_0 dataset, and L_2 is any combination of L_0 and/or L_1 datasets into an integrated dataset.

Computing and communication systems enable us to access and integrate data over widening spatial, temporal, and thematic extents. This process of data integration can be expected to increase for the indefinite future across scientific disciplines. A similar type of process, often called data fusion, has been used in military and intelligence communities for decades albeit for obviously different purposes (Basir and Shen 1996). However, the conduct of science through the scientific method has certain key aspects that require attention and may be benefited from a method that helps to:

1. Ensure the self-consistency of source data.
2. Ensure the self-consistency of composite data (i.e., integrated) derived from multiple source data.
3. Reproduce analyses, using source or composite data, conducted at time t_1 again at time t_2 when some or all source data components involved may have had changes made to correct errors or anomalies.
4. Decompose a composite data object to enable the identification of components presenting inconsistencies with respect to other data components enabling the unambiguous identification of the components that have changed and their effect on the time t_2 analysis while preserving the ability to reproduce the time t_1 result.
5. Preserve the source material and traceability forward to derived data objects as well as the backward traceability to the parent data objects for integrity checking as well as proper attribution of authorship or ownership.

A methodology to support these requirements should ensure a minimal level of self-consistency of fundamental digital objects, such as files, according to the conventions of the given scientific discipline, "domain-specific standards," being addressed as well as enabling the declaration of the dependencies of an integrated (i.e., composite) or derived digital object on its components. This idea of dependency is similar to the notion of functional dependency found in the relational data model as described by (Ullman 1980) but pertains to data files rather than the attributes (or parameters) within a file or database. For example, attributes may span multiple data files. Such a method must also define the means by which self-consistency is achieved, the way it is verified and how these features are provided for the integrated data (i.e., composite) data object. To meet these objectives, we need a new basis for quality control predicated on a formal model. A preliminary version of such a model is presented.

To begin with, a notion of *levels-of-data* (Level 0, Level 1, Level 2) is useful to establish a precise hierarchy and nomenclature for discussing the differences between source data and derived data in its various forms. The hierarchical relationship among these levels is shown in Figure 3.2 and described here. This hierarchy provides a framework for describing the heritage of data products resulting from processing that combines data owned by others. In particular, the earth science community has used similar notions regarding *levels-of-data* for decades as an aid in declaring the degree of calibration and corrections, aspects of quality control, that have been applied to a particular data file (cf. e.g., Kidwell 1997). In the following discussion, the term digital object is used as a generalization for any type of digital data of which the most common, to date, is the familiar data file. However, the ideas presented in this paper apply to any data locus whether in memory or within communication systems that are without a representation in a file system. Therefore, the term ADO, is used to emphasize the implementation independence of the approach described here. This term has been introduced previously in Helly et al. 1999 and 2002.

Level 0 (**L_0**) data are simple digital objects in the sense that they are not composed from other digital data objects and are internally consistent in terms of purpose, measurement

Table 3.2

Valid transitions between data levels (0-prohibited, 1-allowed).

FROM	TO			
		L_0	L_1	L_2
	L_0	0	1	0
	L_1	1	1	1
	L_2	1	0	1

units, projections, coding, and so forth. This is the original, master copy as received from data author and is differentiated from the other levels for that reason. $\mathbf{L_0}$ data may be in any format since it must accommodate proprietary data formats for digital objects such as geographic information system coverages (e.g., ARCINFO™), image file formats, DBMS databases, and other proprietary formats such as SAS™, NetCDF, HDF, and so forth.

Level 1 ($\mathbf{L_1}$) data are simple digital objects that result from any modification of an $\mathbf{L_0}$ digital object and distinguished by the fact they directly back (one-to-one) to a single $\mathbf{L_0}$ predecessor. Typical modifications include conversion of units, nomenclature, map projections, and so forth in preparation for further integration. Nominally, such modifications are done to regularize or normalize the individual digital objects for subsequent analysis or further integration. However, any modification, even a common and seemingly minimal change such as compressing blanks from a file, makes the resultant file a new digital object. It is different from the $\mathbf{L_0}$ object provided by the data submitter and can contain errors introduced during the processing that produced the new ADO. To facilitate portability and universal access the recommended format is a flat ASCII file.[1] It is, however, impractical to insist on this convention as described earlier. $\mathbf{L_1}$, therefore, must permit non-ASCII (e.g., binary) formats with that format defined and documented in associated metadata and its heritage from an $\mathbf{L_0}$ object described.

Level 2 ($\mathbf{L_2}$) digital objects are derived from more than one $\mathbf{L_1}$ file and are composite (or compound) digital objects. The definition of this level is driven by the need to identify composite digital objects to emphasize the unique Quality Control (QC) required to ensure consistency across the files being integrated. To ensure maximum transportability of data, flat-ASCII files with publicly defined formats, seem preferable. However, for the same reasons as in $\mathbf{L_0}$ and $\mathbf{L_1}$, the $\mathbf{L_2}$ definition permits proprietary data formats.

Transitions between Levels

Since the transformation of data from one level to another should be controlled and verified, it is useful to formalize these transitions and associate wiht them specific kinds of QC processing. Therefore, we define the allowable transitions between the data levels in Table 3.2 and illustrate them in Figure 3.3.

[1] This is a file encoded in ASCII with parameters organized into columns and individual observations as rows (i.e., records). Not all data are amenable to this format, particularly hierarchical data, and it is presented as a goal not a rule.

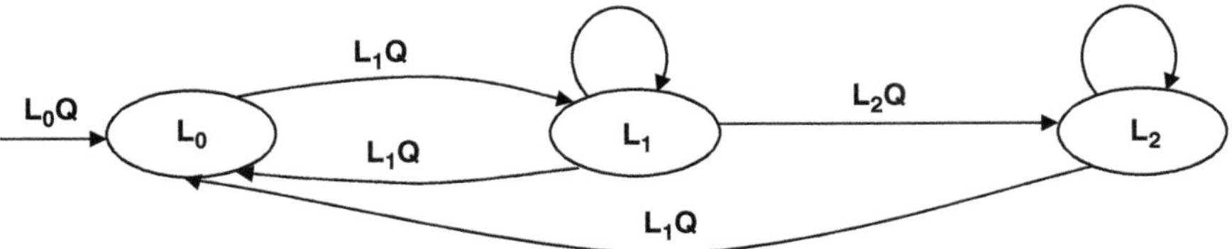

FIGURE 3.3
Transition diagram depicting relationship of data levels and ADRC processing.

The prohibited transitions, the cells with 0 in them, are perhaps the most enlightening. For example, an $\mathbf{L_0}$ object can never become another $\mathbf{L_0}$ object. However, it is possible for infinite chains of heritability through $\mathbf{L_1}$ and $\mathbf{L_2}$ objects differentiated only by their simple or compound derivation. An $\mathbf{L_1}$ or $\mathbf{L_2}$ digital object represents both the effort of the source data author(s) as well as the value-added by the postprocessing of the data. Both efforts represent legitimate intellectual and processing contributions and both deserve recognition from an intellectual property perspective. It is also true that the $\mathbf{L_1}$ and $\mathbf{L_2}$ objects require curation and therefore they are allowed to be cast into an $\mathbf{L_0}$ object. The challenge is to maintain the heritage of the object in an effective manner.

3.1.2.3 Quality Control Processing

Quality Control (QC) is generally an arbitrary, domain-specific process of anomaly detection and correction. To this general conception let us add reporting and refer to the entire process as Anomaly Detection, Reporting, and Correction (ADRC). Metadata (Michener et al. 1997) plays a central role in the ADRC process and is a widely discussed topic with some of the most active and authoritative work in the library community exemplified by the Dublin Core (14) and the Resource Description Framework (RDF) (15) efforts, the long history of spatial data (16) and now with the advent of numerous XML efforts to encode metadata of all varieties. A detailed discussion of metadata is beyond our immediate purpose and the merits of various alternatives will not be discussed here. For our purpose, however encoded, we require metadata content sufficient for accurately processing and comprehending the data, finding the data in a search, and for the publication and auditing of the data; particularly in support of the ADRC processing. An example of a type of sufficient to this purpose, called BKM, is described in Helly et al. (1999). Figure 3.3 depicts the relationship of the QC levels to the levels of data.

Level 0 QC ($\mathbf{L_0Q}$) is ill-defined and to a large degree implicit in the experimental and data acquisition methods used by the author of the data. It is recommended that data authors perform the equivalent of $\mathbf{L_1Q}$ (see below).

Level 1 QC ($\mathbf{L_1Q}$) involves the use of standard univariate statistical descriptions including ranges, first and second moments, percentiles and frequency distributions such as those described by Tukey (1977) and found in PROC UNIVARIATE of the Statistical Analysis System™ (13). This is done to characterize the data and provide a basis for the extraction of thematic information based on the content of the data. For example, the *.log and *.lst files produced by SAS, or the equivalent in other software systems, may be subsequently processed themselves to extract data such as species names from which a species list may be built. This level of processing is also suitable for producing a standardized report in preparation for integration into an $\mathbf{L_2}$ data object.

Level 2 QC ($\mathbf{L_2Q}$) is used to vet the consistency and coherency of the integrated result of multiple $\mathbf{L_1}$ data files. For example, although two files may have successfully passed through $\mathbf{L_1Q}$ and are therefore asserted to be individually self-consistent, it is possible that data anomalies may emerge when such data are merged or concatenated. An example would be a measurement of the same parameter taken with two different instruments having inconsistent calibration. In such a case the measurements would reflect a correct pattern of measurement variability within a reasonable range but would ambiguously estimate the actual value of the parameter. The resolution of such discrepancies typically comes down to expert opinion and various methods adjustment as it would with a person with two inconsistent timepieces.

Logs and listings extracted during QC processing, as described later, can be stored along with the $\mathbf{L_1}$ and $\mathbf{L_2}$ data themselves as part of the metadata by bundling them in the archive file. Nonquantitative data such as lists of species names, geographical names such as station identifiers and placenames can be extracted for comparison with gazetteers (i.e., standard lists of geographic placenames) and other authoritative data.

3.1.3 Functions Needed for the Publication of Scientific Data

In order to support the concepts of the previous sections, we have identified and described a set of basic functions for the publication of scientific data (REF). These are reproduced here in Table 3.3. Many modern systems contain at least a subset of these functions but the ensuing discussion of the HIS incorporates them all so we will refer back to this table throughout the exposition of the HIS design.

3.2 Building the Hydrologic Information System Digital Library

The HIS digital library design is based on a federated design strategy that has the following three key design principles:

1. A set of data servers are configured and managed as peers as depicted in Figure 3.5.
2. The storage of data is based on the concepts of ADOs with each ADO's attendant metadata stored in particular interoperable file representation (*.mif) and we think in terms of (ADO, mif) pairs.
3. The system implements the set of functions described in Table 3.3.

The resulting federated data network provides internet-based access to the collective data resources of the entire CUAHSI HIS federated system through common programming and user interfaces, a common metadata schema and controlled vocabulary to enable network-wide searching for and retrieval of data. It provides this user-level consistency while enabling the storage and management of heterogeneous, multisource data.

The digital library technology is based on work supported by NSF's National Science Digital Library program and will be integrated with the GEON system through jointly managed interfaces. The HIS federated digital library will provide interfaces to commonly used analytic applications such as GIS and visualization tools.

3.2.1 Conceptual Design

The conceptual design for a generalized Federated Digital Library (FDL) system must bridge-the-gap between empirical data sources and applications. The gap is the ability to obtain the important metadata information and the discovery and acquisition of the

Table 3.3

Functions for controlled publication of scientific data. (Modified from Helly 2000.)

Function	Purpose
User registration	Assignment of a user ID and password to a given user while acquiring the user's email address and related contact information. This is used for auditing data access and communication with users.
Data acquisition	Data contribution or submissions (i.e., "upload") and the entry of, at least, a minimal set of metadata. This initiates the automatic creation of a unique name for the ADO and a transportable metadata file which is bundled within the ADO.
Search and retrieval	A search system providing for spatial, temporal, and thematic (e.g., keyword) queries based on the content of the metadata.
Deletion control	The ability to delete an ADO must be tightly controlled to prevent the arbitrary deletion of data that users have copied. In a manner analogous to journal articles, one should not unpublish data. Errata can be accommodated by publishing a revision (i.e., new version) of the data. An important special case to consider is the editorial peer-review process. This requires confidentiality and the ability to remove an ADO if not accepted for peer-reviewed publication. A looser deletion policy would allow deletion of data if it had never been copied.
Assignment of persistent names	The persistent name (i.e., accession number in Figure 3.1) of an ADO is used within the data repository to access the ADO, to monitor updates of previously published ADOs, identify the retrieval of ADOs by users, notify users of anomalies or issues related to an ADO, establish precedence by publication date, and enable citability in other publications.
Quality control policy and methods	This function can exist (or not exist) to varying degrees. It is exemplified by peer-review or nonpeer-review and by anomaly detection and reporting. It must be explicitly stated. Some investigation is beginning on approaches to semiautomating QA/QC for specific types of data.
Access control	Access control enables the data contributor to specify a password known only to him or her and that may be subsequently provided to other users to access the contributed ADO. This enables data submitters to independently control access to their published data. Any user attempting to retrieve a password-protected ADO from the system must obtain that password from the contributor of the data.
Traceability of data heritage	A mechanism for establishing the heritage of data contained within an ADO is required to inform the user of the measured, derived, or computed nature of the data. This is also essential to preserving rights in intellectual property analogous to claims of copyright or trademark.
Bulk data transfer	The ability to upload or download large numbers of ADOs and *.mif files into from the system without manually selecting each individual object.

data. As shown in Figure 3.4, this can be viewed as a hierarchy with the empirical data sources at the bottom and the analytical applications layer at the top. In the middle of these two, is the federated digital library layer providing the data and computing services needed to ingest, organize, search, and retrieve digital objects and cast them into the appropriate representation required by a particular data model.

The term data model is heavily over-loaded with meanings. There is no single, precise definition of the term but in a very general sense, data models are used to store, in a computer system, the information needed to enable computing applications to be built to investigate conceptual models. Conceptual models provide a basis for the design of corresponding data models. For example, the mass reservoirs and fluxes of the standard diagram of the hydrologic cycle have properties that must be represented in a mathematical model to be run on a computer to compute the mass balance of the cycle using empirical

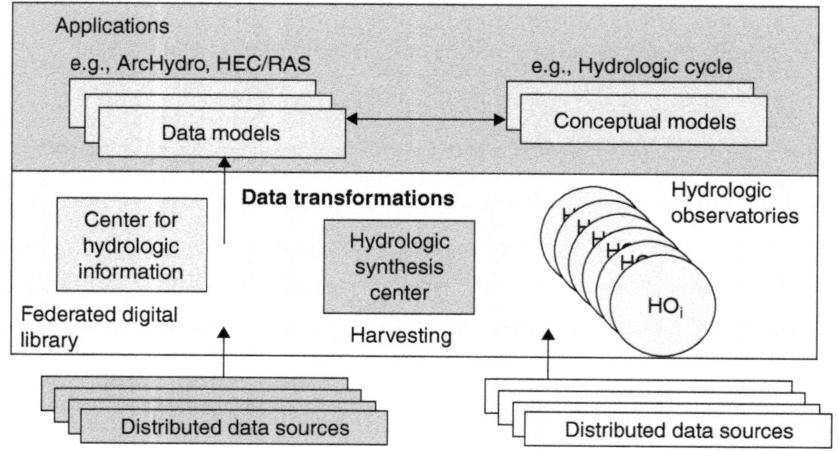

FIGURE 3.4
The hierarchical structure of the HIS digital library design is shown here with data sources listed at the bottom, components of the CUAHSI system in the middle as federated network nodes and end-user applications at the top corresponding to data models used by application software in support research related to corresponding conceptual models of nature.

data. The representation of these properties is the data model for that mathematical model. However, there are data models at various levels of computer system architecture ranging from the hardware representation of numbers to those used to design database management system applications. The reader is encouraged to consult standard computer science textbooks for more on the interesting and diverse topic of data models.

Within the FDL, data are organized into collections and the collections can be arbitrarily defined by the managers of an FDL node. For example, a Neuse River Hydrologic Observatory might have one collection for all of its data or a set of collections broken into useful categories such as atmosphere, surface water, and groundwater or by major subbasin. Essentially any type of collection may be defined according to the needs of the researchers.

3.2.2 System Architecture

The system architecture for an HIS digital library network node is depicted in Figure 3.5. A node in a communications network is a locus where some type of information processing takes place. It may be information production, consumption, or switching. It is important to recognize that, although we talk here mainly about the properties of a single digital library network node, the network we keep referring to would be made up of a set of computers located at hydrologic observatories and laboratories. The fact that they are *federated* means that they operate with a common governance structure and set of standards, conventions, and procedures. The main infrastructure components are the Storage Resource Broker (SRB) middleware and a metadata catalog running on a server node using the Linux operating system. The harvesting process listed in the center is a set of programs written in various languages including Perl and shell scripting languages (e.g., Bash) to extract data and metadata from sources and generate the (Boutilier 1999) pairs (see Figure 3.7). At the time of this writing, these programs are still undergoing development so a detailed enumeration and description of them would be quickly outdated. The interested reader can refer http://cuahsi.sdsc.edu for the latest version of these codes. However, there is a code called adoCreator.pl used to produce the pairs (Boutilier 1999) and a separate set of codes to

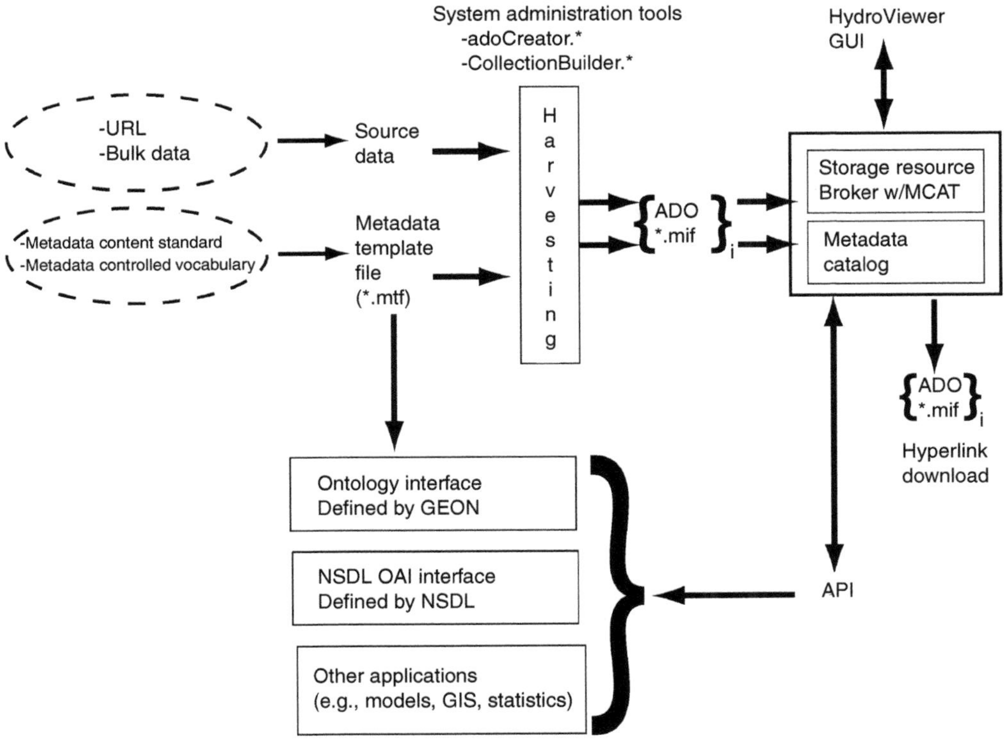

FIGURE 3.5
Generalized system architecture of federated digital library nodes. The host computer for this configuration may range from a laptop to supercomputer-scale servers using exclusively open-source software. The preferred metadata catalog DBMS is Postgres although the proprietary DBMS Oracle can also be used.

install and configure the digital library server node corresponding to the CollectionBuilder tool listed in Figure 3.5. Once created, the ADOs are loaded into the SRB and the metadata contained in the *.mif files is loaded into the metadata catalog using a loading code.

3.2.2.1 External Interfaces

External interfaces are provided through both a REST and a SOAP protocol to support the API shown in Figure 3.5. Both interfaces provide http-based access to web services with responses to the http-based queries and commands encoded in XML. Here is an example URL that searches for collection=CUAHSI_NEUSE and hence finds all ADOs. Please note that these particular URLs are meant to be illustrative of the syntax used and they will not work as written at the time you read this. The naming and exact construction is still evolving.

```
http://cuahsi.sdsc.edu/DL/beta/cgi-bin/rest-
server/rest.cgi?verb=FindADO&collection=CUAHSI_NEUSE&title=
&keywords=&creator=&coverage=&date=&type=
```

The REST service has been expanded and now includes: MoveADO, FindADO, and GetADOurl. Also, the FindADO verb now also returns URLS to the discovered objects.

The URL is in the <record> <header> for each object. Example URLs for the three REST verbs are:

```
http://cuahsi.sdsc.edu/DL/beta/cgi-bin/rest-
server/rest.cgi?verb=MoveADO&collection=CUAHSI-HIS-
Neuse&identifier=CUAHSI_Neuse_20040909120134_20040909120134.
mif&targetLocation=ftp://dotnet.sdsc.edu/Cuahsi

http://cuahsi.sdsc.edu/DL/beta/cgi-bin/rest-
server/rest.cgi?verb=FindADO&collection=CUAHSI-HIS-
Neuse&title=&keywords=&creator=&coverage=&date=&type=

http://cuahsi.sdsc.edu/DL/beta/cgi-bin/rest-
server/rest.cgi?verb=GetADOurl&collection=CUAHSI-HIS-
Neuse&identifier=CUAHSI_Neuse_20040909120134_20040909120134.mif
```

The external interfaces avoid the need for interaction with any client software, such as HydroViewer that we describe later. These external interfaces are the principal means for integrating with other data systems and the major method for interoperability. The interoperability needed for federation is provided internally by the SRB software using a federated, system-level metadata catalog and by an application-level metadata catalog synchronization strategy that harvests the metadata contained in the *.mif files from network nodes in the federation. The federation mechanism makes it possible for each digital library network node to "know" about the contents of the data holdings of each other federated network node. With this information each node can independently check for new metadata, information it does not have, and then obtain the *.mif files from the other nodes and update its local application-level metadata catalog.

3.2.2.2 Storage Resource Broker

The SDSC SRB is client–server middleware that provides a standardized interface for managing heterogeneous physical data resources over a network. SRB, in conjunction with the Metadata Catalog (MCAT), provides a way to access datasets and resources based on their attributes rather than their names or physical locations and unique physical and logical features and commands or drivers. It is important to note that in the HIS digital library, we differentiate the metadata catalog shown in Figure 3.5 from the MCAT described here. The MCAT is a "system" metadata catalog for the internal administration of the SRB while the HIS metadata catalog is an "application" metadata catalog.

3.2.2.3 HydroViewer Graphical User Interface

As an example of a potentially large class of client applications, we present some of the details about one that we have developed called HydroViewer. HydroViewer is a standalone Java application designed to provide search, retrieval, and activation of ADOs contained within the digital library. It provides the user with the ability to select from the various data collections within the federated digital library, such as the Neuse River or Santa Margarita River collections, to retrieve and view the metadata for any given ADO and to select one or more ADOs for transfer to the user's workstation for local utilization.

Referring to Figure 3.6, the large window is partitioned into three subwindows. The upper-left window is where the user selects the collections to be included in a search and to specify search parameters in terms of spatio-temporal and thematic criteria. The lower window

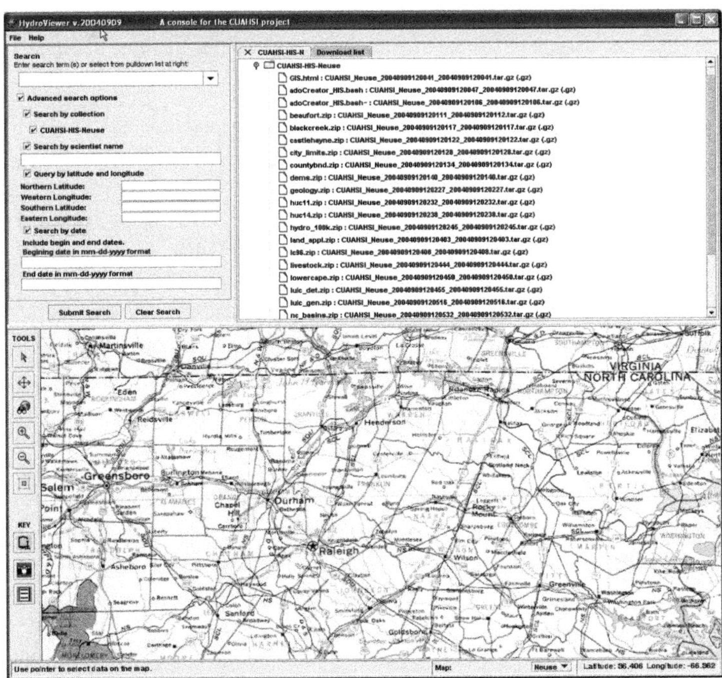

FIGURE 3.6
HydroViewer graphical user interface. This is a Java application that represents the larger class of potential graphical user interfaces (GUIs) that can be built on top of the basic digital library node infrastructure.

contains a basemap where the user can specify a geospatial search using the graphical tools listed on the left margin of the window. The map is also used to display the location of the data returned by the search. The upper left window returns the list of ADOs satisfying the search criteria and also provides access to the ADO metadata and activation by right clicking on a given object.

3.2.3 Arbitrary Digital Objects

In our approach, the basic objects to be managed and published in digital library collections are computer-system files that we refer to as ADOs, *arbitrary digital objects*. Just as the name implies, ADOs can be any type of digital object ranging from ASCII data to MPEG video and anything else that can be stored in a computer file system. For example, the entire contents of a DataBase Management System (DBMS), to the extent that it can be exported into a file, can be stored.

The ADOs are typically stored using an archive file format such as *.tar or *.zip. The reason for doing this is that it enables the original file, which might be named something like rivers.txt that is unlikely to be unique in the world, to be repackaged into something that is unique across the relevant population of digital library collections without altering its original name or structure. The archive file formats have other advantages. For example, when a set of files, such as those containing individual fields surveys over a year, logically comprise a single data object they can be stored together in one ADO.

Since ADOs contain one or more data files with arbitrary file formats and internal logical structure, their contents cannot be directly searched. Searching is performed using a catalog

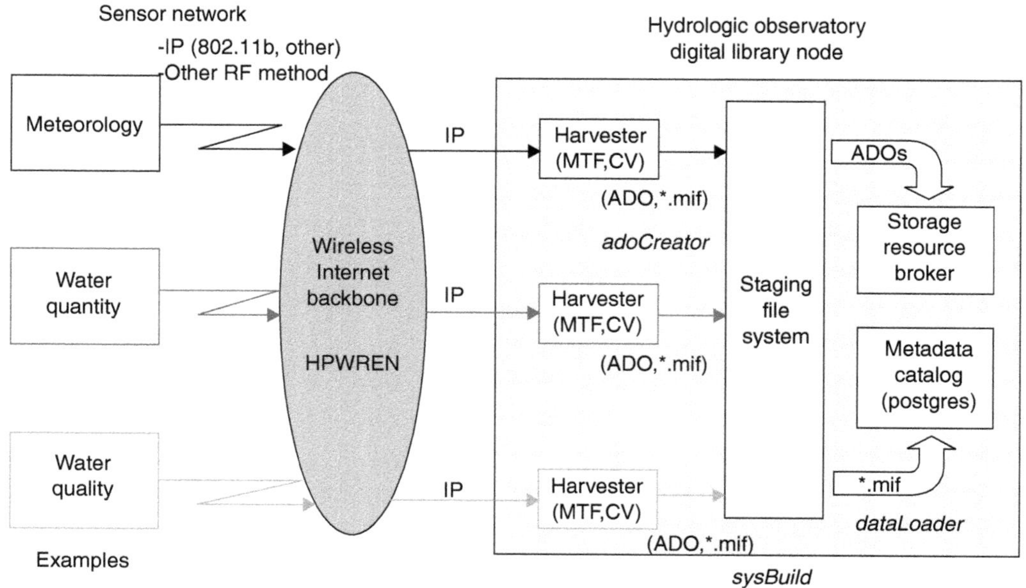

FIGURE 3.7
Sensor network integration with a node of the HIS digital library. HPWREN is a wireless communications network supported by the National Science Foundation under the direction of Hans–Werner Braun, National Laboratory for Applied Network Research, University of California, San Diego.

of metadata based on information provided by the data contributor during the data publication process. This metadata is stored in *.mif (metadata interchange format) data files separately. The metadata are used to populate a metadata catalog to enable the efficient search and retrieval of the ADOs from a digital library.

3.2.4 Metadata

Perhaps one of the most frustrating and tedious aspects of building digital library systems are the description and generation of metadata (Figure 3.7). However, without it these systems cannot operate. In our design, metadata provides the search space for data queries as well as the configuration control for the structure of the metadata catalog and the specification of the parameter set for the harvesting of data and metadata. Our metadata approach (Helly et al. 2002), contains two key components: (1) the metadata template file (*.mtf) and (2) the metadata interchange file (*.mif). The current definitions of these can be found on the HIS web site (http://cuahsi.sdsc.edu).

The metadata template file (*.mtf) is the file that defines the content of any given *.mif file. It defines the metadata schema according to a canonical, block-oriented format such that there are blocks (i.e., sets of rows) for the collection level and blocks for each type of ADO. Canonical means standard or accepted and is often used in computer science to connote the standard definition or reference configuration or form. Our canonical definitions are published on the web site and will eventually be published in a separate collection of system administration ADOs. We refer to these as canonical because they define a standard minimum set of parameters whose nomenclature may not be arbitrarily altered by an individual user. These canonical parameters are assumed by the query codes and related applications to enable predefined queries to be executed reliably. This dependency on the canonical parameters is a necessary but not sufficient condition to ensure a complete and consistent search; we also need a controlled vocabulary.

In order for queries to be reliable, it is necessary that the keywords and phrases used as search criteria are the same as those used by the data producers. This requires a common dictionary, thesaurus for words (i.e., controlled vocabulary) and an additional syntax for phrases. If this is not available and adhered to by data producers and query developers, then there is no guarantee that any given search will return all of the appropriate digital objects. In other words there is no guarantee of the completeness or accuracy of any given search.

To some degree, this ambiguity in search terms can be compensated for using the synonymy and semantic mappings such as is often attempted with ontologies. An ontology has come to mean, in this context, the set of information necessary to properly interpret ambiguous relationships between concepts related to data. It is an abused term that has its roots in philosophy and the interested reader is encouraged to consider its roots in contrast to how it is used today. However, these are at best heuristics that do not provide surety of search completeness. They are not a substitute for controlled vocabulary and metadata content definition.

References

Basir, O.A. and H.C. Shen (1996). Modeling and fusing uncertain multi-sensory data. *Journal of Robotic Systems* 13: 95–109.

Boutilier, C. (1999). Knowledge representation for stochastic decision processes, in *Artificial Intelligence Today. Recent Trends and Developments*, M.J. Wooldridge and M. Veloso, Eds., Berlin, Germany, Springer-Verlag, pp. 111–152.

Helly, J. (1998). New concepts of publication. *Nature* 393: 107.

Helly, J., T.T. Elvins et al. (1999). A method for interoperable digital libraries and data repositories. *Future Generation Computer Systems*, Elsevier 16: 21–28.

Helly, J. and N.M. Kelly (2001). Collaborative management of natural resources in San Diego Bay. *Coastal Management* 29: 117–132.

Helly, J., T.T. Elvins et al. (2002). Controlled publication of digital scientific data. *Communications of the ACM* 45: 97–101.

Kidwell, K. (1997). *NOAA Polar Orbiter Data User's Guide, Tiros-N, NOAA-6, NOAA-7, NOAA-8, NOAA-9, NOAA-10, NOAA-11, NOAA-12, NOAA-13 and NOAA-14*. Suitland, MD, U. S. Department of Commerce, National Oceanographic and Atmospheric Administration, National Environment Satellite, Data, and Information Service, National Climatic Data Center, Climate Services Division, Satellite Services Branch, E/CC33.

Michener, W.K., J.W. Brunt et al. (1997). Nongeospatial metadata for the ecological sciences. *Ecological Applications* 7: 330–342.

National Science and Technology Council (1996). "Foundation for America's information future. Washingon, DC, The White House, p. 131.

Tukey, J.W. (1977). *Exploratory Data Analysis*. Reading, MA, Addison-Wesley.

Ullman, J.D. (1980). *Principles of Database Systems*. Potomac, MD, Computer Science Press.

e

al

re.

4

Hydrologic Metadata

Michael Piasecki

Contents

4.1 Introduction to Metadata

The word "meta" has its origin in the Greek language and means *with, after, about, between,* or *among* [1,2]. It is typically used as a prefix meaning "one level of description higher," and has found its way into many words that we are familiar with though slightly different meanings emerge. For example, when used in *metabolism* or *metamorphosis* it refers to changes or alterations, while its use in *meta-mathematics* or *meta-ethics* indicates an interest in the concepts and results of the named discipline. There is also a widespread use of the word *meta* in the field of Chemistry, where it is used to differentiate an isomer or a polymer related to a specified compound like in *metaldehyde,* or to denote an oxyacid, which is a lower hydrated form of an anhydride or a salt of such an acid, as in *metaphosphoric acid.*

The use of the word *meta* in Metadata obviously attempts to introduce the concept of "about" with respect to data and in its shortest definition means "data about data." Although this is a very concise way of defining it, other definitions have emerged that attempt to describe better what the concept of metadata really entails. A more "big picture" way of thinking about metadata is as "the sum total of what one can say about any information object at any level of aggregation" [3]. Another attempt to capture the gist of metadata has been suggested by Batchelder [4] who states that "metadata is information about a thing, apart from the thing itself." There are several other definitions that attempt to capture a "deeper" meaning of metadata, and while the discussion is ongoing on how to best define metadata, let us settle for the shortest version "data about data" for brevity sake. As you will see in the following sections, we will not lose anything about the concept of metadata by doing so, and perhaps it should be left to you as a readers' prerogative to define your own version of metadata and what it will mean to you in the end.

Over the past few decades the term metadata has emerged as an important concept to describe information objects and it would be unthinkable not to talk about metadata in the context of making the information age a reality. However, metadata is nothing new; in fact, metadata permeates our daily life at about every level and all of us use metadata constantly be it at home, at school or work, or during off-hours. The reality is that we are constantly bombarded with metadata information otherwise we would not know what kind of juice is in that container, what fabric this sweater is made of, and where to make turns when driving to a location you have not been before. Any attempt of our brain to bring structure to complexity is in fact a manifestation of metadata of some sort. This entails the labeling of containers holding frozen food, the organization of books in a bookshelf by topic and author, the fact that you like all knifes, spoons, and forks organized in this drawer and that dish-rack, and even appears when trying to put the Lego pieces into the yellow box in the left shelf while the PlayMobil figures should go into the red bin in the closet.

From the above examples it is clear that metadata is ubiquitous in our life and that it appears in many different forms and very different levels. A label on the frozen food container is a written (or human readable) form of metadata about an object as is the labeling of folders, or as is the author, title, publisher, ISBN number, year, edition, category, and location identifier of a book in a library. All these pieces of information tell us of what is inside, without us actually having to probe the contents to find out. This is a concept we tend to automatically embrace (not all of course, there is a saying that order is for the feeble minded only, while the genius masters the chaos) whenever we have to bring order to a large collection of things that are alike. On the other side we use an internal metadata framework that is stored and constantly altered in our brain cells only, that is, a very personal system that tends to work for us as an individual but not for others even if they live in the same household or are colleagues at the workplace; who does not know the situation in which somebody asks you to go down into the basement and find a certain item in the storage locker among heaps of boxes, shelves, and bags without an apparent (or human readable) ordering system leaving oneself at a loss, while the one who packed the room would know instantly where to go and look. The problems are manifold as (1) you are unable to figure out what the geospatial reference system is and you discover that what was meant to be "in the back of the room" turns out to be "the front of the house," (2) you may be told that you will find a certain item in a "box," which switches your search mode into what you think a box is, while in fact in your vocabulary you would need to look for a container, because you differentiate between cardboard (for a box) and plastic (for a container), or (3) your definition of what "rug" means is not comparable to what the other person associates with it. You could also try to discern what the ordering system is, just as you could inspect the book shelf of a friend to find out the cataloguing and shelving system, if you would have enough time at your hand. This is typically not the case, prompting the need to develop metadata approaches that permit you to better understand how the object you are looking at or for is described and then starting a well-informed search that will lead you to your objective in a reasonable amount of time.

Much more could be written about the use and philosophy of metadata. There are many more examples that one can draw from his/her personal life and even larger entities like corporations, government institutions, or research communities like the Hydrologic

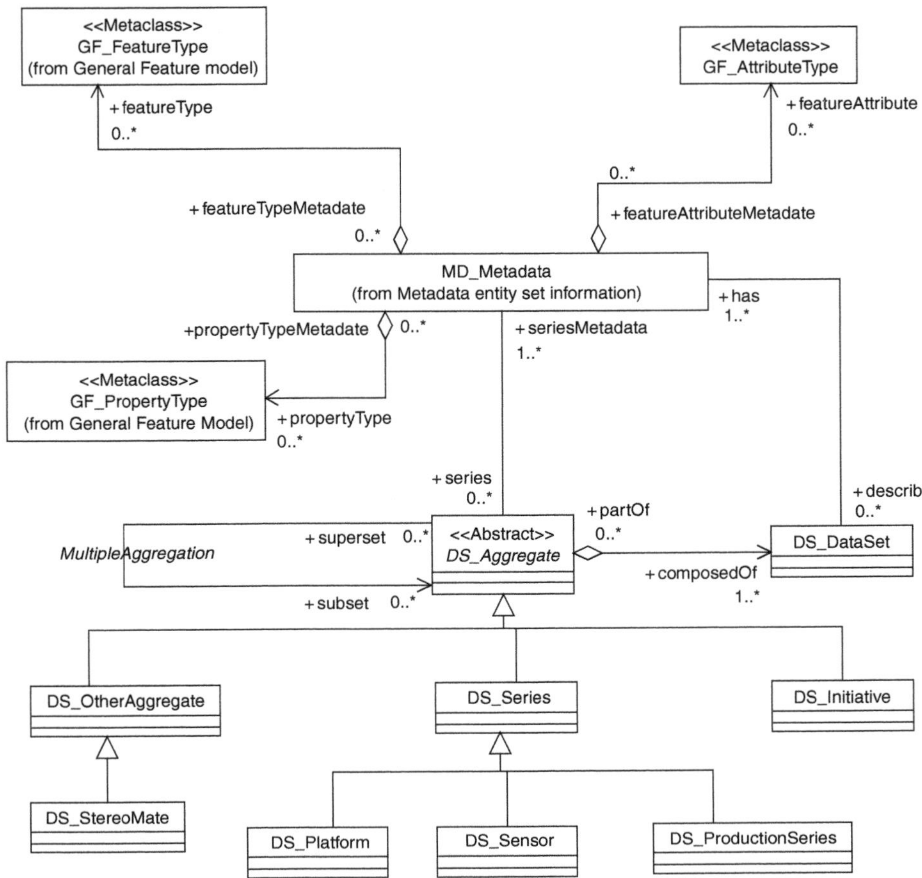

FIGURE 4.1
A generic metadata application.

Sciences community, for example, to fully map the complexity of the metadata world. It is sufficient here to stress the obvious need to use metadata in our daily life and its ubiquitous application everywhere we turn, and also to appreciate the underlying and implied difficulties we encounter when attempting to apply our personal metadata system across our individual boundaries (Figure 4.1). Just imagine then the difficulties that arise when trying to coordinate the metadata needs (and sometimes habits) among many individuals within a community like the Hydrologic Sciences, or even across communities. These thoughts will accompany us in the following sections where we will try to bring some order to the metadata world (a meta-metadata concept of sorts if you like) by outlining some of the approaches that have been undertaken to overcome the problems we tried to allude to in the receding paragraphs.

4.2 Definition of Metadata Categories

In the previous section we tried to point out some of the problems that arise when using metadata systems by using some, admittedly simple, examples from our daily life. It is therefore helpful to shed some light on what metadata is supposed to deliver and how it can be organized and categorized according to the specific purpose it is used for. If we look at an information object regardless of the physical or intellectual form it takes, be it a book, a dataset, or an image, then we can identify three distinct features. These are content, context, and structure, all of which need to be addressed or reflected through metadata. Content relates to what the object contains or is about and is obviously intrinsic to the information object, that is, if we desire to read a book about Huckleberry Finn then we would expect this particular story to be in the text and not any other story by Mark Twain. Context on the other side addresses the many "w's," for example, the who, what, why, where, and how aspects that are associated with the object's creation and as such can be categorized as extrinsic. In our small example that would address the author, date of print, publisher, ISBN number, and so on. Finally, structure provides information about a formal set of associations within or among individual information objects. In our example of a book, this would target the table of content, which would tell us that the book is organized into a preface, a foreword, an epilog, and a number of parts, each containing a number of chapters, a glossary, and an acknowledgment section.

The role of structure has been growing as computer processing capabilities become increasingly powerful and sophisticated. Information communities are aware that the more structured an information object is, the more that structure can be exploited for searching, manipulation, and interrelating with other information objects. Capturing, documenting, and enforcing that structure, however, require specific types of metadata. For example, an Internet resource provider might use metadata to refer to information being encoded into HTML metatags for the purposes of making a Web site easier to find. Individuals digitizing images might think of metadata as the information they enter into the header field for the digital file to record information about the image, the imaging process, and image rights. A hydrological science data archivist might use metadata to refer to the systems and research documentation necessary to run and interpret a stream of data coming from a stream gauge containing raw (or sensor) data. Metadata is also critical in personal information management and for ensuring effective information retrieval and accountability in recordkeeping — something that is becoming increasingly important with the rise of electronic commerce and digital government. In all these diverse interpretations, metadata not only identifies and describes an information object, but it also documents how that object behaves, its function and use, its relationship with other information objects, and how it should be managed.

In view of these aspects it is clear that metadata is essential in the development of networked digital information systems, but it also does not come as a surprise that the conception of metadata functionality is rather broad. In order to bring structure to the metadata realm it is helpful to break metadata down into distinctive subcategories that serve a specific task. There are a number of categorizations published in literature, [3,5,6] one of which is shown in Table 4.1.

The first block, Content or Search, is focused on descriptive metadata elements that are used to search and find the dataset that you are looking for including some generic elements that are ubiquitous to any dataset. The second and third blocks, Syntactic + Semantic or Use, contain metadata elements that you typically will not use for a search but which are necessary to describe how the information object is configured or structured so that it is clear to the user how it can be processed for further use, that is, for downloading, fusion, visualization, analysis, or any other action that a user might want to perform on the dataset. Although the above attempt of categorization is not a cut-and-dry operation because some metadata elements could also appear in a different block, it is helpful in recognizing that metadata serves a number of different purposes. Notice that these categorizations should not be understood as a mandatory step in defining a community metadata profile nor

TABLE 4.1

Metadata categorizations.

Type	Subcategory	Definition	Examples
Content (Search) (Piasecki and Bermudez [5])	General	Generic to all information resources, for example, the core of standards	Publisher/collector When collected Metadata version
	Specific	Specific to a certain type of information resource	Subclasses y/n Geospatial y/n Lineage y/n
Syntactic (Use)	Content part	Info about the structure of the information resource	Unit system Reference/datum Locator (URL)
	Syntactic part	Description of structure itself	Formats Sequence/order Relations/structure
Semantic (Use)		Description of the data contained in information resource	What purpose Number of data Units

do they represent a rule framework that needs to be adhered to. Rather, they keep the organization of metadata manageable and allow a selective targeting of certain metadata type groups for changes, improvements, adaptations, and extensions.

4.3 Metadata: Problems and Standardization

As we have learnt in the previous section, metadata can mean of lot of different things to different people. This leads to an extremely diverse metadata universe that suffers from uncoordinated metadata descriptions.

It is no surprise therefore that initiatives have emerged in recent years that try to deal with this situation by coordinating efforts to establish common standards. These standards would help an individual or group to set up a metadata profile that (1) would fit their data or information object description's needs, and also (2) make sure that the group uses descriptive elements that are understood by all involved. One can say that an attempt is being made to develop a common "language" so that everybody is on the same communication track. This objective is also supported by the recognition that many information objects share very similar features. For example, a geospatial reference need (latitude and longitude, for instance) is common for many datasets originating from very diverse user communities. Medical statistics that are linked to regions or cities, cadastral departments in government institutions that manage property delineations, and the hydrologic community, which needs to know about rainfall, wells, stream flows, or land use characteristics are all information objects that need to be associated with a point or region somewhere on the surface of the earth. In this section we will introduce standards published by the following five entities:

- Federal Geographic Data Committee, FGDC
- International Organization for Standardization, ISO
- Dublin Core Metadata Initiative, DCMI

- Global Change Master Directory, GCMD (NASA)
- ADEPT–DLESE–NASA, ADN

4.3.1 Federal Geographic Data Committee

Allthough the name says this is a committee and not a standard, the Federal Geographic Data Committe (FGDC) publishes the "Content Standards for Digital Geospatial Metadata," or CSDGM [7], whose latest version (v2) is officially referred to as CSDGM Version 2 - FGDC-STD-001-1998. The FGDC is a 19 member interagency (United States only) committee composed of representatives from the Executive Office of the President, Cabinet-level, and independent agencies charged with developing metadata standards within the United States as part of the National Spatial Data Infrastructure (NSDI) initiative. It is interesting to observe that the development of the standard is the result of a presidential executive order. In April 1994, the then President Clinton signed the executive order 12906, "Coordinating Geographic Data Acquisition and Access: The National Spatial Data Infrastructure." Section 3, Development of a National Geospatial Data Clearinghouse, paragraph (b) states: "Standardized Documentation of Data, . . . each agency shall document all new geospatial data it collects or produces, either directly or indirectly, using the standard under development by the FGDC, and make that standardized documentation electronically accessible to the Clearinghouse network." In other words, this standard is open to everyone who wants to use it and the document can be downloaded freely from the FGDC Web site (see Reference 7).

The scope of the standard is (citation from Reference 7):

> The standard was developed from the perspective of defining the information required by a prospective user to determine the availability of a set of geospatial data, to determine the fitness the set of geospatial data for an intended use, to determine the means of accessing the set of geospatial data, and to successfully transfer the set of geospatial data. As such, the standard establishes the names of data elements and compound elements to be used for these purposes, the definitions of these data elements and compound elements, and information about the values that are to be provided for the data elements. The standard does not specify the means by which this information is organized in a computer system or in a data transfer, nor the means by which this information is transmitted, communicated, or presented to the user.

The last part of this statement is important because it alludes to the fact that the user will only receive a document that represents the metadata concept through graphics and text. How this standard, or any other for that matter, is implemented in a "machine-readable" format is left to the end user. In case of the CSDGM, the OpenGIS Consortium (OGC) has taken on the task to publish the standard in machine readable format. At the time of writing this book (Summer 2004) the OGC was publishing the GML, or Geographic Markup Language [8], using the eXtensible Markup Language, XML [9], in fact, the Schema implementation, for encoding the standard. In other words, the CSDGM plus XML Schema makes the GML.

As part of the ongoing efforts, the FGDC has also promoted the development of the so-called mappings to other standards, namely the ISO standards, in order to provide a means to comply with other widely used standards. In fact, at the time of writing this manuscript FGDC had announced plans that the third version of the CSDGM would be fully compliant with ISO standard 19115:2003 making the FGDC standard interchangeable with the ISO 19115:2003. In addition, the FGDC has used a number of sub-committees to define more domain specific community profiles, for instance, for marine and coastal, soil, ground transportation, cartographic, and cadastral datasets to name just a few. Many of these community profiles were in different stages of development (Summer 2004) and it remains to be seen how far they evolve in future.

4.3.2 International Organization for Standardization 19115:2003

The International Standards Organization, ISO, combines under its umbrella many of the national standards organizations, like the DIN (Deutsche Industrie Norm) for Germany or the ANSI (American National Standards Institute) for the United States to name just a few. Obviously, metadata standards are only a very small fraction of norms that are being addressed by the ISO as its scope is much bigger. Inside the ISO, technical committees take on the task to discuss and formulate norms and standards for the various disciplines and areas. The technical committee responsible for the geographic information series of standards (the 19xxx series) is the ISO/TC211 Geographic Information/Geomatics group that currently has 58-member organizations divided into participating and observing members. It also maintains a list of external and internal liaisons to other groups like, not surprisingly, the Open GIS Consortium (external) and other technical committees inside ISO. The scope is (we cite here from Reference 8):

> Standardization in the field of digital geographic information. This work aims to establish a structured set of standards for information concerning objects or phenomena that are directly or indirectly associated with a location relative to the Earth. These standards may specify, for geographic information, methods, tools and services for data management (including definition and description), acquiring, processing, analyzing, accessing, presenting and transferring such data in digital/electronic form between different users, systems and locations. The work shall link to appropriate standards for information technology and data where possible, and provide a framework for the development of sector-specific applications using geographic data.

The ISO 19115:2003 was moved from draft to official recommendation status in 2003, which is indicated by the ":2003" extension. This however does not mean that the TC211 has put this standard to rest, rather, the members continue to expand and update the standard based on the recommendations and feedback they receive from the user groups, using the standard for their purpose. It is also important to note in this context that while the 19115 is the "Metadata Standard," it makes reference to other standards in the 191xx group or series of standards. For example, the TC211 deemed it important enough to develop a separate standard for Temporal Schema (ISO 19108), which can be referenced when using the ISO 19115:2003. It is perhaps best to view the ISO 19115:2003 as the umbrella standard that contains many of the metadata packages, which geographic data user communities might want to use with links to typically smaller and specialized norms for inclusion. In Table 4.2 we have listed the other norms of the 191xx series that are indispensable when using the ISO 19115:2003 [9].

The above list is not complete, in fact there are currently some 40 norms, or projects as the TC211 likes to call them, being worked on that carry the prefix *Geographic Information*. This is a rather long list of course, and as it turns out some of the projects overlap with others or are specializations of a more generic one. Some of these overlaps are certainly due to the fact that different initiatives started at different points in time and that some of them are in fact dormant or that the workgroup has been disbanded. This hampers the ease of use of the 191xx series, and on the other side it provides a rich source of metadata concepts for almost anything that can be geographically referenced.

The ISO 19115 is organized into Unified Modeling Language (UML) packages each of which contains one or more metadata entities (UML classes) that contain elements (UML

Table 4.2

Some associated norms from the 191xx family.

Name	Title
ISO 19106[a]	Geographic Information — Profiles
ISO 19107[a]	Geographic Information — Spatial Schema
ISO 19108:2002	Geographic Information — Temporal Schema
ISO 19109[a]	Geographic Information — Rules for Application Schema
ISO 19110[a]	Geographic Information — Methodology for Feature Cataloging
ISO 19111:2003	Geographic Information — Spatial Reference by Coordinates
ISO 19112[a]	Geographic Information — Spatial Reference by Geographic Identifiers
ISO 19113:2002	Geographic Information — Quality Principles
ISO 19114[a]	Geographic Information — Quality Evaluation Procedures
ISO 19117[a]	Geographic Information — Portrayal
ISO 19118[a]	Geographic Information — Encoding
ISO 19130[a]	Geographic Information — Sensor and Data Models for Imagery and Gridded Data
ISO 19136[a]	Geographic Information — Geography Markup Language (GML)
ISO 19139[a]	Geographic Information — Metadata Implementation Specs

[a] To be published.

attributes) which identify the discrete units of metadata. It is important to separate the terminology in UML (packages–classes–attributes) from the one that is used in the ISO (section–entity–element). The ISO norm provides an example metadata application that we show here for demonstration purpose, Figure 4.2.

In this case you can see the central metadata block (ISO 19115) MD_Metadata that has a number of other metadata- or entity-elements (coming from other standards) attached to it, like the General Feature norm ISO 19109. The package Metadata Set Entity contains an entity called MD_Metadata that is mandatory for all datasets. As an entity can be an aggregate of other entities, the MD_Metadata itself is an aggregate of 11 other entities that are listed below and shown in Figure 4.2. These are

- MD_Identification
- MD_Constraints
- DQ_DataQuality
- MD_MaintenanceInformation
- MD_SpatialRepresentation
- MD_ReferenceSystem
- MD_ContentInformation
- MD_PortrayalCatalogueInformation
- MD_Distribution
- MD_MetadataExtensionInformation
- MD_ApplicationSchemaInformation

Any of the above entities (classes) may be an aggregation of a set of other entities (classes) or just contain a number of one or more metadata elements. It is interesting to note that the entity MD_Metadata has its own set of elements, even though its extent is largely being defined by the other entities. It is perhaps best to think of this structure as a hierarchical system into which the metadata entities (classes) and their elements are organized. You should also be familiar with the UML to better understand the involved UML syntax so

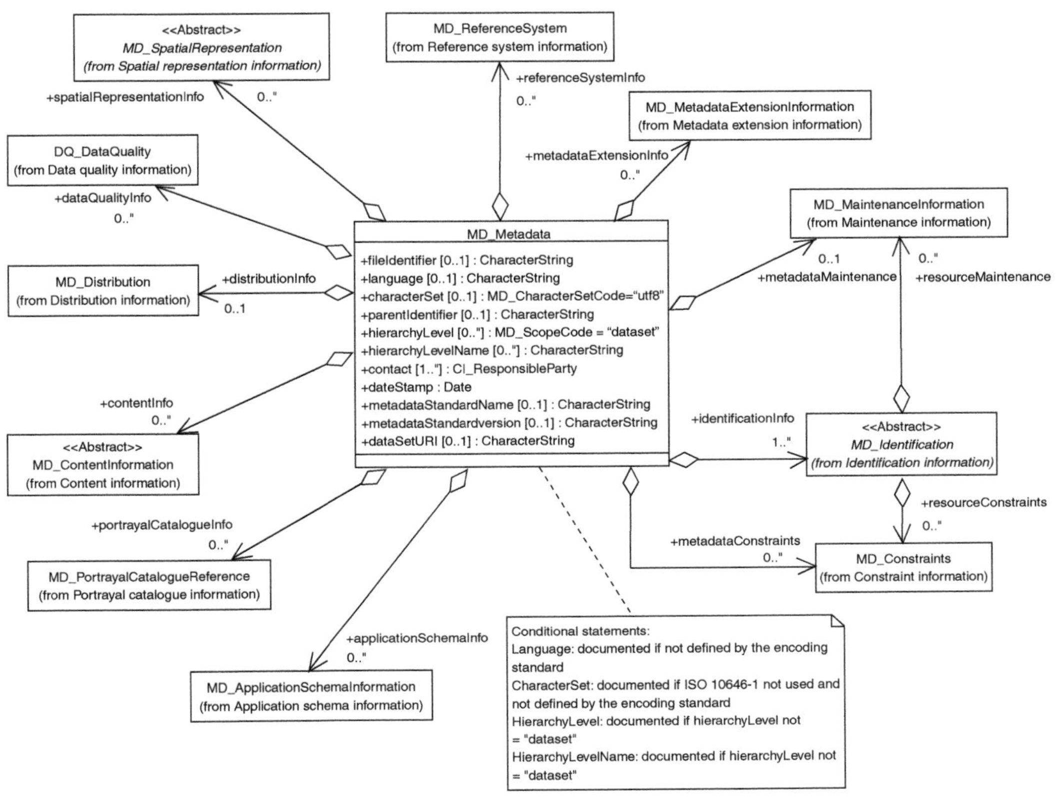

FIGURE 4.2
Metadata entities and elements making up the MD_Metadata Set.

that you know how to discover cardinalities and aggregation rules between entities and elements. If not then it is a good time for you to take a small detour to brush up on your knowledge of UML.

The ISO 19115:2003 documentation provides a detailed UML class diagram for all of the 11 entities and also provides a so-called Data Directory in which all entities and their elements are being listed with their proper syntax and a short description of what they are intended for.

Last but not the least we should also mention that the makers of the ISO 19115:2003 had a number of guidelines to give to the users. First of all, the TC211 realized that a standard that is to be used by a large and diverse community cannot possibly contain all possible profiles up-front. This task is much too daunting and they decided instead to provide a general framework only rather than going into details. To this end a set of rules has been established that will allow a specific user community to extend the current collection so that it fits its own needs which is a crucial aspect for community acceptance. Second, knowing that only a subset of all possible elements would be used for any community specific dataset, the TC211 decided to define a "core" set of metadata elements that should by ubiquitous for any geographic dataset description. This set encompasses 26 specific elements that have been identified from a number of the aforementioned entities, as shown in Table 4.3 [9]. Notice that the TC211 went even further by mandating the use of 7 of them (dark gray), while the other either have "optional" (15) or "conditional" (4, highlighted in light gray) status. The optional elements are deemed critical (that is why they are part of the core set)

Table 4.3

Metadata core elements as defined in ISO 19115:2003.

Parent class (entity)	Core elements	Parent class (entity)	Core elements
MD_Metadata	*fileIdentifier*	MD_Resolution	*equivalentScale*
MD_Metadata	*metadataStandard Name*	MD_Resolution	*Distance*
MD_Metadata	*metadataStandard Version*	CI_Responsible Party	*Role*
MD_Metadata	*language*	MD_Format	*Name*
MD_Metadata	*characterSet*	MD_Format	*Version*
MD_Metadata	*dateStamp*	CI_Citation.	*Title*
MD_DataIdentification	*geographicBox*	CI_Date	*Date*
MD_DataIdentification	*geographicIdentifier*	CI_Date	*dateType*
MD_DataIdentification.	*language*	CI_Online Resource	*Linkage*
MD_DataIdentification	*characterSet*	EX_Extent	*Ex_Temporal Extent*
MD_DataIdentification	*topicCategory*	EX_Extent	*EX_GeographicExtent*
MD_DataIdentification	*spatialRepresentation Type*	MD_Reference System	*Ref. SystemIdentifier*
MD_DataIdentification	*abstract*	LI_Lineage	*Statement*

Table 4.4

DCMI metadata core elements.

Type	Core elements	Type	Core elements
Element	*Contributor*	Element	*Publisher*
Element	*Coverage*	Element	*Relation*
Element	*Creator*	Element	*Rights*
Element	*Date*	Element	*Source*
Element	*Description*	Element	*Subject*
Element	*Format*	Element	*Title*
Element	*Identifier*	Element	*Type*
Element	*Language*		

but do not have to be used in case there is no real need to do so. The conditional elements become mandatory if a certain condition is fulfilled, typically the demand to use either one or another element.

It should be clear that a user community is at liberty to change the *optional* or *conditional* status of any element to *mandatory* if it is deemed necessary. It is not possible however to change the mandatory elements.

4.3.3 Dublin Core Metadata Initiative

The DCMI [10], has its roots in the library sciences but has grown to encompass more diversity of the years. The DCMI is an open, international, consensus-driven association of metadata practitioners motivated by the desire to make it easier for people to find information resources, using the Internet that currently has some 50 participating institutions or individuals from around the world. Like the ISO 19115:2003 the DCMI has a set of core elements but unlike the DCMI these do not have mandatory status, rather, they are all only recommended. Just to give you a comparison to the ISO 19115 we have listed the DCMI core elements in Table 4.4.

As you can see, there is quite an overlap in what the TC211 members and the DCMI creators thought should be part of a core metadata set, like "Title," "Date,"

"Description = Abstract," "Language," "Type = topicCategory," "Publisher = Role," and also "Source = Linkage." The ISO 19115:2003 is a little more specific; the number of core elements (26 ISO vs. 15 DCMI) suggests this. Yet it is in a sense comforting to observe that despite the fact that many individuals contribute to these efforts there appears to be an unofficial consensus on what type of metadata elements need to be included in any core metadata set.

There are also 35 additional elements or element-refinements that are either "recommended" or "conforming." The difference in DCMI terminology between these two classifications is that the former is deemed useful for resource discovery across domains or disciplines, whereas the latter has only a demonstrated need for a specific community without necessarily meeting the stricter criteria of usefulness across domains. There is also a set of 12 DCMI maintained vocabulary terms that are recommended for a controlled vocabulary. This refers to the need to agree on a specific term to describe a certain data instance so that one does not end up with a dozen or so different terms attempting to describe identical types of data. As mentioned earlier, this is termed "semantic heterogeneity" a fundamental problem in metadata generation to which we will return a little later again in this chapter. Finally, the DCMI also provides 17 different encoding schemes to make sure that numbers, pointers, dates, geographic names, points in space, and other time references are clearly identified using a common format. Interestingly enough, several of these encoding schemes have been taken from the ISO standards, emphasizing the influence and utility that the ISO standards have on other initiatives around the globe.

The DCMI has a much smaller number of metadata elements compared to the ISO and also the FGDC. While this certainly somehow limits the applicability of the standard, it is also its strength; the DCMI is very easy to grasp and one can quickly put together a subset of elements that suites a specific need. On the other side, it does need some work in case you cannot find what you are looking for, in which case new elements need to be created that are not part of the standard anymore. Here lies the danger when using the DCMI: one community might take off with many new elements that are not part of another community, which almost inevitably lead to occurrences of semantic heterogeneity, which in turn makes discovery of resources across domains very difficult. This problem has in fact received much attention as of late and efforts are underway to deal with this lack of interoperability between metadata descriptions. One should also mention that the DCMI currently has no abstraction model, that is, it is not formally represented using, for example, UML. At the time of writing this chapter, however, efforts underway to cast the DCMI collection into UML but there is no proposed date for publication.

4.3.4 Global Change Master Directory

The Global Change Master Directory was started (actually renamed from NASA Master Directory, NMD) by NASA in the early 1990s in a continuing attempt to bring some order to the type of data NASA programs were collecting. Part of this effort was the creation of the Directory Interchange Format, or DIF, which is now (Summer 2004) at its ninth version. The DIF is used to create directory entries that describe a group of data and consists of a collection of fields, which detail specific information about the data. Like the other metadata standards the DIF set has a number of mandatory elements (highlighted in light grey) that are required and a number of recommended elements, as shown in Table 4.4; the others expand upon and clarify the information some of which are free text fields while others require the use of controlled keywords.

If we take a closer look at the list then we recognize, that many elements are quite similar when compared to the ISO and DCMI core sets. This may reconfirm the fact that there exists an agreement across the disciplines and standards about what elements are considered absolutely crucial and as such must be required for any dataset. The DIF set is, like the DCMI, without a formal representation (like UML), and also does not come with a standard encoding scheme, like XML or OWL. Instead, the DIF elements are published as a free text list with only some guidelines addressing the syntax of the element, the order in which it has to appear, and how long the character string can be when creating an entry. This requires special tools that are designed to parse this specific format for the information and represents one of the shortcomings of the GMDC, because there is neither a formal way of extending the standard nor can the standard be used on the World Wide Web (WWW). On the other side, like the DMCI, the set is relatively clear and limited in its complexity, and subsets can easily be derived. Also, Appendix 2 of the DIF set (suggested keyword lists) is a very rich source for agreed upon keywords that can be used as a seed list by any other standard to ensure a controlled vocabulary for a variety of disciplines, among them is Hydrology.

4.3.5 Mapping of Metadata Standards

As we have seen in the above paragraphs, not only are core elements part of all four introduced standards, but they also map very well into each other. There are of course some discrepancies as far as the coverage is concerned (the core sets vary slightly in length and composition) but low and behold we can conclude that the similarity of the core collections clearly establishes a need to include some 10 to 15 descriptive metadata elements regardless of the origin and the domain these descriptions will be used for, that is, these core sets represent a global standard.

It is clear that not all standards provide the same coverage, and also that some of the standards are not readily usable for applications that involve the WWW, that is, where we need machine readability so that parsers can discover data resources across the Internet. In view of these aspects, it should be clear that the ISO 19115:2003 stands out among the four presented standards: first, because it has the largest coverage; second, because it has a formal representation through the use of UML; third, because it has clearly defined interface that allows user groups to extend it to meet the groups specific descriptions needs while maintaining ISO compliance; and fourth, because it has been encoded in machine readable formats like XML and OWL. It is no surprise therefore that we devoted the largest amount of space to introduce the ISO 19115:2003.

The prominence of the ISO 19115:2003 has been recognized by essentially all other metadata standard organizations, the manifestation of which can be seen in the attempt of all non-ISO standards to provide mappings to the 19115 norm. In other words, each standard tries to make sure that its own collection of elements can be found in the ISO 19115:2003 as well, that is, the claim for ISO compliance is being made. This is also true for other metadata standards we did not talk about like the Spatial Information Council of Australia and New Zealand (ANZLIC) standard that is very close in its definitions to the ISO standards. In addition, the FGDC is planning to make the ISO 19915:2003 norm its official recommendation for version 3 of its content standard (CSDGM) toward the end of 2004. The only exception to this group of standards is the ADN (ADEPT/DLESE/NASA) metadata framework published by the Digital Library for Earth Science Education, DLESE [11], initiative. Even though the ADN set shows 20 required elements (the set is in fact quite close in its core element selection to the other standards), no crosswalk has been provided for the ISO 19115:2003 or any other of the aforementioned standards so far. The ADN has been realized however using XML schemas and because of this fact the collections can be parsed using the WWW tools, which distinguishes it from the other non-ISO standards.

In short, it is fair to say that the ISO 19115:2003 norm is on its way to become the global reference metadata standard that should be used, or at least complied with, when developing new community metadata profiles.

4.4 Hydrologic Metadata

At present there is no comprehensive metadata profile that has been developed for the hydrologic community, even though some smaller efforts have been undertaken to fill this gap based on individual needs of various agencies (like HYDROML for USGS). There are, in fact, numerous "pieces" of hydrologic metadata, some of which conform to metadata standards while others are the result of free-wheeling collections of whatever metadata the authors deemed suitable. In short, the field of Hydrology has never experienced a focused effort to develop an overarching community metadata profile. However, recent efforts by the Consortium of Universities for the Advancement of Hydrologic Sciences Incorporation, or CUAHSI, are aimed at developing a hydrology metadata profile [12]. The task of doing so, however, is not entirely unproblematic because hydrology is a large field with many subareas, like geo-hydrology, hydro-climatology, hydro-paleontology, and eco-hydraulics to name just a few. From these word constructs it becomes clear that hydrology has interests in meteorology, ecology, as well as in the subsurface fields like geology making it an inherently cross-disciplinary field. In short, when developing a metadata profile for hydrology one needs to take into account the definitions, or semantics, that other communities use to describe their datasets, which puts an additional difficulty to the task. A very good starting point is to explore the content of the small "pieces" we talked about earlier and to gauge what other metadata hydrology related communities have been developing to suit their needs. This will reduce the development time of the CUAHSI hydrologic profile. Also, it allows extraction of the details of the descriptions through an interface that permits these potentially disparate description frameworks to work together. In this regard it is particularly valuable to identify conceptualizations of metadata that are both human readable and machine readable.

4.4.1 Related Markup Languages

Ecological Metadata Language, EML [12], is a metadata specification developed by the ecology discipline and for the ecology discipline. It is based on prior work done by the Ecological Society of America and associated efforts (Michener et al. 1997, ecological applications). EML is implemented as a series of XML document types that can be used in a modular and extensible manner to document ecological data. Each EML module is designed to describe one logical part of the total metadata that should be included with any ecological dataset. The EML project is an open source, community oriented project dedicated to providing a high-quality metadata specification for describing data relevant to the ecological discipline. The project comprises of voluntary project members and project decisions are made by consensus. Individuals who invest substantial amount of time and make valuable contributions, like the development of the EML schemas, writing documentation, and helping with maintenance to the development and maintenance of EML, have the chance to be invited to become EML project members. The EML is in its 2.0.1 version and also has a validation service that provides an application against which a user can test or validate whether a certain metadata instance is EML conform.

The Earth Science Markup Language, ESML [13] is a metadata specification developed through a joint project between NASAs Earth Science Technology Office and the Information Technology and Systems Center at the University of Alabama. This effort concentrates on the problem that earth science data is archived and distributed in many different formats varying from character format, packed binary, "standard" scientific formats to self-describing formats. The resulting heterogeneity causes data-application interoperability problems for scientific tools for which ESML is a possible solution. ESML is an interchange technology that enables data (both structural and semantic) interoperability with applications without enforcing a standard format within the Earth science community. The current (August 2004) list of file formats supported includes; GRIB, HDF-EOS, netCDF, generic Binary, generic ASCII, and NEXRAD LII. Users can write external files using these ESML schemata to describe the structure of the above data file formats. Also, the development group provides an ESML library (for parsing the description and decoding them) that can be used for applications development. As a result, software developers can now build data format independent scientific applications utilizing the ESML technology. Furthermore, semantic tags can be added to the ESML files by linking different domain ontologies to provide a complete machine understandable data description. This ESML description file allows the development of intelligent applications that can now understand and "use" the data.

The OpenGIS Consortium Geography Markup Language, GML [14], is largely based on the family of the ISO 19100 norms we talked about earlier and serves as the conduit of implementing these ISO norms in machine readable formats like XML. In fact, version 3.0 (v3.0; published in January 2003) has been implemented using XML schemas (just like the EML), which are available for free download from the OGC Web site. The scope of the GML v3.0 is to provide a venue for the modeling, transport, and storage of geographic information including both the spatial and nonspatial properties of geographic features. The specification defines the XML schema syntax, mechanisms, and conventions that:

- Provide an open, vendor-neutral framework for the definition of geospatial application schemas and objects.
- Allow profiles that support proper subsets of GML framework descriptive capabilities.
- Support the description of geospatial application schemas for specialized domains and information communities.
- Enable the creation and maintenance of linked geographic application schemas and datasets.
- Support the storage and transport of application schemas and datasets.

In addition to the ISO norm compliance (improved in v.3.0) the OGC has also developed a number of geographic features (two-dimensional linear, complex, nonlinear, three-dimensional geometry), with temporal properties, and spatial and temporal reference systems to extend the applicability of the GML, which has ballooned the current version to eight times the size of the previous version (2.1.2). Although, while the inclusion of these added-value geographic features metadata blocks make GML an intriguing alternative for describing geographic features, it does not contain a specific hydrology section or any other domain specific block.

While these markup or metadata languages (EML, ESML, GML) are by no means the only ones around (ADN too is available as XML Schema), they represent some of the most prominent concepts having the largest scope for the geosciences field. Among those three, the GML is perhaps the best starting point as it is ISO compliant and also adds some

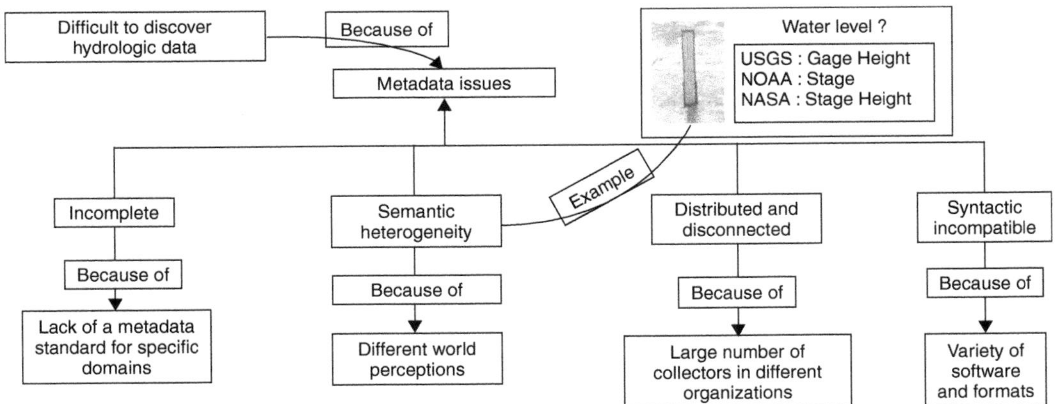

FIGURE 4.3
Interoperability problems in hydrology.

additional components (the "features" list) to the extent content. Either one of the above implementations of metadata standards (that is the transition from a conceptualization in UML to a machine readable implementation like XML Schema), however, is not readily usable for hydrology as many of the descriptive terms and hydrologic features are not implemented in the sets provided. In fact, there are several challenges that exist when creating a domain specific profile, which we will highlight in the next section.

4.4.2 Challenges in Creating Domain-Specific Metadata Profiles

As mentioned at the beginning of this chapter it is difficult for a community to agree on a common description for datasets because individuals tend to organize and describe their data collections in many different ways. As a result, a number of interoperability problems emerge that make the search and exchange of information objects extremely difficult, if not impossible, as shown in Figure 4.3.

One of the main problems that scientists face when trying to discover information objects is the lack of adequately uniform metadata descriptions of data, that is, the absence of a common community metadata profile. In other words, even though the data exists somewhere, it is not described at all, and hence cannot be discovered. Another problem is that of semantic heterogeneity. What we mean by this is the use of different keywords to describe the same thing, for example, the use of Stage Height, Stage, and Stage Height to identify water level measurements. Another problem is the presence of many datasets that are similar in nature but reside in a distributed and disconnected environment that is unable to provide sufficient communication between the various data sources. Finally, a vexing problem is the use of different data formats and software products to process data, which has hampered the exchange of data within a community and across community boundaries. While we will not elaborate on all nuances that one can encounter, some of these problems are perhaps best explained by using the above example.

Water elevation measurements in streams are one of the most crucial and significant data collections that hydrologists (not all but certainly surface water hydrologists) might be interested in. The U.S. Geological Survey operates a vast network of stream gages (12,000 plus) that are distributed over the entire United States. USGS refers to these measurements as "gage height," which is a keyword by which these measurements can be found in their National Water Information System (NWIS) data base. The USGS, however, is not

Table 4.5

Metadata core elements in the GCMD DIF.

Parent field	Core element	Comments
Entry_ID	*Entry_ID*	Must be first, use only once, max 31 chars
Entry_Title	*Entry_Title:*	Use only once, max 160 chars
Parameters	*Category:*	Entry must be from list: Parameter Valids
Parameters	*Topic:*	Entry must be from list: Parameter Valids
Parameters	*Term:*	Entry must be from list: Parameter Valids
Parameters	*Variable:*	Entry must be from list: Parameter Valids
Parameters	*Detailed_Variable:*	Free text, max 80 chars
ISO_Topic_Category	*ISO_Topic_Category:*	Repeat possible, must select entries from ISO 19115 MD_TopicCategory Code
Data Center	*Data_Center_Name:*	Use only once, must be from list: Data Center Valids
Data Center	*Data_Center_URL:*	Use only once, max 160 chars
Data Center	*Data_Set_ID:*	Repeat possible, max 80 chars
Data Center Personnel	*Role:*	Fixed to: Data Center Contact
Data Center Personnel	*First_Name:*	Max 80 chars
Data Center Personnel	*Middle_Name:*	Max 80 chars
Data Center Personnel	*Last_Name:*	Max 80 chars
Data Center Personnel	*Email:*	Max 80 chars
Data Center Personnel	*Phone:*	Max 80 chars
Data Center Personnel	*FAX:*	Max 80 chars
Data Center Contact	*Address:*	Max 80 chars, multiple lines possible
Data Center Contact	*City:*	Max 80 chars
Data Center Contact	*Province_Or_State:*	Max 80 chars
Data Center Contact	*Postal_Code:*	Max 80 chars
Data Center Contact	*Country:*	Max 80 chars
Summary	*Summary:*	Use only once, max 80 chars/line, no more than 30
Metadata Name	*Metadata_Name:*	Comes from ISO 19115 MetadataStandard_Name
Metadata Version	*Metadata_Version:*	Comes from ISO 19115 MetadataStandard_Version

the only entity collecting water surface elevation measurements, in fact, various state level departments and also the US Corps of Engineers have datasets addressing the same measurement, yet they refer to it as "water depth," "stage," or "water level," as summarized in Table 4.5 (from Reference 15).

The consequence of this keyword-hodgepodge is clear: if you were to search for all datasets that measure water surface elevation by typing into a search engine the string "gage height," you would only get the datasets from the USGS, but not the others, simply because you did not use the correct or entire list of possible keywords. This is of course a relatively simple problem that a human brain can easily resolve. A machine, however, cannot see the equivalency of these terms and will miss a substantial subset of data you might be interested in. Also, if you think along the lines of "googleing" for water elevation data using the keyword "Stage" you will find that the machine has no means to differentiate between the dataset you are interested in and a platform for the performing arts. Both of these variations are referred to as "semantic interoperability" (or better lack thereof) and can be, in layman's terms, described as a people speaking different dialects. The solution to this problem is twofold: first, it is necessary to agree on a so-called "controlled vocabulary" (the agreement to use one specific term and not many) for any community. Second, one needs to realize and accommodate the fact that not all entities controlling data holdings will be willing to change keywords in order to conform to a specific controlled vocabulary. In other words, provisions need to be made to accommodate already existing keyword selections by implementing technologies that can automatically resolve semantic conflicts by establishing the equivalency of certain keywords, a thesaurus of sorts if you like.

Other hindrances to perfect interoperability concern the data structure itself. It is not uncommon for researchers to spend much of a day's work reformatting datasets, because the format of a certain received dataset is not the format they need to further process the data. This is referred to as "syntactic interoperability" (or again lack thereof) and is likely to pose a substantial time sink when retrieving datasets. The hydrologic community does not have a preferred format, rather, it has remained a fact of live that many different formats abound. This typically leaves it to the individual researcher to figure out how to get the data in a format that he needs it in. Some communities have made great strides to overcome this problem and as a result data formats like SEED (for seismology), GRIB (oceanography NOAA), or netCDF (for atmospheric sciences) have emerged over the years that are widely accepted as reference data formats. In the next section you will learn more about another data format (Hierarchical Data Format, HDF5), which has received widespread acceptance. In view of the broad scope of datasets that are relevant for the hydrologic and its neighboring communities, it is fair to assume that there will always be a plethora of different formats the hydrologic community has to content with resulting in a significant challenge when attempting to define a hydrologic metadata profile.

Last but not the least, there is a challenge to bring the community together to agree on concepts, standards, and protocols that should be used for hydrology. This is a process that requires a substantial amount of time because a community specific profile needs to mature over time before it receives a widespread acceptance. On one side, much of this acceptance will depend on the degree of involvement of the community in developing the standard. On the other side, acceptance will also depend on the degree of ease-of-use because too much metadata requirement bears the danger of becoming tedious while too little will hamper the utility of providing metadata to start with. This calls for a change in attitude toward the provision of metadata (each individual must understand the importance of it) and also the necessity to provide guidance, information about, and tools for metadata creation that will make it easier for researchers to actually attach metadata descriptions to their datasets.

4.4.3 Overview of CUAHSI Community Profile

The Hydrologic Information System (HIS) of the Consortium of Universities for the Advancement of Hydrologic Science Incorporation (CUAHSI), is in the process of developing a community metadata profile based on the ISO 19115:2003 metadata norm [16]. This is to ensure compliance with a metadata norm that is likely, as mentioned earlier, to attain reference status in the near future and also to facilitate an increased level of interoperability between related or neighboring communities that share certain dataset interests. However, common standards such as the ISO 19115:2003 are quite general in nature and typically lack domain specific vocabularies, complicating the adoption of standards for specific communities. In order to permit growth and maturation over time, it is imperative that the profile is implemented such that it can be easily extended or otherwise modified as time passes. While several possibilities exist to implement the profile in machine readable form, the Web Ontology Language (OWL) was chosen because of its specific properties that set it apart from technologies like XML schema or RDF schema (more about this can be found in Reference 17).

The profile consists of a two pronged approach. The first prong is the use of the ISO 19115 norm (implemented in OWL) that provides the backbone of the metadata profile by identifying the metadata entities and their properties (the latter make up what is called the metadata-tags, that is, the values that eventually need to be supplied to describe a specific dataset). The selection of these entities and their properties is based on what is available in the metadata packages of the ISO 19115:2003 (as shown in Section 4.3.2) plus what needs to be added to the profile by extending the ISO 19115:2003 (the extensions are carried out based on the rule sets of the norm). The second prong concerns the development of keyword list or controlled vocabularies that are specific to the hydrologic community, because the ISO 19115:2003 norm contains only a very rudimentary top-level keyword set. In other words, terms like lake, watershed, hydrography, water body, stream, and so on are not present and must be developed. While one can think up a hydrology thesaurus, it is better to work with an already exiting thesauri like the GCMD, or wordNET, to name a few. These can serve (and have served) as a good starting point to identify the necessary hydrologic controlled vocabularies. These too have been implemented in OWL, which permits the merger and interaction with the ISO 19115:2003 based metadata skeleton to form the CUAHSI community profile.

The CUAHSI metadata profile will eventually contain a substantial number of individual metadata blocks that comprise of a Core Block (mandatory for all CUAHSI datasets) plus any number of data specific metadata blocks that can be added to the Core Block according to need. For brevity sake, we will only address the CUAHSI Core Block here. The ISO 19115 norm identifies a set of 24 Recommended Metadata elements, as shown in the left panel of Figure 4.4. ISO recommended metadata are classified as either M (Mandatory), O (Optional), or C (Conditional), while all other ISO noncore metadata elements are optional by default. The red labeled elements have been chosen for the CUAHSI Core profile, which is composed of almost all ISO-core elements (except those that are left black) and other non-ISO core elements (e.g., Data legal Constraints). The CUAHSI Core elements are set to (M) Mandatory by default, that is, all of these elements must be included in any dataset description.

While the use of a standardized metadata element structure is an important aspect when developing a community profile, that is, the use of the ISO 19115:2003 norm, the identification of a controlled vocabulary (CV) is equally essential. This is due to the fact that the ISO 19115 norm does not provide a rich list of terms that can be used for any community profile. As mentioned earlier (see Table 4.6) some of the interoperability problems stem from the fact that different members of a community use different terms to describe a certain phenomena or measurement. As a result, the CUAHSI–HIS effort is spending a considerable amount of effort to select a CV that can be used for describing datasets.

When identifying a hydrologic CV it is best to start from an already established set of hydrologic terms. There are several sources that lend themselves to be used for this purpose, for example, the GCMD [18], UNESCO International Glossary of Hydrology [19], or WordNET [20] to name just a few. These sources can then be used to expand and merge terms from other glossaries to form a more complete and useful hydrologic CV. The CUAHSI effort has resulted in a first cut at specific hydrologic controlled vocabulary (approximately 80 terms organized in a thematic taxonomy), part of which is shown in the right panel of Figure 4.4. In early 2005 the status of the hydrologic CV was still temporary and in development, but it is expected that the CV will mature in the following years.

All of these collections exist in ASCII format and are easily accessible for download. These ASCII formats however are too simple in their representation and there is no formal implementation of any of the glossaries that could be used by application to parse and deduce meaning from these terms via the Internet. Because of the need to (1) keep the CV flexible, (2) update it easily, (3) maintain readability and access to the outside world, and (4) formalize it in a way that facilitates interoperability, the hydrologic CV is being developed and implemented in the web ontology language (OWL). Because an ontology is essentially a hierarchy of objects and their properties, it is an ideal application for categorizing a glossary of terms that can be divided along thematic areas going from more general to more specific. The use of OWL carries the prospect of providing a technology

CUAHSI Core Elements
From ISO Recommended Base Set
Dataset title (M)
Dataset reference date (M)
Dataset language (M)
Dataset topic category (M)
Abstract describing the dataset (M)
Metadata point of contact (M)
Metadata date stamp (M)
Distribution format (O)
Dataset responsible party (O)
Additional temporal extent information (O)
Lineage (O)
On-line resource (O)
Metadata standard name (O)
Metadata standard version (O)
Metadata language (C)
Geographic location by four coordinates (C)
Geographic location by geographic identifier (C)
Dataset character set (C)
Metadata character set (C)
Spatial resolution of the dataset (O)
Additional vertical extent information (O)
Spatial representation type (C)
Reference system (O)
Metadata file identifier (O)
CUAHSI Core Elements
From the Rest of ISO Set
CUAHSI - thematic - Keywords
Dataset legal Constraints
Dataset Security Constraints
Topography
Barometric Altitude
Bathymetry
Geotechnical Properties
Land Use Classes
Soil Classification
Landforms
Vegetation
Snow/Ice Cover
Satellite Imagery
Map
Watershed Area (Drainage Area)
Digital Elevation Model
Hydrology
Hydrologic Model
Runoff
Direct Runoff
Base Runoff
Surface Runoff
Surface water
Discharge or Flow
Water Depth
Stage Height
Base Flow
Pressure
Water Yield
Ground water
Ground Water Flow
Safe Yield
Infiltration
Hydraulic Conductivity
Drainage
Moisture
Wells
Aquifer

FIGURE 4.4
CUAHSI metadata elements and keyword list.

platform that can be used to overcome and resolve the problem of semantic heterogeneity in future.

4.4.4 Future Outlook

Metadata is like interest — it accrues over time as the user community continues to revisit the profile trying to expand and upgrade it based on the feedback that is being received. Carefully designed metadata profiles result in the best information management in the short- and long-term. If thorough, consistent metadata has been created, it is possible

Table 4.6

Various keywords used to identify gage measurements.

Measurement	Organization	Link
Gage height	USGS	http://waterdata.usgs.gov/nwis/discharge
Gage	State of Colorado-Division of Water Res.	http://dwr.state.co.us/Hydrology/Ratings/ALAWIGCO02.txt
Water depth	Montana Natural Resource Informat. Sys.	http://nris.state.mt.us/wis/
Stage	U.S. Army Corps of Engineers, New Orleans	http://www.mvn.usace.army.mil/eng/edhd/Wcontrol/miss.htm
Water level	Delaware River and Bay PORTS	http://coops.nos.noaa.gov/dbports/AllWL_db.html

to conceive of it being used in many unforeseen ways to meet the needs of nontraditional users, for multiversioning, and for data mining. The resources and intellectual and technical design issues involved in metadata development and management are far from trivial. Some key questions that must be resolved by information professionals as they develop digital information systems and objects include:

- Identifying which metadata schema or schemas should be applied in order to best meet the needs of the information creator, repository, and users.
- Deciding which aspects of metadata are essential for what they wish to achieve, and how granular they need each type of metadata to be — in other words, how much is enough and how much is too much. There will always be important tradeoffs between the costs of developing and managing metadata to meet current needs, and creating sufficient metadata that can be capitalized upon for future, often unanticipated uses.
- Leaving provisions in the design and implementation of the metadata profile that ensures ease and flexibility of expansion and also the incorporation of new technology like the advent of the Semantic Web.
- Paying close attention to the need of interoperability because availability of datasets within a community is too narrow a scope, rather it needs to be realized that other communities might have an interest as well. The realization that datasets of a specific community in the end are just part of something much bigger is of utmost importance. In other words, it is essential to acquire or maintain an out-of-the-box view when implementing metadata profiles for which the use of an international standard is just one step.

What we know is that the existence of metadata will prove critical to the continued physical and intellectual accessibility and utility of digital information resources and the information objects that they contain. In this sense, metadata provides us with the Rosetta-Stone that will make it possible to decode information objects and transform their information into knowledge.

References

1. Dictionary Definition of Meta, HostingWorks, accessed July 2004. http://hostingworks.com/support/ dict.phtml?foldoc=meta.
2. Definition of Meta, Word Reference.com, accessed July 2004. http://www.wordreference.com/definition/meta-.htm.
3. Gilliland-Swetland, Anne J., Introduction to metadata information, accessed July 2004. http://www.getty.edu/research/conducting_research/standards/intrometadata/2_articles/index.html.
4. Batchelder, Ned, Metadata is nothing new, accessed July 2004. http://www.nedbatchelder.com/index.html.
5. Piasecki, Michael and Bermudez, Luis, HydroML: Conceptual development of a hydrologic markup language, IAHR Congress, August 24–29, 2003, Thessaloniki, Greece.
6. Cornillon, Peter, OPeNDAP: Accessing data in a distributed, heterogeneous environment, NSF-Sponsored CyberInfrastructure Workshop, October 31, 2003.
7. Federal Geographic Data Committee, FGDC, Content standards for digital geographic metadata, CSDGM, version 2.0, accessed July 2004. http://www.fgdc.gov/.
8. International Organization for Standardization, ISO/TC211 Geographic information/Geomatics, accessed July 2004. http://www.isotc211.org/.
9. International Standard ISO 19115:2003, *Geographic Information Metadata*, 1st ed., ISO Copyright Office.
10. Dublin Core Metadata Initiative, DCMI, accessed July 2004. http://dublincore.org/documents/.
11. Digital Library for Earth Science Education, DLESE, ADN metadata framework version 0.6.50, accessed July 2004. http://www.dlese.org/Metadata.
12. Ecological Metadata Language, EML, version 2.0.1, accessed July 2004. http://knb.ecoinformatics.org/software/eml/.
13. Earth Science Markup Language, ESML, version v3.0.1, accessed August 2004. http://esml.itsc.uah.edu/index.jsp.
14. OpenGIS consortium geography markup language, GML, version v3.00, Implementation Specification, OGC document 02-023r4. http://www.opengis.org/docs/02-023r4.pdf.
15. Bermudez, Luis and Piasecki, Michael, (2003), Achieving semantic interoperability with hydrologic ontologies for the WEB, IAHR Congress, Saloniki, Greece, September 2003.
16. CUAHSI Metadata Elements version 1.0, Metadata specifications for the hydrologic community, accessed February 2005. http://loki.cae.drexel.edu:8080/web/how/me/metadatacuahsi.html.
17. Bermudez, Luis and Piasecki, Michael, (2004), Metadata community profiles for the semantic web, *GeoInformatica Journal*, September 2004 (submitted).
18. Global Change Master Directory (GCMD), accessed February 2005, http://gcmd.gsfc.nasa.gov/index.html.
19. UNESCO, International glossary of hydrology, accessed February 2005. http://webworld.unesco.org/water/ihp/db/glossary/glu/HINDENT.HTM.
20. WordNet, a lexical database for the English language, accessed February 2005. http://wordnet.princeton.edu/.

5

Hydrologic Data Models

Benjamin L. Ruddell and Praveen Kumar

CONTENTS

5.1 Data Models

5.1.1 Contemporary Challenges in Hydrologic Data Management

In recent years there has been an explosion in the availability of earth science data from satellites, remote sensors, and numerical computations. This avalanche of data brings unprecedented opportunities for the study of environmental processes, and has led to new difficulties in organizing and communicating data for study. In many earth science studies, including hydrology, a majority of time and investment is now spent wrestling data into a useful form. Everyone who has experience migrating and integrating geospatial data for study can tell stories of the troubles involved: incompatible files, differences in projection, and the bugs and glitches of a pantheon of tools and libraries to handle all the different data types. Finding the right data on the Internet can be time consuming. Once the data is located, the piece of information of interest is often buried in a huge file, of which only a small portion is required. File sizes are now routinely measured in Gigabytes. Specialized knowledge is required to handle each of the file formats involved, and the process of data conversion is manual. When the data has been assembled, its quality is often questionable and unverifiable. This learning curve is steep, and knowledge gained in the process is rarely applicable to the next project.

The need for data models that are specialized to handle the requirements of scientific data has been discussed for more than a decade. In 1995, IBM released a paper outlining a framework for scientific data models in anticipation of the flood of scientific data generated by NASAs Earth Observing System (EOS) [1]. This paper explains that, for the purposes of scientific visualization and analysis, new data models are needed to handle this EOS data.

> Traditional methods of handling scientific data such as flat sequential files are generally inefficient in storage, access or ease-of-use for large complex datasets, i.e., >1 GB in size, particularly for such applications. Relational data management systems are also deficient because the relational model does not accommodate multidimensional, irregular or hierarchical structures often found in scientific datasets nor the type of access that associated computations require.... Therefore, there is a need for some type of data (base) model that possesses elements of a data base management system but is oriented towards scientific datasets and applications. ... The implementation of such a model effectively decouples the management of and access to the data from the actual application...
>
> *Scientific data models for large-scale applications* [1]

As the magnitude of the data and scale involved increases, so do the troubles caused by older data management technology. As a consequence, the wealth of earth science data generated by environmental sensors, and by re-analysis and fusion of these data into new datasets, has been chronically under-studied. Much of its potential lies dormant because the right people are unable to efficiently access, integrate, analyze, and share the data. Examples of underutilized large-scale hydrologic environmental datasets include:

- Satellite images of the earth's surface (particularly EOS data)
- RADAR, atmospheric, and precipitation measurements
- Weather data and atmospheric soundings
- Geologic, soil, and aquifer mapping
- Surface and groundwater quantity and quality monitoring
- Supercomputer weather and climate model outputs
- Stream network hydrology and hydrography
- Ecology and vegetation coverage
- Snow and ice cover
- Storm-track data

These datasets are all different views of the earth's environmental system, measuring different aspects of that system. These datasets are *geospatial*, meaning that they are registered and projected across the earth's surface (geo), and distributed in space (spatial). Moreover, an understanding of the earth's environment and hydrology requires the integrated study of different elements of its systems. The challenge of contemporary earth scientists, then, is to access and integrate this huge diversity and volume of data, and make it talk to us via visualization and quantitative analysis. Better methods of data integration, documentation, and communication are needed, and are now being developed. Because of the centrality of the hydrologic water cycle to the earth's climate and ecology, hydrology can be a valuable "common denominator" and point of integration for diverse environmental data. The hydrologic data model is central to the emerging paradigm of scientific communication and study, and will be discussed in this chapter.

The goal of organizing and sharing high-quality large-scale environmental data is now in reach with the advent of geodata modeling. Geospatial data models are able to address the issues of scale and heterogeneity inherent in environmental datasets. Issues of scale arise when datasets of significantly different spatial resolution or extent are combined, such as when a local drainage basin is being analyzed using regional climate model data and continental-scale remote sensing images. Only a small portion of one dataset is required, and a well-designed data model allows efficient access to just what is needed. Issues of heterogeneity arise when data concerning environmentally related parameters are measured and aggregated in incompatible forms, such as when land-use data is stored on a raster grid aggregated by county, and that data must be combined for analysis with vector hydrologic data aggregated by watershed. An appropriate data model provides a framework for relating these heterogeneous data.

5.1.2 What Is a Data Model?

The definition of *data model* varies from source to source. It varies because it is a very practical and an applied concept — the form and to some degree the theory behind it will depend on the application it is designed for. Here are two definitions for a data model.

> A formal method of describing the behavior of the real-world entities. A fully developed data model supports entity classes, relationships between entities, integrity rules and operations on the entities
>
> *ESRI GIS glossary* [2]

> A data model says what information is to be contained in a database, how the information will be used, and how the items in the database will be related to each other.
>
> *Free On-Line Dictionary of Computing* [3]

A data model can also be understood as a:

- Standardized structure that organizes data
- Way of "modeling" and making sense of the relationships between data
- Common language for communication of data
- Enforcer of data quality and integrity
- Organizer of metadata: what is this data? Where is it from?
- Pictorial, visual representation of the structure and relationships of data and metadata

Data models were first used in private industry to track orders, customers, jobs, and development programs. They were created because information-driven business depends on the abilities of its employees at many locations to easily and accurately communicate and track information. Intranet databases are the core of these systems, and the data models describe the form in which business information is to be communicated and stored. In the business world, data modeling has successfully improved data management and efficiency.

Data model theory has been developed to describe the form and role of data models in their diverse contexts. The focus of this section is on object-oriented relational data modeling, which is used by geodatabases. It is important to understand the meaning of the terms *data, metadata, meta-model,* and *meta-metamodel,* which combine to form the data model concept.

Data	The raw alphanumeric information, which is the subject of analysis.
Metadata	*Meta* means "about," so *metadata* means "about-data." The alphanumeric descriptors of a specific set of data. The categorization, tag, or label on a specific data series. For example, metadata on a streamflow time series might include "cfs" and "Porcupine Creek at mile 3.4."
Metamodel [4]	The blueprint of the metadata. It lists requirements and data types for tags on a specific type of data. For example, the metamodel of a streamflow time series might require "units" as a string datatype and "gauge name" as a string datatype.
Meta-metamodel [4]	The schematic relationships between different collections of data and metadata. An integrated picture of how different metadata/data constructs relate to each other.

A data model operates on all four of these levels. It encompasses the data, descriptions of data, requirements for how data is described and packaged, and the relationships between the metamodel groups (see Figure 5.1). The data model structure is application-driven and tool-driven. Application-driven means that a data model is implemented for a real-world

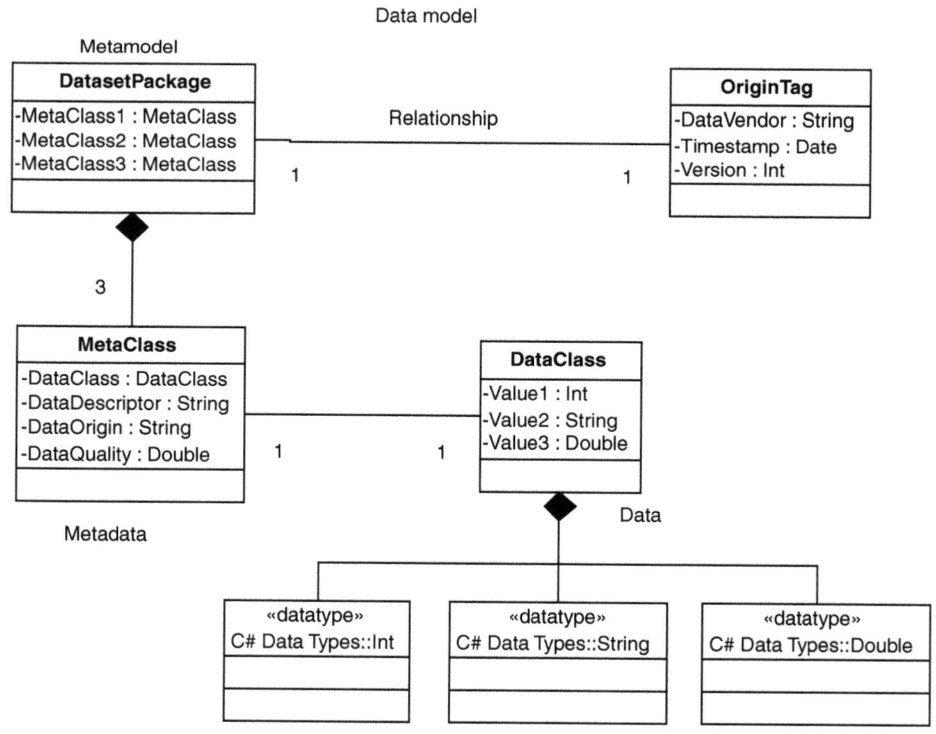

FIGURE 5.1
An example of a data model structure for a data package provided by a vendor. An integer, string, and double aggregate to form a DataClass. A DataClass is associated with a MetaClass, which describes the information in the DataClass. Three MetaClasses are included in a DatasetPackage, which is the highest-level class in this model. Each DatasetPackage is associated with one OriginTag, with vendor information about the dataset.

use, such as modeling surface hydrology or a supply chain. It is designed for a specific purpose, and its metamodel reflects an effort to describe real-world objects such as stream reaches and drainage catchments. Tool-driven does not mean that a data model can only work with one piece of software. It means, however, that data models are designed in light of the capabilities and limitations of software tools. Geographic Information Systems (GIS) is the key technology-driving tool in the design of geodata models.

The requirements of industry-standard Database Management Systems (DBMS) and data repositories have an important influence on data model architecture, because the data model is implemented inside a DBMS and repository. Data models are independent from a specific DBMS technology or software application, but model design must be compatible with database standards. A powerful attribute of data-modeled datasets is that their data and metadata is extremely portable, and can be exported to and used by any software that understands the data model, independent of the DBMS, repository, platform, or software in use. This portability creates a tremendous opportunity for the application of generic visualization and data mining tools that can read a data-modeled dataset.

An effective data model solution answers the challenge of data management by effectively and efficiently organizing datasets and their metadata, enforcing quality requirements on the data and metadata, and tending the relationships between data objects. It is important to keep a practical perspective when designing a data model. We model data to make it useful to us, so how do we decide what is useful? What do we want to do with the data, and what metadata and relationships do we need to follow if we want to do that? The goal is to make a model as tight, efficient, and specific as possible, while providing flexibility and extensibility so that additional future modeling can be built alongside and on top of the old model, without the need to convert and migrate data to a new structure. By making data management more streamlined and standardized, data models save time and money. The biggest benefit of a hydrologic data model is its ability to standardize the communication of data across a distributed system, such as the Internet. More details on distributed, standards-based data communication are discussed later in the chapter.

5.2 Geodata Models

5.2.1 What Is a Geodata Model?

The GIS systems are becoming ubiquitous to science and industry. They are powerful and useful because they are able to process and visualize spatial data. The GIS has moved from being a tool for the conversion and visualization of map data, to being a full-service solution to data management and analysis. The GIS systems are most useful in the field of hydrology.

A geodata model is a data model, which incorporates spatial data readable by a GIS application along with associated attribute data and metadata. It is designed to represent the data and relationships important for a specific geospatial application. A hydrologic geodata model might contain elements describing a spatial hierarchy of watersheds, streams, lakes, aquifers, and inputs, flows, and abstractions of water from the system. An archaeological geodata model would contain categories for different types of artifacts, buildings, roads, and their locations within a dig. Geodata models are a total solution for describing the spatial layout of objects, describing the objects with data and metadata, and documenting the interactions and relationships between objects. Widespread application of GIS geodata models has the potential to bring to the environmental sciences the benefits of distributed access to a well-formed data.

5.2.2 Geodata Standards, Web Services, and the Internet

With the adoption of standard hydrologic data models, distributed access to data is becoming more streamlined and reliable. Vendors of data are able to store and dispense their data products in a format that everyone can access. For example, the United States Geological Survey (USGS) has adopted the ArcHydro hydrologic data model for its National Hydrographic Dataset (NHD). Government, academic, and industry interests are moving to create a geospatial data infrastructure combining data-modeled databases with distributed Internet-based access. This infrastructure will include web portals (gateways to data) that maintain catalogs of distributed data. Web portals function by taking query-based requests for data, scanning metadata of registered data sources to locate the proper information, and brokering the communication of that data to the client.

Metadata standards underlie the communication of data in distributed systems. They are transparent to the user of a database, data model, and web service, but they make communication of queries and data possible by standardizing the structure and meaning of metadata. Currently the most important metadata standards for geospatial data are the Federal Geographic Data Committee (FGDC), Content Standard for Digital Geospatial Metadata, and the International Organization for Standardization (ISO) 19115 Metadata Standard. Data interoperability requires that geospatial data models are compatible with these metadata standards. Extensible Markup Language (XML) is now universally utilized to communicate structured metadata.

Geodata models are a key to web-based spatial data interoperability because they describe what is contained in a database, so the portal's search engine can make good decisions about the relevance of data. Data modeled databases also allow query-based data extraction, meaning that only the relevant data and metadata needs to be returned with a query. This feature makes a modeled database ideal for supplying web portal-based data services. Via a web portal, a huge online national hydrologic database can be scanned for relevant data, queried, and those particular results may be returned in a standard, self-describing model format. Alternatively, the relevant data may be extracted from several different online servers, compiled using the proper data model, and returned as one complete and comprehensible unit. The FGDC has already implemented a prototype, the National Geospatial Data Clearinghouse, which utilizes the National Spatial Data Infrastructure (NSDI). The NSDI is a network of data gateways that send requests for data to participating Internet servers.

The emerging framework for web-based distributed data access is being implemented using *web services*. Web services are not limited to spatial data exchange. Rather, they are a broad architecture for the exchange of, and access to, all types of data by all sorts of interconnected devices. By adopting open standards for data description, query, and communication, a web service decouples the format of data and data management methods from the access to that data. In other words, a data vendor/server can store and manage their data any way they want to, but make the data accessible through a standards-compliant web service (see Figure 5.2). Web services publicly describe their capabilities and content using the Web Services Description Language (WSDL) and the Universal Description, Discovery and Integration standard (UDDI). Requests for information are phrased using the XML-based Simple Object Access Protocol (SOAP). ESRI, Inc., which has invested in the implementation of web-based data communication standards, describes the capabilities of web services:

> Web services provide an open, interoperable, and highly efficient framework for implementing systems. They are interoperable because each piece of software communicates with each other piece via the standard SOAP and XML protocols. This means that if a developer "wraps" an application with a SOAP API, it can talk with (call/serve)

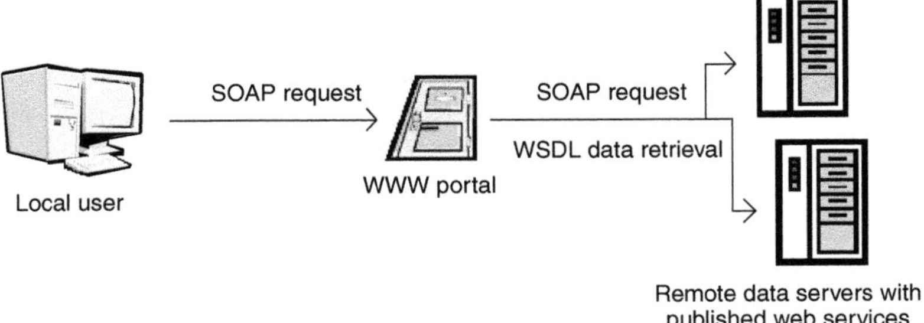

Figure 5.2
A web services framework. SOAP protocol is used to request data from a web portal, and the portal queries its registered servers for that data. Available data content on those servers is described using WSDL protocols.

> other applications. Web services are efficient because they build on the stateless (loosely coupled) environment of the Internet. A number of nodes can be dynamically connected only when necessary to carry out a specific task such as update a database or provide a particular service.
>
> *Spatial Data Standards and GIS Interoperability, ESRI* [5]

5.2.3 Open GIS: Extensibility and Interoperability

Open GIS is the principle of building GIS tools and data standards so that their functionality is accessible to all users. The Open GIS Consortium (OGC), www.opengis.org, was formed to promote open GIS standards and interoperability. Along with the NGDC and participating academic and industry groups, the OGC has successfully developed and disseminated spatial data standards. The US Federal agencies participate in the OGC standards via the NGDCs requirements. Open GIS benefits all parties involved in spatial data analysis. ESRI, Inc., vendor of the leading GIS applications, has actively participated in this effort. ESRI has adopted the Open GIS approach because of these benefits:

> An open GIS system allows for the sharing of geographic data, integration among different GIS technologies, and integration with other non-GIS applications. It is capable of operating on different platforms and databases and can scale to support a wide range of implementation scenarios from the individual consultant or mobile worker using GIS on a workstation or laptop to enterprise implementations that support hundreds of users working across multiple regions and departments. An open GIS also exposes objects that allow for the customization and extension of functional capabilities using industry-standard development tools.
>
> *Spatial Data Standards and GIS Interoperability, ESRI* [5]

Put simply, Open GIS makes it possible for everyone to share data more easily, and for software developers to produce more powerful and flexible tools for working with the data. ESRIs ArcGIS software is built on an open object-oriented component model called the ArcObjects™. The ArcObjects are a successful experiment in Open GIS interoperability. The ArcObjects are compliant with the Component Object Model (COM) developed by Microsoft, and the Object Model Diagram (OMD), and component libraries are publicly available.

By making the ArcObjects available to the public, ESRI has opened the door to the extension and customization of its software to meet special requirements. Every component of the ArcGIS software may be modified, extended, and included in new implementations.

This modification is achieved using either the integral Visual Basic scripting capability or by building an independent application using an object-oriented programming language. These ArcObject libraries may be referenced and accessed from Windows^{TM} based program development environments, and used together with other libraries. The Open GIS platform, and future evolutions of the ArcObjects and related technologies, will be an important element of the developing geospatial data infrastructure.

5.2.4 The ESRI Geodatabase: A Foundation for GIS Data Modeling

Data storage and management within ArcGIS is accomplished using a geographic database (geodatabase). The geodatabase is a structured format for a GIS-enabled database, which can be implemented inside an industry standard DBMS. The geodatabase specifies a format of storage of spatial data structures in database objects, which can be read and understood by a GIS tool. A relational hierarchy of tables of spatial data and associated attributes and metadata is specified. A massive amount of spatially descriptive data stands behind each point on a map, including projection information, location coordinates, and feature metadata. The geodatabase organizes this spatial data so the GIS understands the features in the database. It is debatable whether the geodatabase is itself a data model because it is not specific to a particular data application. Rather, it is an open framework that supports the development of data models based on inheritance of its GIS data structures.

The ESRI promotes the geodatabase framework by listing its major benefits [6]:

- Centralized management of a wide variety of geographic information in a DBMS
- Large data management in a continuous integrated environment
- Full support for multiuser editing and advanced versioning environment
- Support for advanced data types and geometry (i.e., the ability to store 3D coordinates, measures, and true curves)
- Faster and more efficient data entry with data rules and relationships
- Create and edit feature-linked annotation
- Create and edit geometric networks
- Relationships with tabular data sources stored in any industry-standard DBMS
- Create and edit topologically integrated features
- Create subtypes and domains for maintaining database integrity
- Support for the development of industry-standard data models

A geodatabase is a collection of object classes that inherit interfaces and attributes from the ArcObjects structures. The basic level of implementation of data storage in a geodatabase is accomplished with *Object Classes* and *Feature Classes*. Object and feature classes are composed of a group of objects of the same class, and a list of attribute data for each object. Object classes contain a specific data object (any class type is acceptable) along with attribute data in fields. Feature classes inherit from object classes, and store geospatial geometric features and their GIS spatial reference data. A feature class has a specific geometry type (such as point, polyline, or polygon) and a specified list of attribute fields. Attribute fields contain one data object type each, for example, "long integer." In a geodatabase, object class tables are normally associated with feature classes via relationships.

Some key capabilities of the geodatabase framework for scientific geodata models include support for topology, data domains, geometric networks, relationships, and validation rules for database features.

Subtypes are categories that distinguish between otherwise identical feature classes. For example, two subtypes under "stream channel" might be "artificial" and "natural." Default values may be specified so that the most common subtype is assumed unless otherwise specified.

Topology is a specification of spatial connectedness within a set of database features. For example, it is necessary that a collection of watershed features share common boundaries, and do not have spatial discontinuity or overlapping. It might also be important that stream channels do not cross each other. Geodatabase topology provides rules to enforce the quality of these spatial relationships.

Data Domains specify a valid range of values in an attribute field in the database. There are two types of data domains. *Coded Value Domains* specify a list of correct values for the field. For example, a "flow channel type" attribute might have a coded value domain of 1 = natural and 2 = artificial, where natural and artificial are the only acceptable values. *Range Domains* specify a valid continuous range of numeric data values. For example, the "tree height" attribute must be set between 0 and 400 feet.

Geometric Networks are organized, topologically related collections of point and line feature classes. Points form junctions and nodes, and the lines connect those nodes as links. Attributes such as flow direction may be assigned to the network. *Sources and Sinks* are points in a geometric network where a parameter is input to or abstracted from the network. Flow direction is assigned based on the distribution of sources and sinks.

Dimensions are measurements of features in the geodatabase space. *Aligned dimensions* measure the true map distance between two points in geodatabase space. *Linear dimensions* record a useful measurement associated with database features (such as a flowpath length), where the distance is not necessarily the true map distance "as the crow flies" between two points.

Relationships record an association between attributes of two database classes. The important attributes of a relationship class are the origin class, destination class, cardinality of each relationship end, and the primary key and foreign keys of indexing. Geodatabase relationships are typically of a one-to-one or a one-to-many *cardinality*. This means that a given record field attribute in the origin class is related to either exactly one, or to many, attribute fields in the destination class. A relationship associates identical index values in two database classes: the primary index key is the origin class index field name, and the foreign index key is the index field name in the destination class. Relationships are accessed by some software to automatically manipulate and retrieve data associated with a given database class.

Validation Rules exist in several forms in the database. Attribute validation rules track the validity of data entries by checking attribute data against data domains. Network connectivity rules specify which type of point and line features may be connected to each other in a geometric network. Relationship rules restrict which subtypes of a feature class may be connected in a relationship, and also the range of cardinality acceptable in a one-to-many relationship (i.e., 1 to 4 storm sewers must be connected to a manhole node). The rules and structure built into the geodatabase framework may be used to manually or programmatically search for and enforce data quality in a geodatabase.

5.2.5 The Structure of a Geodatabase

The geodatabase is a complicated object structure (see Figure 5.3), and its important elements will be explained here. Its critical structural elements are the workspace, row, object and feature classes, object and feature datasets, relationship classes, rules, cursors, query filters, and name objects.

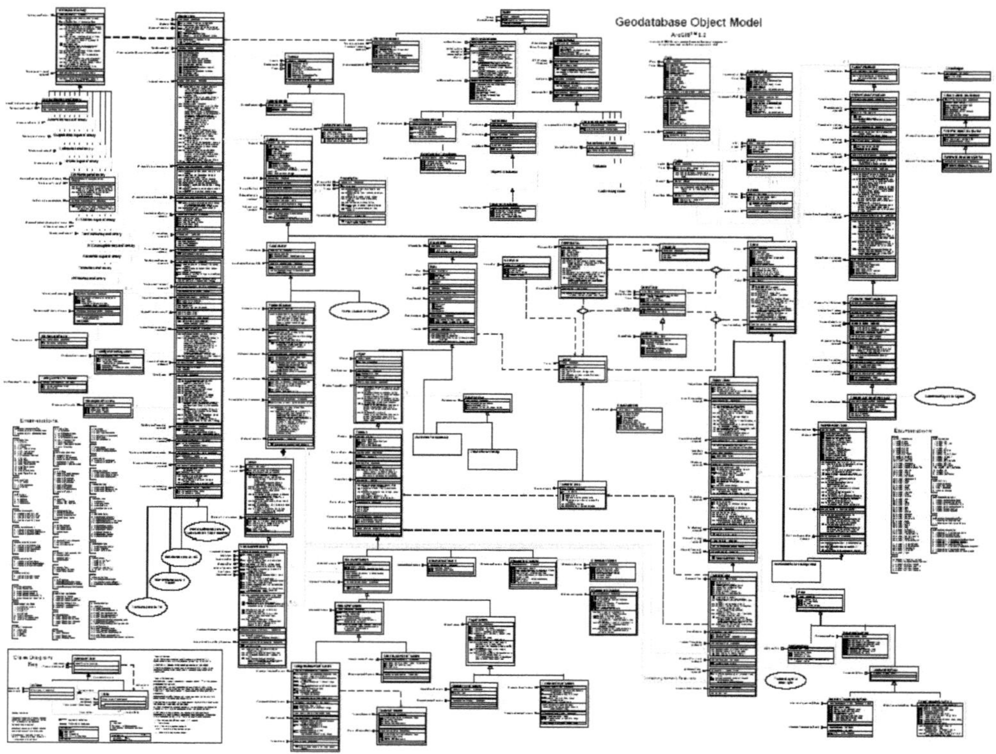

FIGURE 5.3
This geodatabase OMD describes part of the COM-compliant ESRI ArcObjects. This figure is intended only to show the intricacy of the object model, and should not be used for technical reference as it is not readable. Full OMD available at ESRI ArcObjects online (9).

Every geodatabase is stored in a *workspace*. This workspace represents the physical storage location of the data in a database repository, using a supported DBMS. Each DBMS has its own unique workspace factory (i.e., AccessWorkspaceFactory, ShapefileWorkspace-Factory) — which is used to access a specific DBMS and generate a standard workspace object to access that database. This workspace object is then used to read and write objects in the database.

Rows are a basic element of the database's tables. The row object contains the methods needed for reading and writing the data at the row-by-row level of a database table. These rows have a description of their fields, and store the actual data for each field. A row object must be retrieved to read or write data.

Feature classes are special types of tables, which store ESRI GIS geospatial feature objects in rows along with attribute data. All objects in a feature class will share the same object type and attribute fields. A more general version of the feature class, from which the feature class inherits, is the *object class*. Object classes store general data that is not geospatial.

Feature datasets are collections of feature classes. A feature dataset specifies the *spatial reference* for all classes within the collection. A spatial reference object contains the projection, datum, and georegistration metadata necessary to locate spatial objects. Object classes are not allowed inside feature datasets because they are not geolocated. Rather, object classes are stored in object datasets.

Relationship classes associate two object classes (feature classes are a specialized type of object class). Their function has already been discussed in some detail. Relationship classes are stored as stand-alone objects in the geodatabase.

Rule classes, including the specialized attribute, relationship, and connectivity rules, are stored in association with object classes, in a similar manner to relationship classes.

Cursors are special objects used to collect from a table specific rows which meet certain filter criteria. *Query Filter* objects take SQL statements and scan a table for rows matching the query, then those rows are returned as a cursor. Using the cursor, the matching rows are read or modified.

Name objects are "lightweight" objects, which contain the title and location data of a large object (such as a workspace or dataset) without the actual data contained in that object. Names may sometimes be used instead of the full-size object, which is efficient when very large objects are being used.

5.3 The ArcHydro Data Model

5.3.1 The History and Purpose of ArcHydro

ArcHydro is the result of years of conceptual work on GIS data structures for surface water resources, merged with the ESRI geodatabase technology to form a relational GIS data model for water resources. It was officially published by ESRI and the Center for Research in Water Resources of the University of Texas at Austin in 2002; the original book on ArcHydro, *ArcHydro: GIS for Water Resources* [7] is available from ESRI Press and contains a thorough discussion of the model and its applications. The input of a variety of nongovernmental agencies government resource agencies, and industry experts was included in the development process. It is an open-source, nonproprietary data model, and its intended purpose is to provide a common data framework for the water resources and hydrology communities. ArcHydro has been successful in this goal, and is coming into widespread use by university and government agencies — including the U.S. Army Corps of Engineers (USACE), which supports it by developing analysis software to work with its datasets, and the United States Geological Survey/Environmental Protection Agency which have begun to distribute the National Hydrography Dataset in an ArcHydro format. From its inception ArcHydro has been intensely practical, and this is a big factor in its success: its structures are designed to support standard methods of hydrologic and hydraulic analysis (such as the USACEs HEC–HMS and HEC–RAS software [8]), and to function within industry-standard database and GIS systems. ESRI has developed a suite of tools to support the data model, which, along with preexisting tools provided in the ArcGIS software, provide a basic level of functionality for building and working with an ArcHydro database.

One design goal of the data model is to provide scalability of access to data based on the aggregation of water resources data by watersheds. In fact, the entire ArcHydro structure is based on a watershed approach to data representation and analysis. Groups of Watersheds are aggregated into Basins. Watersheds and stream networks are related using a common outlet point, where all streams in the watershed combine and exit. Network topology enforces a reasonable arrangement of streams and nodes, where water flows "downhill" toward the watershed outlet. Stream channel geometry is stored along with the stream network, describing the shape of river bottoms and flood plains. Hydro Response Units describe properties of the land surface related to hydrologic processes. Monitoring points provide a geolocation for associated time series data. All this data may be accessed using relational query in a database, and viewed using the ArcGIS or other compatible visualization software.

The ArcHydro is a work in progress. It is designed for the representation of surface water resources, but development is underway to expand its scope to include ecological, groundwater, and atmospheric water cycle processes for a more complete and integrated perspective on water resources. These future developments will not alter the basicArcHydro framework, but will rather build on top of the existing technology to extend its capabilities. ArcHydro will be the basis of other creative adaptations that extend its value in new and unanticipated ways. Software tools designed to function with the basic ArcHydro framework will therefore continue to function with future extended versions of the database, but new tools will be needed to capitalize on new developments. Because ArcHydro is designed for the developing paradigm of GIS and information technology, its current implementation will eventually become outdated. However, the basic concepts underlying its design, and lessons learned from its implementation, are likely to persist as a classic study in water resources science and hydroinformatics.

5.3.2 The Structure of ArcHydro

The ArcHydro surface water data model is composed of four thematic feature datasets, namely *Network, Hydrography, Drainage,* and *Channel*, and two object classes, *Hydro Event* and *Time Series*:

- Network — points and lines connected as a pathway for water flow
- Hydrography — raw hydrologic data input from GIS maps and tables
- Drainage — drainage areas defined by topography and urban hydraulics
- Channel — profile lines and cross sections describing the 3D geometry of stream channels
- Hydro Event — describes extra properties of HydroFeatures not included in class attributes
- Time Series — indexed data describing time-varying parameters for any HydroFeature

These model packages are integrated to provide the necessary functionality.

Each hydrologic feature in the database is assigned a unique *HydroID* integer to identify it. The HydroID is the fundamental index of the ArcHydro model. Database tools maintain a table of assigned HydroIDs, the *HydroIDTable*, to prevent duplication. Along with the HydroID, each feature is assigned a *HydroCode* string descriptor, which permanently identifies the feature, normally in a format that is comprehensible to a human user. Database relationships between features depend on HydroID to identify associated objects.

The network structure is perhaps the most complicated and important element of ArcHydro. It is comprised of an ESRI geometric network (see Figure 5.4: Network Schema UML). Its two components, the HydroEdge (line) and HydroJunction (point) combine in this network to form a topologically integrated flow network. The HydroEdge class inherits from the ESRI ArcObjects class "ComplexEdgeFeature" and has two subtypes, *Flowline* and *Shoreline*. Flowlines are watercourses, and Shorelines are the edges of large bodies of water. The HydroJunction class inherits from the ESRI ArcObjects class "SimpleJunctionFeature." Using the geometric network and descriptive attributes of the network classes, a thorough description of stream networks is possible. Attributes describe the length of the edge, distance to the next sink on the network, flow direction along an edge, and area drained

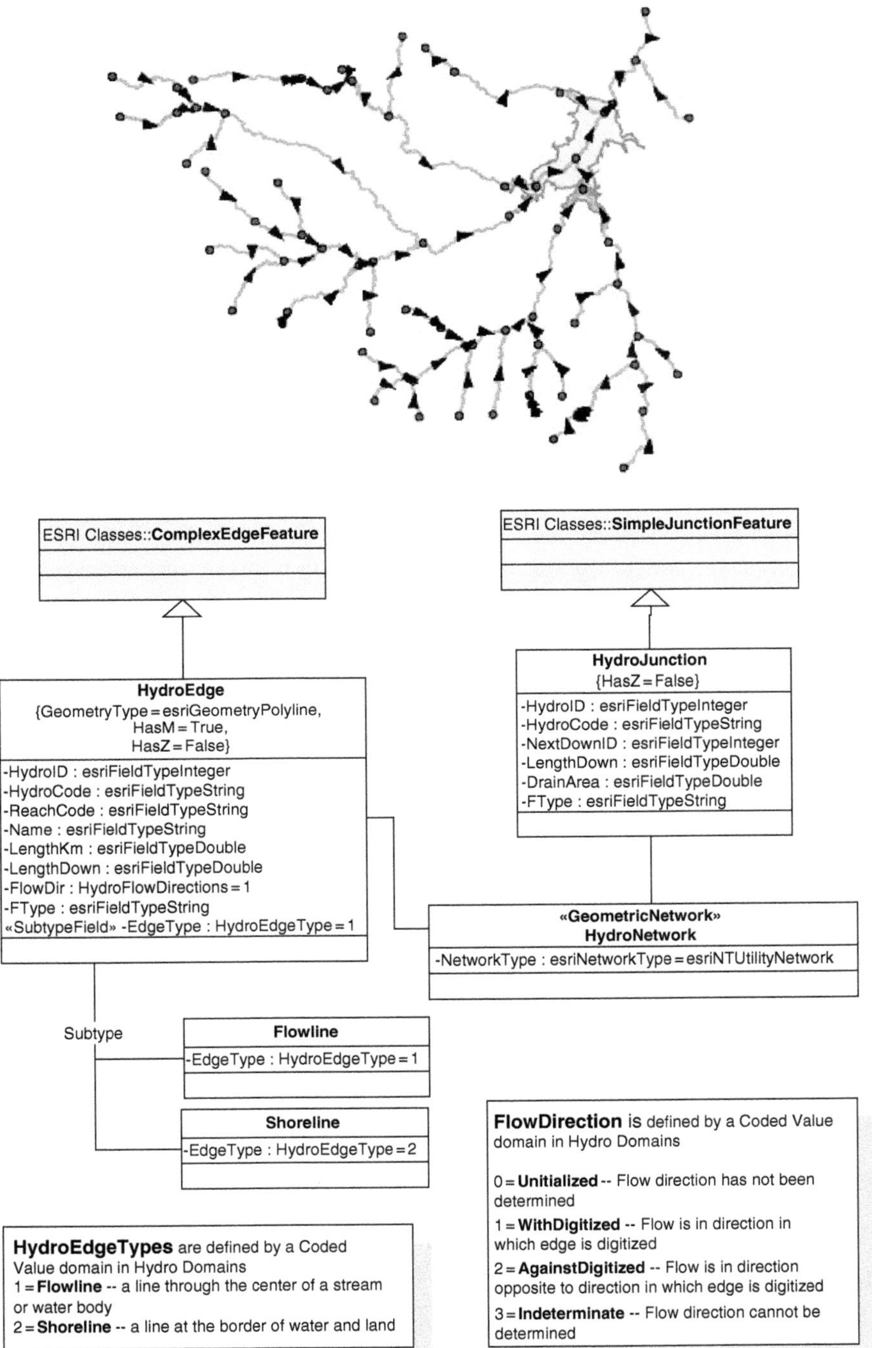

FIGURE 5.4
Network schema UML. This ArcHydro network visualization shows junctions, stream reaches, flow directions, and a water body [7].

through a particular HydroJunction. The HydroJunction's NextDownID attribute contains the HydroID of the next junction downstream on the network. Using the NextDownID, a user can traverse the network upstream and downstream, tracing the path of flow along the network and accessing any interesting features or channel geometries along the way.

A special attribute, the ReachCode, is included with HydroEdge features; the ReachCode refers to the 14-digit code applied to U.S. streams in the NHD and used by the USEPA in discharge permitting procedures. It is the *de facto* standard designation for streams in the United States and is utilized in the ArcHydro data model. The first eight digits of the ReachCode correspond with the USGS Hydrologic Unit Code (HUC). The HUC is an index for a specific hydrologic region/basin in the United States containing many watersheds and stream reaches. The last six digits of the ReachCode indicate a specific stream reach within the HUC. The NHD ReachCode attribute indexes each stream in the United States according to its basin and uniquely identifies it within that basin, and is included in the ArcHydro data model to facilitate interoperability with national datasets.

The hydrography package is inspired and informed by attention to existing regional and national datasets in the United States, including the database of river structures maintained by the USACE and the NHD. The ArcHydro data structure is able to represent all data types recorded in these large-scale datasets (see Figure 5.5: Hydrography Schema UML). The *HydroResponseUnit* is a flexible polygon feature class allowing the description of spatially distributed parameters affecting the interaction of the land surface with the atmosphere and

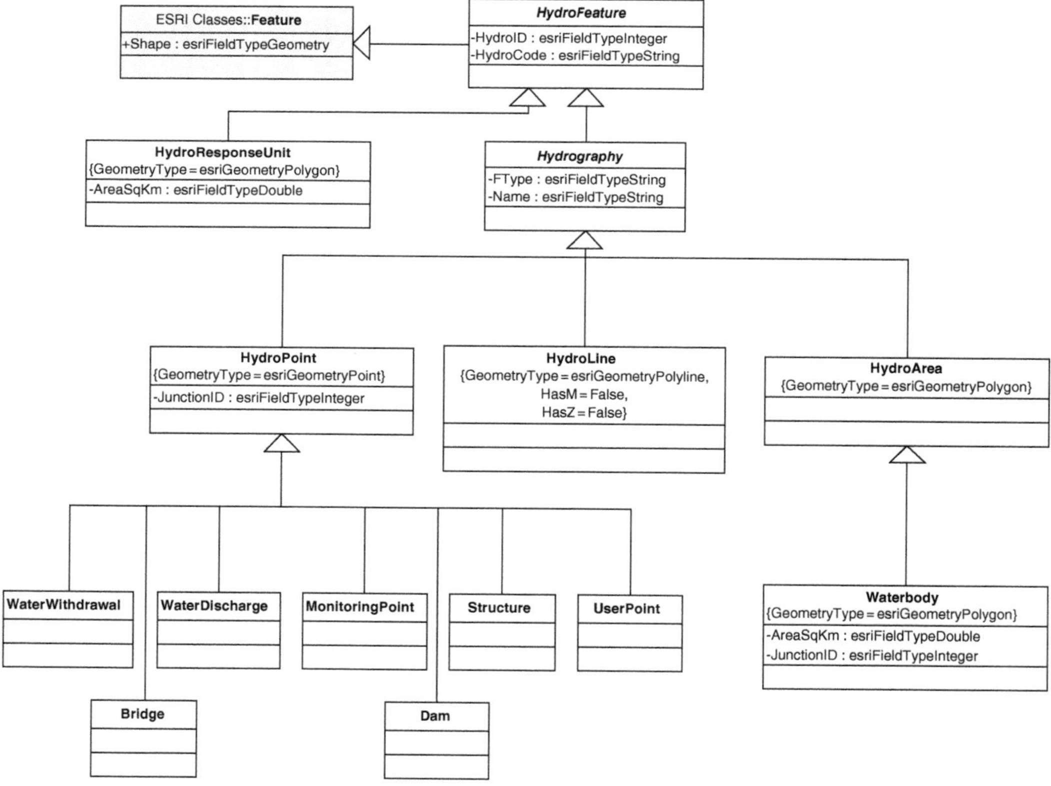

FIGURE 5.5
Hydrography schema UML. This image shows a layout of monitoring points, water withdrawl points, water quality stations, and water discharge points along with network and drainage features [7].

groundwater. For example, SCS land cover values could be stored for a NEXRAD rainfall grid, and used to model runoff in a watershed. A *Waterbody* is a polygon feature class representing a lake or reservoir. Several *HydroPoint* classes represent specific features along the stream network. *Dam, Structure, Bridge, WaterWithdrawl,* and *WaterDischarge* classes represent human structures that affect water flow. The *MonitoringPoint* class represents any location in a watershed where hydrologic data is recorded. MonitoringPoints have a relationship with records in the Time Series class, where associated data is stored, and with the HydroJunction class, if the MonitoringPoint is gauging data at an in-stream location on the network. *UserPoint* classes represent any point not covered by the hydrography framework. *HydroPoint, HydroLine,* and *HydroArea* represent other types of hydrographic features, such as administrative boundaries. The *FType* attribute of Hydrography features is derived from the NHD, and stores the NHDs name or descriptor for each feature.

The Drainage feature dataset (see Figure 5.6: Drainage Schema UML) is based on the "raindrop model"; in other words, when a drop of rain falls somewhere, where does it go? It flows into Catchments, Watersheds, and Basins, and then through the stream network to an outlet point/sink. *Basins* are administrative boundaries formed for the purpose of large-scale water resource management, such as the Lower Mississippi River Basin (This basin is identified by the USGS Hydrologic Unit Code 08), and are the basic units of organizational data packaging. *Watersheds* are a subdivision of basins, normally based on drainage into a particular tributary of the basin. *Catchments* are not defined by administrative conditions, but rather by a set of physical, hierarchical formulae. One such formula could specify that each catchment drains a single stream, with the stream being defined as draining at least 1000 Digital Elevation Model (DEM) cells. Two other feature classes, *DrainagePoint* and *DrainageLine,* are implemented to facilitate DEM-based watershed delineation and analysis within the geodatabase (read on for more details).

An important relationship is established between the Watershed and HydroJunction classes. Each Watershed has a "JunctionID" attribute, corresponding to a HydroJunction's HydroID attribute. Therefore each Watershed is "attached" to the stream network at a specific point. This relationship enables a user to trace flows from a watershed downstream via the stream network, or to retrieve watersheds contributing flow to a junction on the network.

The channel geometry feature dataset is not like the other datasets; it is not "flat." Whereas other datasets represent map-like arrangements of hydrologic features, the channel describes the 3D shape of the stream channel and its floodplains (see Figure 5.7: Channel Schema UML). By assembling the elements of channel geometry, an intricate 3D Triangular Irregular Network (TIN) representation of a channel may be constructed from a Digital Terrain Model (DTM) or exported to modeling software (such as HEC–RAS). Channel geometry is composed of *CrossSection* and *ProfileLine* classes. Both are registered according to a ReachCode. CrossSections are transverse 2D "slices" of a channel geometry arranged perpendicular to the direction of flow, composed of many *CrossSectionPoint* objects which record the x and y coordinates (CrossM and Elevation attributes) of points along the channel section. CrossSections are located at a distance along the stream reach profile, and this location is recorded in the ProfileM attribute. ProfileLines store the location of three important channel features: BankLines, StreamLines, and Thalwegs.

Hydraulic modeling requires additional parameters such as channel roughness or levee location associated with channel sections. The *HydroEvent* class provides a structure for the storage of such parameters (see Figure 5.8: Hydro Event UML). *HydroPointEvents* and *HydroLineEvents* store user-specified attribute data associated with a particular stream reach (indexed by ReachCode), such as channel roughness. However, events are not limited to just hydrologic channel parameters, and can record offsets or administrative data associated with the channel. Specific attributes other than "measure" are not predefined in the data

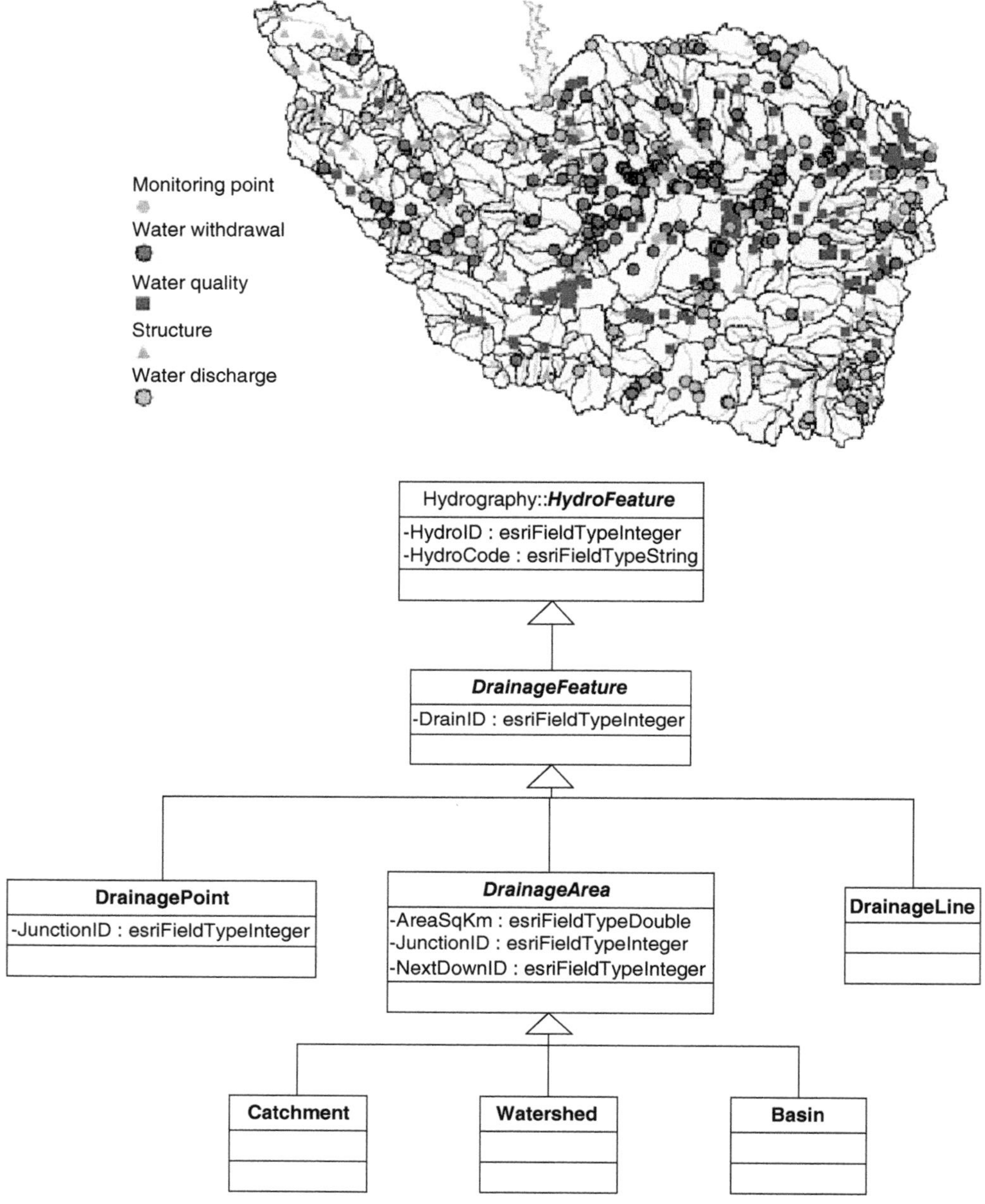

FIGURE 5.6
Drainage schema UML. The ArcHydro drainage model is built around Watershed, Basin, Catchment, Line, and Point features [7].

model because the variety of parameters that might be needed is too great to anticipate in the design of the data structures.

The *Time Series* object class is a wonderfully simple and powerful data structure (see Figure 5.9). It stores data records indexed by HydroID (FeatureID), time series type (TSTypeID), time of record (TSDateTime), and data value (TSValue). Any time series of data may be stored in association with any feature in the database possessing a HydroID, although the time series is meant to be used with the related MonitoringPoint feature class.

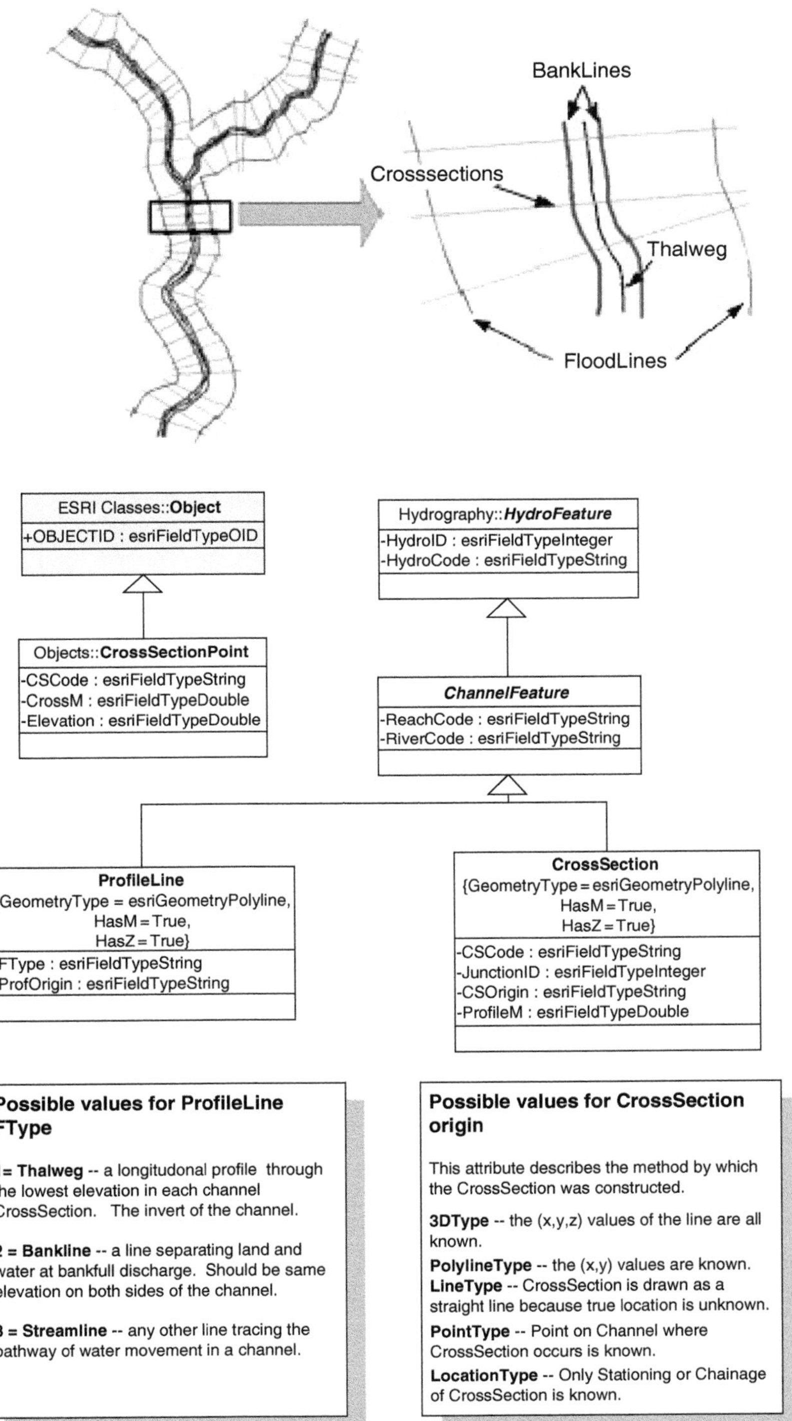

FIGURE 5.7
Channel schema UML. The channel model is a series of channel geometry descriptive cross-sections, combined with profile lines, which run parallel to the streamflow [7].

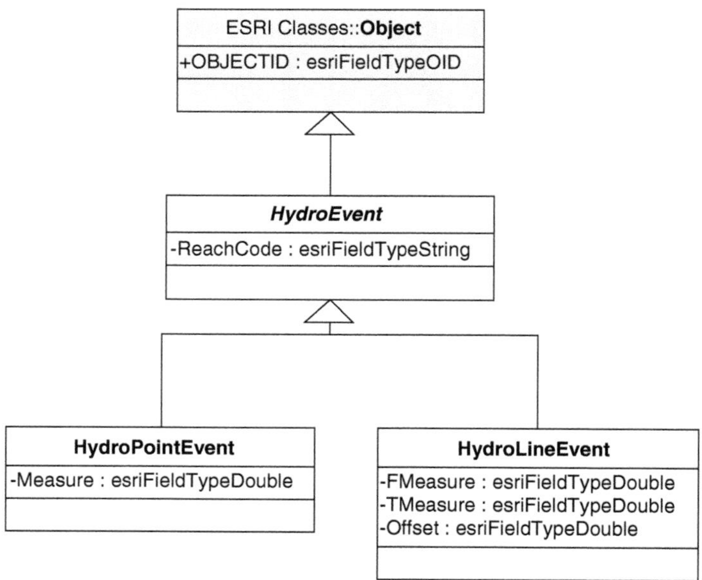

FIGURE 5.8
Hydro event UML. Hydro Events describe "extra" data in association with channels [7].

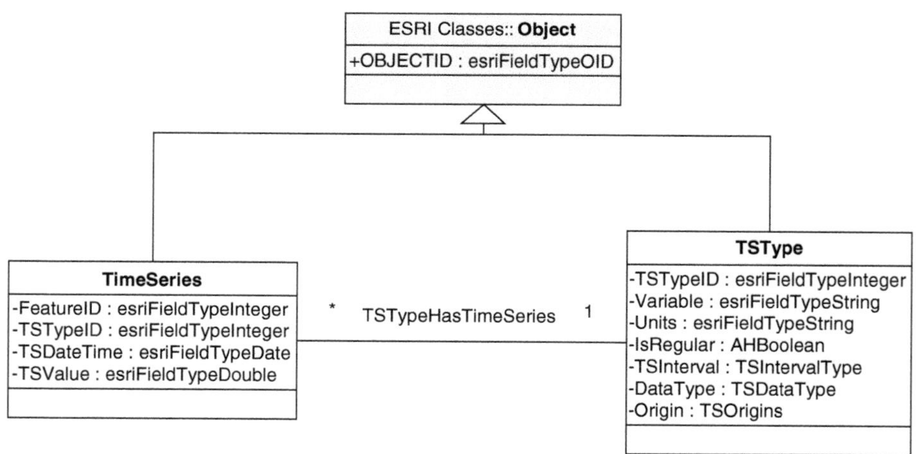

FIGURE 5.9
Time series schema UML. Time-varying data is recorded in the Time Series table, and the descriptive metadata for that time series is stored in the associated TSType index table [7].

Data types are described in the *TSType* object class and indexed with a TSTypeID integer. TSType's six attributes include the units of record, a string description of the data are listed along with whether the time series follows a regular time interval, what that interval is, the type of time series (instantaneous, cumulative, incremental, average, maximum, minimum), and the origin of the data (recorded or generated). Time Series records may be queried and sorted according to feature, type, time, and value.

References

1. Treinish, Lloyd A. *Scientific Data Models for Large-Scale Applications*. Yorktown Heights, NY: IBM Thomas J. Watson Research Center, 1995.
2. ESRI online, www.esri.com.
3. Howe, Denis, Ed. *The Free On-Line Dictionary of Computing*. http://www.foldoc.org/.
4. Tannenbaum, Adrienne. *Metadata Solutions*. Boston, MA: Addison-Wesley, 2002.
5. ESRI, Inc. *Spatial Data Standards and GIS Interoperability*. Redlands, CA: ESRI, Inc., 2003.
6. ESRI, Inc. *Working with the Geodatabase: Powerful Multiuser Editing and Sophisticated Data Integrity*. Redlands, CA: ESRI, Inc., 2002.
7. Maidment, David R. *ArcHydro: GIS for Water Resources*. Redlands, CA: ESRI Press, 2002. pp. 141–166, 177–187. Figures used with permission.
8. The United States Army Corps of Engineers Hydrologic Engineering Center. Software publications at: http://www.hec.usace.army.mil.
9. ESRI, Inc. *ESRI Geodatabase Object Model Diagram*. July 2004. http://arcobjectsonline.esri.com/ArcObjectsOnline/Diagrams/Geodatabase%20Object%20Model%201.pdf.

6

The Modelshed Geodata Model

Benjamin L. Ruddell and Praveen Kumar

Contents

6.1 Modelshed Framework

6.1.1 A Data Model for Remote Sensing, Fluxes, and Earth Science Modeling

Beyond the territory of hydro data models lies the diverse field of environmental Earth science modeling. The Earth's built and natural environments are represented in models; these models are used to store data and simulate environmental processes. Contemporary computer models come in two varieties: numerical matrix-based models and datasets that use raster grids and object-oriented models based on the interaction of objects, with data stored in vectorized table formats. Geographic Information System (GIS) can spatially register, visualize, and access both types of data, and are an ideal platform for the integration of spatial Earth science models (see Figure 6.1). The integration of raster and vector data is a challenge for GIS systems, and must be substantively addressed by environmental data models.

In the most general sense, Earth science data can be organized in GIS data model using a combination of points, lines, polygon areas, volumes, associated time series data, and raster images. The environment is not static, so environmental parameters (e.g., water, energy) are in flux and must be represented by an environmental model. The model structure should be sufficiently rigid and specific to maintain useful relationships between model

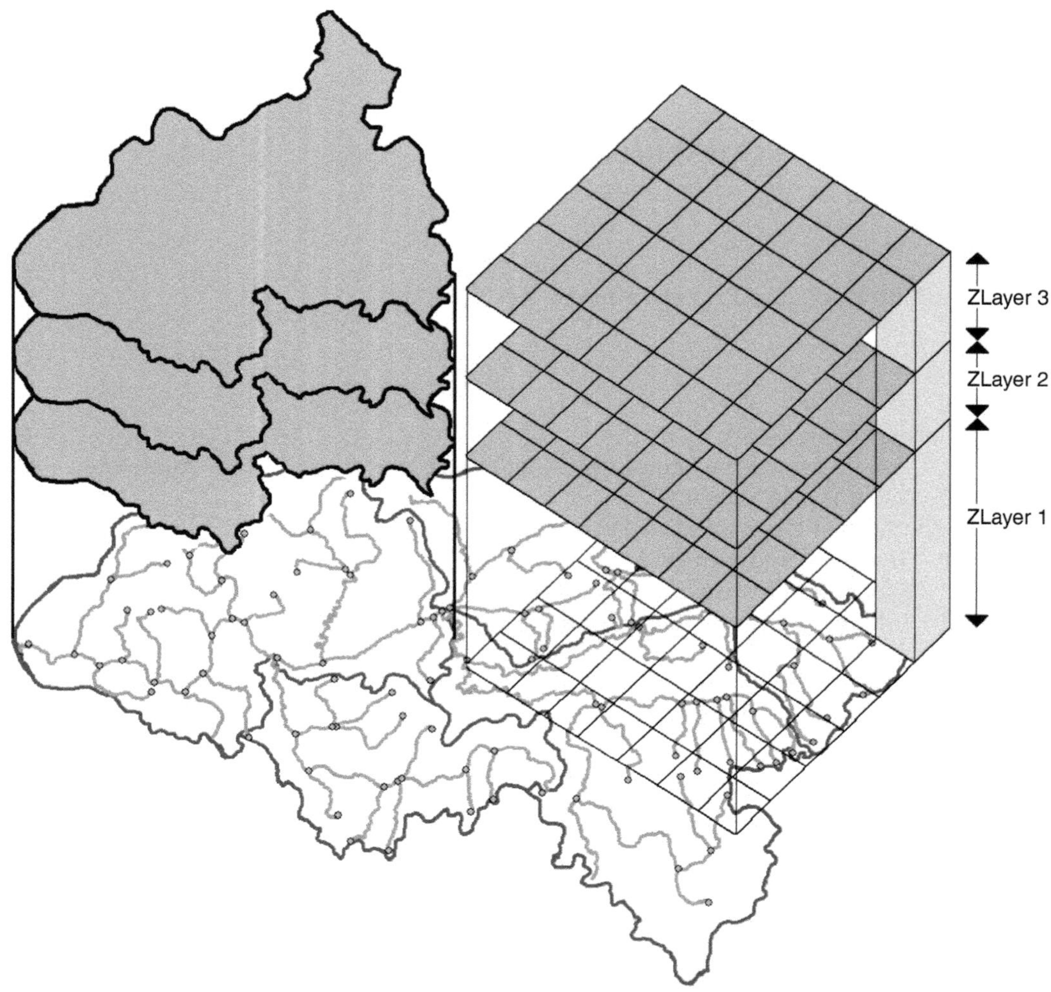

FIGURE 6.1
A visualization of the Modelshed geodata model framework, including grid and watershed-based Modelshed type domains, vertical ZLayer indexing, and the associated hydrographic data in an ArcHydro-compatible model.

classes; however, it should be sufficiently flexible to accept a broad range of environmental datasets for diverse applications, storing metadata relevant to each. The Modelshed framework is a generalized GIS data model for the organization and modeling of diverse geospatial data. It represents point, line, area, and volumetric database objects in three dimensions, and stores time series and flux data in association with all model features. It provides data structures to facilitate the geospatial analysis of time-indexed raster datasets and the integration of raster data with the vector structures of the data model.

The design of the Modelshed framework is approached from the perspective of environmental modeling, rather than the data storage for a particular application. Therefore, the framework focuses on what environmental models have in common, and on ways to allow different types of models to operate in a GIS environment, with GIS data structures. Modeling is often conducted in a numerical, grid-based environment, and the Modelshed framework puts a priority on representing and interoperating with this sort of modeling.

The challenge here is to represent rigorous grid-based applications in the vector-space of a GIS data model, and to do so efficiently, and to provide structures that allow transfer of numerical data in and out of the database environment. Another challenge is to generalize the model so that generic tools and codes may be applied to many different environmental applications.

The Modelshed framework is implemented using the Environmental Systems Research Institute (ESRI) geodatabase technology, and being built on this technology, is compatible with all software tools designed to work with industry-standard DBMS (DataBase Management System) and the OpenGIS standards of the geodatabase. Some tools commonly used to access a Modelshed-compatible data include ArcGIS™, MSAccess™, MSExcel™, Oracle™, and Extensible Markup Language (XML) readers. The Modelshed framework is implemented on top of the ArcHydro data model and uses some of ArcHydro's objects and attributes. ArcHydro is discussed extensively in Chapter 5. Tools designed for use with ArcHydro remain compatible with a Modelshed-extended database, because Modelshed does not modify the underlying ArcHydro structure. Some functions of the ArcHydro tools, particularly the HydroID assignment utility, will function with Modelshed features. However, it is important to note that the Modelshed framework is conceptually independent from ArcHydro, and could be implemented as a stand-alone data model or on top of another data model. Modelshed is implemented on top of ArcHydro to leverage ArcHydro's data structures and gain access to ArcHydro-compatible datasets, which are becoming highly available in the hydrologic and GIS community. As an extension to ArcHydro, Modelshed adds powerful capacities to an already valuable data model: three-dimensional (3D) indexing of spatial features, modeling of fluxes between objects, storage of statistical time series data, integrated access to four-dimensional (4D) raster datasets, and new relationships between some features in the database. The Modelshed framework will remain compatible with future extensions of the ArcHydro data model, assuming that the existing ArcHydro data structures are not altered.

Whereas ArcHydro is a data model specialized for the hydraulic and hydrologic description of the land surface using contemporary hydrologic datasets, Modelshed is much more general and will accept nonhydrologic geospatial datasets. Whereas in ArcHydro the objects are specific to hydrology, Modelshed has metadata structures which describe the meaning of groups of database features and index them according to the type. It is therefore a vehicle for the integration of loosely related environmental datasets. Ecology, geology, atmospheric science, sociology, agriculture, climatology, and many other disciplines may store datasets using the Modelshed framework. These diverse datasets have a common language to converse and relate. Tools developed for visualization, analysis, and data transfer of Modelshed-compatible databases will work with any specific application, in any branch of geospatial science. Hydrology will benefit from this integration because of the relationship between ArcHydro and Modelshed, and the usefulness of the ArcHydro model and the watershed paradigm as a frame of reference for many other geospatial sciences.

The Modelshed framework is proposed as a generalized 4D data model for the integration and common study of diverse geospatial and Earth science data, for the modeling of environmental fluxes in space and time, and for the development of generic visualization, data mining, analysis, and data transfer tools. It can:

- Store data for diverse spatio-temporal applications and phenomena
- Address issues of scale, heterogeneity, and resolution
- Build on top of existing data models (e.g., ArcHydro) to leverage existing data structures and tools

- Establish new relationships
- Model environmental fluxes
- Connect raster data and numerical models with object-relational data models

6.1.2 What Is a Modelshed?

A Modelshed is a volumetric spatial model unit, registered in three dimensions by a GIS, with which time-varying data, model fluxes, spatial relationships, and descriptive metadata are associated. It is a generalized conceptual structure for the representation of model features in GIS. A modelshed can be formed from any point, line, or area that has a unique Feature ID (HydroID in ArcHydro). It can have multiple associated time-series' of data, and multiple types of fluxes. The resulting structure has the capabilities of a 3D time-varying matrix, but exists in relational, tabular data model space and is not constrained to rectangular grids and model resolutions (see Figure 6.2: The Modelshed framework UML).

The Modelshed framework is composed of six elements:

- Modelshed feature classes (ModelPoint, ModelLine, ModelArea, and others) indexed by ModelshedType
- 3D indexing with ZLayer
- Statistical data storage in StatisticalTS
- Flux modeling with FluxLink, FluxType, and FluxRecord
- 4D Raster cataloging in RasterTSIndex
- New relationships established with ModelshedLink, AreaLink, and OrthogonalLink

Modelsheds are indexed by ModelshedTypeIDs. A ModelshedType describes a group of Modelsheds and collects them together. ModelshedType stores metadata describing many specific layers of model features, and distinguishes these types from each other. In this way only the generic geometric shapes of a GIS model, the points, lines, and areas, are needed to implement the abstract Modelshed class. The points, lines, and areas represent a diversity of features in model space, but they are all described using the common ModelshedTypeID index. For example, a Modelshed type could be called "watershed," "rain gauges," or "climate grid," and be comprised of the appropriate features.

A ModelshedType must be composed of a single feature class. A ModelshedType record specifies this feature class in the ModelshedClassName attribute. This attribute contains the name of an implementation feature class, including ModelArea, ModelPoint, ModelLine, and ArcHydro classes and custom classes that may take part in the Modelshed framework. Using the ModelshedType index, a user may traverse from any link or data record to the ModelshedType table, and locate the class where model features are stored. If a native ArcHydro feature class is to be used in association with Modelshed features, it must be registered as a ModelshedType, just like features stored in the ModelArea, ModelPoint, and ModelLine feature classes.

The Modelshed abstract class is the root of the Modelshed framework. It inherits from the ArcHydro class HydroFeature. The essential element of this inheritance is the HydroFeature's HydroID, which serves as the unique database ID for all features in the Modelshed database. Model time series data, flux data, and other model relationships are linked on this unique ID.

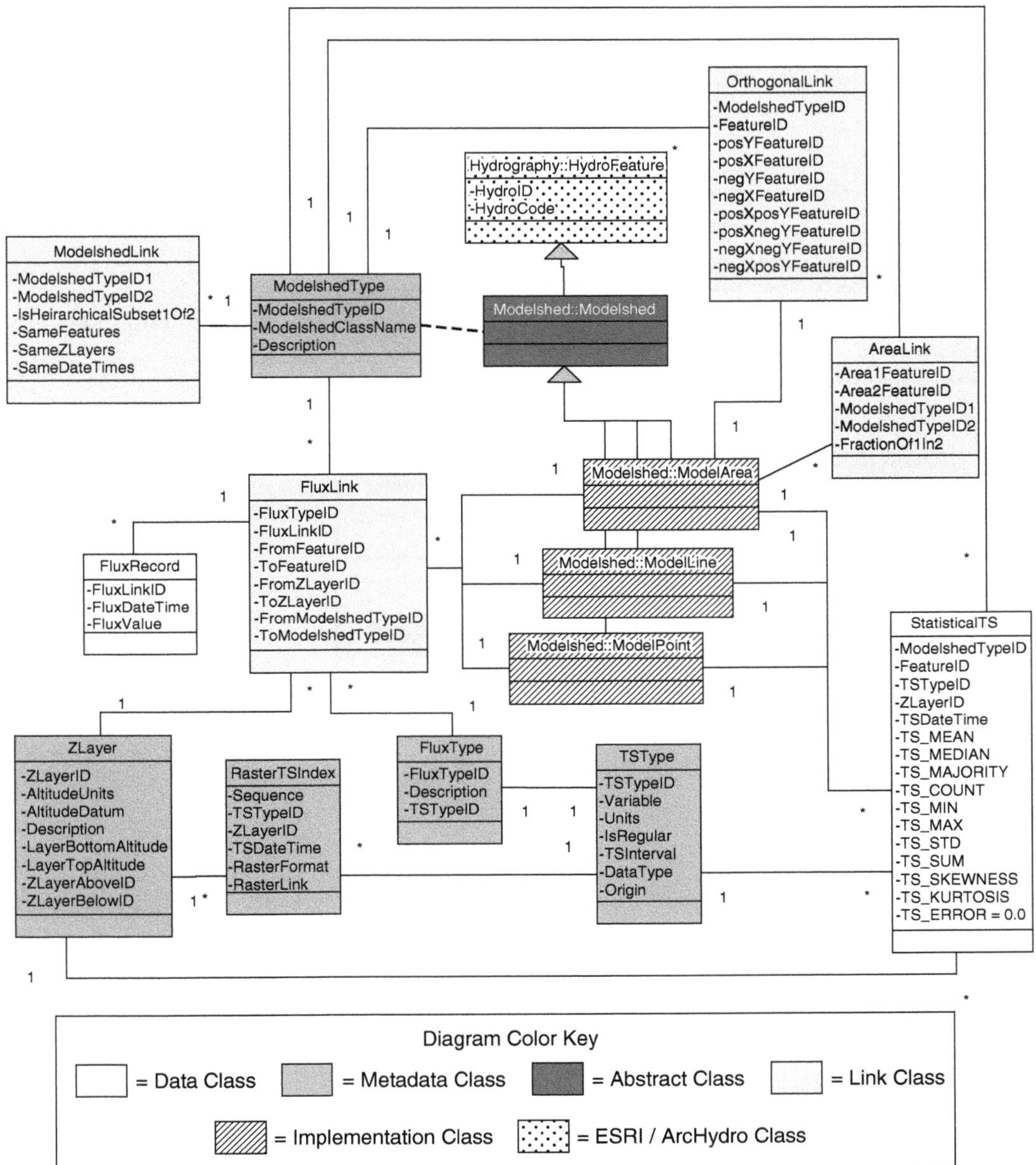

Figure 6.2
Modelshed framework UML and data dictionary. The Modelshed geodata model combines spatial features inheriting from the ArcHydro HydroFeature with a network of metadata classes.

6.2 The Modelshed Geodata Model Structure

6.2.1 Temporal and Vertical Dimensions

A limitation of GIS data model is the difficulty in representing time-varying and 3D datasets. While visualization of 4D GIS data is an ongoing area of research, the storage

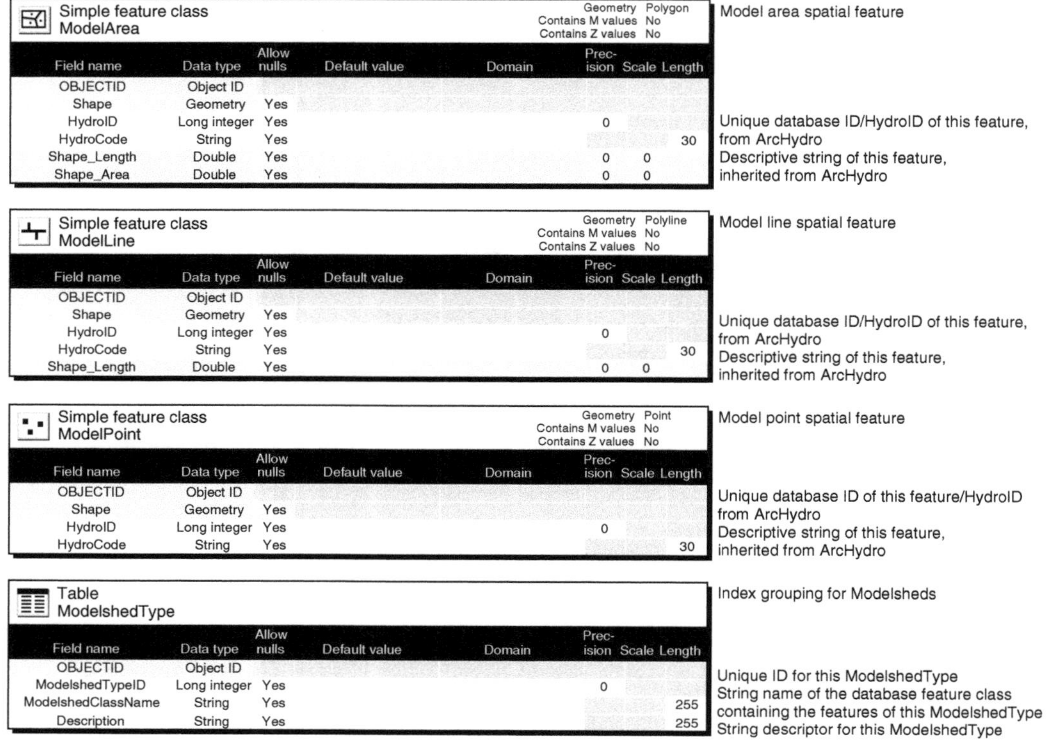

Simple feature class ModelArea — Geometry: Polygon; Contains M values: No; Contains Z values: No — Model area spatial feature

Field name	Data type	Allow nulls	Default value	Domain	Precision	Scale	Length
OBJECTID	Object ID						
Shape	Geometry	Yes					
HydroID	Long integer	Yes			0		
HydroCode	String	Yes					30
Shape_Length	Double	Yes			0	0	
Shape_Area	Double	Yes			0	0	

Unique database ID/HydroID of this feature, from ArcHydro
Descriptive string of this feature, inherited from ArcHydro

Simple feature class ModelLine — Geometry: Polyline; Contains M values: No; Contains Z values: No — Model line spatial feature

Field name	Data type	Allow nulls	Default value	Domain	Precision	Scale	Length
OBJECTID	Object ID						
Shape	Geometry	Yes					
HydroID	Long integer	Yes			0		
HydroCode	String	Yes					30
Shape_Length	Double	Yes			0	0	

Unique database ID/HydroID of this feature, from ArcHydro
Descriptive string of this feature, inherited from ArcHydro

Simple feature class ModelPoint — Geometry: Point; Contains M values: No; Contains Z values: No — Model point spatial feature

Field name	Data type	Allow nulls	Default value	Domain	Precision	Scale	Length
OBJECTID	Object ID						
Shape	Geometry	Yes					
HydroID	Long integer	Yes			0		
HydroCode	String	Yes					30

Unique database ID of this feature/HydroID from ArcHydro
Descriptive string of this feature, inherited from ArcHydro

Table ModelshedType — Index grouping for Modelsheds

Field name	Data type	Allow nulls	Default value	Domain	Precision	Scale	Length
OBJECTID	Object ID						
ModelshedTypeID	Long integer	Yes			0		
ModelshedClassName	String	Yes					255
Description	String	Yes					255

Unique ID for this ModelshedType
String name of the database feature class containing the features of this ModelshedType
String descriptor for this ModelshedType

FIGURE 6.2 (Continued)

of 4D data is now within reach, thanks to the geodatabase's capacity for relational indexing. Time series records may be indexed with any database object, and vertical indexing can add the third dimension to a formerly flat map.

The StatisticalTS object class is the all-purpose data repository of the Modelshed framework (see Figure 6.3: StatisticalTS Objects). Any Modelshed may have associated data in the StatisticalTS table, indexed by FeatureID (HydroID in ArcHydro), TSTypeID, ZLayerID, time, and ModelshedTypeID.

Data records in a database are simplifications of conditions in the model space; they are discrete aggregations of parameters that continuously vary across space and time. A variety of statistical parameters is included in a StatisticalTS record to allow a more thorough quantitative description of the variation and the error inherent in data collection and modeling. For some applications, such as ensemble runoff modeling, the statistical variability of data is critical to the generation of a probability distribution for model results. For the aggregation of continuous raster data according to Modelshed areas, a statistical summary provides a repository for a thorough description of what that raster contained both for continuous and discrete raster data files. Continuous files represent the variation of one parameter across the raster surface, while in discrete rasters each cell records an integer value that is associated with a specific descriptive category for that cell (e.g., a DEM is continuous, while a land use raster is discrete).

The Modelshed geodata model framework assimilates the ArcHydro TSTypeID index as its own data parameter description metadata object. The TSTypeID is well suited to all sorts of data types. Because the TSTypeID is essential to the function of the Modelshed

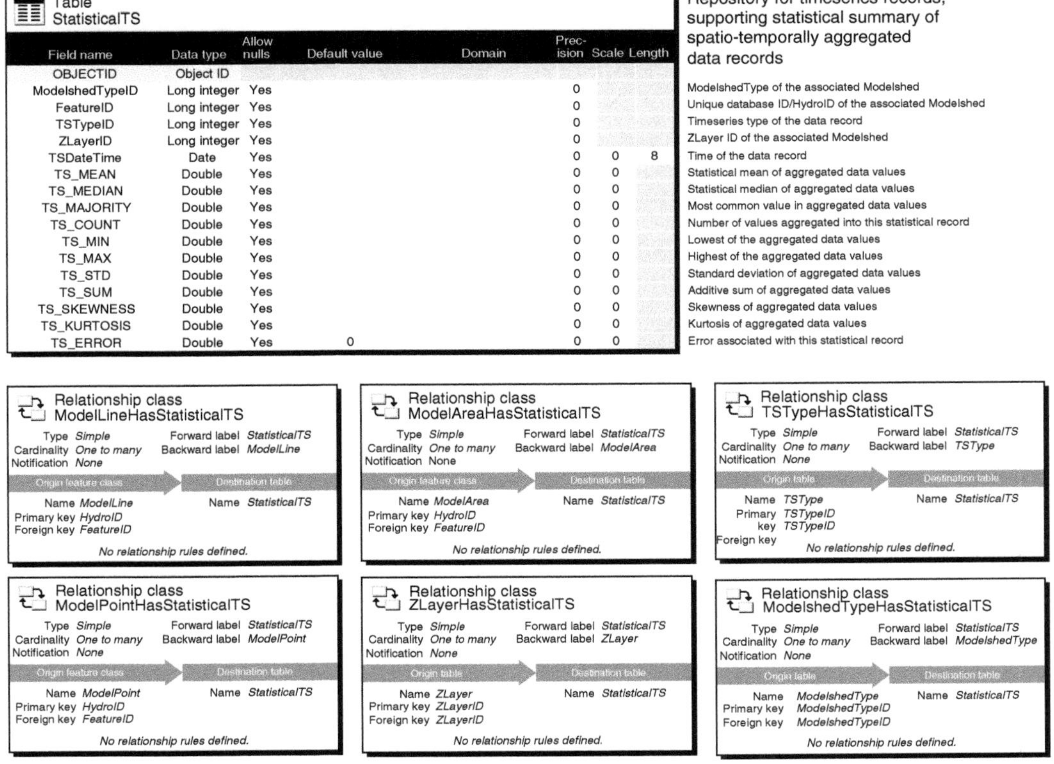

Table StatisticalTS

Repository for timeseries records, supporting statistical summary of spatio-temporally aggregated data records

Field name	Data type	Allow nulls	Default value	Domain	Precision	Scale	Length	Description
OBJECTID	Object ID							
ModelshedTypeID	Long integer	Yes			0			ModelshedType of the associated Modelshed
FeatureID	Long integer	Yes			0			Unique database ID/HydroID of the associated Modelshed
TSTypeID	Long integer	Yes			0			Timeseries type of the data record
ZLayerID	Long integer	Yes			0			ZLayer ID of the associated Modelshed
TSDateTime	Date	Yes			0	0	8	Time of the data record
TS_MEAN	Double	Yes			0	0		Statistical mean of aggregated data values
TS_MEDIAN	Double	Yes			0	0		Statistical median of aggregated data values
TS_MAJORITY	Double	Yes			0	0		Most common value in aggregated data values
TS_COUNT	Double	Yes			0	0		Number of values aggregated into this statistical record
TS_MIN	Double	Yes			0	0		Lowest of the aggregated data values
TS_MAX	Double	Yes			0	0		Highest of the aggregated data values
TS_STD	Double	Yes			0	0		Standard deviation of aggregated data values
TS_SUM	Double	Yes			0	0		Additive sum of aggregated data values
TS_SKEWNESS	Double	Yes			0	0		Skewness of aggregated data values
TS_KURTOSIS	Double	Yes			0	0		Kurtosis of aggregated data values
TS_ERROR	Double	Yes	0		0	0		Error associated with this statistical record

Relationship class ModelLineHasStatisticalTS
Type *Simple* Forward label *StatisticalTS*
Cardinality *One to many* Backward label *ModelLine*
Notification *None*
Origin feature class → Destination table
Name *ModelLine* Name *StatisticalTS*
Primary key *HydroID*
Foreign key *FeatureID*
No relationship rules defined.

Relationship class ModelAreaHasStatisticalTS
Type *Simple* Forward label *StatisticalTS*
Cardinality *One to many* Backward label *ModelArea*
Notification *None*
Origin feature class → Destination table
Name *ModelArea* Name *StatisticalTS*
Primary key *HydroID*
Foreign key *FeatureID*
No relationship rules defined.

Relationship class TSTypeHasStatisticalTS
Type *Simple* Forward label *StatisticalTS*
Cardinality *One to many* Backward label *TSType*
Notification *None*
Origin table → Destination table
Name *TSType* Name *StatisticalTS*
Primary key *TSTypeID*
Foreign key *TSTypeID*
No relationship rules defined.

Relationship class ModelPointHasStatisticalTS
Type *Simple* Forward label *StatisticalTS*
Cardinality *One to many* Backward label *ModelPoint*
Notification *None*
Origin feature class → Destination table
Name *ModelPoint* Name *StatisticalTS*
Primary key *HydroID*
Foreign key *FeatureID*
No relationship rules defined.

Relationship class ZLayerHasStatisticalTS
Type *Simple* Forward label *StatisticalTS*
Cardinality *One to many* Backward label *ZLayer*
Notification *None*
Origin table → Destination table
Name *ZLayer* Name *StatisticalTS*
Primary key *ZLayerID*
Foreign key *ZLayerID*
No relationship rules defined.

Relationship class ModelshedTypeHasStatisticalTS
Type *Simple* Forward label *StatisticalTS*
Cardinality *One to many* Backward label *ModelshedType*
Notification *None*
Origin table → Destination table
Name *ModelshedType* Name *StatisticalTS*
Primary key *ModelshedTypeID*
Foreign key *ModelshedTypeID*
No relationship rules defined.

Figure 6.3
The StatisticalTS data dictionary.

framework, it is considered as an original Modelshed class for the purposes of model explanation.

The ZLayer object describes a vertical layering index. A Zlayer is a horizontal layer in GIS space with a specified top and bottom altitude, measured from a datum. ZLayers are indexed using the ZLayerID integer. This integer ID is used to vertically index StatisticalTS data records, link Modelsheds using FluxLinks, and index raster datasets. If a group of ZLayers is stacked in a vertical sequence, the ZLayers above and below the layer of reference are recorded by ZLayerID, allowing a user to traverse a stack of Zlayers (see Figure 6.4: ZLayer Objects). ZLayers may be indexed with reference to any datum, such as "NAD1983," or "ground surface," "average treetop height." Positive altitudes are above the datum in elevation, and negative altitudes are below. It is not necessary that a ZLayer represents atmospheric layers; it might describe a soil layer or an aquifer.

ZLayer is a critical component of the Modelshed framework because it indexes the third dimension of the data model. The GIS surface features describe the 2D position and form of an object, and the ZLayer index projects that object vertically in space. All database features may be projected vertically, including lines, areas, and points. These vertical projections may combine to represent complex 3D structures. For example, a combination of points, lines, and areas projected below the ground surface could represent wells, aquifers, and boundary areas between aquifers. A point projected into the atmosphere could represent the path of a weather balloon, and associated StatisticalTS records could record data at intervals along the path.

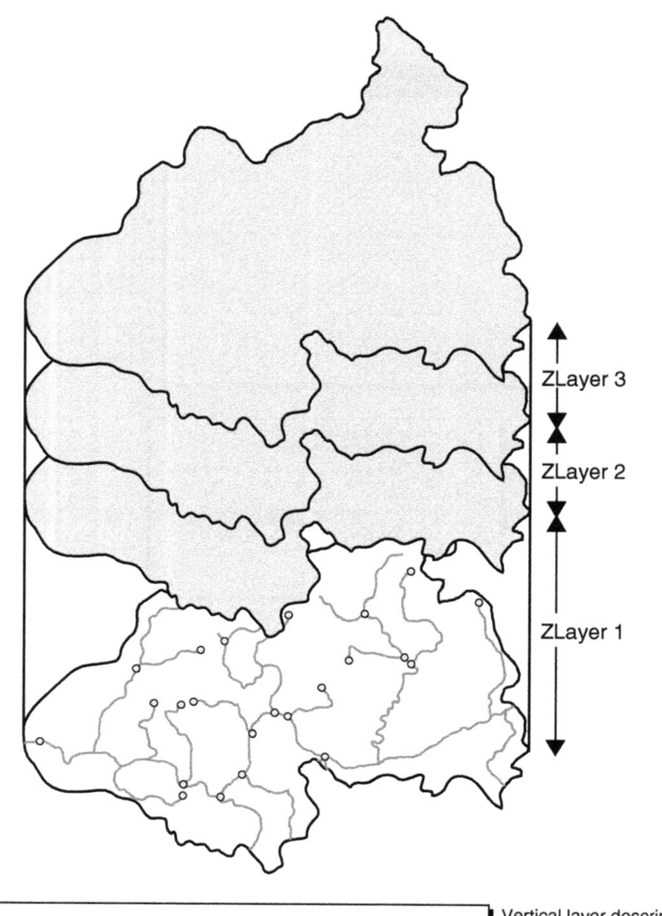

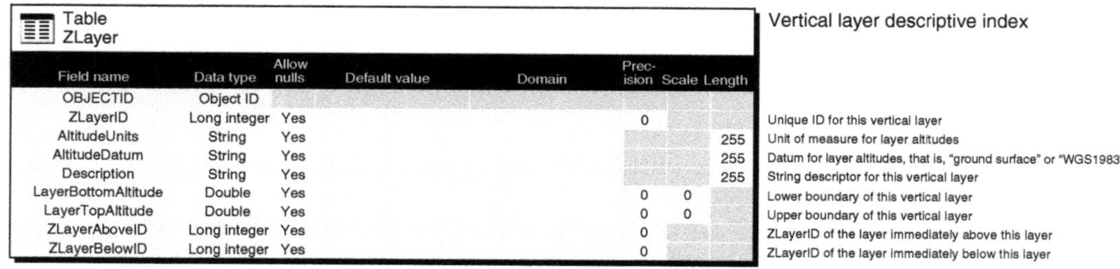

Table ZLayer — Vertical layer descriptive index

Field name	Data type	Allow nulls	Default value	Domain	Precision	Scale	Length	
OBJECTID	Object ID							
ZLayerID	Long integer	Yes			0			Unique ID for this vertical layer
AltitudeUnits	String	Yes					255	Unit of measure for layer altitudes
AltitudeDatum	String	Yes					255	Datum for layer altitudes, that is, "ground surface" or "WGS1983"
Description	String	Yes					255	String descriptor for this vertical layer
LayerBottomAltitude	Double	Yes			0	0		Lower boundary of this vertical layer
LayerTopAltitude	Double	Yes			0	0		Upper boundary of this vertical layer
ZLayerAboveID	Long integer	Yes			0			ZLayerID of the layer immediately above this layer
ZLayerBelowID	Long integer	Yes			0			ZLayerID of the layer immediately below this layer

FIGURE 6.4
A ZLayer visualization and data dictionary. ZLayer relationships are shown in Figure 6.2.

6.2.2 Modeling Fluxes with FluxLinks

In the natural world, fluxes flow from one place to another, crossing space and time. In the Modelshed framework these "places" are represented by database objects (see Figure 6.5: Flux Objects). Fluxes may be established between any two database features with a FeatureID (equivalent to the HydroID in ArcHydro). A flux could record the movement of any parameter. Examples include matter and energy, perhaps in the form of water, sediment, wildlife, and radiation. These quantities may move from place to place. The amount of a parameter in a certain place at a certain time is already recorded in the StatisticalTS table. The FluxRecord object class tracks the movement of a specific value of that parameter from

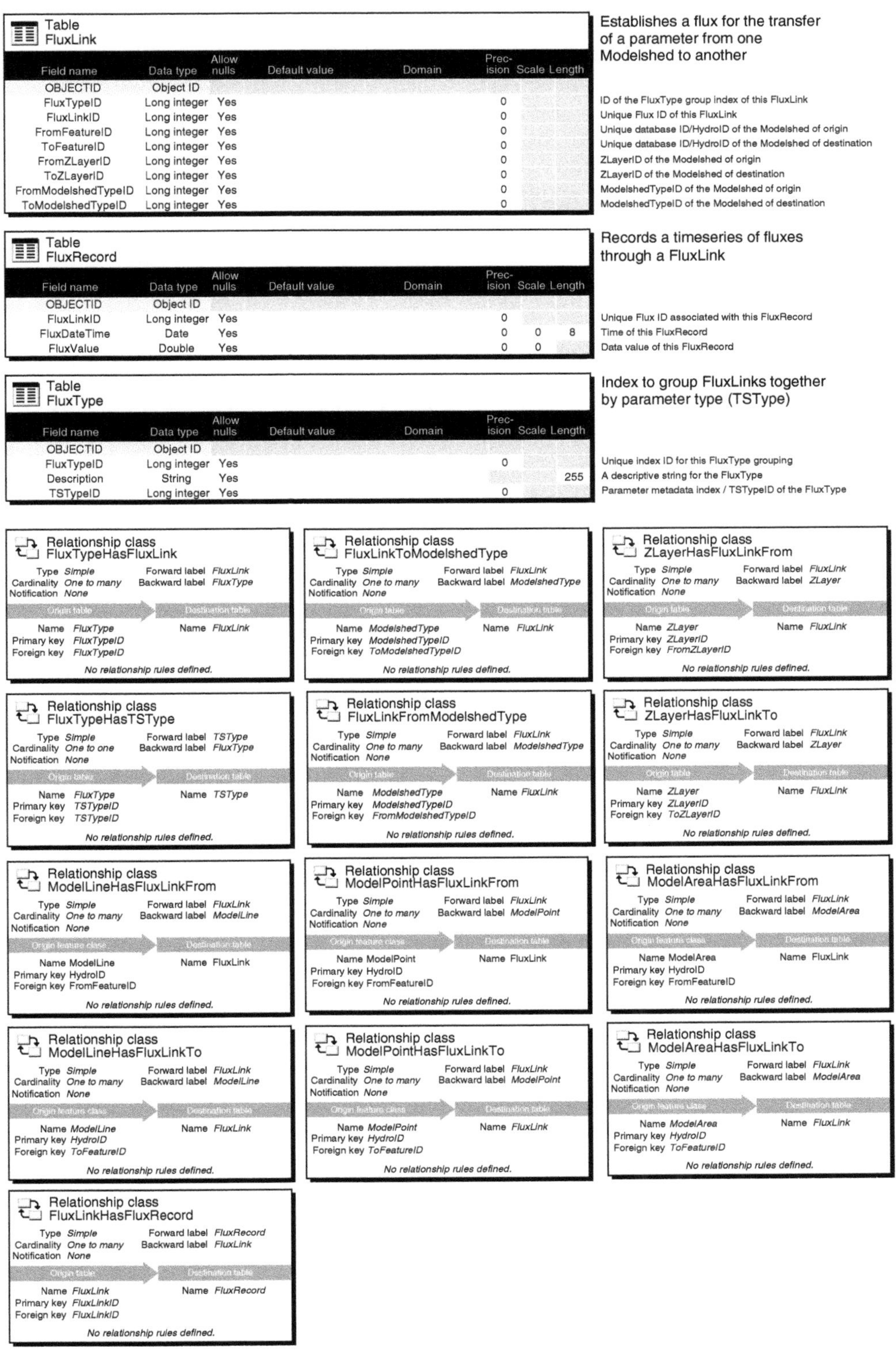

Table FluxLink — Establishes a flux for the transfer of a parameter from one Modelshed to another

Field name	Data type	Allow nulls	Default value	Domain	Precision	Scale	Length	Description
OBJECTID	Object ID							
FluxTypeID	Long integer	Yes			0			ID of the FluxType group index of this FluxLink
FluxLinkID	Long integer	Yes			0			Unique Flux ID of this FluxLink
FromFeatureID	Long integer	Yes			0			Unique database ID/HydroID of the Modelshed of origin
ToFeatureID	Long integer	Yes			0			Unique database ID/HydroID of the Modelshed of destination
FromZLayerID	Long integer	Yes			0			ZLayerID of the Modelshed of origin
ToZLayerID	Long integer	Yes			0			ZLayerID of the Modelshed of destination
FromModelshedTypeID	Long integer	Yes			0			ModelshedTypeID of the Modelshed of origin
ToModelshedTypeID	Long integer	Yes			0			ModelshedTypeID of the Modelshed of destination

Table FluxRecord — Records a timeseries of fluxes through a FluxLink

Field name	Data type	Allow nulls	Default value	Domain	Precision	Scale	Length	Description
OBJECTID	Object ID							
FluxLinkID	Long integer	Yes			0			Unique Flux ID associated with this FluxRecord
FluxDateTime	Date	Yes			0	0	8	Time of this FluxRecord
FluxValue	Double	Yes			0	0		Data value of this FluxRecord

Table FluxType — Index to group FluxLinks together by parameter type (TSType)

Field name	Data type	Allow nulls	Default value	Domain	Precision	Scale	Length	Description
OBJECTID	Object ID							
FluxTypeID	Long integer	Yes			0			Unique index ID for this FluxType grouping
Description	String	Yes					255	A descriptive string for the FluxType
TSTypeID	Long integer	Yes			0			Parameter metadata index / TSTypeID of the FluxType

Relationship class FluxTypeHasFluxLink
Type Simple; Cardinality One to many; Notification None; Forward label FluxLink; Backward label FluxType
Origin table: Name FluxType; Primary key FluxTypeID; Foreign key FluxTypeID — Destination table: Name FluxLink
No relationship rules defined.

Relationship class FluxLinkToModelshedType
Type Simple; Cardinality One to many; Notification None; Forward label FluxLink; Backward label ModelshedType
Origin table: Name ModelshedType; Primary key ModelshedTypeID; Foreign key ToModelshedTypeID — Destination table: Name FluxLink
No relationship rules defined.

Relationship class ZLayerHasFluxLinkFrom
Type Simple; Cardinality One to many; Notification None; Forward label FluxLink; Backward label ZLayer
Origin table: Name ZLayer; Primary key ZLayerID; Foreign key FromZLayerID — Destination table: Name FluxLink
No relationship rules defined.

Relationship class FluxTypeHasTSType
Type Simple; Cardinality One to one; Notification None; Forward label TSType; Backward label FluxType
Origin table: Name FluxType; Primary key TSTypeID; Foreign key TSTypeID — Destination table: Name TSType
No relationship rules defined.

Relationship class FluxLinkFromModelshedType
Type Simple; Cardinality One to many; Notification None; Forward label FluxLink; Backward label ModelshedType
Origin table: Name ModelshedType; Primary key ModelshedTypeID; Foreign key FromModelshedTypeID — Destination table: Name FluxLink
No relationship rules defined.

Relationship class ZLayerHasFluxLinkTo
Type Simple; Cardinality One to many; Notification None; Forward label FluxLink; Backward label ZLayer
Origin table: Name ZLayer; Primary key ZLayerID; Foreign key ToZLayerID — Destination table: Name FluxLink
No relationship rules defined.

Relationship class ModelLineHasFluxLinkFrom
Type Simple; Cardinality One to many; Notification None; Forward label FluxLink; Backward label ModelLine
Origin feature class: Name ModelLine; Primary key HydroID; Foreign key FromFeatureID — Destination table: Name FluxLink
No relationship rules defined.

Relationship class ModelPointHasFluxLinkFrom
Type Simple; Cardinality One to many; Notification None; Forward label FluxLink; Backward label ModelPoint
Origin feature class: Name ModelPoint; Primary key HydroID; Foreign key FromFeatureID — Destination table: Name FluxLink
No relationship rules defined.

Relationship class ModelAreaHasFluxLinkFrom
Type Simple; Cardinality One to many; Notification None; Forward label FluxLink; Backward label ModelArea
Origin feature class: Name ModelArea; Primary key HydroID; Foreign key FromFeatureID — Destination table: Name FluxLink
No relationship rules defined.

Relationship class ModelLineHasFluxLinkTo
Type Simple; Cardinality One to many; Notification None; Forward label FluxLink; Backward label ModelLine
Origin feature class: Name ModelLine; Primary key HydroID; Foreign key ToFeatureID — Destination table: Name FluxLink
No relationship rules defined.

Relationship class ModelPointHasFluxLinkTo
Type Simple; Cardinality One to many; Notification None; Forward label FluxLink; Backward label ModelPoint
Origin feature class: Name ModelPoint; Primary key HydroID; Foreign key ToFeatureID — Destination table: Name FluxLink
No relationship rules defined.

Relationship class ModelAreaHasFluxLinkTo
Type Simple; Cardinality One to many; Notification None; Forward label FluxLink; Backward label ModelArea
Origin feature class: Name ModelArea; Primary key HydroID; Foreign key ToFeatureID — Destination table: Name FluxLink
No relationship rules defined.

Relationship class FluxLinkHasFluxRecord
Type Simple; Cardinality One to many; Notification None; Forward label FluxRecord; Backward label FluxLink
Origin table: Name FluxLink; Primary key FluxLinkID; Foreign key FluxLinkID — Destination table: Name FluxRecord
No relationship rules defined.

FIGURE 6.5
The FluxLink data dictionary including its relationships.

one place to another at a specific time. A FluxRecord lists the ID of the FluxLink associated with the data, the time of the flux, and the value of the flux.

The FluxLink object class establishes a permanent relationship between two Modelsheds. FluxLinks are indexed by a FluxLinkID. FluxLinks are grouped and described using a FluxTypeID which is described in the FluxType metadata object class. FluxType includes a string description of this collection of fluxes and a TSTypeID describing the data parameter being passed in the flux. There should be at least one FluxLink per FluxType connecting any two Modelsheds. In other words, a FluxLink represents a pathway for the transfer of a particular parameter between two Modelsheds. FluxLinks are indexed by the FeatureID (HydroID in ArcHydro), ZLayerID, and ModelshedTypeID of both Modelsheds involved in the flux.

A combination of StatisticalTS data and FluxRecord data allows total accounting for fluxes of parameters within the database. This is the equivalent of nodes and links in a computational network, where Modelsheds are nodes and FluxLinks are links. Continuity and balance equations may be established, accounting for the moves in and out of nodes, and by which path it travels, and how much the stored quantity at a node increases or decreases at a given time step. For a given Modelshed and parameter TSType, given that the value of a parameter in the Modelshed at a given time is recorded in the time series table (StatisticalTS), and given that all fluxes to and from the Modelshed are recorded in the FluxRecord table, a parameter is conserved (equation 6.1).

$$\text{Timeseries}^{T} + \sum^{T+1} \text{FluxRecord} = \text{Timeseries}^{T+1} \tag{6.1}$$

Likewise, conservation of a parameter and computation of the average residence time of a parameter in a modelshed may be computed from the same data. Conservation is established by balancing the value of a parameter gained in a Modelshed via a FluxLink against the value of the same parameter lost by the other Modelshed at a given time step. The average residence time is computed by dividing the value of a parameter in a Modelshed at a given time step by the sum of outgoing FluxRecords.

The FluxLink concept is essentially redefines vectors within data model space. In coordinate space, a vector consists of a magnitude and a direction within that coordinate system. In data model space, a vector can be represented by the combination of FluxLink and FluxRecord classes. Vector directionality follows the FluxLink from one Modelshed center to another. The magnitude of this vector is established by the value and sign of a FluxRecord.

6.2.3 Linking Remote Sensing Raster Data with Vector Data: Query-Based Analysis

The RasterTSIndex class lists rasters available by a link to their location external to the geodatabase. Rasters are indexed by DateTime, TSType, and Zlayer (see Figure 6.6: RasterTSIndex Objects). The class is a functional link to external data that allows automated raster processing and data ingestion into the StatisticalTS structure of the Modelshed framework. This catalog keeps a track of the source of geoprocessed statistics and data in the database. The class is simple but very powerful and it can organize and catalog raster images, which do not possess their own time, data type, or layering metadata. It can link to files from distributed sources on a network or the internet. It refers to rasters, including those in formats not readable by ArcGIS.

Because the rasters are not stored in the database itself, data integrity is a concern. However, the benefit of this approach is a dramatically reduced database size for remote sensing applications and the elimination of the need to create raster structures within an

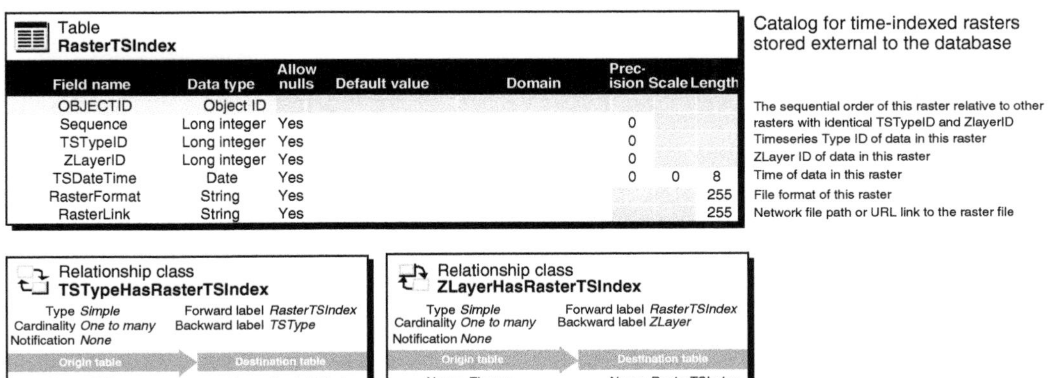

Table
RasterTSIndex

Catalog for time-indexed rasters stored external to the database

Field name	Data type	Allow nulls	Default value	Domain	Precision	Scale	Length	
OBJECTID	Object ID							
Sequence	Long integer	Yes			0			The sequential order of this raster relative to other rasters with identical TSTypeID and ZlayerID
TSTypeID	Long integer	Yes			0			Timeseries Type ID of data in this raster
ZLayerID	Long integer	Yes			0			ZLayer ID of data in this raster
TSDateTime	Date	Yes			0	0	8	Time of data in this raster
RasterFormat	String	Yes					255	File format of this raster
RasterLink	String	Yes					255	Network file path or URL link to the raster file

FIGURE 6.6
The RasterTimeSeriesIndex data dictionary.

otherwise vectorized database. On contemporary workstations, this benefit is very important when working with raster datasets greater than 1 GB in size. The 32-bit GIS workstations that do most GIS work today support only 2 GB maximum file sizes, and a database with large rasters stored internally will quickly exceed this limit. Another benefit of this layout is that raster data stored external to the database can be edited and accessed by nonGIS software. If a database is going to be communicated, it makes sense to store only the data that is associated with database features at an effective spatial resolution that matches the level of detail found in other parts of the database. The Modelshed geodata model handles raster data by keeping an index of raster source files stored outside the database, and ingesting that raster data as needed by converting it into statistical vector data associated with database features.

More important than tracking where rasters provide a way to ingest the data into the database and access it in relationship with Modelshed features. This is the most important function of the RasterTSIndex class. By indexing a time series of rasters by time series type, time, and vertical layer, it is possible to automatically geoprocess those rasters and extract data that is associated with database area features. ModelArea features may be overlaid on the raster files, and the data in the rasters may be summarized by those areas and entered as a time series record in the database. Underlying this approach to data management is the idea that the value of data in raster files derived from its association with real-world objects. The rectangular cells and digits in a raster image are not physically meaningful themselves. However, when that data is associated with a "real-world" object in the data model, such as an ArcHydro watershed, it becomes meaningful.

When raster data is summarized and stored in relationship with model features, the most immediate benefit is query-based access and visualization of that data. Formerly, the information was accessible to a user only by viewing the raster image or by geoprocessing. Now, that information is available with a simple sort or join on a database table, and the desired data may be extracted in relationship to the relevant database features. The SQL query approach to analysis is intuitive, simple, and widely understood. Relational data mining software can now explore the raster data. Features may be selected and manipulated in GIS software based on associated raster statistics. Often what is needed is not an entire stack of time-indexed rasters, but rather just a time series of data for a few database features. With raster data summarized and relationally indexed using the StatisticalTS class, it is

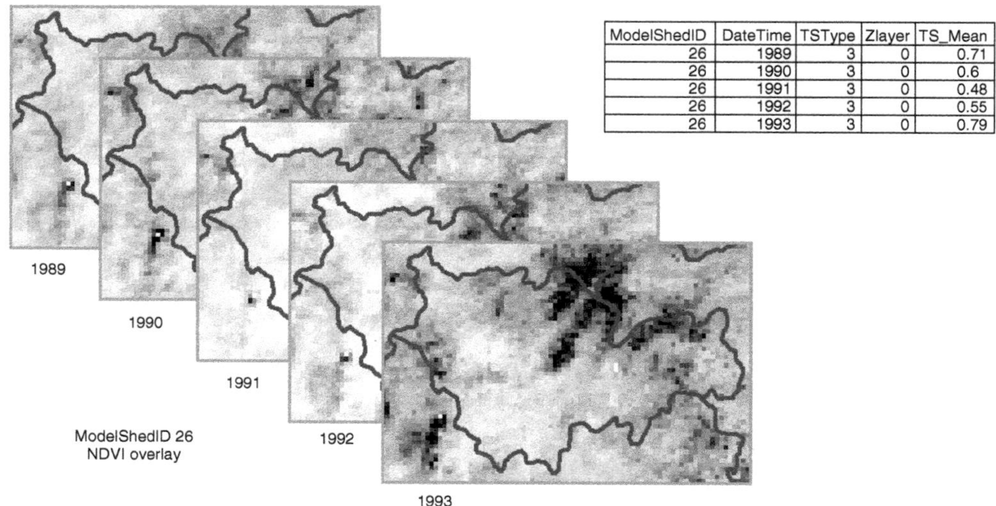

ModelShedID	DateTime	TSType	Zlayer	TS_Mean
26	1989	3	0	0.71
26	1990	3	0	0.6
26	1991	3	0	0.48
26	1992	3	0	0.55
26	1993	3	0	0.79

FIGURE 6.7
A time series of raster images is overlaid on a Modelshed feature, ID# 26, and the aggregated values are statistically summarized and stored in the StatisticalTS table.

simple to extract the right time series and package it for export or analysis (see Figure 6.7: Raster stack analysis into StatisticalTS).

6.2.4 New Relationships: ModelshedLink, AreaLink, and OrthogonalLink

The ModelshedLink object class establishes an association between two ModelshedTypes (see Figure 6.8: ModelshedLink Objects). The association indicates whether one type is a hierarchical subset of another (e.g., in ArcHydro basins are subdivided into watersheds). Equal spatial and temporal resolution is also indicated in the ModelshedLink. Modeling is greatly simplified when the input datasets have the same resolution. In the Modelshed framework, resolution is gauged by comparing the features, ZLayers, and times associated with the ModelshedType. If all three are the same, it can be said that the two Modelshed-Types have the same spatial and temporal resolution, and simplifying assumptions may be applied.

AreaLinks are established between any two areas or polygon features in the database. The link is formed based on the FeatureIDs of the two features (HydroID in ArcHydro) and is indexed according to the ModelshedTypeID of the two features (see Figure 6.9: AreaLink Objects). The purpose of an AreaLink is to establish a natural, structural relationship between the features of two otherwise unrelated Modelshed types, based on the degree to which features overlap. This relationship makes sense in geospatial applications, where the most powerful relationship between objects is established by the vertical pull of gravity.

Qualitative and quantitative linkage is accomplished using the FractionOf1In2 attribute. This attribute is just what it sounds like: it records the fraction of Area1 that overlaps Area2 in a 2D GIS projection of the two areas. AreaLink records are normally created in pairs: a link is established from feature A to feature B, and also from feature B to feature A. In this way the link may be traversed in both directions. Logical relationships may be derived from this link. For example, if the fraction of A in B is 1 and the fraction of B in A is <1, we know that B is larger than A and entirely contains A. This is an example of a test for

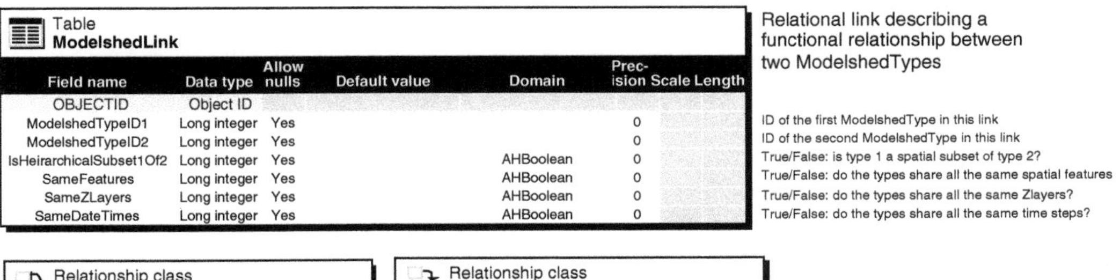
Table
ModelshedLink

Relational link describing a functional relationship between two ModelshedTypes

Field name	Data type	Allow nulls	Default value	Domain	Precision	Scale	Length	
OBJECTID	Object ID							
ModelshedTypeID1	Long integer	Yes			0			ID of the first ModelshedType in this link
ModelshedTypeID2	Long integer	Yes			0			ID of the second ModelshedType in this link
IsHeirarchicalSubset1Of2	Long integer	Yes		AHBoolean	0			True/False: is type 1 a spatial subset of type 2?
SameFeatures	Long integer	Yes		AHBoolean	0			True/False: do the types share all the same spatial features
SameZLayers	Long integer	Yes		AHBoolean	0			True/False: do the types share all the same Zlayers?
SameDateTimes	Long integer	Yes		AHBoolean	0			True/False: do the types share all the same time steps?

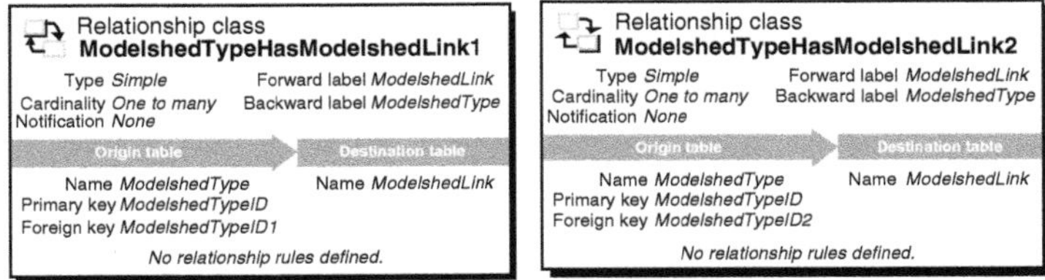

FIGURE 6.8
The ModelshedLink data dictionary.

hierarchy; in the ArcHydro framework, feature A could be a Watershed, while feature B could be a Basin.

One advantage of this approach is that once the AreaLink is established, a user can traverse the link and retrieve related areas without geoprocessing (see Figure 6.10). Simple modeling may be accomplished by processing data associated with features in one Modelshed type according to area-weighted averages, and registering the weight-averaged data with features in the other Modelshed type. For example, if vegetation land cover raster statistics are aggregated according to a watershed ModelshedType, and that data is needed for the analysis of an ecoregion ModelshedType, an AreaLink between the watershed and ecoregion types allows area-weighted access to the data according to ecoregions. Another example of AreaLink application is found in the vertical association of aquifers, watersheds, and atmospheric zones. Using AreaLinks between these features, rainfall could be measured in atmospheric zones (perhaps these are NEXRAD rainfall grid cells), then abstracted into the underlying (and thus related) watersheds, stream networks, and aquifers.

Orthogonal Links may be established between topologically adjacent curvilinear (four-faced) ModelArea features that share the same Modelshed type. They replicate in vector space the functionality found in matrix computations, by linking each cell to the others around it and permitting the user to traverse a curvilinear grid in vector space (see Figure 6.11: OrthogonalLink Objects). Curvilinear systems are not limited to square grids. In fact, some of the most interesting applications of curvilinear grids use irregular and curved edge lines (see Figure 6.12). For example, geographically projected grids are in fact made up of curved longitudinal and latitudinal boundaries. Another application of OrthogonalLink lies in the establishment of flow paths for groundwater and surface water transport. A group of powerful groundwater models use flow paths to describe the movement of groundwater, and these flow paths may be represented by a curvilinear grid. OrthogonalLink allows a program to traverse the flow path to downstream (posX) cells and parallel (posY) cells.

6.2.5 Dynamic Modelshed Features

To this point, all the structures of the Modelshed geodata model have been formed from temporally and spatially static modelshed features, with associated time series data. Another

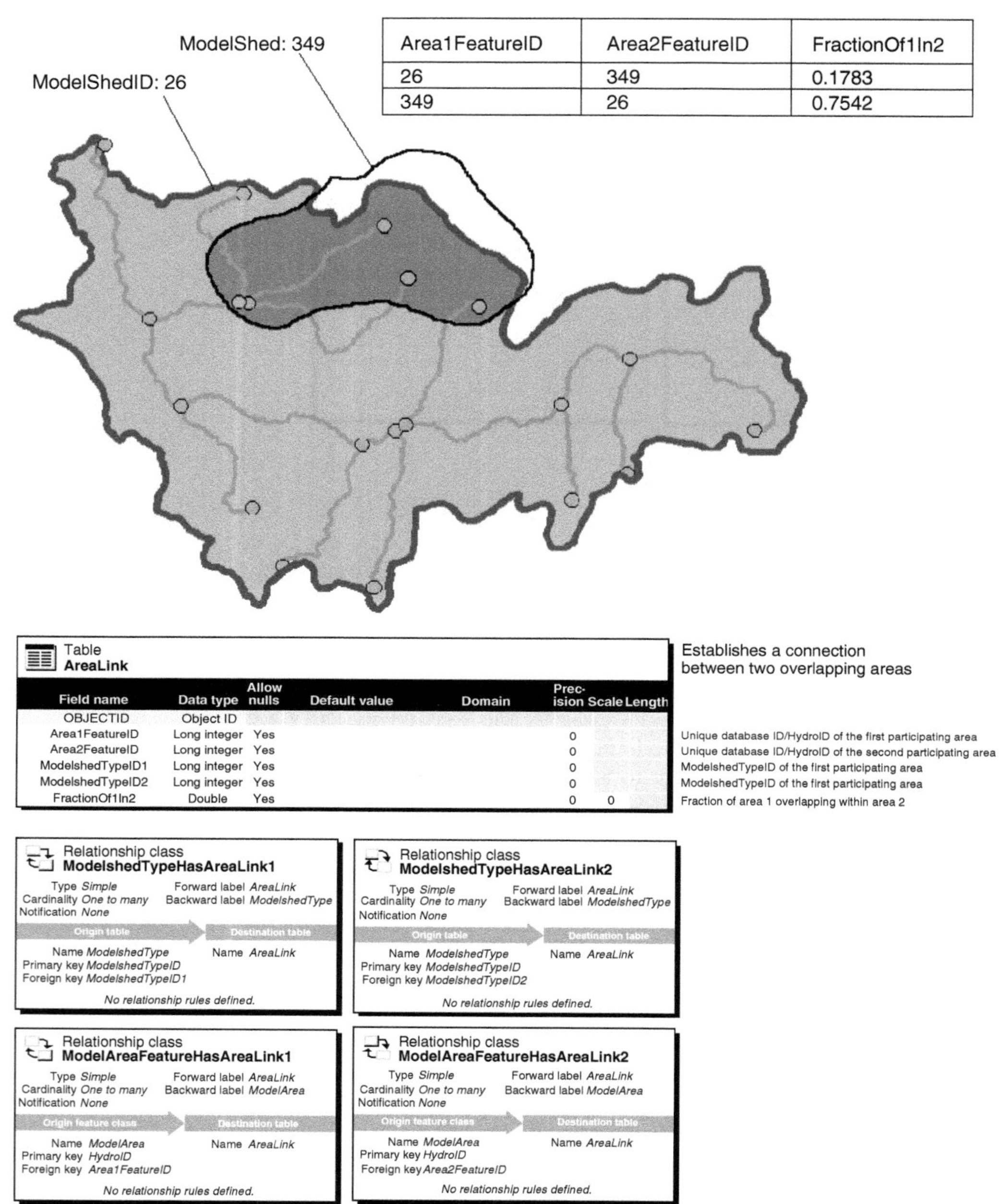

FIGURE 6.9
AreaLink objects.

approach is possible, one in which it is the features themselves that move in time. Such features are dynamic, because they are able to change shape and location in time within the database. Time series data is still associated with such a feature as a means of storing attribute data (see Figure 6.13: Dynamic Feature UML Objects). In this way the Modelshed geodata model is able to track the motion and evolution of spatially dynamic features and their associated attribute data.

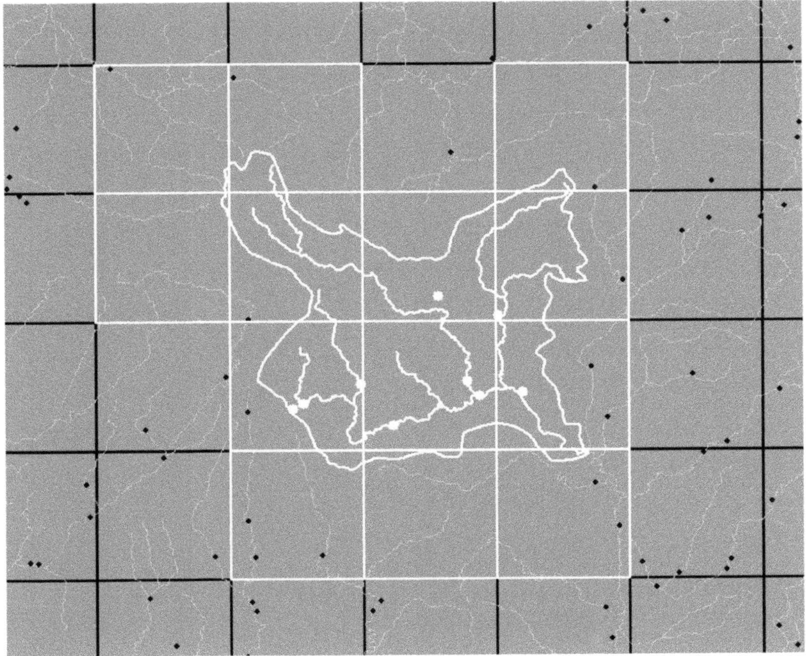

Figure 6.10
Using AreaLinks, overlapping model areas may be queried and selected directly from the database, without the need for spatial processing. Watershed and grid ModeshedTypes are overlaid and selected in this figure.

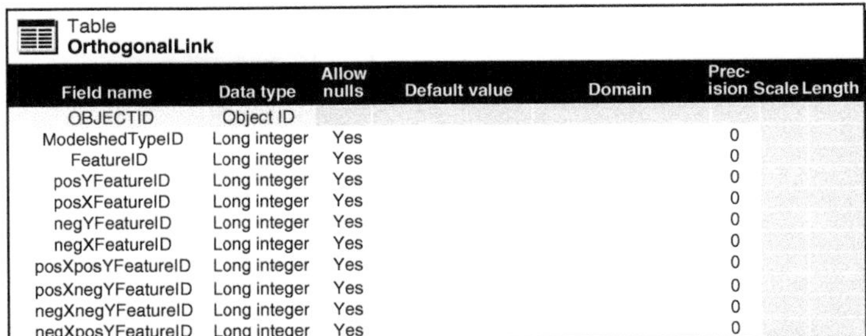

Table
OrthogonalLink

Relational link connecting polygon features in curvilinear grids to their neighbors

Field name	Data type	Allow nulls	Default value	Domain	Precision	Scale	Length	
OBJECTID	Object ID							
ModelshedTypeID	Long integer	Yes			0			Shared ModelshedTypeID of features in this link
FeatureID	Long integer	Yes			0			Unique database ID/HydroID of this feature
posYFeatureID	Long integer	Yes			0			Unique database ID/HydroID of feature in Y+ direction
posXFeatureID	Long integer	Yes			0			Unique database ID/HydroID of feature in X+ direction
negYFeatureID	Long integer	Yes			0			Unique database ID/HydroID of feature in Y- direction
negXFeatureID	Long integer	Yes			0			Unique database ID/HydroID of feature in X- direction
posXposYFeatureID	Long integer	Yes			0			Unique database ID/HydroID of feature in X+Y+ direction
posXnegYFeatureID	Long integer	Yes			0			Unique database ID/HydroID of feature in X+Y- direction
negXnegYFeatureID	Long integer	Yes			0			Unique database ID/HydroID of feature in X-Y- direction
negXposYFeatureID	Long integer	Yes			0			Unique database ID/HydroID of feature in X-Y+ direction

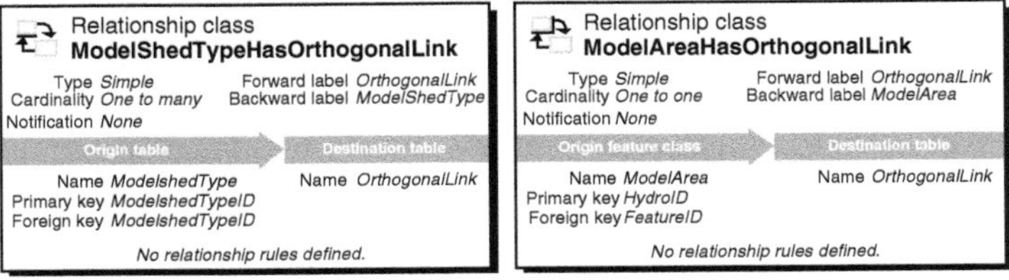

Figure 6.11
The OrthogonalLink data dictionary.

Tracking an object as it evolves through different spatial configurations requires another index, called the DynamicFeatureID index. Each dynamic modelshed feature shape still has a unique HydroID to identify it within the database, but now all the feature shapes representing spatio-temporal evolution of one dynamic feature share the same

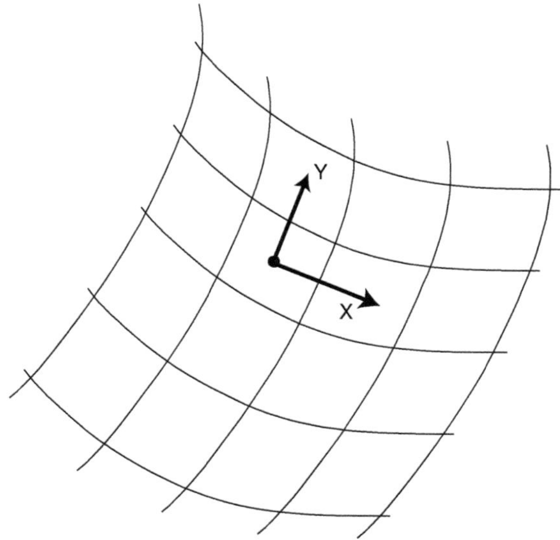

FIGURE 6.12
OrthogonalLink form associations between eight adjacent polygons in a curvilinear grid.

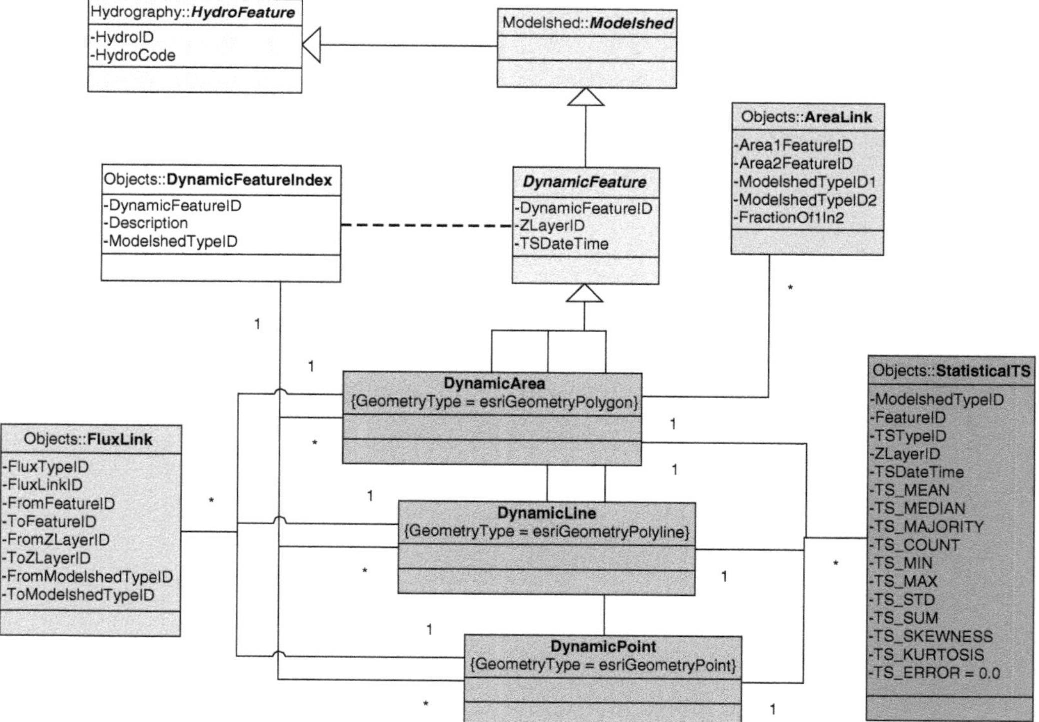

FIGURE 6.13
The Dynamic Modelshed feature UML and data. Dynamic features relate to Modelshed metadata classes just like static features, and possess similar relationships. The DynamicFeatureIndex keeps track of dynamic features.

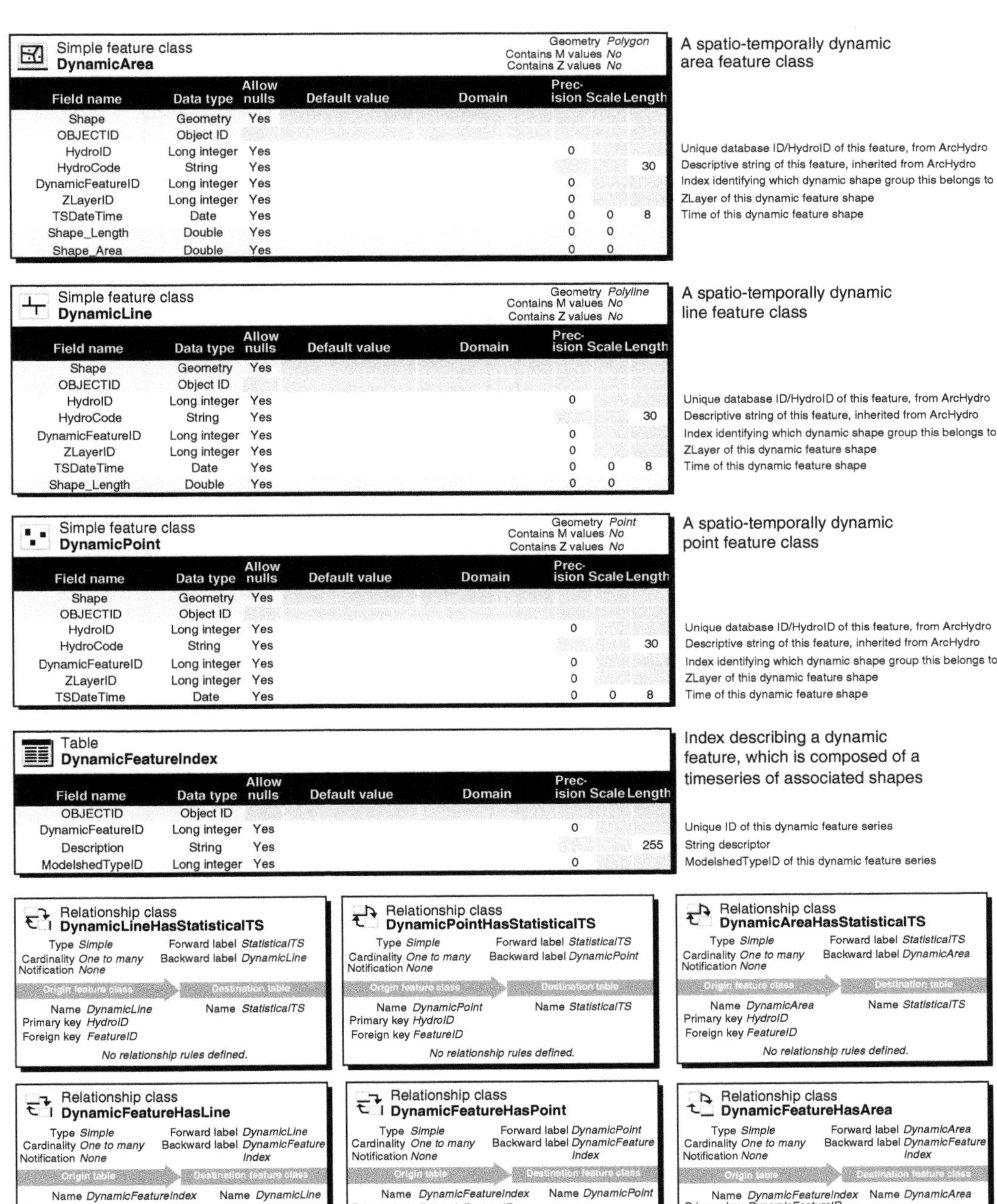

Simple feature class **DynamicArea** — Geometry *Polygon*; Contains M values *No*; Contains Z values *No*

A spatio-temporally dynamic area feature class

Field name	Data type	Allow nulls	Default value	Domain	Precision	Scale	Length
Shape	Geometry	Yes					
OBJECTID	Object ID						
HydroID	Long integer	Yes			0		
HydroCode	String	Yes					30
DynamicFeatureID	Long integer	Yes			0		
ZLayerID	Long integer	Yes			0		
TSDateTime	Date	Yes			0	0	8
Shape_Length	Double	Yes			0	0	
Shape_Area	Double	Yes			0	0	

Unique database ID/HydroID of this feature, from ArcHydro
Descriptive string of this feature, inherited from ArcHydro
Index identifying which dynamic shape group this belongs to
ZLayer of this dynamic feature shape
Time of this dynamic feature shape

Simple feature class **DynamicLine** — Geometry *Polyline*; Contains M values *No*; Contains Z values *No*

A spatio-temporally dynamic line feature class

Field name	Data type	Allow nulls	Default value	Domain	Precision	Scale	Length
Shape	Geometry	Yes					
OBJECTID	Object ID						
HydroID	Long integer	Yes			0		
HydroCode	String	Yes					30
DynamicFeatureID	Long integer	Yes			0		
ZLayerID	Long integer	Yes			0		
TSDateTime	Date	Yes			0	0	8
Shape_Length	Double	Yes			0	0	

Unique database ID/HydroID of this feature, from ArcHydro
Descriptive string of this feature, inherited from ArcHydro
Index identifying which dynamic shape group this belongs to
ZLayer of this dynamic feature shape
Time of this dynamic feature shape

Simple feature class **DynamicPoint** — Geometry *Point*; Contains M values *No*; Contains Z values *No*

A spatio-temporally dynamic point feature class

Field name	Data type	Allow nulls	Default value	Domain	Precision	Scale	Length
Shape	Geometry	Yes					
OBJECTID	Object ID						
HydroID	Long integer	Yes			0		
HydroCode	String	Yes					30
DynamicFeatureID	Long integer	Yes			0		
ZLayerID	Long integer	Yes			0		
TSDateTime	Date	Yes			0	0	8

Unique database ID/HydroID of this feature, from ArcHydro
Descriptive string of this feature, inherited from ArcHydro
Index identifying which dynamic shape group this belongs to
ZLayer of this dynamic feature shape
Time of this dynamic feature shape

Table **DynamicFeatureIndex**

Index describing a dynamic feature, which is composed of a timeseries of associated shapes

Field name	Data type	Allow nulls	Default value	Domain	Precision	Scale	Length
OBJECTID	Object ID						
DynamicFeatureID	Long integer	Yes			0		
Description	String	Yes					255
ModelshedTypeID	Long integer	Yes			0		

Unique ID of this dynamic feature series
String descriptor
ModelshedTypeID of this dynamic feature series

FIGURE 6.13 (Continued)

DynamicFeatureID. This allows all of a dynamic feature's shapes to be queried from the database and displayed or analyzed as a group. Along with a DynamicFeatureID, each feature shape has a ZLayer and DateTime attribute to track the time associated with this shape and the vertical extents of the shape. With a static modelshed feature, a time-series is realized as a series of points on a graph; with a dynamic modelshed feature, the

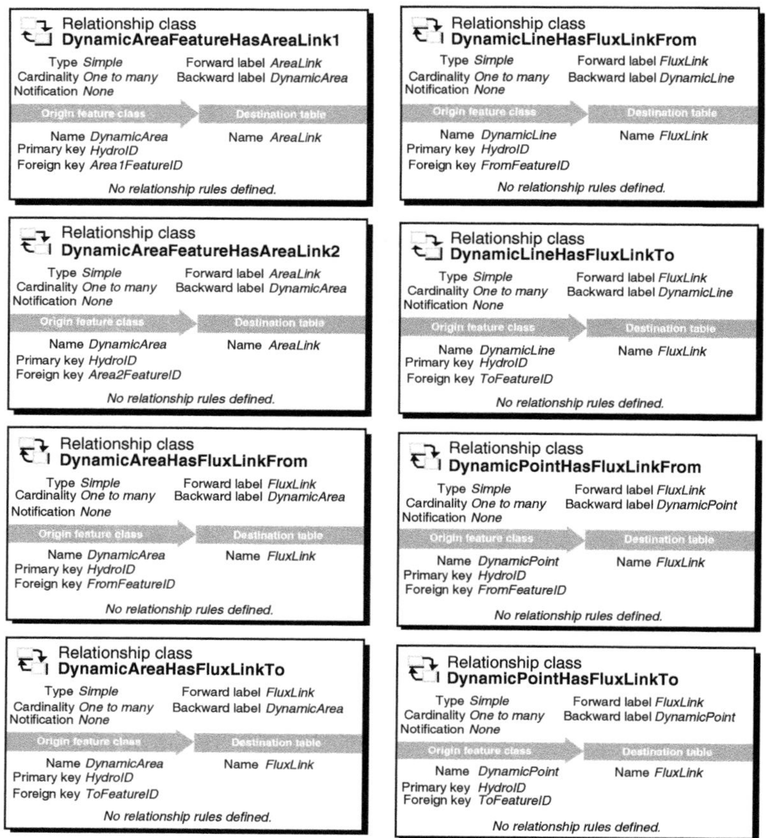

FIGURE 6.13 (Continued)

time series becomes a sequence of shapes that track the object as it moves across the map (Figure 6.14: spatio-temporal evolution of a dynamic feature). These shapes may be either points, lines, or areas, and a dynamic feature may not transition from one to another. Thus, a dynamic feature may be used to represent atmospheric phenomena such as storms and clouds, a plume of contaminants in an aquifer, the progress of a weather front contour through the atmosphere, or the motion of a particle or aircraft through space and time. Much like with the static Modelshed feature, StatisticalTS records may be associated with each shape in the dynamic feature, since each shape has a unique HydroID.

Multiple feature shapes may be assigned to a dynamic feature at the same time step, each with a distinct ZLayer index. Thus, dynamic features are able to model horizontal "slices" through spatially dynamic objects. A good application for dynamic Modelshed features would be a thunderstorm cell. The cell will develop and move in time, shifting in altitude and shape. The footprint of the cell in each 500 ft Zlayer altitude zone is recorded, and radar reflectivity and rainfall measurements from the cell may be recorded in associated StatisticalTS records.

AreaLink and FluxLink relationships may be established with dynamic features the same way as static features. Since all these Modelshed relationships depend on the unique feature identifier HydroID, any feature with a HydroID may participate. AreaLinks may be built between a dynamic thunderstorm feature and the underlying static watershed features,

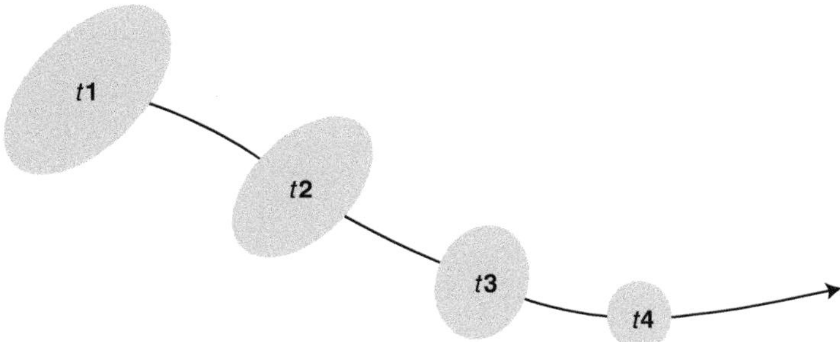

Figure 6.14
The Spatio-temporal evolution of a dynamic feature as it changes shape and location through time. Each shape has its own HydroID and spatial data, but they share a DynamicFeatureID.

or between two dynamic feature shapes. This means that the AreaLinks will change at each time step as dynamic features shift with relation to other database features. FluxLinks may be established between that thunderstorm feature and those underlying watersheds, to record the transfer of water from the clouds to the ground. Like AreaLinks, new FluxLinks must be continually established as features shift in space and time.

An important difference between static and dynamic relationships lies in the number of features and the associated size of the database. Since a dynamic feature lasts for a limited number of time steps and are usually replaced with other dynamic features, each having its own spatial data, indexes, AreaLinks, and FluxLinks, the amount of data stored will quickly explode. This is a limitation of the concept. Thus, at this point the use of dynamic features will be useful for the close study of small-scale dynamical environmental systems over relatively short timescales.

Section II

Managing and Accessing Large Datasets

7

Data Models for Storage and Retrieval

Michael J. Folk

CONTENTS

7.1 Survey of Different Types and Uses of Data

As hydrology intersects with so many different scientific domains, the data that the hydrologist works with comes from a huge variety of sources and in a daunting assortment of formats. First, there is metadata, as described in this section, which is typically not large but can be quite complex. The metadata is typically followed by the rest of the data — the digital objects to which metadata refers — which can be very large and equally complex. To organize data for efficient storage and access, it is a good idea to come up with a conceptual model of the type of data we wish to organize.

An apt way to get an idea of the challenges in organizing hydrologic data is to look at some examples. In the following examples, we will see that hydrologic data can be characterized in a number of ways:

- There is a wide variety of datatypes, and a wider variety of data models and formats
- Some objects are extremely large, and promise to become larger in the future
- Some objects can be quite complex

- The number of objects can be quite large
- The way the data is accessed can vary greatly, and often it is difficult to find tools for accessing the data easily or efficiently

Figure 7.1 through Figure 7.4 are a few examples from the XMS suite of modeling systems of the Environment Modeling Research Laboratory at Brigham Young University [1]. XMS supports more than 50 different models, each dealing with its data in its own special way. The XMS software has been adapted to support the formats and the data models for all these models. In order to avoid writing software to accommodate every model's data format, the project looks for a common format and data model that can encompass all of these, as well as future models.

7.1.1 Remote Sensing

Hydrology increasingly deals with remote-sensed satellite data. A range of remote-sensed data types is exemplified by NASA's Earth Observing System Data and Information System (EOSDIS) [2], which is the data and information system supporting the study of global climate change. At the heart of this research is the study of the global connections among

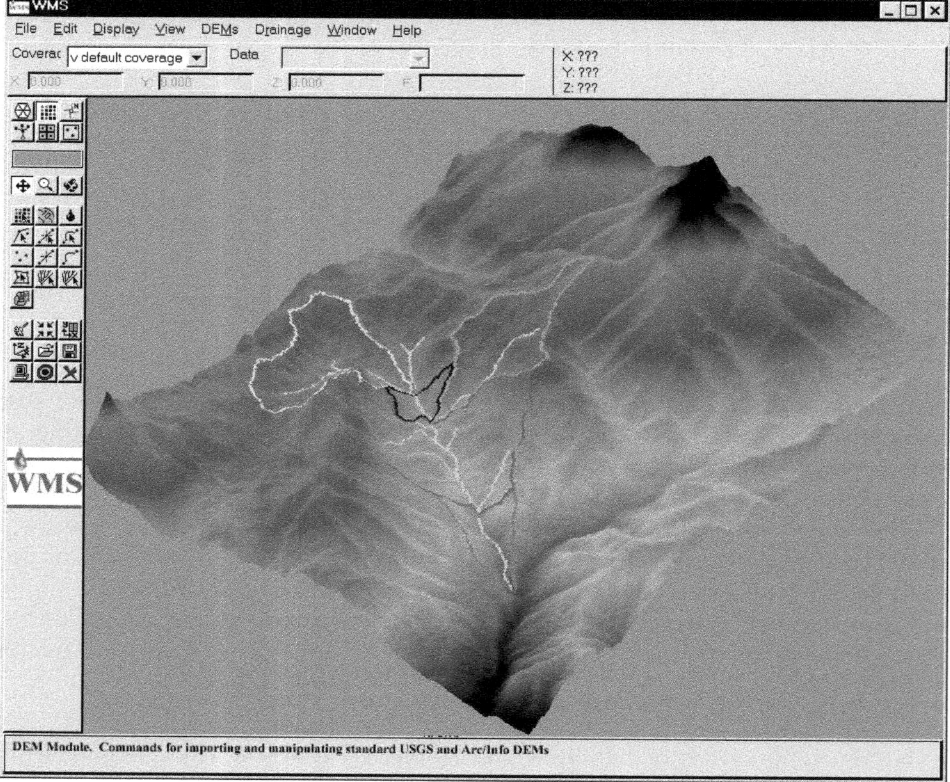

FIGURE 7.1
Visualization of a watershed. Data includes water bodies and rivers, and elevation. How should this data be modeled and organized? (Courtesy Norman L. Jones, BYU Environmental Modeling and Research Laboratory, 2003.)

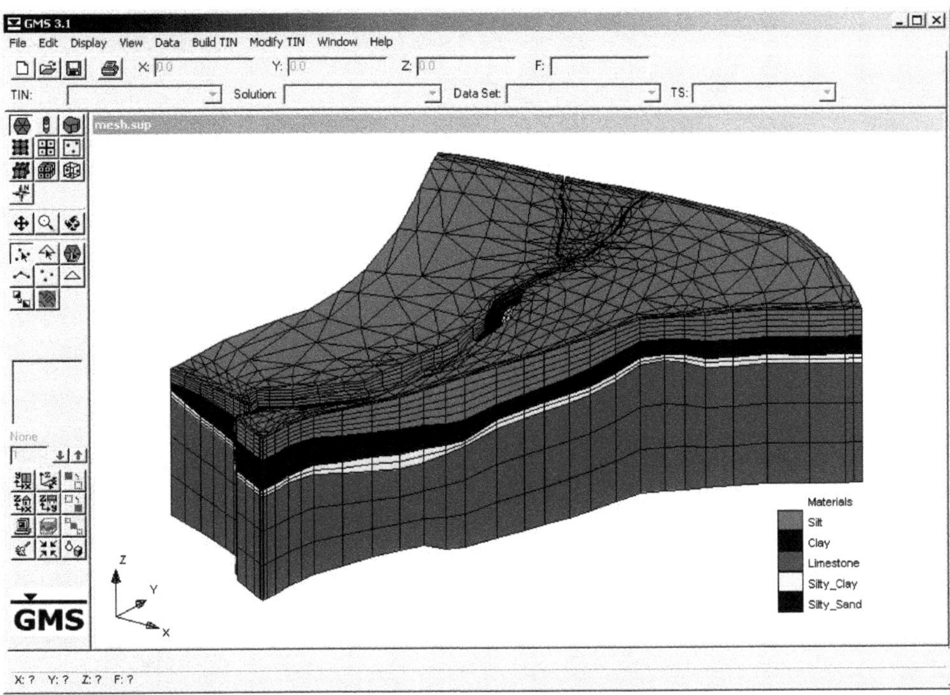

FIGURE 7.2
Visualization of groundwater simulation, showing different geologic layers in an acquifer. How will the information that describes the different layers of materials be stored? What structure is used to describe the geometry? How will this structure be organized and stored for easy access and subsetting? (Courtesy Norman L. Jones, BYU Environmental Modeling and Research Laboratory, 2003.)

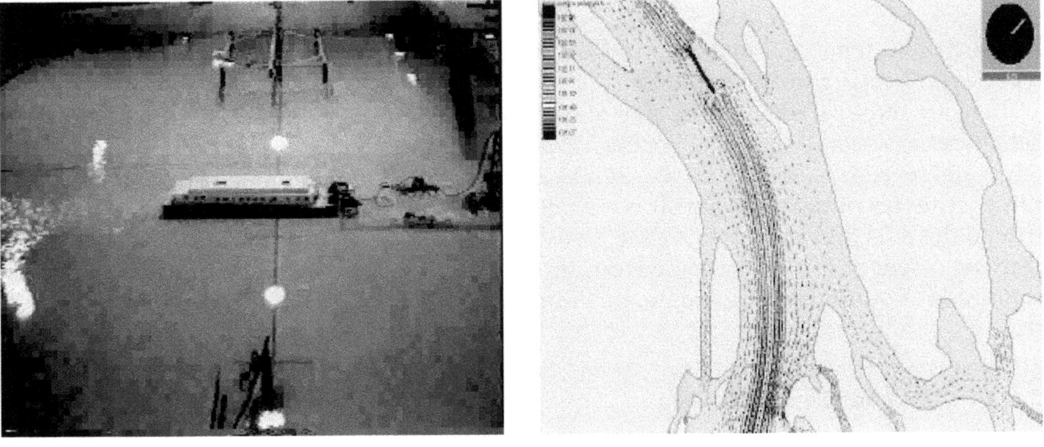

FIGURE 7.3
A physical model emulates selected sections of the river and model barges are pulled through these sections. At the same time, the SMS software uses computer models of the moving barges. The computer models are cheaper than the physical models, and they allow the engineers to analyze a wider variety of scenarios. What kinds of data do we see in the computer model, and how will it be organized and stored? (Courtesy Norman L. Jones, BYU Environmental Modeling and Research Laboratory, 2003.)

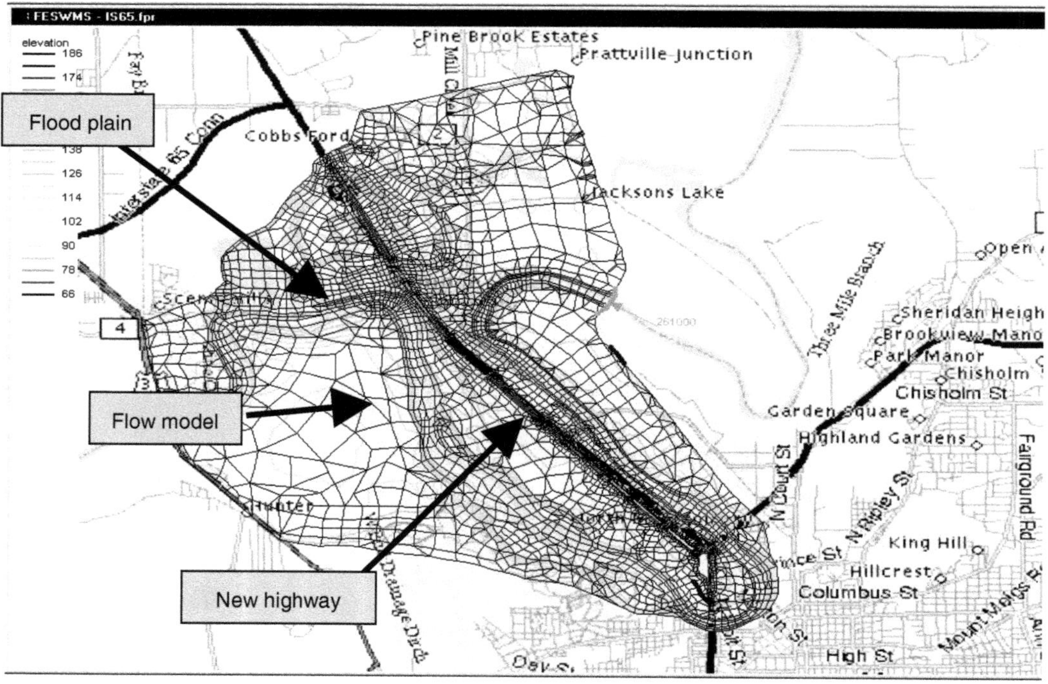

FIGURE 7.4
This application of SMS involved the design of a new highway running through a flood plane, through the center of a long, winding bend in the river. There was a concern that the highway would become a flow barrier during the annual flood stage for the river. SMS was used to simulate the flow conditions at flood stage, helping to ensure that the relief openings placed beneath the highway were large enough to pass the flood water beneath the highway and prevent flooding of the city. What are the different types of data involved in this simulation? Since these data are likely to come from different sources, how might they be described and organized so that they can be integrated for this study? (Courtesy Norman L. Jones, BYU Environmental Modeling and Research Laboratory, 2003.)

the earth's systems, as illustrated in Figure 7.5. Data about these systems are collected in a variety of ways, mostly by satellite.

There are three satellites at the core of EOS, named Terra, Aqua, and Aura.[1] Each of these satellites has several sensors collecting dozens of different kinds of data about layers of the atmosphere, oceans, snow and ice, and land. Between them, the satellites generate almost three terabytes of data per day. It is a challenge to describe all these data in a way that is true to the data and at the same time gives us some kind of a common view. Although the primary users of EOS data come from the global climate research community, there are hundreds of other uses of the data, including many hydrologic applications.

7.1.2 Common Geospatial Data

As a final set of examples, there is a small number of geographic data types that form the core of many geographical applications, which these are illustrated in Figure 7.6. Figure 7.6(a) is a *Digital Orthophoto Quadrangle* (DOQ) [3]. It is essentially an aerial photograph, consisting of rows and columns of pixels, and is also geo-rectified, which means that it has been

[1] http://terra.nasa.gov/, http://aqua.gsfc.nasa.gov/, http://eos-chem.gsfc.nasa.gov.

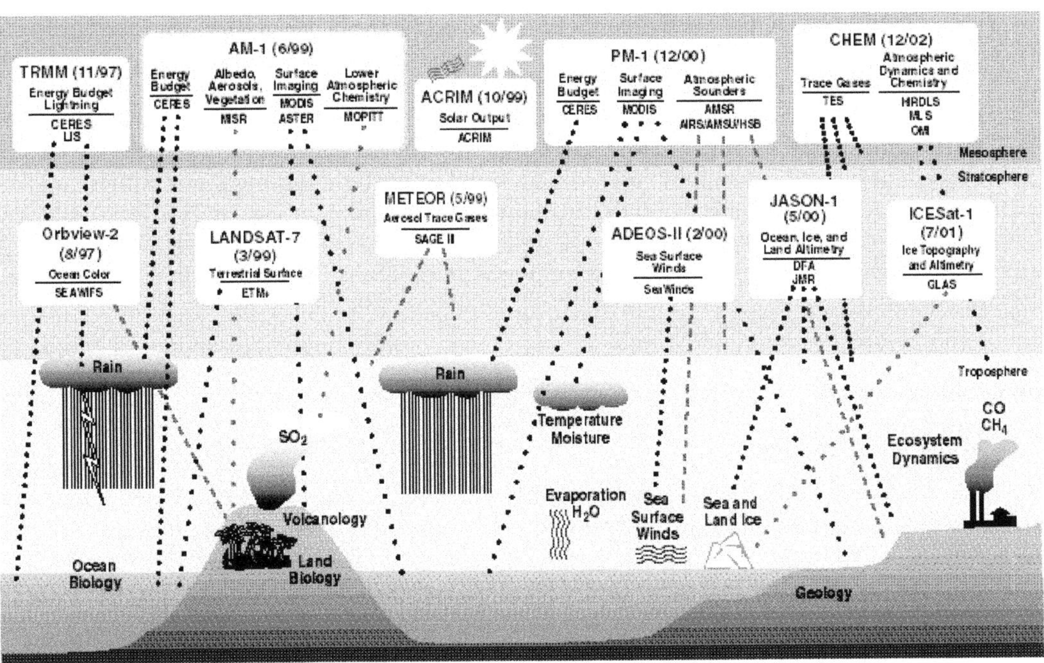

FIGURE 7.5
A varied constellation of satellites collects data about the earth. NASA's EOSDIS' three satellites and multiple instruments organizes its data according to three models, simplifying data storage, access, interoperability, and sharing for its many users.

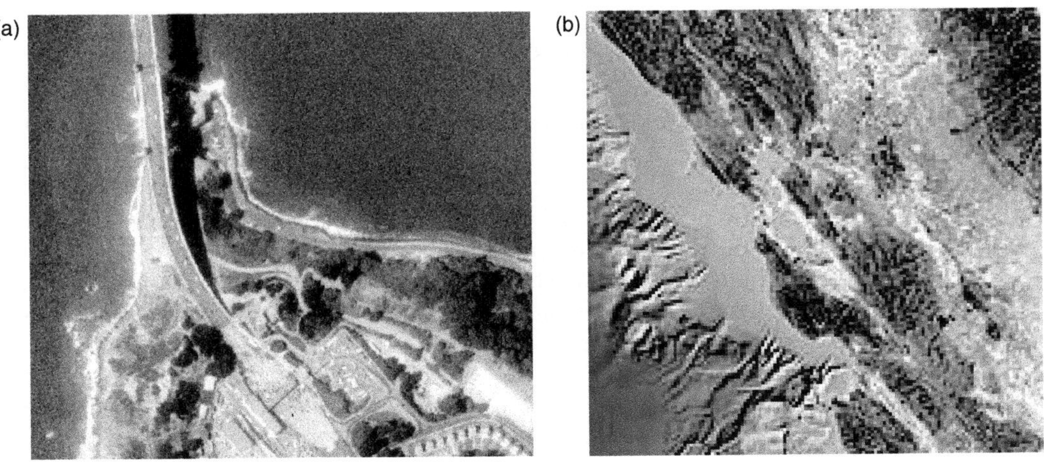

FIGURE 7.6
(a) DOQ showing entrance to Golden Gate Bridge. (b) DEM showing the Bay Area and ocean floor off the California coast. DOQs and DEMs are relatively simple objects, but can be very large.

corrected to remove distortions to maintain spatial accuracy. More importantly, it means that other kinds of information such as streets and elevation information, can more readily be combined with this picture. Figure 7.6(b) is a rendering of a *Digital Elevation Model (DEM)* [4], which is essentially an array of sampling of elevations at regularly spaced intervals. DEMs and DOQs must also include information needed to interpret the meaning of each pixel, such as the upper left and the lower right corners, the spatial resolution, and the type of number used to represent each pixel.

A more complex geospatial example is a *Digital Line Graph (DLG)* [5], which consists of a vector data representing cartographic information. DLGs contain a wide variety of information depicting geographic features (hypsography, hydrography, boundaries, roads, utility lines, etc.). A "large-scale DLG" contains nine categories of data. DLGs are derived from hypsographic data (contour lines) using USGS 7.5 min, 15 min, 2 arc sec (30 by 60 min), and 1:2 million-scale topographic quadrangle maps. Like DOQs and DEMS, DLGs come in different forms.

7.2 Who are the Users?

We have seen in these examples that different kinds of data are available, in different forms and combinations, for different purposes, and by different types of users, including:

- Discipline scientists, who generally have an intimate knowledge of data and are often interested in large collections of similar data.
- Multidisciplinary scientists, who need to access and integrate data from many sources.
- Application specialists, such as farmers, land-use planners, earthquake engineers, and civil engineers, who also need to integrate the data, often in circumstances in which there is incomplete information.
- Students and teachers, who typically do not need large collections, but need good data examples and tools.
- The general public, who have little knowledge or resources and hence need the information in easily digested form.
- Politicians and decision makers, whose needs can be unpredictable, eclectic, and urgent.

Each category has specific needs with respect to access, metadata, formats, and tools. Indeed the variety of uses of these data, particularly their uses in combination, is what makes modeling and organizing the data challenging and interesting. Some scientific data is used by just one segment of the user community. Other data might span many different communities.

Some are just occasional users, such as people browsing the web looking for an image of a hurricane or a volcano. For these users, data should be easy to view and interpret using common tools such as a browser. Others have to do science or other extensive work with the data, and for them accuracy, completeness, efficient storage, and access can all be important.

7.3 Gathering, Using, and Archiving Data

It should be clear that one challenge to managing data is in finding ways to accommodate the tremendous variety of datatypes. To understand the challenges to modeling and

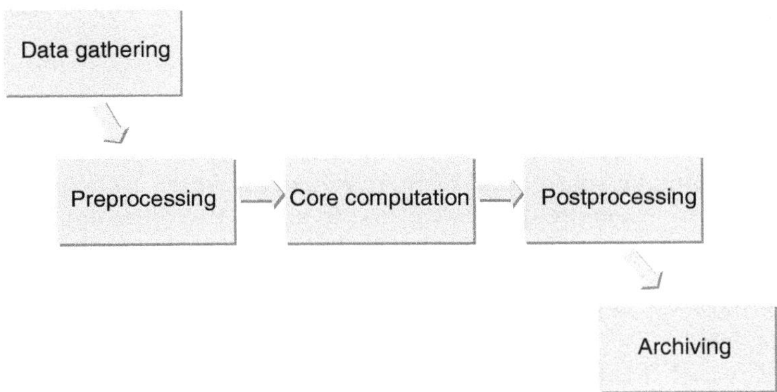

FIGURE 7.7
The complete process, from a data perspective.

organizing data, we also need to think about how the data are gathered, used, and in some cases archived. The complete process can be illustrated as in Figure 7.7.

Data gathering can come from instruments, archived data, databases, or simulations. It may come from one or many local or remote sources and may be necessary to search for. Our worry is about how to get it, when to get it, scheduling or synchronizing data transport, and getting space and permission for staging it.

Preprocessing may involve generating data, loading restart files, cleaning or reorganizing the data, changing its format, integrating it with other data, and creating files or databases that it will be written to.

Core computations occur after the data is collected, cleaned, and integrated. This may involve a lot or a little I/O. Partial I/O may be required, meaning that only subsets or subsamples of the data are read in and written out. We may have to reorganize the data as it is read or written, where I/O may be constant or "bursty." The data may be streamed in, meaning that the computing process must be able to keep up with whatever data rate it is given. The computation may stream the data out to another process, such as visualization postprocess. In parallel computing environments, I/O may be evenly distributed or clumpy, and may go to one or many destinations and go through other processes. The file system that stores the data may be capable of serial or parallel access. There may be steering, whereby results are sent to another process (or person), which can change the computation in real time.

Postprocessing occurs when the primary data analysis is completed. It could be visualization, a process of annotating the data, or producing derivative datasets in new forms or formats. Postprocessing may involve splitting up the data into parts that go to different places. It may involve sub-sampling or subsetting and require annotation or other extra information to be added to the data.

Archiving occurs when the data is to be used for later access. The data may need to be archived locally or at some remote location, or perhaps duplicated at more than one location. The archive, as well as the dataset itself, may be large or small; many or a few, or none of the objects may need to be archived. The data may need to be reorganized, put into a new format, and annotated. The archive may be a long-term archive, where access is expected to occur at some future time, or an active archive, where access is expected to occur presently and frequently.

Of course, all of these steps can overlap. For example, streaming input is a part of data gathering, preprocessing, and core computation.

7.4 Data Management Challenges

The applications and data sources we have seen add up to a proliferation of types of data objects and metadata, data formats, and the accompanying applications and software. How is a person to make sense of all of this? What can we do to help make it possible to share this data among many different users, to access the same data in different software packages or on different computer systems? How can we make it possible to combine different types of data easily and efficiently? What can we do to save users from writing separate preprocessing, analysis, visualization, and other software for each of these types and formats of data?

Furthermore, we see that the datasets are becoming very large, and the complexity of the data is increasing. One satellite in EOS can generate a terabyte (trillion bytes) of data per day, and it will not be long before petabytes (thousand trillion bytes) are common. How are we to collect, store, and interpret data of such size and complexity? Data are often archived in one place, and then needed somewhere else. More commonly, just a portion of the data may be needed somewhere else. How can we transport data quickly from place to place, or access only those portions that we need, or avoid moving it altogether? Even when data is local, the process of getting it into and out of a computer can be difficult. How can we access all or part of a dataset efficiently and easily?

We can address these challenges in several ways, as illustrated in Figure 7.8. *Data models* provide a common way to conceptualize the content and the use of data, *file formats* are developed that represent the content we want to store and use, *standard application programming interfaces (API)* provide a way for the software written to query and access the data, and these three *enable common applications and software* to be created.

7.4.1 Data Models

7.4.1.1 Data Model

We need to identify some common ways to think about this variety of data. For example, at a certain conceptual level a DOQ and a DEM are both just different images of the same

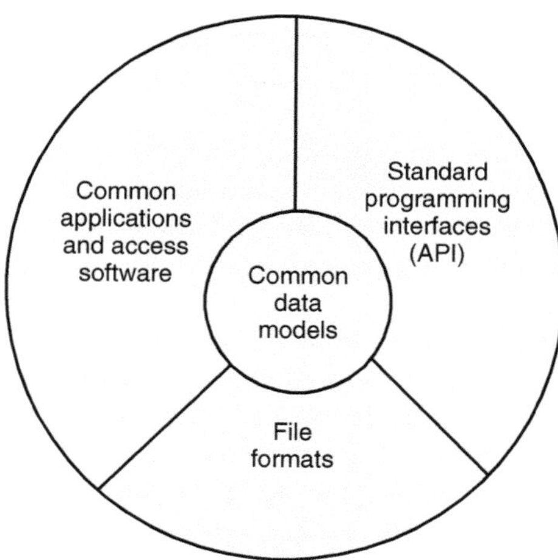

FIGURE 7.8
Four-part approach to addressing important data management challenges.

piece of land. A DOQ is an image showing the things that one sees on the land, with each pixel representing a very small part of the whole image. A DEM is also an image, possibly over the exact same land area, but in this case each pixel represents an elevation. So we can think of both as "images," each described by several rows of elements, together with some information that helps us locate the corresponding land area geographically. Once we describe DEM and DOQ in terms of a common concept (image) and metadata content (geolocation and other information), we can also think about common operations that we might want to perform on these objects. These conceptual tools — the way we describe objects within some application domain, their attributes, and allowable operations — are collectively called a *data model.*

A good data model and implementation enables us to address many questions raised at the beginning of this section. It helps us to conceptualize applications to operate on similar data objects coming from different sources, to share data among different users, to combine different types of data more easily and efficiently, and it can save users from writing a separate preprocessing, analysis, visualization, and other software for each type and format of data.

7.4.1.2 File Format

A data model provides a conceptual view of objects, but does not address the need to organize the objects in ways that facilitate the operations we need to perform. The choice of a file format can be of critical importance for storing, accessing, and managing these objects by computers. We examine data formats in detail in Chapter 8.

7.4.1.3 Programming Model and API

A data model describes the attributes and the access methods associated with objects, but do not say how the applications can actually access them. An API describes the calling conventions by which a program (e.g., a scientific application or visualization software) accesses the data. An API should reflect in a meaningful way the query and the access methods that are a part of the data model. File formats for storing scientific data can be very complex, especially if they effectively support services such as fast access and querying, scalable storage, and subsetting. By providing an interface between the application and the storage format, an API hides from an application the format and other implementation details involved in getting to the actual data.

An API must also accommodate the needs of anticipated applications with regard to programming languages and platforms. This is referred to as the language *binding* of the API. It is not unusual to require that an API be implemented in more than one language, because the programs that will use the API occur in more than one language. For instance, scientific analysis codes are commonly in Fortran, whereas tools for visualization and other operations are commonly in C, C++, or Java, or perhaps all four. Because of this multilingual use of an API, it is often important that the general API and programming not be defined in a way that is too language specific. A well designed API can often be extended to new language bindings as the need arises.

7.4.2 Example — HDF-EOS

We have described three basic components to address many data management challenges: a good data model, a standard file format, and an implementation in the form of a programming model and API. The Earth Observing System described earlier (Figure 7.5) provides a good example of how these components can be defined and used. The instruments on EOS satellites collect dozens of different types of raw data, and many more data products are derived from the initial data. Faced with the daunting tasks of creating software for managing, analyzing, viewing, and archiving this variety of instrument data and derived products, the EOSDIS project looked for common features and requirements among the different products. When viewed from the standpoint of data structures, metadata, and access requirements, it found strong similarities among many of them, and EOSDIS was able to define three data models that covered most of the data products and data types. Most of the raw and derived data, it found, could be described either as point datasets from field observation studies, datasets from the swaths of satellite borne sensors, or data sets on gridded map projections. Hence, three models were defined, and these were called *point*, *swath*, and *grid*.

The next task for EOSDIS was to determine how to actually organize and access these objects. NASA had examined a number of file formats that are used for storing scientific data, and determined that the Hierarchical Data Format (HDF) had data structures that could be used to store any of the three types of data. Once the data model and storage format were available, EOSDIS created a programming model, API, and software library, called HDF-EOS, to provide uniform access to the diverse data products stored in the format. The HDF-EOS software library allows a user to query and access the contents of a file by earth coordinates and time if there is a spatial dimension in the data.

7.4.2.1 HDF-EOS Swath Data Model

To understand the process of creating these models and profiles, we focus on the HDF-EOS swath structure. Recall that a data model consists of three types of information: a conceptual description of objects, their attributes, and allowable operations.

Swath concept

Figure 7.9 illustrates how a satellite scans a portion of the earth to create a swath. Figure 7.9(a) shows an instrument taking a picture of the earth, and in Figure 7.9(b) we see how the swath is a series of scan lines perpendicular to the ground track of the satellite as it moves along that ground track. In essence, a swath is a time sequence of records, where each record consists of one or more scan lines recorded by instruments.

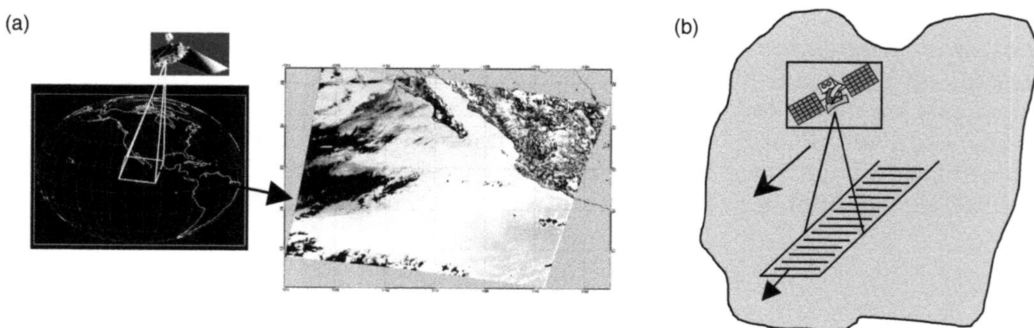

FIGURE 7.9
Illustrations of a swath.

Swath data and attributes

In the HDF-EOS swath data model, a standard Swath is made up of four primary parts: data fields, geolocation fields, dimensions, and dimension maps:

Data fields are the main part of a Swath from a science perspective. All the other parts of swath are there to provide information about the data fields or to support particular types of access to them. Data fields usually contain the raw data (scan lines) taken by the sensor, or parameters derived from that data. Typically, two-dimensional arrays can be used for this data, but they can have as few as one dimension or as many as eight.

Geolocation fields allow the Swath to be accurately tied to particular points on the Earth's surface. The Swath model requires the presence of at least a time field or a latitude/longitude field pair.

Dimensions define the axes of the data and geolocation fields by giving them names and sizes.

Dimension maps define the relationships between the data and geolocation fields. Sometimes there is a one-to-one correspondence between axes and scan lines, but often this is not the case. There may, for instance, be only one geolocation value for every ten scan lines, and dimension maps provide this information.

Swath operations

Swath operations are what make it possible for scientific applications to define, access, and query swath objects. The HDF-EOS model supports the following operations.

Access — Open, create, and close swath objects or collections of swaths

Definition — Set key parameters that define a swath

Input/Output — Read/Write a swath or its metadata

Inquiry — Return information about a swath

Subset — Read partial data from specified region of a swath

These operations are fully defined in Reference 6.

Swath format

The HDF4 file format was used initially for implementing the swath profile, then later a second implementation was done with HDF5. The implementation is illustrated in Figure 7.10. In a later section, we take a closer look at this implementation.

Swath programming model and API

The data model describes the types of operations we wish to perform on EOS swath data. The HDF-EOS swath programming model, API, and library are a software implementation of the swath operations described earlier, which enables scientists, tool builders, and others to create applications that use the swath data. The programming model describes the steps taken to write and read HDF-EOS swaths:

- **Open** the HDF-EOS file
- **Create** or **Attach** to a specified Swath object

- **Define** key object features
- **Query** about objects
- **Read** or **write** in part or all of a swath object
- **Read** or **write** attributes of a swath object
- **Detach** from the Swath object
- **Close** the HDF-EOS file

The API provides routines that are called by a program to perform these operations. A large number of routines are available for working with HDF-EOS swaths. There are too many routines to list here, but Table 7.1 illustrates the routines available for querying about a Swath object. As the table shows, there is a C and a Fortran binding for the HDF-EOS Swath API.

7.5 Summary

In this chapter, we illustrate the great heterogeneity of data that hydrologists must deal with, and list some of the challenging characteristics of this data. Objects can be very large, numerous, complex, represent quite different models, and come in a daunting diversity of

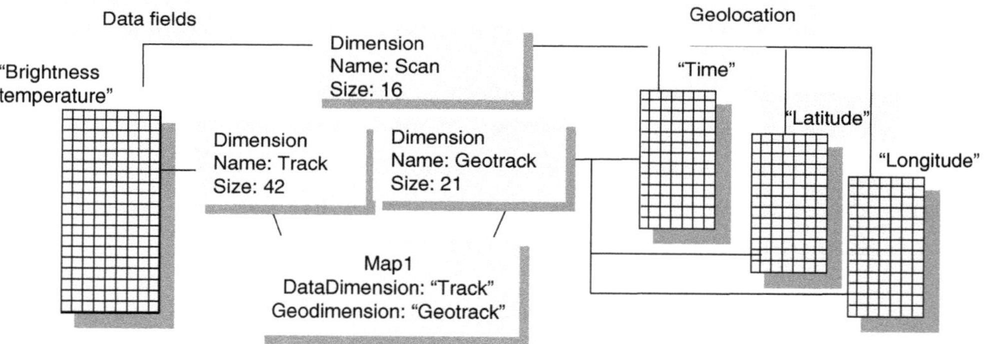

FIGURE 7.10
File structures used for HDF-EOS swath.

TABLE 7.1
Some inquiry routines for the HDF-EOS Swath API

Routine name		
C	Fortran	Description
SWinqdims	swinqdims	Retrieves information about dimensions defined in swath
SWinqdatafields	swinqdflds	Retrieves information about the data fields defined
SWinqattrs	swinqattrs	Retrieves number and names of attributes defined
SWdiminfo	swdiminfo	Retrieves size of specified dimension
SWmapinfo	swmapinfo	Retrieves offset and increment of specified geolocation mapping
SWattrinfo	swattrinfo	Returns information about swath attributes
SWfieldinfo	swfldinfo	Retrieves information about a specific geolocation or data field
SWinqswath	swinqswath	Retrieves number and names of swaths in file

formats. We see that the types of users and uses, of hydrologic data can be equally varied, and yet these different constituencies often deal with the same data.

Working with digital hydrologic data typically involves several steps, including data gathering, preprocessing, core computations, postprocessing, and archiving. Each of these presents data management challenges.

One way to help deal with the heterogeneity, size, and complexity of hydrologic data is to step back and identify common data models. A data model provides a common way to conceptualize the content and use of data. We look at some examples of data that at first seem quite different, but in fact share a great deal in common, and can be described with the same data model. Data models are abstractions that can be implemented in different ways, so it is useful to define common APIs and programming models, which provide a way for software be written to query and access the data. Having common data models, APIs, and programming models can simplify our ability to share data, to combine data from different sources, and to build and share software for working with data.

A good example of a common data model, together with an API and implementation, is HDF-EOS, the software library and tools created for NASA's Earth Observing System. HDF-EOS makes it possible to work with dozens of different kinds of data in a common way. We examine the HDF-EOS swath data model in detail, describing the conceptual structure, the conceptual operations to be provided, and their implementation in the HDF-EOS API.

References

1. Environmental Modeling Research Laboratory (EMRL) at Brigham Young University. http://www.emrl.byu.edu/home.htm.
2. Earth Observing System Data and Information System (EOSDIS). http://spsosun.gsfc.nasa.gov/eosinfo/Welcome/index.html.
3. USGS Western Region Geography Web site. What is a DOQ? http://geography.wr.usgs.gov/doq/.
4. USGS Rocky Mountain Mapping Center Web site. USGS digital elevation model information. http://rockyweb.cr.usgs.gov/elevation/dpi_dem.html.
5. EROS Data Center Web site. USGS Digital Line Graph. http://edc.usgs.gov/guides/dlg.html.
6. Taaheri, E.M. and D. Wynne. An HDF-EOS and data formatting primer for the ECS project. 175-WP-001-002. NASA EOSDIS Core System Project, March 2001.

8

Data Formats

Michael J. Folk

Contents

8.1 Formats and Abstraction Layers

A data format is a method for representing data in digital storage. Digital storage could consist of a persistent medium such as tape or disk, or a transient medium such as a computer's memory or a network. A data format incorporates a model for data representation, such as a specification of how numbers are to be represented and stored. A data format may also define standard methods for storing, reading, and inquiring about data that is stored in the format.

We define a format as follows: a format specifies the organization of information, at some level of abstraction, as contained in one or more byte streams that can be exchanged between

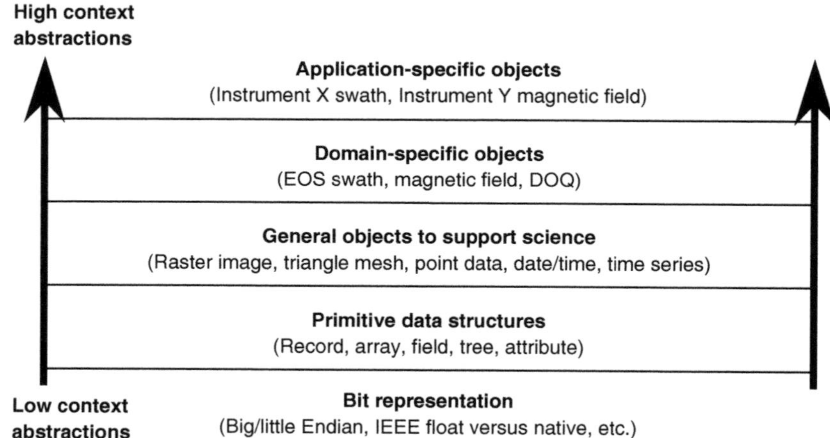

FIGURE 8.1
Low context and high context data models.

systems. We focus here on formats designed for scientific data exchange, access, and archiving. We exclude discussion of other types of containers, such as relational databases, although data types that are described here often apply equally well to all databases.

Scientific data formats, or formats that support scientific data, are specified with paper, or electronic, documents describing the meanings, layout, and relationships of objects that the format recognizes. These documents define what is, and by omission what is not, a part of the format.

We have seen in the preceding section that we can describe data models in terms of data formats, creating a layered view of data formats and models. When we do this, the distinction between a data model and format becomes blurred. An HDF–EOS 5 Swath is a data model, but it is also a format in the sense that it describes how to represent swath data in HDF5. HDF5 in turn corresponds to a data model, whose objects are the building blocks used to implement HDF–EOS 5. And so it goes. The representation of a given object can consist of a number of layers, where the objects become increasingly primitive as one works down through the layers. Each successive layer can be thought of as the "format" for the layers above.

Figure 8.1 shows one way to think about layers of formats. At the lowest level there are bits in a digital store, such as tape, disk, or memory. At this level, formats (bit representation) are specified for simple atomic data types like numbers and text. Another higher level format may describe primitive data structures (record, array, etc.) that consist of collections of the atomic data types. Out of these are constructed more complex objects that have general use and meaning in science, such as images, meshes, date, and time. The formats for these are described as composites of the layer below, the primitive data structures.

As we travel up the layers in Figure 8.1, the scientific context becomes increasingly specific, hence the bottom layers are labeled "low context" and the top is "high context." The layered approach to dealing with formats has a number of advantages:

- Each layer provides information structures that can be used as building blocks to create structures at higher layers.
- In object terms, a more specialized object can inherit the semantics and methods of the more general objects.
- This inheritance can have enormous advantages for software engineering, as it may permit the higher levels to re-use methods at the lower levels.

- For users, data and tools can be meaningfully shared at the highest layer of common semantics, without having to be concerned about the underlying complexity.

8.2 Concepts of Data File Formats

We have seen how data formats occur in the context of modeling and representing scientific data. In this section we look at the features of formats themselves. A good understanding of formats will be helpful when we need to decide upon a format to use for a particular application. Or, given that we must use a particular format, it will help us understand how to use that format effectively.

8.2.1 File Access

The way a file is organized has a lot to do with how we intend to use it; so, before we look at file organization, let us think a little about how we access files. File access methods can be divided into four general categories: whole-file, sequential, direct, and indexed.

By way of example, think about a document that you might find on your desk, such as a photograph, a short paper, or a book. But instead of the document being on your desk, you receive the document in an email in the form of a file. Typically when you get the file, you use some application to load the file into your computer and you work with it there.

Suppose the file contains a photograph. You probably have an application that will open and display the picture so that you can view it. In this case, you access and use the whole file at once. This is *whole-file* access.

Suppose instead that your file contains an important five-page document that you need to proofread carefully for typographical errors. You open the file, you start reading at the beginning, and read straight through, without skipping a single word, to the end. This kind of file processing is called *sequential access.*

As a third example, imagine that you receive a very large document, and you only need to read the section that starts on the 317th page. In this case, with any luck your application allows you do jump directly to page 317 and start reading. This is called *direct access* — it means that any object in the file can be accessed without having to access other objects.

A variation on direct access occurs when the file is organized in such a way that you cannot really tell what page the material is on that you want to read — you cannot go directly to a certain page, although you might like to. If you are lucky, there is a table of contents, or index, in the file that you can consult, and it tells you just what page to go to. When you access a certain part of a file by first consulting an index, this is called *indexed* access.

Both direct and indexed access are examples a *random* access to objects. Random access means that any object in a file is as likely to be access as any other.

All of these examples describe situations in which a human accesses a file, albeit with the help of a computer. In the world of hydroinformatics, all of these types of access are needed. In some cases they involve humans, but in many instances they just involve computers.

8.2.2 File Structures

The structure of a file refers to how it is organized. We describe four categories of file structures: simple, record structured, indexed, complex. Not surprisingly, these structures correspond loosely to the types of access described in the previous section.

All of the structures we describe here contain optional headers records. A *header* is a record placed at the beginning of a file that is used to store information about the contents of the file, such as content metadata and structural metadata.

8.2.2.1 Simple Files

Simple files have very little structure. They consist typically of a repeated list of values. For example, a file containing a stream of integers that correspond to measurements from a single sensor would qualify as a simple file. Simple files are easy to understand, and it is usually easy to implement readers and writers for these files. A good example of a simple file is the USGS DOQ [1], which consists of a text header, followed by an image consisting of a number of lines of samples (pixels). Since the size of each line and the number of lines is given in the header, it is easy to write programs to read or write a DOQ.

Simple files can be accessed sequentially (e.g., processing a stream of integers), as whole files (e.g., viewing a DOQ), or directly (e.g., getting the 475th reading from an instrument).

8.2.2.2 Record Structured Files

A *record structured* file is a sequence of records. A *record* is an aggregate structure that could come in many forms. A record could be an image, or it could be something more complex and heterogeneous, such as a set of measurements. The units in a heterogeneous record are called *fields*. For example, an experiment might take measurements from a number of instruments at each of a number of time steps. A record could be stored for each time step, with one field holding the time of the measurements, and another field for each of the measurements. The form of records can vary a great deal. Fields may be optional, and they may be fixed or variable in size.

Record structured files are good for repeated, predictable data.

Whole-file access is possible with record structured files, as long a the file can fit into a computer's memory. Increasingly in hydrologic applications, this is not the case, as number of measurements can far exceed the capacity of a computer to hold them all.

Sequential access is possible, and common, with record structured files, as long as there is some way to tell when one record ends and the next one begins.

Direct access to records can be achieved if the size of records is fixed, and then only if the sequence number of the desired record is known. In this case, we can compute the byte offset of the record we are interested in, jump to that offset, and start reading. However, if the records are variable-length, there is no way to know where a particular record begins, so the only way to get to the record we want is to start at the beginning and read through all records that precede the record of interest.

Another way that fixed-length record files can facilitate direct access is through the use of record keys. A *key* is a field that is the target of a search. If the records are sorted by key field, then the records containing a certain value of the key can be located by performing a binary search on the file.[1]

8.2.2.3 Indexed Files

An *indexed file* is one for which one or more data structures (indexes) exist for finding records or other objects of interest in a file. Indexes can be part of a file, or they can be in a separate file, part of a database, or elsewhere.

[1] A binary search is performed by starting with the middle record of a file, determining which half of the file the record of interest lies, then repeating the process with that half of the file until the record of interest is found.

Indexes are useful when there is no easy way to know the location of objects that we might be searching for.

As an example, suppose that you collect data at varying time intervals from a variety of instruments, collecting one variable-length record per time interval. Each record includes a time-stamp field, giving the time when its measurements were made. Also suppose that you know your applications will require random access to records based on the times when measurements were made. If the records are stored end-to-end in the file, then direct access is not possible, for two reasons. The records are variable length, so you cannot compute the offset of a given record. But even if the records were fixed length, the time intervals are not even, so it is not possible to calculate which record you need. This is a good case for building an index. A second, fixed-length record file could be created, wherein each record has two fields, the first field containing the time stamps, the second containing the byte offsets of the corresponding records. If this file were sorted based on the values in the time stamp field, it would be relatively easy to do a binary search on this file for the time stamp of interest.

There can be more than one index for a file, providing alternate views. For instance, in the example above there could be one index that is sorted by time stamp and other indexes sorted according to other criteria. Each index provides a different view of the ordering of the records, even though the records themselves are ordered in just one way, and hence facilitate quick random access to records.

We have described indexes built from fixed length records. What happens to this kind index if we build it, then later append a new record to the original file? Or what if we delete a record from the file? Or what if the binary search is too slow for our needs? As it turns out, searching is such an important part of computing that many different types of index structures have been created to address these and other requirements. These are beyond the scope of this treatment, but two basic approaches are worth mentioning: tree structures and hash tables.

Tree-structured indexes involve data structures that enable records to be added and deleted efficiently. There are many types of trees, but one that is very common among file structures is the B-tree [2,3].

A *hash table* is an index in which keys are mapped to positions in the index by means of an operation called a *hash function*. As we saw earlier, if we know the position of a record, we can jump directly to that record, without even having to do a binary search. Hence, hash table indexes provide a very fast way to look up values [3].

8.2.2.4 Complex Files

Complex files are files whose organization can have many different structures and storage methods, including combinations of those described above. These files are usually sufficiently complicated that users generally do not write their own software to access them; instead, some third-party software is generally available to access these files.

Complex files are useful when the data, metadata, or both are complex or might vary from one instance to another. Complex files are very common in scientific computing, including hydrologic applications, because they provide needed flexibility and efficiency.

It is not unusual for a file format to begin as a simple or format, then evolve into a complex format as new features are added. As the code for accessing such a format also evolves, the result can be unwieldy, ill-structured software that is hard to maintain and use. And as applications try to integrate data stored in many different complex formats, these problems multiply.

Because of these problems, there are efforts to develop robust, general purpose complex formats that meet the needs of scientists, but avoid problems of proliferating formats and software degradation. We discuss these formats in detail later.

8.2.2.5 Multiple vs. Single File

A *multiple-file* structure is one in which the objects of interest are broken into pieces and the pieces are stored in different files.

Multiple files can help deal with size problems when a dataset is larger than what can conveniently be handled by a technology. For example, many file systems can only store files that are 2 GB or smaller. If a dataset is larger than 2 GB, it may be accommodated on such a file system by breaking it into several files that are smaller than 2 GB.

Multiple-file structures can be useful when the separate files perform different functions. For example, it is not unusual to put metadata in one file and raw data in a separate file, then access the metadata through one process and the raw data through a separate process. In some instances it may even be convenient to store the different members in entirely different places.

Although multiple files can be very useful, they have the drawback that they must remain associate with one another, an yet file systems do not automatically keep them together. Thus, without careful file management, it may be possible to lose one of the parts of a multiple file, perhaps rendering the other parts unintelligible.

8.2.3 Other Format Options

File structures are not the only thing that must be considered in selecting or designing a file format. Here we examine four others:

1. Encoding — text vs. binary
2. Access software — library vs. a user's code
3. Standardization — standard vs. *ad hoc* formats
4. Dealing with multiple objects — multiple files vs. a single file
5. Ownership — proprietary vs. open

8.2.3.1 Text vs. Binary Encoding

In Figure 8.1 we see that, at its lowest level of abstraction, a file format is just a collection of primitive objects. In scientific computing, there are quite a few different types of primitive objects. These include text objects, such as annotations and the names of variables and parameters, and numbers in many sizes and types, such as integer, floating point, and complex numbers. Data structure information may also be included in files — pointers and bit-maps, for instance. There are also special encodings of all of these primitive objects, such as data compression or encryption schemes.

Each of these primitive objects is represented by a series of bits, and these bits are arranged to represent information according to some encoding scheme. The way that this encoding is done can be divided into two general classes: text and binary.

A *text encoding* represents primitive objects using characters, such as the characters in the ASCII seven-bit character set [4], or one of the UNICODE schemes [5].

A *binary encoding* represents primitive objects in the same way that they tend to be represented in a computer. Text is still represented in a text encoding such as ASCII, but numbers are commonly presented as in some fixed, nontext format. Integers are typically one-, two-, four-, or eight-byte bit patterns representing binary integers. Floating point numbers are generally represented using the IEEE floating point encoding scheme, and so forth.

Many factors determine whether a binary or pure text encoding scheme is more suitable for a file format. These included human readability, compatibility with system utilities, computer readability, file size, performance requirements, and format complexity.

Human readability

Whereas text files can be displayed and read with any text editor, or can easily be printed, all of the data in a binary file, except that which is represented in text form, requires translation by software. On the other hand, a great deal of data is not meant for direct human consumption. A 24-bit image, for instance, is meant to be interpreted and displayed by a computer, with humans never looking at the pixel values that they are comprised of.

Compatibility with system utilities

Most operating systems provide useful utilities for working with text. For instance *sort*, *grep*, and *wc* are powerful UNIX utilities that are designed to work with text files only. Equivalent tools for binary files often must be implemented individually for a specific binary format.

Computer readability

The values stored in a text file that are not character values must be decoded in order to be understood by computers. This is not difficult, but it is an extra encoding or decoding step, and can add some extra time to an I/O operation. In many case this decoding is not required for binary values.

There is another aspect to computer readability, however, that can make text formats sometimes better than binary formats. Not all computers have the same architectures, nor do all computers represent binary numbers in the same way. When files are moved between two computers that use different binary representations, the numbers may need to be translated.

One common difference is that byte order is handled differently on different machines, This is referred to as a machine's *endianness*. *Big endian* computers store the most significant byte of any multi-byte data object (such as a four-byte integer) at the lowest memory address, and *little endian* machines store the least significant byte at the lowest memory address. When files are written directly from the memory of a big endian machine, they are typically written in big endian order in the file, and vice versa for files written from little endian machines. This difference may not be difficult to accommodate, but does require knowledge and intervention by applications.

Another difference is in the fundamental way that numbers are represented. This can require complex number conversions, and even then there are likely to be problems with accuracy. Although text is generally superior to binary in this regard, it is the case that not all computers represent text in the same way either, so some translation may sometimes be required when moving text files from one computer to another.

File size

Text encoding can often result in a greater space requirement, especially when very large or accurate values are stored. The integer −1,000,000,000 requires 11 or more bytes to store as text, but can be stored with four bytes as binary integer. Hence, a text file with many large numbers can easily be twice the size of the corresponding binary file. Of course, a text file that has mostly very small numbers can be relatively small.

Compression can often be used to mitigate the difference between text and binary storage, but then the values stored would need to be decoded.

Accuracy

Text representations can easily be made as long as required to represent a value accurately, as long as the value can be represented by a series of digits. (Pi, for instance, cannot be represented with digits.) Standard binary encodings are generally fixed in length, and thus have a finite range of size and accuracy.

I/O and transfer time

If there is a large amount of data, the greater size of a text file can appreciably affect the time it takes to get it into and out of a computer, or to transfer it over a network. The larger size of text can appreciably increase I/O and transfer time simply because more bytes have to be moved. In addition, numeric values stored as text usually need to be translated during input and output operations, which can take extra time.

Parallel I/O

Parallel read and write operations, in which a number of processors do I/O simultaneously to different parts of a file, is far more manageable when it is easy to predict the sizes of those elements that are being accessed. Again, binary files tend to be better in such instances because elements are fixed size, whereas in text files elements tend not to be.

Direct access capability

We have seen how the choice of a file structure should correspond to the types of uses that are anticipated, and the kind of data being represented. For instance, we have seen that fixed length record files and indexed files can improve our ability to access data directly and to randomly search for records. Because values in binary files are of a fixed size, binary files are more easily organized into fixed length records than are text files. Text files can be organized with fixed length records, but the result can be that records are much larger than they would otherwise need to be. Similarly, indexed files can be either binary or text, but the resulting text files can be much larger and not as easily converted to a form that computers easily deal with. This is particularly the case for complex index structures, such as search trees.

How to decide?

So what is the conclusion, text or binary? There are clearly advantages on both. If files needs to be read primarily by humans, or if someone will need to write software to read and write them, and if file size is not a major concern, text is good. If size, complexity, or random access are required, or if the data is not meant for direct human consumption binary is good. Another thing to think about is whether a data format that initially seems simple and small might evolve into one that is more complex and large. Text may initially seem like the best solution, but as the size or complexity of datasets increases, the disadvantages of text may outweigh its benefits.

8.2.3.2 Library vs. User-Code

So far we have characterized formats according to way they are structured, their expected modes of access and use, and whether they have binary or text encoding. Another important characteristic that distinguishes file formats is the degree to which third party software is available or required for accessing them. There is software readily available for accessing many formats, but some formats are just documented and you have to implement your own readers. Sometimes you have a choice. We call formats that are generally accessed through third party software *formats emphasizing interfaces*. At the other extreme are formats that are explicitly described, so that an application can contain its own software to decode the format. We call these *formats emphasizing data structure definitions*.

Formats emphasizing interfaces

Many formats are characterized primarily in terms of the interface that is used for interacting with files. This could be a graphical user interface (GUI), which enables humans to interact with the data, or it could be or an application programming interface (API), which enables application programs to interact with the data. If data in the format is to be accessed through a GUI, it may be sufficient to describe the format in terms of the types of objects that appear on a screen.

If a format is to be accessed through an API, it will likely be described in terms of a library of routines (or methods) that application programs can call.

Interfaces generally hide the actual structure of the underlying data, leaving this to associated software implementations or other specification documents not officially part of the format standard. For example, the format CDF (Common Data Format) [6] has an API for storing and accessing arrays, and there is an implementation of this API so that the API can be called by software to access data in CDF. This adds an abstraction level to the format, in which an application does not have to be concerned with how the bits are actually organized.

Indeed, the use of an API specification for a format makes it possible to organize the corresponding data in more than one way. HDF5, for example, can store its objects in a number of ways, and indeed provides facilities for applications to replace its default layouts with ones of their own choosing. This point reveals one of the great advantages of interface based formats, and that is that an interface can hide and manage a great deal of complexity. It would take an individual years to write software to take advantage of all of the capabilities of a format such as HDF5, but the availability of a software interface can allow an individual to take advantage of their powerful features with relative ease.

Although an interface can make your job much easier by providing access methods, storage options, and other features that would be hard to create on your own, there is no guarantee that a GUI or API will be easy to use. The powerful features of many interface-based formats can result in large libraries or applications that require work to set up and to learn how to use.

An interface may not be able to do all that you need with your data; it may be a general tool or library that you adapt to your needs. In this case, you may need to do quite a bit of work making this happen. And in this case your job is significantly simplified if the tool or library is open source.

Formats emphasizing data structure definitions

Some formats emphasize descriptions of the objects and other structures that constitute the format. This can be done using normal English prose, but it is also common to use data description languages (DDLs) for this task. DDLs are generally simple, flexible, machine-interpretable, languages that can be used to describe a wide variety of object types, layouts, and relationships. For example, the EAST DDL, specified in an ISO standard [7], can describe a sequence of repeating records where part of the record structure can vary from record to record, and where a wide variety of standard and custom data types are supported at the bit level. XML [8] is a general-purpose markup language that, in combination with appropriate Document Type Definitions (DTD) or XML Schema, can describe almost any character based format, and is therefore a type of DDL.

DDLs either include specification of the associated data to the bit level, to allow generic DDL tools to manipulate the data, or they provide pointers where this information may be obtained.

8.2.3.3 Interface vs. Data Structure Based Formats

Formats emphasizing data structure definition may, or may not, include an API as part of their specification. Formats emphasizing interfaces may, or may not, include data structure definitions as part of their specification. The usefulness of interfaces and structure definitions depend upon the user context. In the context of data analysis and processing, a clear API that allows users to quickly and conveniently access the data is very helpful. These users are typically not concerned with details of exactly how data is stored; they wish to be able to translate the logical structures of the data into something they can easily manipulate. But for use in an archive system, the situation may be quite different. The data is likely to be useful long after the current software used to read or write it is obsolete. In this context a complete and detailed description of the location and meaning of each bit of the data format is essential so that future generations of scientists and software can effectively use the archive.

Sometimes it is best to have both an API and a DDL for a format. It is common to use a single format for exchange, access, and sometimes for archiving as well even though it may not be optimized for each of these roles. To address these roles, it is common to add an API component to a DDL based format to facilitate access. Likewise, it is common to develop a DDL to describe the underlying structure of an API based format that otherwise does not go to the bit level.

8.2.3.4 Standard vs. Ad Hoc Formats

Ad hoc formats

Traditionally, scientific applications were often written by one or a few individuals, working in a relatively closed environment with data that they would produce and only they would use. In this environment, it was easy for teams to develop their own data formats, specific to their data and their data only. The ad hoc formats and software developed in these environments have the advantages that they can be optimized for efficient storage and for the kinds of access that will be made to the data.

Furthermore, a data model can be created that exactly fits the needs of the application. This common approach has many advantages, and is often very compelling at the beginning of a project, when enthusiasm is high, ideas are plentiful, and the task looks simple. But risks also come with the ad hoc approach. The invention of a new format, and the creation of software to read and write the format take resources, not to mention the need to maintain the software by its creator for as long as it is needed. In addition, the use of an ad hoc format may make it much harder to share data or software among different applications, especially if the format is very complex.

Figure 8.2(a) illustrates the situation that arises when projects develop their own format and tools. These so-called stovepipe solutions create their own formats, their own access software, and their own tools in such a way that none of the parts created by one project can be shared with the others.

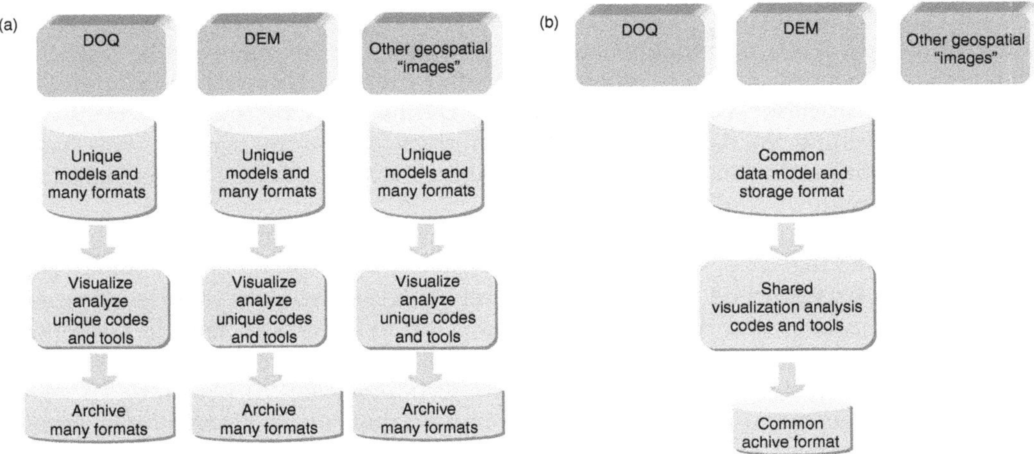

FIGURE 8.2
Stovepipes vs. common storage.

Standard formats

The disadvantages of ad hoc formats become clearer as the community of participants in a particular research endeavor grows, as the need to share data across teams becomes common, and as the benefits of sharing software become evident. And so, we often see an emphasis on developing standard formats and software — communities get together and agree to develop (or adopt) a common format, and to use the format in standard ways, eliminating the need to invent, implement, and maintain different versions of software that all do essentially the same thing. These advantages are illustrated in Figure 8.2(b).

There are a few disadvantages to the use of standards. Because standards are generally more broadly defined than ad hoc solutions, they typically do not fit any one application exactly as well as an ad hoc solution might, so a certain amount of work may be required to adapt a specific data model to that of the standard, and even then the fit may not be perfect. Also, by offering more general solutions, a standard implementation may not offer the same efficiency of storage or access as a special purpose ad hoc solution. Finally, the software that implements the standard may be bulkier or more complex than the ad hoc solution, and may take some work to adapt to an individual application.

8.2.3.5 Proprietary vs. Open

Another criterion for comparing file formats and access software is where they fit on the spectrum from proprietary to open. There is of course a continuum between the two, and most technologies fall somewhere between the extremes of "completely proprietary" and "completely open."

Proprietary formats and software

The term *proprietary* can mean many things, but in a general sense a technology is proprietary if it is owned by some individual or organization, and if there are restrictions on how the technology may be used. The reasons for these restrictions are generally commercial — by maintaining ownership of a technology, one can make money from its sale or use — but they may simply stem from a desire to control the quality and scope of the technology.

In the case of a file format, proprietary restrictions may apply to the format itself, or to software used to access information stored in the format. There may even be restrictions on reverse engineering activities, prohibiting anyone from publishing a description of the format or the software used to access it. It is sometimes the case that software is freely available for reading the data in a proprietary format, but not for writing. This has the benefit that information can be copied and distributed freely, with constraints placed only on those who produce the data.

Proprietary technologies have the benefits that they can be tightly controlled by their creators, insuring that all of the components are exactly appropriate for one another, and lowering the costs and need for systems integration. This can also make them less expensive to implement, revise, and maintain. Proprietary technologies also tend to concentrate expertise in one place, that of the proprietor, and as a result users may be less likely to be required to understand the complexities of the technologies.

Government agencies use the term COTS (commercial off-the-shelf) to refer to proprietary items (in this case, technologies) that can be bought ready-made, without having to expend the time and cost of building them from scratch [9]. Because they have to compete with other commercial products, and because the lead-time in developing commercial products is necessarily short, COTS technologies are also more likely to take advantage of advancements in technology. COTS technologies also tend to be supported well, since user satisfaction is important to the their success.

Proprietary technologies are not without their risks and costs. By definition their usage is restricted, and some users may be scared off from using them for fear of legal ramifications. We have seen this happen with formats in particular, in cases in which a company might wish to use a particular file format for internal purposes, but fear the consequences if the file format, or access software, inadvertently leaks out. Even when licensing restrictions are well-understood, the need to accommodate licensing restrictions can significantly impact system architectures that incorporate proprietary technologies. Proprietary software can have the disadvantage that users cannot make small changes necessary to adapt it to local needs, or repair minor flaws in the software.

An often-expressed concern about proprietary formats is that they may change significantly or be discontinued because of marketplace changes or strategic changes in the goals of the companies that own them. This can be a problem for the many scientific endeavors that require computing systems and applications to work over several decades. Also, many scientific applications can make very good use of historical data covering very long time periods. In such cases, it is important to them that data does not become inaccessible because a company goes out of business or changes its mind about a format.

On the other hand, highly successful proprietary formats can create such a huge following that their persistence over time is almost guaranteed. PDF seems like one such format.

Open formats and software

The term *open* can also have many meanings and nuances, but fundamentally open technologies are flexible, adaptable, and available. An open system is defined in [9] as

> a collection of interacting software, hardware, and human components, designed to satisfy stated needs, with the interface specification of components:
>
> - Fully defined
> - Available to the public
> - Maintained according to group consensus
>
> in which the implementations of components are conformant to the specification.

Open formats and software are less likely to change over time without agreement from a broad base of users. Because they are fully defined and publicly available, vendors and others are free to implement improved versions or versions that are more appropriate for specific circumstances, at the same time maintaining the interoperability required by the specification. If a software implementation is also *open source*, meaning that the source code is available for anyone to examine, improvements can be made through community contributions. Members of the community can fix bugs or add enhancements and offer these back to the others in the community.

Open system software is often free or inexpensive, especially when it is the product of a scientific community. This not only makes it available to a wider range of possible users than many proprietary systems, but it also promotes good will among its users and thereby fosters participation in its support and maintenance.

Open formats and software are particularly valuable in support of long-term preservation of data. Because they are fully, publicly defined and maintained by and for the community of users, their existence over time does not depend solely on a proprietor's marketing strategy.

There are risks and costs associated with open systems. Whereas the commercialization and marketing can help sustaining proprietary technologies, there may be less interest in selling a free, open technology to a community whose interest might decline over time for a variety of reasons. Open software is often free, which means the resources to implement, maintain, and support it must come from some source other than product sales. These resources are often voluntary, and can slip away as responsibilities and interests change. Resources may come from a consortium of organizations, a large federal agency, or some other institution whose own resources or priorities could change.

Also, because of the mandate to serve the needs of a broad community, open software and formats may be more general and hence more complex than their more narrowly targeted proprietary counterpart. This could impact on the cost of implementation and maintenance, and could make it harder to tune them for specific requirements.

Open technologies are not necessarily free technologies — large, complex open systems may be implemented at great cost and sold accordingly. Licensing restrictions on open systems can inhibit use, even when those restrictions are designed to encourage free distribution of the software.

Finally, with respect to long-term preservation, being open does not guarantee long-term existence of either a format or software. If either is difficult to maintain because it is poorly implemented or overly complex, if other open technologies supercede it, or if its usefulness lessens over time, the interest and resources required to maintain it can easily wane.

8.2.4 Format Examples

In this section, we describe some common formats that some sectors of the hydrology community is likely to find of interest. These formats illustrate many of the characteristics of formats that are described in earlier sections. They are netCDF, SDTS, ESRI Shapefiles, and XML.

8.2.4.1 netCDF

netCDF (network Common Data Form) is an interface-oriented format developed for the atmospheric sciences community but widely used in many branches of science. The netCDF library also defines a machine-independent format for representing scientific data.

NetCDF is a good example of a *de facto standard* — it is not currently endorsed by any national or international standards body, but is so widely used and well supported that many scientific communities treat it as a standard.

The netCDF data model was designed for applications in atmospheric science, but has very broad applicability. It is simple and flexible, and directly reflects the types of data and data access that many scientists need. The netCDF documentation [10] describes the netCDF data model as follows:

> A netCDF dataset contains *dimensions, variables,* and *attributes,* which all have both a name and an ID number by which they are identified. These components can be used together to capture the meaning of data and relations among data fields in an array-oriented dataset. The netCDF library allows simultaneous access to multiple netCDF datasets which are identified by dataset ID numbers, in addition to ordinary file names.
>
> A netCDF dataset contains a symbol table for variables containing their name, data type, rank (number of dimensions), dimensions, and starting disk address. Each element is stored at a disk address which is a linear function of the array indices (subscripts) by which it is identified. Hence, these indices need not be stored separately (as in a relational database). This provides a fast and compact storage method.

NetCDF data is organized in a way that can provide fast direct access to individual variables or parts of variables. A great deal of scientific data is collected as a sequence of records, for example a sequence of measurements over time, one set of measurements per time step. netCDF accommodates this in a natural way by storing data so that it can be efficiently accessed by record.

As an open, free standard, netCDF has gained a large following among many scientific communities, and has spawned a large number of software tools and applications. It enjoys particularly widespread use in the earth sciences, hydrology among them.

For more information about netCDF, visit http://my.unidata.ucar.edu/content/software/netcdf/.

8.2.4.2 SDTS

The *Spatial Data Transfer Standard (SDTS)* is an open standard DDL-based multiple-file format designed primarily for exchanging spatial data among federal agencies. Mandated by the Federal Government and developed over the course of many years by the U.S. Geological Survey (USGS), SDTS is used to distribute spatial data by many federal agencies, such as the USGS, the U.S. Census Bureau, and Army Corps of Engineers.

SDTS is an international standard format, having been ratified by the National Institute of Standards and Technology (NIST) as a Federal Information Processing Standard (FIPS 173). SDTS is well and rigorously documented [11–13], and has a rich web site at http://mcmcweb.er.usgs.gov/sdts/index.html.

Whereas the netCDF data model is independent of any particular science, SDTS is specifically designed for spatial data. Within that context, however, the SDTS data model is quite general, and can accept virtually any kind of spatial data. There are three levels to the data model, a conceptual model, a collection of spatial features, and profiles. The *conceptual model* defines generic objects, such as points, lines, polygons, and grids, which are used as the building blocks for creating spatial features. The descriptions of these objects include both the geometry (graphic depiction), the topology (connectivity and spatial relationships) of spatial features, and other attributes. *Spatial features* are the real-world phenomena (a road, a lake) that the generic objects describe.

Because the number of features defined for SDTS is enormous and growing, any given application is bound to have trouble dealing with all of the different types of objects it might encounter in an SDTS file. For this reason, a third level of specification is available for SDTS files, called a *profile*. A profile provides rules that restrict the contents of an SDTS file to a particular type of spatial data. One profile, the *Topological Vector Profile* (TVP) [14], includes spatial datasets in which vector features are represented together with geometry and topology. Examples of data that can be distributed using the TVP are the Digital Line Graph datasets [15] and Census TIGER files.

We have already described four conceptual levels for SDTS, but we are not yet done. Since SDTS is as DDL-based format (there is no standard API for accessing data in SDTS), there is a rigorous specification of the SDTS format. The SDTS format specification is based on the ISO 8211 [16] encoding standard, a media-independent standard that provides the full syntax and semantics necessary to describe SDTS records, fields and subfields right down to the byte level, and in a machine readable form.

An important difference between SDTS and netCDF has to do with their relative emphasis on access vs. content. Whereas netCDF is designed to provide efficient direct access to data, the design of SDTS pays no attention to this capability. STDS is a *transfer standard*, and concentrates on capturing all information it can in the most rigorous way possible. The SDTS format is a sequential format, and does not permit direct access. This tradeoff on the part of SDTS was by deliberate — the SDTS designers assumed that users who needed efficient direct access could convert their data from SDTS into some other format when direct access was desired.

8.2.4.3 ESRI Shapefiles

The *shapefile* format is a multiple-file binary format for storing certain kinds of geospatial data. The shapefile format was originally developed by Environmental Systems Research Institute (ESRI) to manage data used by its commercial GIS software. Although ESRI controls the format, it does so in a way that keeps the format open and serves the broad community of users of geospatial data. Here is how the ESRI Web site describes the format and its evolution:

> The Shapefile spatial data format is Open and published by ESRI. A shapefile stores non-topological geometry and attribute information for the spatial features in a data set. Shapefiles can support point, line, and area features. This file format is growing in adoption and capability to accommodate more complex spatial data; ESRI will announce and publish changes as they are developed. ESRI encourages developers and users to create interchange capabilities to both read and create shapefiles [17].

The shapefile specification is format-based. The specification [18] provides all of the information needed to write software to create and read shapefiles. As a result, a number of libraries and other tools have been written and are freely available for accessing shapefile data and for translating between shapefiles and other formats.

It is interesting to contrast shapefiles with SDTS. Although both are geospatial data formats, their intended uses are quite different. Whereas SDTS is designed to store and transmit complete information about geospatial phenomena, shapefiles are designed for very fast direct access to individual records, by tools that do not need a lot of the information that SDTS provides. As a result shapefiles are organized quite differently, and omit much of the information that can be found in SDTS. This also means that the shape file format can be simpler, making it is easier to write software to read and write shapefiles.

A shapefile is a good example of an indexed file. It consists of three different files, a main file, an index file, and a dBASE table. The *main file* is a variable length record file containing the actual data. Each record describes a shape, or feature, in terms of its vertices. The number of vertices for a given shape can vary enormously, from one vertex (a point) to thousands. The *index file* is a fixed length record file wherein each record contains the offset and length of the corresponding shape record in the main file.

The main file and index file contain the bare bones information needed to get features and to get them quickly. Other information, such as attributes of features, are contained in the *dBASE file*, whose format is defined by a standard called DBF.

Shapefiles provide us a good example of appropriate file design. The design emerged from a proprietary need, but was offered as an open standard, which not only helped users get their data into and out of the format, but also gave the community a useful standard for non-ESRI applications. The design is simple and effective for what it is meant to do. ESRI has evolved the format over time, but in a way that avoids feature bloat and does not compromise the core goals of the format. And, by adopting the DBF format for metadata, ESRI demonstrates the value of not inventing yet another format when it is not necessary.

8.2.4.4 XML

XML (Extensible Markup Language) is a single file[2] text format for describing the structure and content of documents. XML was developed by the World Wide Web Consortium (W3C) [8] to provide a rich, web-friendly format that would add semantic information to web-based documents. XML has been extraordinarily successful in achieving this goal, spawning may outstanding tools and applications, and XML has become the lingua franca for web documents that need to contain more information than simple HTML documents can provide. Uses of XML are described in detail in Chapters 1–4. XML is examined here in the context of its characteristics as a file format.

XML software is available on all major systems for working with XML, making it highly portable and useful for heterogeneous systems with multiple users and multiple purposes. It is an excellent language for sharing data across systems, across conceptual domains (different users, communities, and uses), and for archiving data.

An XML is both a markup language and a format. A markup language is a language that describes the structures in a document. That is, a document described in XML contains not just the normal content, but also the structure. What makes XML particularly interesting, however, is that it does more than just describe the structure of powerful capabilities to describe the meaning of objects within a document by means of machine readable tags.

It is possible to create XML profiles that describe standard structures and tags for particular application domains using mechanisms called DTDs and Schema, which are described in Chapters 2 and 11. This capability makes XML effective for storing a wide range of data types, from database schema to web services to scientific data, and we find XML playing a central role in virtually every computational discipline.

This emphasis on semantics makes XML well-suited for storing scientific and engineering metadata. Unlike NetCDF, SDTS, and shapefiles, which are designed with application domains in mind, XML is independent of any particular scientific or other domain. Nevertheless, what makes XML interesting here is that the uses of XML overlap significantly with those of some scientific formats, especially formats like netCDF and SDTS. Information that might otherwise be stored in SDTS or netCDF has a reasonably likelihood of instead (or in addition) being stored in the format used for XML.

[2] There have also be proposals for multiple-file collections that combine XML with other content-specific data files.

Hydrologists are particularly interested in the fact that DTDs and Schema have been used to create XML profiles for scientific applications, including earth science applications. One such profile is ESML (Earth Science Markup Language) [19], an XML profile for describing earth science data. ESML can store data, or it can be used as a meta-format, to describe data in other scientific formats, such as netCDF. See Chapter 4 for more details on ESML.

Like SDTS, XML is a text-base sequential format, and hence the disadvantages of text formats apply, particularly those involving efficiency, scalability, storage of complex objects, and algorithms involving complex access:

- Space-utilization can be even worse than for other text-based formats because XML tags and other structural metadata can consume a large amount of space, so it generally is not a good choice for representing large arrays of numbers and structures.
- Since access is sequential, complex data structures such as search structures (e.g., search trees) are not easily dealt with in XML — the only easy way to find an item in an XML file is by starting at the beginning and reading until the object is found. Other structures used in scientific computing, such as unstructured grids, also do not fit well with the XML model.
- XML does not allow direct selection from or to portions of objects, such as portions of images.
- Scatter operations, in which a collection of object elements is distributed to many parts of a file, are not possible in XML.

There are ways to mitigate the text-vs.-binary problems. Because certain XML symbols are repeated frequently, certain compression techniques can decrease file-size enormously. Also efforts are underway to define a version of XML that can contain binary data in those sections where large arrays of binary numbers occur [20]. Another approach that is being investigated is to store metadata in XML and simultaneously to store raw data in some other format.

8.3 Summary

This chapter covers characteristics of data file formats. We see that formats can be described at many levels, ranging from a low-level specification of how bits are stored to how high-level conceptual objects are organized. Each level builds upon those below. The higher the level, the more it relates to the context of a particular application and the less concern there needs to be with the underlying, bit-level complexity.

Some basic concepts of file formats are examined to help us think about the appropriateness of individual formats for particular uses, recognizing that no single format is best for all uses. File format concepts include file access operations and file structures. File structures can be categorized as file, record structured, indexed, complex, and multiple-file. Other file format options include text vs. binary, human vs. computer readability, compatibility with system utilities, accuracy or number representation, I/O and transfer time, ability to support parallel I/O, and direct access capability.

Some formats are defined by APIs, so that the underlying format may not even be known by a user. Others are defined by the data structures they contain. Some formats are ad hoc, while others are endorsed as standards. The relative advantages and disadvantages of proprietary vs. open formats are covered in some detail.

The chapter concludes with an investigation of four format examples, drawn from formats that occur frequently in scientific computing and hydroinformatics: netCDF, SDTS, ESRI Shapefiles, and XML. These formats are described and evaluated in the context of the many format features that have just been listed.

References

1. Part 2 Specifications: Standards for Digital Orthophotos. National mapping program technical instructions. U.S. Geological Survey National Mapping Division, 1996. http://rockyweb.cr.usgs.gov/nmpstds/acrodocs/doq/2DOQ1296.PDF.
2. Bayer, Rudolph and Edward M. McCreight. Organization and maintenance of large ordered indexes. *Acta Informatica*, 1:173–189, 1972.
3. Folk, Michael, Bill Zoellick, and Greg Riccardi. *File Structures, and Object Oriented Approach with C++*, 3rd ed. Reading, MA: Addison-Wesley, 1998.
4. International Organization for Standardization. ISO 7-bit coded character set for information interchange. 1991. http://www.iso.org/iso/en/ISOOnline.frontpage.
5. International Organization for Standardization. Universal multiple-octet coded character set (UCS). 2003. http://www.iso.org/iso/en/ISOOnline.frontpage.
6. CDF, the Common Data Format. National Space Science Data Center. http://nssdc.gsfc.nasa.gov/cdf/cdf_home.html.
7. ISO 15889:2003. Space data and information transfer systems — data description language — EAST specification. 2003. http://www.iso.ch/iso/en/ISOOnline.frontpage.
8. World Wide Web Consortium Web site, containing extensive information about XML: http://www.w3.org/XML/.
9. Oberndorf, Patricia. COTS and Open Systems. SEI monographs on the use of commercial software in government systems. Carnegie Mellon Software Engineering Institute, Pittsburgh, PA 15213-3890, February 1998. http://www.sei.cmu.edu/cbs/papers/monographs/cots-open-systems/cots.open.systems.pdf.
10. Unidata. *NetCDF User's Guide Chapter 2: Components of a NetCDF Dataset*. http://my.unidata.ucar.edu/content/software/netcdf/guidec/guidec-7.html.
11. SDTS Task Force. *The Spatial Data Transfer Standard — Senior Management Overview*. http://thor-f5.er.usgs.gov/sdts/articles/pdf/senmgr.pdf.
12. SDTS Task Force. *The Spatial Data Transfer Standard — Guide for Technical Managers*. http://thor-f5.er.usgs.gov/sdts/articles/pdf/mgrs.pdf.
13. SDTS Task Force. *The Spatial Data Transfer Standard — Handbook for Technical Staff*. http://thor-f5.er.usgs.gov/sdts/articles/pdf/staff.pdf.
14. SDTS Task Force. *Topological Vector Profile: A Profile of SDTS (FIPS 173) that Deals with Fully Topologically Structured Vector Data*. http://mcmcweb.er.usgs.gov/sdts/SDTS_standard_oct91/index_3.html.
15. SDTS Task Force. *DLG-3 SDTS Transfer Description*. http://thor-f5.er.usgs.gov/sdts/datasets/tvp/dlg3.
16. ISO 8211:1994. Information technology — specification for a data descriptive file for information interchange. 1994. http://www.iso.ch/iso/en/ISOOnline.frontpage.
17. ESRI Web site for Shapefile Information: http://www.esri.com/software/opengis/openpdf.html.
18. ESRI Shapefile Technical Description. An ESRI White Paper. J-7855. Environmental Systems Research Institute, Inc. 1998.
19. ESML Web site: http://esml.itsc.uah.edu/.
20. XML Binary Characterization Working Group public page: http://www.w3.org/XML/Binary/.

9

HDF5

Michael J. Folk

Contents

In this chapter, we take an in-depth look at the HDF5 format and library. Using HDF5 as an example, our goal is to provide a detailed understanding of a data format and its core software.

9.1 What Is HDF5?

The HDF5 is one of the handful of complex, general-purpose scientific data format in widespread use today. HDF5 is used in hundreds of different scientific and engineering applications, including many that impact on hydrology. It is an interface-oriented format, with open source software and an open specification. The HDF5 user communities are best described in terms of their format requirements, rather than discipline needs.

The following data management challenges have influenced the design, implementation, and support of HDF5.

Size and complexity of data:

- *Large objects, many objects.* Many formats, including earlier versions of HDF, already have difficulty handling very large objects and very large numbers of objects. The size and number of objects we need to store will increase in future, and formats need to be able to handle this.
- *Variety of data types and structures.* Scientific and engineering applications deal with a large variety of different data types and structures, and we need to accommodate these in persistent storage, in files.
- *Metadata flexibility.* Applications need to store metadata in a variety of forms, sometimes mixing different kinds of metadata in one container.

Availability of data:

- *Moving data.* The increasing distribution of data makes it necessary to move data from place to place. This can be facilitated by formats and software that are portable, and by formats with features such as compression that can decrease transmission time.
- *Data sharing — technologies.* Data is shared among those within one discipline, across disciplines, among different tools, and over time. Data sharing can be facilitated by formats that are self-describing, that are in common use, that are free and open, and for which software is portable.
- *Data sharing — community standards.* Equally as important as the technical specifications of a format and software is the extent to which communities determine how to use it to their advantage, and create environments that facilitate this common usage.

Ease of access:

- *Availability of applications and tools.* To a large extent, the usability of data is determined by the availability of applications and tools to create, access and view data. Complex formats, in particular, are of little value to most users without good applications and tools.
- *Cross-platform software support.* Because data, as well as applications and tools, often needs to be available in many different computing environments, it is important

to have software that works across platforms. Again, this is particularly important for complex formats.

- *Direct and partial access*. Direct access to individual records or objects, as well as partial access to objects become important when the size of objects or collections grows. Formats, as well as software, need to provide direct and partial access to meet the performance needs of users.

Efficient I/O and storage:

- *Fast I/O*. There are many situations in which fast access is required. We have seen how direct and indexed access can provide fast retrieval of objects in files. Partial access can improve access when only a small portion of an object is desired. And on systems with many processors that work in parallel, parallel I/O can make it possible to write several parts of an object or file simultaneously.
- *Efficient storage*. Very large collections can cost a great deal in storage. Formats that store data efficiently by using data compression, binary encoding, and other means, can save storage costs, I/O time, and network transfer time.

HDF5 addresses these challenges in a number of ways, as illustrated in Figure 9.1.

9.1.1 HDF5 Data Model

In Chapter 8, data models are described as ranging from "low context" to "high context." The HDF5 data model falls in the lower two portions of that continuum — it stores bit representations and primitive data structures, and is essentially domain-independent. Higher-context structures, such as images, tables, and unstructured grids, are constructed from these building blocks.

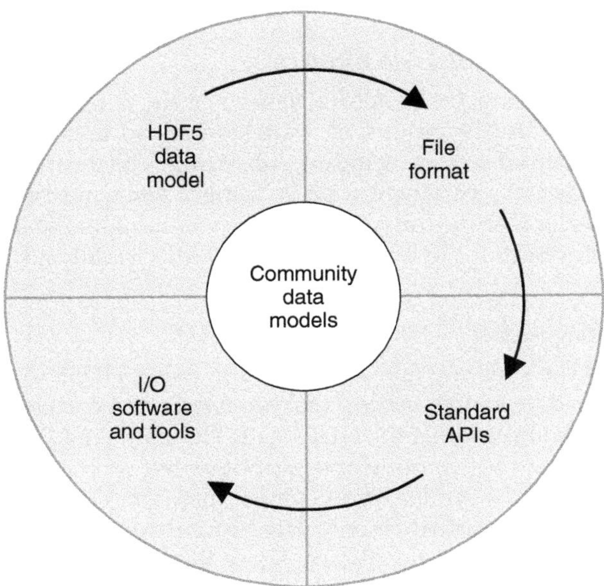

FIGURE 9.1
HDF5 solution to data challenges.

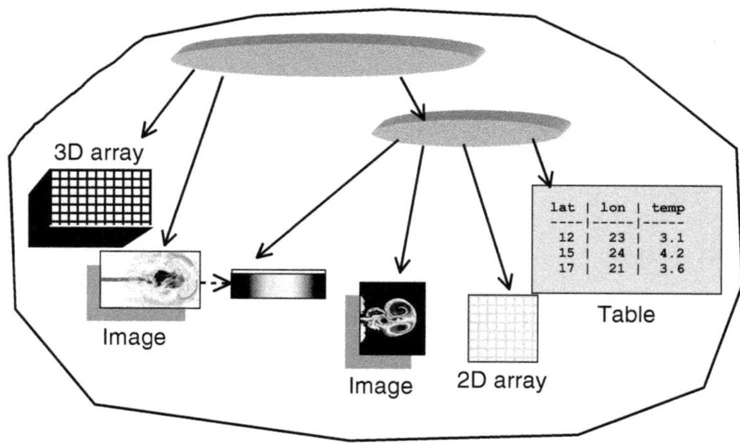

FIGURE 9.2
Example of HDF5 file.

A data model is described in Chapter 7 as "the way we describe objects within some application domain, their attributes, and allowable operations." Since HDF5 is domain-independent, the HDF5 model supports general data structures that are needed to organize and represent science and engineering data. These include multidimensional arrays (datasets), attributes, and a grouping structure. Figure 9.2 has a example of an HDF5 file, where images, multidimensional arrays, tables, etc., are created from the basic structures. It also shows that objects of different types can be mixed together as needed. Typically, there will be metadata stored with objects to indicate what type of object they are.

The HDF5 model also specifies that these structures can be organized and stored in ways that facilitate efficient and convenient storage and access. Allowable operations include those for creating, reading, writing, and querying objects, as well as operations to control *how* objects are stored and accessed. These are described in detail below.

9.1.1.1 Storage Model — HDF5 File Format

The HDF5 format implements the structural aspects of the data model. It determines how the objects (datasets, datatypes, groups, etc.) are stored, and uses structures that facilitate fast direct access and partial access, including subsetting. The format also allows objects to be organized or encoded in a variety of ways to achieve efficient storage and access in different circumstances. Because the format has so many options and features, it is necessarily complex, requiring applications to use the HDF5 API for virtually all access.

9.1.1.2 Programming Model — HDF5 Library and API

Figure 9.3 illustrates the software layers involved in using HDF5. At the bottom layer is an HDF5 file or other data source. Above that are two layers corresponding to the HDF5 software library, which implements the HDF5 API. The *low level API* concentrates on basic I/O: opening and closing files, reading and writing bytes, seeking, etc. This corresponds to the lowest level in the low context/high context diagram. HDF5 provides a public API at this level so that users can write their own drivers for reading and writing to places other than those already provided with the library. Those that are already provided include the UNIX standard input/output file interface ("stdio"), and MPI–I/O, an interface for parallel systems.

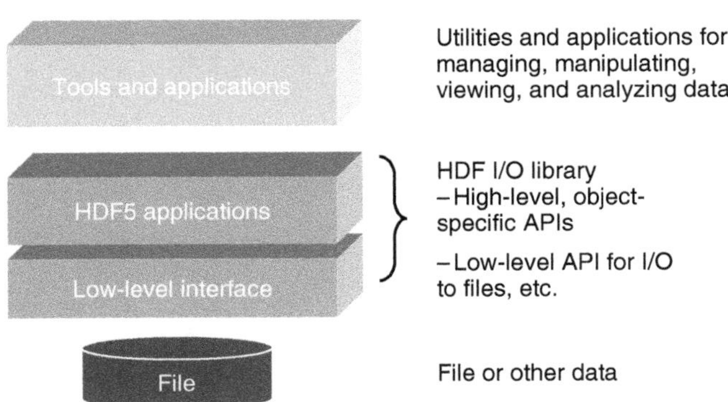

Figure 9.3
HDF5 software layers.

The next level is the *high-level, object-specific HDF5 interface*. This would be the next level up in the low–high context continuum. This is the API that those who develop HDF5 applications use most frequently. This API is used to create a dataset or group, read and write datasets, extract subsets of datasets, and query about a file or the objects it contains.

9.1.1.3 *Community Data Models*

HDF5 is so general that any particular type of data can be stored in it in many different ways — it is as easy to build stovepipes with HDF5 as with any homegrown format. For this reason, HDF5 users are encouraged to organize and access their files in standard ways. The first step in this process is to develop common data models within a community of users. This means agreeing on (1) the conceptual structures that best represent the data that the community will store and share, (2) the operations that will be required for the community to read, write, and query the data, (3) how to represent and organize data within HDF5 files, and (4) the structural and content metadata needed to capture the information that is to be shared. The HDF–EOS example cited in Section 9.2.3 exemplifies this process.

9.1.1.4 *Applications and Tools*

At the top level of the software stack in Figure 9.3 are applications, software tools, and APIs created for specific applications or application domains. This software corresponds to the higher levels of the low–high context diagram. As with any complex format, the ultimate usefulness of HDF5 is greatly influenced by the software that is available at this layer. Some of this software directly implements community data models, in the same way that the HDF5 API implements the HDF5 data model. Examples of software at this level are the HDF–EOS API that supports NASAs EOS swath, grid, and point datatypes. On the other hand, some of these applications do not involve high-context data models at all. They might be tools such as HDFView, a Java-based application for displaying and editing HDF5 files. Or, they might be applications whose use and data are restricted and do not need to be shared.

9.2 HDF5 Data Model — Drilling Down

An HDF5 file is a container for storing scientific data. HDF5 can store two types of primary objects: groups and datasets.

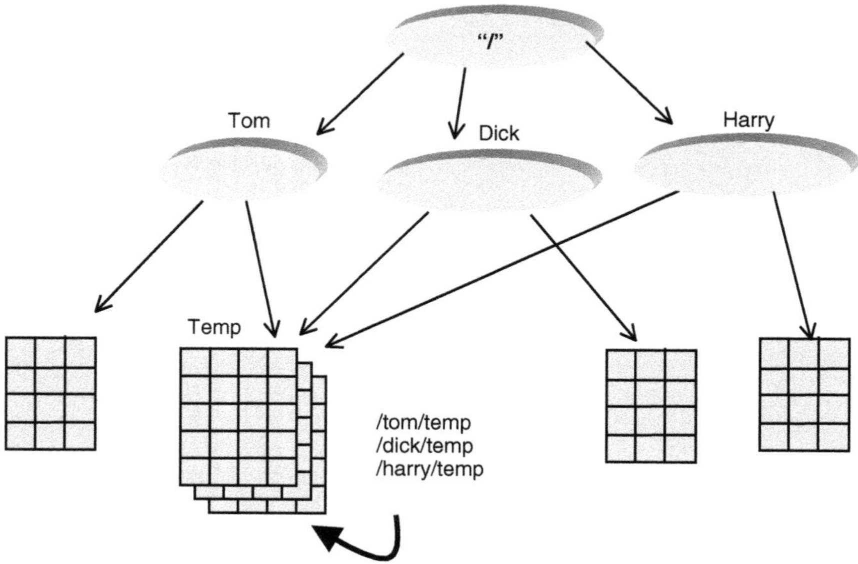

Figure 9.4
HDF5 group example. The oval shapes represent groups, and the rectangular structures represent datasets. Objects in HDF5 files can be identified by their pathnames. As this example illustrates, some objects may have more than one pathname.

9.2.1 Group

An *HDF5 group* is a structure for describing collections of related objects. A group may contain zero or more HDF5 objects, including other groups. Every HDF5 file starts with a *root group*, and all objects can be located in terms of their relation to the root group. The group structure is similar to directory structures found in most operating systems, except in HDF5 there is no requirement that the structure be strictly hierarchical on non-cyclical. This means that groups and members of groups can be shared. HDF5 objects are identified and located by their pathnames, as illustrated in Figure 9.4.

In addition to its member objects, a group also can contain *attributes*, which are structures that let an application attach information of the form "name = value" to an object. Attributes are described in detail in the next section.

9.2.2 Dataset

An *HDF5 dataset* is an object composed of a data array, together with metadata to describe the array and its properties. Figure 9.5 illustrates the components of a dataset:

Data array. A *data array* is a multidimensional array of identically typed data elements distinguished by their indices. The actual data may be organized in the file in a variety of ways, as described below.

Metadata. An *HDF5 dataset metadata* includes a description of dataset's datatype, dataspace, storage properties, and attributes.

Datatypes. An HDF5 dataset's *datatype* specifies how to interpret a data element. HDF5 defines atomic datatypes and compound datatypes. Datatypes can be saved to disk as objects that are called "named datatypes."

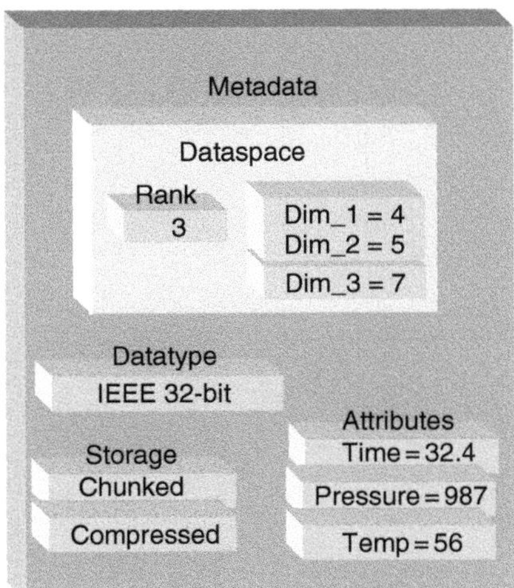

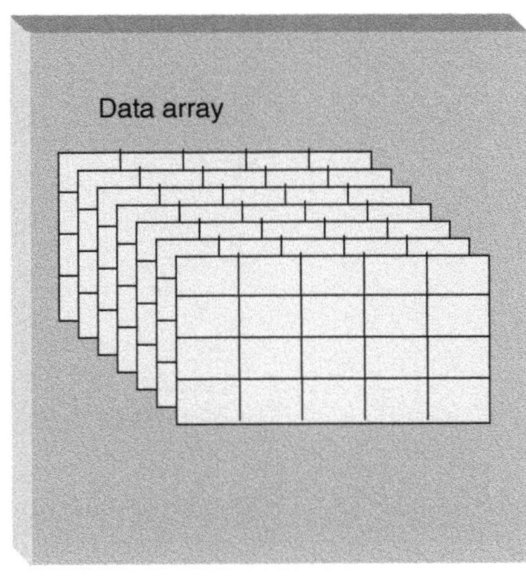

FIGURE 9.5
Dataset components consist of the raw data in the form of a multidimensional array, together with metadata.

Atomic types are distinguished by the fact that they are always dealt with in their entirety. It is not possible to read or write part of an atomic type. They include common *integer* and *floating point* numbers, as well as user-defined integer and floating point types. (The HDF5 format fully describes datatypes, so virtually any type of integer or floating point value can be represented.) *Variable-length types* can be defined, including character strings, but also including variable-length strings of any other atomic type. *References* are a datatype that refer to other objects in an HDF5 file. *Region references* refer to regions within the multidimensional array of a dataset. *Enumeration* types are types whose range of values include nominal values (e.g., names) that are mapped to integers. *Array* datatypes are multidimensional arrays that cannot be subdivided.

A *compound datatype* is one in which a collection of several datatypes are represented as a single unit, similar to a *struct* in C. For instance a compound type may consist of a string (e.g., a name), followed by an integer (e.g., age), and a float (e.g., weight). Members of compound types may be atomic or compound. The parts of a compound datatype are called *members*. The members of a compound datatype may be of any datatype, including another compound datatype. It is possible to read members from a compound type without reading the whole type.

Dataspace. A dataset *dataspace* describes the dimensionality of the dataset. The dimensions of a dataset can be fixed (unchanging), or they may be *unlimited*, which means that they are extendible — they can grow larger. Properties of a dataspace consist of the *rank* (number of dimensions) of the data array, the *actual sizes of the dimensions* of the array, and the *maximum sizes of the dimensions* of the array. Figure 9.6 illustrates a dataset with dimensionality 4×6.

A dataspace can also describe portions of a dataset, making it possible to do partial I/O operations on portions of datasets, called *selections*, in one read (or write) operation. Since I/O operations move data between two places, data transfer functions require two dataspace arguments: one describes the application memory dataspace or subset thereof, and the other describes the file dataspace or subset thereof. Currently selections are limited to sub-arrays with the same rank as the dataset, their unions, and lists of independent points. Several sample cases of selection reading/writing are shown in Figure 9.7.

Storage properties. Although a data array is conceptually a contiguous rectangular array, it can be stored physically and transferred in different ways. The actual storage may be

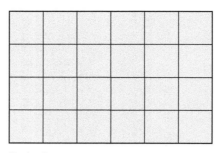

FIGURE 9.6
Dataset with dataspace whose rank is 2 and dimensions are 4 and 6.

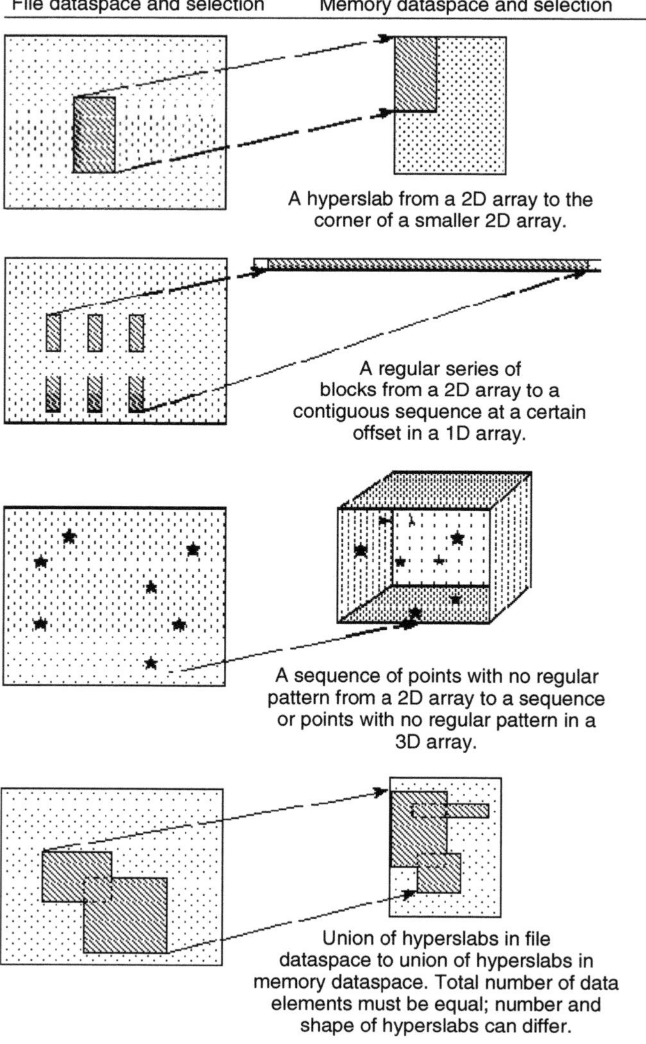

File dataspace and selection
Memory dataspace and selection
A hyperslab from a 2D array to the corner of a smaller 2D array.
A regular series of blocks from a 2D array to a contiguous sequence at a certain offset in a 1D array.
A sequence of points with no regular pattern from a 2D array to a sequence or points with no regular pattern in a 3D array.
Union of hyperslabs in file dataspace to union of hyperslabs in memory dataspace. Total number of data elements must be equal; number and shape of hyperslabs can differ.

FIGURE 9.7
Mappings between file dataspaces and selections, and memory dataspaces and selections.

a set of chunks, which may be compressed, and the access may be through different storage mechanisms and caches. *Storage properties* of a dataset can include maps between the conceptual array of elements and the actual stored data. Other storage properties include a fill value that can indicate that no data has been written to some portion of a dataset, and error detection encodings, such as a checksum.

Compressed storage has obvious advantages in terms of space use and data transfer time. Chunked storage involves dividing the dataset into equal-sized "chunks" that are stored separately. Chunking has three important benefits. (1) Chunking makes it possible to achieve good performance when accessing subsets of the datasets, even when the subset to be chosen is orthogonal to the normal storage order of the dataset. (2) With chunking large datasets can be compressed without sacrificing I/O performance, even when accessing subsets of the dataset. (3) Chunking makes it possible efficiently to extend the dimensions of a dataset in any direction.

Attributes. HDF5 attributes are small named datasets that are attached to primary datasets, groups, or named datatypes. Attributes can be used to describe the nature and the intended usage of a dataset or group. An attribute has two parts: a *name* and a *value*. The value part contains one or more data entries of the same datatype. The HDF5 format and I/O library are designed with the assumption that attributes are small datasets. Hence attribute operations are scaled down versions of the dataset operations — attributes cannot be extended, chunked or compressed, and partial I/O is not available.

9.2.3 Example: HDF–EOS Swath

To illustrate the use of HDF5, we look at the HDF–EOS 5, which uses HDF5 as its underlying format. Groups are represented with ovals, and datasets are represented with rectangles.

First, let us look at the top level of an HDF–EOS 5 file, illustrated in Figure 9.8. Under the root group, the group /HDF–EOS INFORMATION contains structural metadata for the HDF–EOS objects stored elsewhere. This is stored in a dataset as a text document in a machine-parsable format called ODL. The other group, /HDF–EOS, contains the collection of swath and point objects in respectively-named groups. The group /HDFEOS/ADDITIONAL/FILE ATTRIBUTES contains global metadata that a specific application adds to the HDF–EOS file.

This structure makes it easy for the HDF–EOS library to find what it needs in an HDF–EOS file. If an application needs a swath, the library knows to retrieve these under

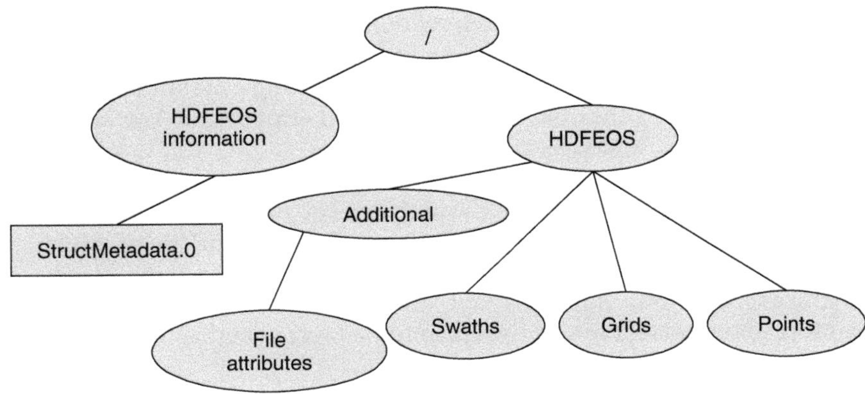

FIGURE 9.8
Top level of an HDF–EOS 5 file.

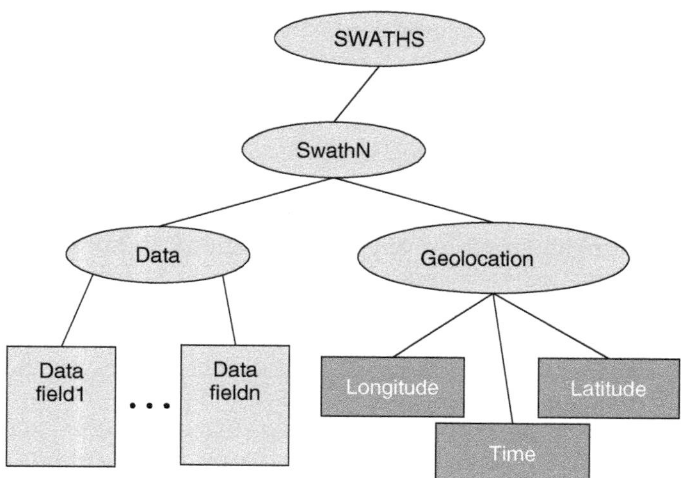

Figure 9.9
HDF–EOS swath structure in HDF5.

the group /HDFEOS/SWATHS. Applications will read and write HDF–EOS objects and metadata using the HDF–EOS 5 API.

If an applications needs to add its own special information to an HDF–EOS file, it can do so by using the normal HDF5 API, and storing that information under the root group. As long as it does not store it under one of the two top-level HDF–EOS groups, the HDF–EOS library will never see it.

Drilling down, Figure 9.9 shows how individual swaths are stored under the "SWATHS" group. There are two subgroups, "Data" and "Geolocation." The Data group contains the instrument measurements, each field stored as a separate HDF5 dataset. The Geolocation group contains a separate dataset for latitude, longitude, and time. For details on how to interpret these objects, see Reference 1.

9.3 HDF5 Library

The HDF5 library provides a relatively rich set of features that enable applications to read, write, query, and otherwise work with objects. Special features of the HDF5 library include the ability to create complex data types, complex subsetting capabilities, control over the way that objects are stored, flexible I/O options, and support for key language models including C, Fortran, Java, and C++.

The HDF5 library has three conceptual layers (Figure 9.10), two of which are exposed to applications. The top layer consists of the high level object API. This is the API that most applications deal with. It is where applications specify and work with groups, datasets, attributes, etc., including full and partial data access involving these objects. It is also the API that establishes and queries about storage properties such as compression and chunking. More about this layer shortly. The middle layer contains the library internals — its functions are not available through a public API.

The bottom layer is the HDF5 *virtual file I/O layer* (VFL), and is one of the special features of HDF5. To understand the VFL, it is important to understand that although the HDF5 file format describes how HDF5 data structures and dataset arrays are mapped to some linear

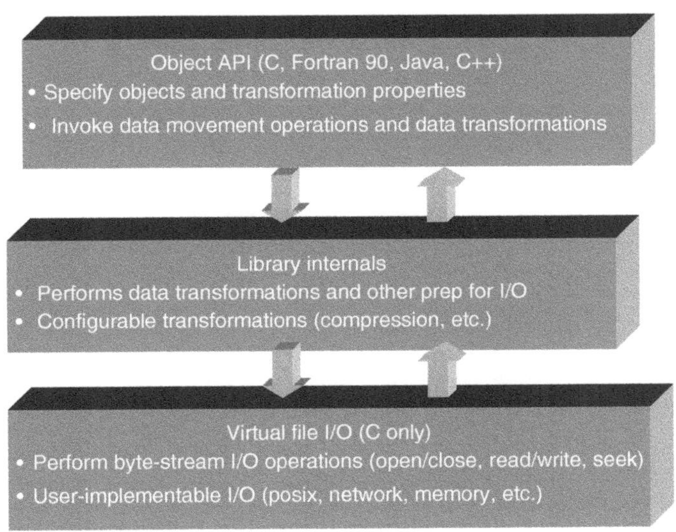

FIGURE 9.10
Structure of HDF5 library.

address space, the format does *not* specify how the address space is mapped onto storage. Generally, that mapping is just to a single file stored somewhere on a disk, but that does not have to be the case. In fact, early users of HDF5 requested the ability to map the format address space onto different types of storage (a single file, multiple files, local memory, global memory, network distributed global memory, a network protocol, etc.) with various types of maps. For instance, some users wanted to be able to handle very large format address spaces on operating systems that support only 2 GB files by partitioning the format address space into equal-sized parts each served by a separate file. Other users want the same multi-file storage capability but want to partition the address space according to the type of data (raw data in one file, object headers in another etc.) in order to improve I/O speeds.

Since the number of storage variations is probably larger than the number of methods that the HDF5 implementers are capable of implementing and supporting, a *VFL API* is available that allows application writers to design and implement their own mapping between the HDF5 format address space and storage, with each mapping being a separate *file driver*, as illustrated in Figure 9.11. The official HDF5 distribution provides a small set of useful file drivers:

- **POSIX.** The default driver, which uses POSIX file-system functions like read and write to perform I/O to a single file.
- **MPI–IO.** The driver of choice for accessing files in parallel using MPI and MPI–IO. It is only predefined if the library is compiled with parallel I/O support.
- **Family.** Partitions large format address spaces into more manageable pieces and sends each to separate storage locations using an underlying driver of the user's choice.
- **Split.** Splits the format address space into metadata and raw data and maps each onto separate storage using the underlying drivers of the user's choice. The meta data storage can be read by itself (for limited functionality) or both files can be accessed together.

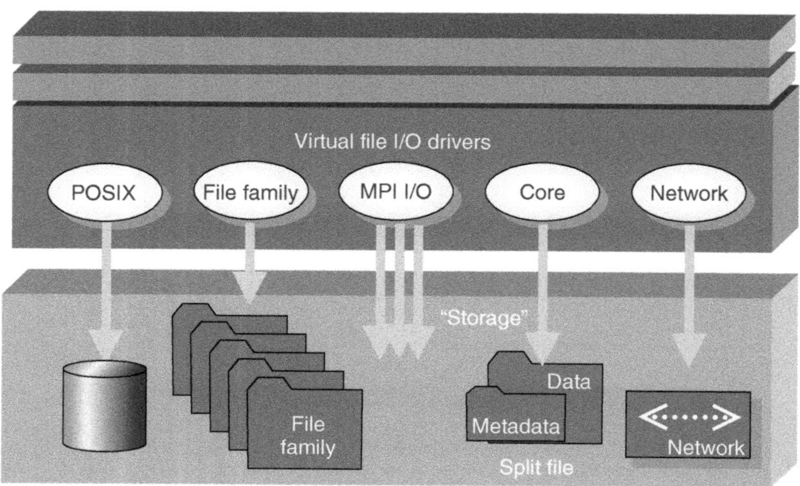

FIGURE 9.11
Examples of Virtual File Layer (VFL) I/O drivers.

(a)

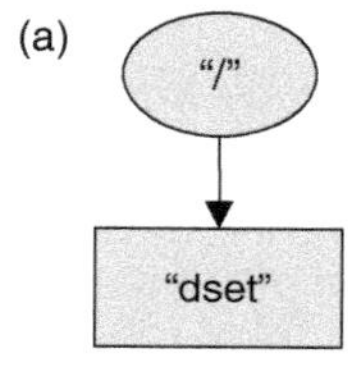

(b)

```
HDF5 "dset.h5" {
      GROUP "/" {
         DATASET "dset" {
            DATATYPE { H5T_STD_I32BE }
            DATASPACE { SIMPLE ( 4,6 ) / ( 4,6 ) }
            DATA {
               1, 2, 3, 4, 5, 6,
               7, 8, 9, 10, 11, 12,
               13, 14, 15, 16, 17, 18,
               19, 20, 21, 22, 23, 24
            }
         }
      }
      }
```

FIGURE 9.12
Sample use of h5dump. A file, shown in (a), consists of the root group and a single 4 × 6 dataset called "dset" with datatype *integer*. The command "h5dump dset.h5" results in the output shown in (b).

9.3.1 A Simple HDF5 Viewer — h5dump

Before looking at the HDF5 API, we describe a simple HDF5 command-line viewer called h5dump. h5dump is a simple, command-line utility that lists the objects in an HDF5 file and their contents. h5dump gives us a simple way to view HDF5 files, and can be especially useful when we are writing programs using the HDF5 library to create HDF5 files.

An example of the use of h5dump is shown in Figure 9.12. If you write programs using HDF5, you will find this to be a very handy utility.

9.3.2 Programming with the HDF5 Object APIs

The HDF5 object API is the one that most applications use, and will be the focus of the rest of this section. This discussion will only scratch the surface in describing how to use the API; further information can be found on the HDF5 web page [2].

9.3.2.1 Naming Conventions

The HDF5 API currently is implemented with C, Fortran 90, Java, and C++ interfaces. The names of C routines begin with prefix H5*, where * is a single letter indicating the object on which the operation is to be performed. (Other languages follow these as closely as possible.) For example:

H5D: **D**ataset interface	e.g., H5Dread
H5F: **F**ile interface	e.g., H5Fopen
H5S: data**S**pace interface	e.g., H5Sclose

9.3.2.2 Defined Datatypes

To achieve portability objects have their own defined datatypes. In the code that follows, these types are used:

hid_t:	object identifiers (native *integer*)
hsize_t:	size used for dimensions (*unsigned long* or *unsigned long long*)
herr_t:	function return value (FAIL(-1), SUCCEED(0))

9.3.2.3 Include Files

There are many definitions and declarations, such as the defined data types, that should be included with any HDF5 program. These definitions and declarations are contained in several *include* files. The main include file is `hdf5.h`. This file includes all of the other files that your program is likely to need. *Be sure to include* `hdf5.h` *in any program that uses the HDF5 library.*

9.3.2.4 Programming Model

The programming model for dealing with an object involves three parts:

- Objects are first opened or created
- Then accessed by read, write, query, and other operations
- Finally closed

Operations on objects have certain dependencies. For example, a file must be opened before a dataset because the dataset open call requires a file handle as an argument. Objects can be closed in any order and reusing a closed object will result in an error.

Property lists

In the following coding examples, we will encounter the use of *property lists*. A property list is a collection of parameters attached to an object that gives the library certain information about how to deal with the object. Some properties are permanently stored as part of the object (e.g., the chunking strategy for a dataset), others are transient (e.g., buffer sizes for data transfer). A common use of a property list is to pass parameters from the calling program to a VFL driver.

Since they contain several pieces of information, property lists can reduce the number of arguments in a function call. They also provide support for unusual cases that arise when creating files, opening files, creating datasets, and reading or writing data. And they allow functionality to be added without affecting applications.

Property Lists are conceptually similar to Attributes. However, Property Lists are information relevant to the behavior of the library, while Attributes are relevant to the user's data and application.

9.3.2.5 Creating an HDF5 File

There are three steps involved in creating an HDF5 file:

- Specify file creation and access property lists, if necessary
- Create a file
- Close the file and the property lists, if necessary

The function used to create an HDF5 file is shown in Figure 9.13(a). The *file creation property list* is used to control file metadata (information about the sizes of file data structures, etc.). For C, specifying a value as `H5P_DEFAULT` uses the default values. The *file access property list* is used to control different methods of performing I/O on files. Specifying `H5P_DEFAULT` uses the default values.

For every function that a program calls to create or open an object, there is a corresponding function that closes the object, releasing the object from control by the program. The function to close a file is described in Figure 9.13(b).

The code fragment shown in Figure 9.13(c) creates an HDF5 file using the property list defaults, then closes the file. If there is a possibility that the file already exists, the user must add the flag `H5ACC_TRUNC` to the access mode to overwrite the previous file's information. Since this file has nothing added to it, using h5dump to view this file would display only the root group:

```
HDF5 "file.h5" {

GROUP "/" {

}

}
```

9.3.2.6 Creating a Group

To create an HDF5 group, use `H5Gcreate`, shown in Figure 9.14. A group may be created in another group by either providing the full path name of the group to the `H5Gcreate` function or by specifying its location relative to a given group. For example, to create the group `Data_new` in the `Data` group, one can use the following call:

```
/* absolute path */

grp_new = H5Gcreate(file, "/Data/Data_new", 0);
```

or

```
/* relative to group */

grp_new = H5Gcreate(grp, "Data_new", 0);
```

Note that the group identifier `grp` is used as the first parameter in the `H5Gcreate` function when the relative name is provided.

9.3.3 Programming with HDF5 High Level APIs

Because of the flexibility of HDF5, the HDF5 library functions present a programmer with many options. The result is that the API can be quite complicated, especially for someone who is just beginning to use HDF5. To simplify programming with HDF5, a set of *HDF5 High Level APIs* has been created [3]. These APIs make HDF5 easier to use by performing more operations per call than the normal HDF5 API. They also encourage standard ways to store certain types of higher level objects, such as images, by enforcing standard representations of such objects in HDF5.

The following high level APIs are available for HDF5 [3]:

- *HDF5 Lite*. This API provides the same basic functionality as HDF5, but is simpler and makes fewer options available.

(a)

```
hid_t H5Fcreate (const char *name, unsigned flags,
                 hid_t create_id, hid_t access_id)

    name          IN:   Name of the file to access
    flags         IN:   File access flags
    create_id     IN:   File creation property list identifier
    access_id     IN:   File access property list identifier
```

(b)

```
herr_t  H5Fclose (hid_t file_id)
    file_id    IN: Identifier of the file to terminate access to
```

(c)

```
 hid_t       file;           /*  file identifier */
/*
* Create a new file using H5ACC_TRUNC access,
* default file creation properties, and default file
* access properties.
* Then close the file.
*/
file = H5Fcreate(FILE, H5ACC_TRUNC, H5P_DEFAULT, H5P_DEFAULT);
status = H5Fclose(file);
```

FIGURE 9.13
Creating an HDF5 file, (a) describes the H5Fcreate function, (b) describes the H5Fclose function, (c) illustrates their use. All example code is in C language.

(a)

```
hid_t H5Gcreate (hid_t loc_id, const char *name, size_t size_hint)

    loc_id      IN: location at which to create the group
    name        IN: name of the group to be created
    size_hint   IN: file space to reserve for names in group. If
                    a non-positive value is supplied, then a
                    default size is used.
```

(b)

```
#include <hdf5.h>
#define FILE "group.h5"

main() {

   hid_t       file_id, group_id;  /* identifiers   */
   herr_t      status;

   /* Create a new file using default properties. */
    file_id = H5Fcreate (FILE, H5F_ACC_TRUNC,
                         H5P_DEFAULT, H5P_DEFAULT);

    /* Create a group named "MyGroup" in the file. */
    group_id = H5Gcreate (file_id, "MyGroup", 0);

    /* Close the group and file. */
    status = H5Gclose (group_id);
    status = H5Fclose (file_id);
 }
```

(c)

```
HDF5 "group.h5" {
GROUP "/" {
         GROUP "MyGroup" {
          }
}
}
```

FIGURE 9.14
Creating a group in HDF5, (a) describes the H5Gcreate function, (b) illustrates how to create a group called MyGroup in an HDF5 file, (c) shows h5dump output for this file.

- *HDF5 Image.* Many applications treat datasets as raster images. Since a raster image can be stored as an array, HDF5 defines a convention for describing datasets as images. This API stores raster images as datasets, together with attributes to help applications interpret them as raster images. It provides functions to read and write images, and also to link raster images to color palettes, where appropriate.
- *HDF5 Table.* Many applications treat datasets as a one-dimensional collection of records, wherein each record is a compound data type — a fixed collection of fields. The purpose of the Table API is to make it easy for applications to interpret datasets as "tables." The Table API has functions to create, store, modify, retrieve, and query tables.
- *HDF5 Dimension Scale.* Sometimes it is useful to associate coordinate systems, and to store scales for each dimension of a dataset. The HDF5 Dimension Scale API makes

it easy to define datasets within HDF5 that can be recognized by applications as scales associated with dataset dimensions.

9.3.3.1 Programming with HDF5 Lite

The HDF5 Lite API consists of higher-level functions, which do more operations per call than the basic HDF5 interface. The purpose is to wrap intuitive functions around certain sets of features in the existing APIs. This version of the API has two sets of functions: dataset and attribute related functions. Currently there are C and Fortran APIs for this library.

The HDF5 Lite programming model is simple: obtain an identifier of the file where we want to use these functions and then call the appropriate function, as illustrated in the following code.

```
/* Create a HDF5 file. */
file_id = H5Fcreate( "test.h5", ...... );
/* Call some High Level function */
H5LTsome_function( file_id, ...extra parameters );
/* Close the file. */
status = H5Fclose( file_id );
```

9.3.3.2 Creating a Dataset Using HDF5 Lite

To create and write a dataset we can use the general function `H5LTmake_dataset`. This function accepts a file id, obtained with the basic HDF5 library function `H5Fcreate` or `H5Fopen`, a dataset name, the number of dimensions in the dataset, an array containing the dimensions, the HDF5 datatype (e.g., `H5T_NATIVE_INT`) for the data, and the data. `H5LTmake_dataset` is defined and illustrated in Figure 9.15.

9.3.3.3 Reading a Dataset with HDF5 Lite

The function for reading a dataset with HDF5 Lite is described in Figure 9.16.

9.3.3.4 Creating Attributes with HDF5 Lite

HDF5 Lite contains functions to create and read attributes. These functions are type specific. That is, there is one function for each of the most common HDF5 datatypes. For example, to write an integer attribute we use the function `H5LTset_attribute_int`. To read an attribute, the steps are similar, except that we use the read functions. To read the previously written attribute we use `H5LTget_attribute_int`. Specifications for these functions are given in Figure 9.17, along with examples of their use.

There are "get" and "set" attribute functions for the following C language datatypes: string, char, short, int, long, float, and double.

9.3.4 Conclusion

The programming information presented in this section should be sufficient to enable you to solve the problem at the end of this chapter. In addition, you will need to learn to compile and run your program with the HDF5 library. Some of the information needed to do this can be found in problem description, but for more detailed information, visit the HDF5 web site.

(a)

```
herr_t H5LTmake_dataset ( hid_t loc_id, const char *dset_name,
                          int rank, const hsize_t *dims,
                          hid_t type_id, const void *buffer )
      loc_id        IN:    Identifier of the file or group to
                           create the dataset within.
      dset_name     IN:    The name of the dataset to create.
      rank          IN:    Number of dimensions of dataspace.
      dims          IN:    An array of the size of each dimension.
      type_id       IN:    Identifier of the datatype to use when
                           creating dataset.
      buffer        IN:    Buffer with data to be written to dataset.
```

(b)

```
#include "H5LT.h"

int main( void )
{
hid_t        file_id;
hsize_t      dims[2]={2,3};
int          data[6]={1,2,3,4,5,6};
herr_t       status;

 /*create an HDF5 file using default properties */
 file_id = H5Fcreate ("ex_lite1.h5", H5F_ACC_TRUNC,
                      H5P_DEFAULT, H5P_DEFAULT);

 /*create the dataset */
 status = H5LTmake_dataset(file_id,"/dset", 2, dims,
                           H5T_NATIVE_INT, data);

/* Close the file */
 status = H5Fclose (file_id);
 return 0;
}
```

FIGURE 9.15
(a) `H5LTmake_dataset` creates and writes a dataset named `dset_name` attached to the object specified by the identifier `loc_id`. The parameter `type_id` can be any valid HDF5 predefined native datatype. (b) A sample program to create and write a dataset whose elements are the native integer type of the local machine (`H5T_NATIVE_INT`).

9.4 Example Problem: Using the HDF5 File Format as IO for an Advection–Diffusion Model

9.4.1 The Advection–Diffusion Model

In nature, transport occurs in fluids through the combination of advection and diffusion. The transport of a substance C in a bidimensional fluid medium referential can be described by the partial differential equation

$$\frac{\partial C}{\partial t} = -u\frac{\partial C}{\partial x} - v\frac{\partial C}{\partial y} + \frac{\partial}{\partial x}\left(K\frac{\partial C}{\partial x}\right) + \frac{\partial}{\partial y}\left(K\frac{\partial C}{\partial y}\right), \tag{9.1}$$

(a)

```
herr_t H5LTread_dataset ( hid_t loc_id, const char *dset_name,
                          hid_t type_id, void *buffer )
      loc_id       IN:   ID of file or group to read dataset within
      dset_name    IN:   The name of the dataset to read
      type_id      IN:   ID of datatype to use when reading dataset
      buffer       OUT:  Buffer with data.
```

(b)

```
 #include "H5HL.h"

int main( void )
{
 hid_t        file_id;
 int          data[6];
 herr_t       status;

 /* open file from example 1 */
 file_id = H5Fopen ("ex_litel.h5", H5F_ACC_RDONLY, H5P_DEFAULT);

 status = H5LTread_dataset(file_id,"/dset", H5T_NATIVE_INT, data);

 status = H5Fclose (file_id);
 return 0;
}
```

FIGURE 9.16
(a) `H5LTread_dataset` reads a dataset named `dset_name` attached to the object specified by the identifier `loc_id` into a memory at location `buffer`. The parameter `type_id` can be any valid HDF5 predefined native datatype. (b) A sample program to read a dataset whose elements are the native integer type of the local machine.

where u and v and are steady velocity components in the 2D space and K is a diffusion coefficient. This equation states that the time variation of the concentration of the substance C at a point depends on two physical processes, advection (first two terms of right-hand side), which models a streaming of infinitesimal elements in a fluid and diffusion (last two terms), which models random motion at microscopic scale.

This equation can be solved numerically using a finite difference scheme. We approximate the spatial domain using a grid of cells, using the notation

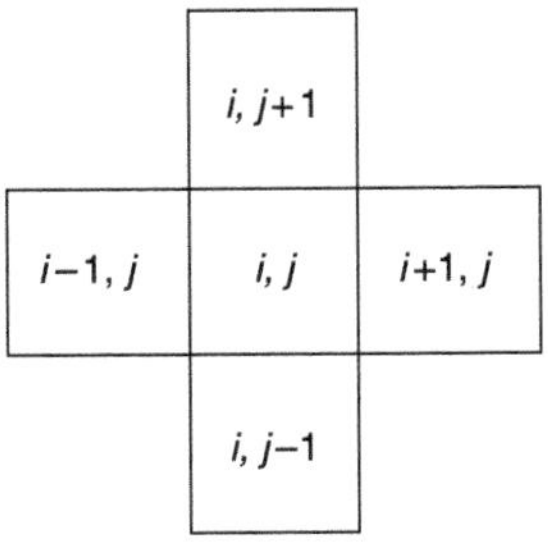

	i, *j*+1	
i−1, *j*	*i*, *j*	*i*+1, *j*
	i, *j*−1	

Using a spatial difference to the left for the advection and a centered difference for the diffusion yields the following explicit scheme, where K was assumed constant.

$$\frac{C_{i,j}^{t+\Delta t} - C_{i,j}^{t}}{\Delta t} = -\,u_{i,j}\frac{C_{i,j}^{t} - C_{i-1,j}^{t}}{\Delta x} - v_{i,j}\frac{C_{i,j}^{t} - C_{i,j-1}^{t}}{\Delta y} + K\frac{C_{i+1,j}^{t} - 2C_{i,j}^{t} + C_{i-1,j}^{t}}{\Delta x} + K\frac{C_{i,j+1}^{t} - 2C_{i,j}^{t} + C_{i,j-1}^{t}}{\Delta y} \tag{9.2}$$

(a)

```
herr_t H5LTset_attribute_int( hid_t loc_id, const char *obj_name,
                              const char *attr_name, int *buffer,
                              hsize_t size)
      loc_id      IN:   Identifier the object (dataset or group)
                        to create the attribute within
      obj_name    IN:   Name of object to attach attribute to
      attr_name   IN:   Attribute name.
      buffer      IN:   Buffer with data to write to attribute
      size        IN:   Size of 1D array (in the case of a
                        scalar attribute, use 1)
```

(b)

```
#include "H5LT.h"

int main( void )
{
 hid_t    file_id;
 hsize_t dims[1] = {5};
 int      data[5] = {1,2,3,4,5};
 herr_t   status;

 /* open the HDF5 file from the previous example */
 file_id = H5Fopen ("ex_lite1.h5", H5F_ACC_RDWR, H5P_DEFAULT);

 /* Create and write the attribute "attr1" on the root group */
 status = H5LTset_attribute_int(file_id,"/","attr1",data,5);

 /* read the attribute back */
 status = H5LTget_attribute_int(file_id,"/","attr1",data);

 /* Close file */
 status = H5Fclose(file_id);
 return 0;
}
```

FIGURE 9.17
(a) `H5LTset_attribute_int` creates and writes a numerical integer attribute named `attr_name` and attaches it to the object specified by the name `obj_name`. The attribute has a dimensionality of 1. The HDF5 datatype of the attribute is `H5T_NATIVE_INT`. (b) A sample program to write and read an attributed or type `int` to dataset `dset`, then to retrieve this attribute.

The subscript in $C_{i,j}^{t+\Delta t}$ denotes the space information in the cell grid, and the superscript denotes the time information. This equation can be easily solved for $C_{i,j}^{t+\Delta t}$. For every step in time, a new value of the solution is obtained explicitly from a linear combination involving the nearest neighbors, which are known either from a previous step or the initial condition.

9.4.2 The Assignment

The assignment will try to model the transport of a pollutant C at the exit of a sewage canal. The computational domain will be a matrix of 20×40, where two cells in the south border identify the sewage canal (Figure 9.18). At the exit of the canal a concentration of pollutant is dumped every time step with a constant flow.

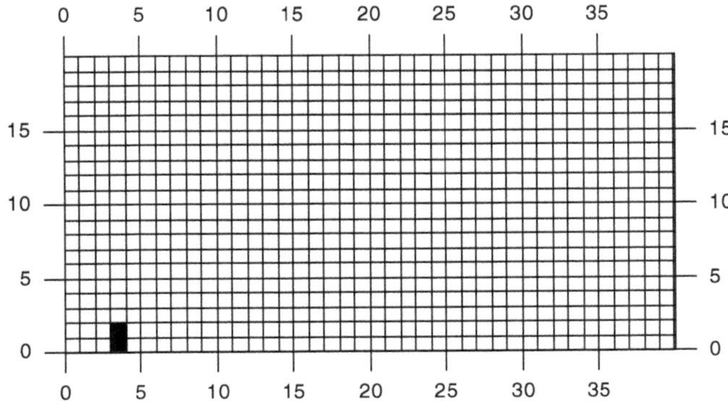

Figure 9.18
Computational domain for simulation.

The added concentration C of the pollutant at this cell is

$$C = \frac{m_p}{V}, \tag{9.3}$$

where m_p stands for the mass of pollutant added and V for the volume of water in the cell. The volume V is obtained knowing the height of water and the space dimensions of the cell. The mass of pollutant m_p can be obtained knowing the flow Q and concentration of the pollutant c_p in the sewer, which are model constants.

$$Q = \frac{V_p}{\Delta t}, \quad m_p = V_p \cdot c_p. \tag{9.4}$$

This added concentration C at this cell is calculated in the GetWaste function of the model.

Your task will be to complete a C++ class ADmodel by adding the data Input/Output routines. The HDF5 file format will be used. We have supplied the header file for the class (ad.h), the implementation of the model (ad.cpp) and the main program (main.cpp). The goal is to implement the functions indicated in the header, which are the following:

- *Start*. This function creates the HDF5 file that is used to save and open the results. It also creates the two groups "initial" and "results." It additionally writes in the group "initial" three datasets of variables that remain constant during the model life cycle. The datasets are: the velocity field (two datasets for each velocity component), and the bathymetry. The dataset names will be: "velocity U" and "velocity V" for the velocity field and "bathymetry" for the bathymetry. You also must define the dimensions of the datasets, which are 2D arrays with dimensions 20 × 40, and pass this array to the call to the HDF5 Lite function that makes the dataset. This function also creates the attributes "time" and "Kdif."
- *GetDispersion*. We define dispersion as the number of cells at each time step that contain the pollutant. You will have to calculate the number of cells that are contaminated by the pollutant. (Hint: to get this value note that the concentration matrix c of values is initialized to zero at the start of the model.) You will use this value as an attribute of the concentration dataset. This function returns an integer value of the dispersion.

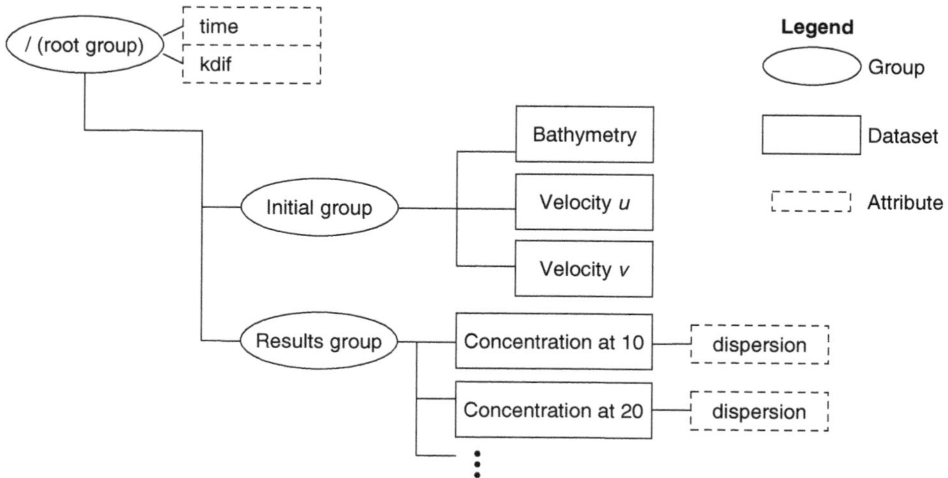

Figure 9.19
Hierarchy of HDF5 file upon completion of simulation.

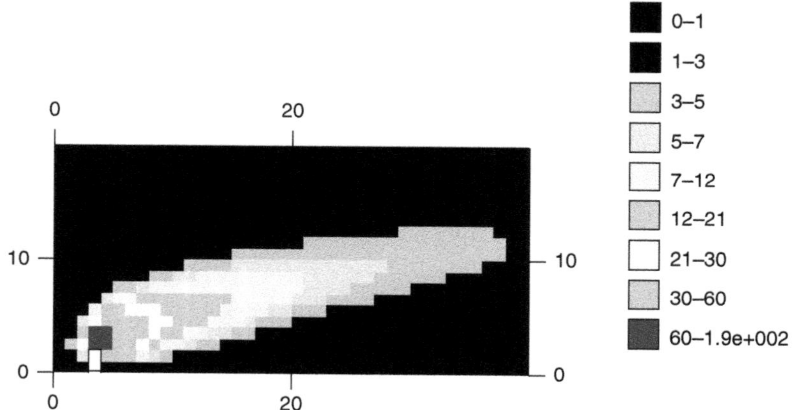

Figure 9.20
Output of a simulation as displayed with the Discrete Map view of HDF Explorer.

- *SaveAt*. This function will write the concentration at a given time step. For each time step an HDF5 dataset must be created. Its name will be a concatenation of the word "concentration at," with the current time of the model (e.g., at time 30, the dataset name will be "concentration at 30." You can use the string manipulation routines of the standard C library to form the string. The function will also save the dispersion as an attribute of the dataset. To do this you will call the GetDispersion function and pass the return value as a parameter to the HDF5 function that writes attributes (note that GetDispersion returns an integer value). Additionally, the attribute "time" must be updated with the current time.
- *OpenAndContinue*. This function will open the HDF5 file previously written, and will read the "concentration" and the current model time.
- *The class destructor*. This function will close the HDF5 file.

The hierarchy of the file upon completion of the program is shown in Figure 9.19.

The file may be viewed using the HDF file viewer "HDF Explorer," which can be downloaded from http://www.space-research.org/. Figure 9.20 shows the output of a simulation

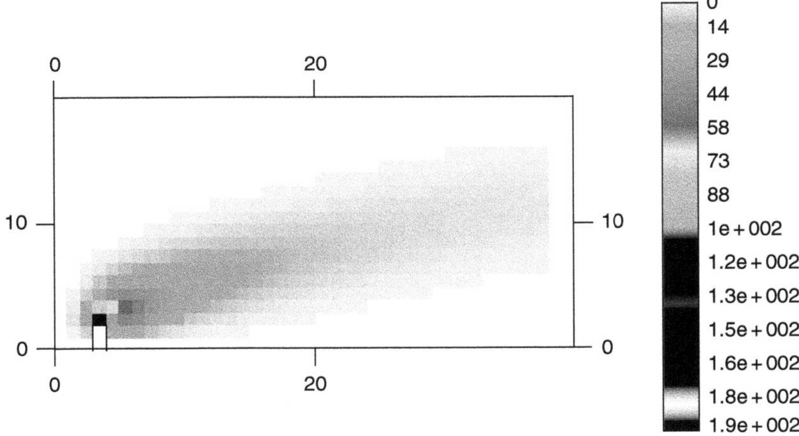

FIGURE 9.21
Output of a simulation as displayed with the Continuous Map view of HDF Explorer. This option allows a user defined palette to be loaded from a file.

as displayed with the Discrete Map view of HDF Explorer. Figure 9.21 shows the output of a simulation as displayed with the Continuous Map view of HDF Explorer.

9.4.3 Implementation Issues

The HDF5 Lite library will be used for IO. See http://hdf.ncsa.uiuc.edu/HDF5/hdf5_hl/ doc/

Code for main(): main.cpp

```
// ************************************************
// *                                              *
// *  main.cpp                                    *
// *                                              *
// *  Sample code used to test the                *
// *   CE498HI part 2 solution                    *
// *                                              *
// ************************************************

#include "ad.h"

//this program runs 2 instances of the ADmodel
//1st a new model is created using the function Start,
//which starts at time 0
//a 2nd run opens the previously saved results

int main()
{
 ADmodel *ad;
 int i;

 //1st run . start a new model
 ad = new ADmodel();
```

```
 ad->Start("ad.h5");

 for (i=0; i< 10; i++ )
 {
  ad->Iterate();
  ad->SaveAt();
 }

 delete ad;

 //2nd run . start a new model, using the saved results
 ad = new ADmodel();
 ad->OpenToContinue("ad.h5");

 for (i=0; i< 10; i++ )
 {
  ad->Iterate();
  ad->SaveAt();
 }

 delete ad;

 return 0;
}
```

Declaration of class ad.h

```
#include "hdf5.h"

#define IDIM 20
#define JDIM 40

#define LAND-55 //land flag
#define RANK 2

typedef double real;

//flow and concentration of substance C at entry point of sewage
struct Waste
{
 real Q;
 real c;
};

//------------------------- ADmodel -------------------------

class ADmodel
{

 //public methods
```

```
public:
 ADmodel();
 void Iterate();

 //private methods and variables
private:
 int cols;
 int rows;

 real h[IDIM][JDIM]; //bathimetry
 real u[IDIM][JDIM]; //horz. velocity
 real v[IDIM][JDIM]; //vert. velocity
 real c[IDIM][JDIM]; //concentration at time t
 real nc[IDIM][JDIM]; //concentration at time t+1
 Waste waste[IDIM][JDIM]; //concentration and flow of polutant

 real dx[ JDIM ]; //spacial step of the model matrix
 real dy[ IDIM ];

 real dt;   //time step in s
 real time; //simulation time in s
 real Kdif; //diffusion coefficient

 //model
 void InitModel();
 real GetWaste( int, int );


 ///////////////////////////////////////////////////////////////
 //TO IMPLEMENT FOR THE ASSIGNMENT
 ///////////////////////////////////////////////////////////////

public:

  //destructor
  //closes the HDF5 file

 ~ADmodel();

 //Start
 //create the file, save initial constant conditions
 void     Start(char* file_name);

 //OpenToContinue
 //Read the model time and concentration
 void     OpenToContinue(char* file_name);

 //SaveAt
 //save concentration at several time steps
 //save dispersion as an attribute
```

```
  void      SaveAt();

 private:

  //GetDispersion
  //calculate the number of cells affected by pollutant
  int       GetDispersion();

  // HDF5 file handle
  hid_t file_id;

};
```

References

1. Zhao, Shen Zhao, Abe Taaheri, Date Ray Milburn. HDF–EOS interface based on HDF5, Vol. 1: overview and examples. NASA Technical paper 175-EMD-001, October 2003. http://hdfeos.gsfc.nasa.gov/hdfeos/Info/Info_docs/HDF-EOS/guides/HDFEOS5.1.6_ug_vol1_oct_2003.pdf.
2. HDF5 Web site, containing tutorials and other documentation. For tutorials, see http://hdf.ncsa.uiuc.edu/HDF5/doc/Tutor/. For other documentation, see http://hdf.ncsa.uiuc.edu/HDF5/doc/.
3. HDF5 high level API. http://hdf.ncsa.uiuc.edu/HDF5/hdf5_hl/.

Section III

Data Communication

10

Web Services

Jay C. Alameda

Contents

In this chapter, we will discuss issues related to coordinating distributed resources, such as the diverse data resources used in hydrology (NASA MODIS [1], Hydrologic Information System, HIS [2], NOAA data, for example, National Operational Hydrologic Remote Sensing Center, NOHRSC [3]). In order to build a base of understanding, we will take a look at some early efforts in distributed object systems (i.e., object-oriented software systems that allow objects to be distributed to multiple computers), to the current approach, web services, that builds on the success of web technologies. We will then look at some of the foundational issues in data and task description using the eXtensible Markup Language, XML. Finally, this will be tied together in efforts to harness vast compute and data resources to build the computing analog of the electric power grid, namely, the *computational and data grid*.

Web services are becoming the defacto communication mechanism for e-commerce and business to business transactions and communications, as evidenced in a recent single-topic issue of *IBM System Journal* devoted to web services and e-commerce [4]. Building on a legacy of distributed object systems that had roots in proprietary systems, web services appear to resolve some of the difficulties that prevented widespread adoption of previous distributed object systems, in part by securing agreement of many parties that typically disagree in fundamental ways. We will start with a consideration of some foundational notions for distributed object systems, survey where we have been with component systems, and then look at web services, some relevant standards, and some patterns of their use.

10.1 Distributed Object Systems

Before we discuss notions relevant to *distributed object systems*, whereby we do not constrain ourselves to interactions on one physical computer, we should review some of the basics, namely, objects, interfaces, and component architectures.

10.1.1 Objects and Interfaces

The notion of an object is really quite simple, namely, an *object* is data plus functions that operate on data [5]. Objects are concrete instances of abstract data types called *classes*. The *data members* within an object are also referred to as *attributes*; similarly, an object also has *member functions*.

A class may have more than one *interface*, which is a set of functions that interact with a family of objects. This particular family of objects that responds to the interface *implements* the interface. As we will see, the interface is particularly important in understanding web services.

Objects are particularly powerful as they provide mechanisms for *inheritance* by providing the ability to *extend* an object class by adding new attributes or member functions to the original object class. This allows the extended class to access the original objects; indeed, the extended class can implement the original interfaces provided by the original objects. Also, there is a large degree of flexibility in this inheritance mechanism to allow parent classes to be overridden or modified by the extended class.

For example, from our own workflow engine's documentation [6], we can look at the inheritance patterns of one class, UriList (one can think of UriList as a directory listing class). In Figure 10.1, we can see the relationship of UriList to other objects, starting at java.lang.Object. Also note that this particular class explicitly implements an interface, "Stateful," which is implemented in a variety of other classes as shown in Figure 10.2. In Figure 10.3, we show that by extending a class ErrorHandlerTask, we can benefit across the board by using the functions from a generic Error Handler in a large number of classes. This is one of the benefits of an object approach. Finally, in Figure 10.4, we illustrate the attributes and functions that are part of UriList.

10.1.2 Component Architectures

In order to take advantage of objects not physically located on the same resource (note that in our discussion earlier we did not make any provisions for object distribution), it is useful to introduce the notion of a *component architecture*. This is an almost deceptively simple notion, that is, a component architecture is a framework for designing and using components, and, they are object class instances whose characteristics are defined by the underlying architecture [5]. One additional feature of a component architecture is that they typically also have a *container*, which is an application that integrates a set of components. Note that this also does not specify "distributed objects" explicitly; however, by providing a mechanism for integrating objects, component architectures have been pressed into service to allow distributed functionality in many circumstances.

Component architectures typically fall into one of the two different programming models. The first such model, *client–server*, has the components to act as the server. A client in this scheme is either a *container of servers* or *proxies (interface)* to the servers. This programming model was dominant in Microsoft's COM+ and CORBA, which we will

ncsa.ogre.core.ant.taskdefs.uri

Class UriList

```
java.lang.Object
  |
  +--org.apache.tools.ant.ProjectComponent
        |
        +--org.apache.tools.ant.Task
              |
              +--ncsa.ogre.core.ant.taskdefs.ErrorHandlerTask
                    |
                    +--ncsa.ogre.core.ant.taskdefs.uri.UriList
```

All Implemented Interfaces:
Stateful

---

public class **UriList**
extends ErrorHandlerTask
implements Stateful

Generic task for listing. The actual operation (local or remote files, resources, methods etc.) is based on the uri schema [see documentation to HandlerMap].

Single directory or patterns can be given using the appropriate attributes, or multiple directories or patterns can be added. In either case, the underlying handler returns a ListResult object encapsulating the list(s) [see under ListResult].

FIGURE 10.1
Illustration of inheritance of classes for class UriList. (Credit: Albert L. Rossi, University of Illinois.)

ncsa.ogre.core.ant.interfaces.extended

Interface Stateful

All Known Implementing Classes:
CheckSysProps, DeserializeList, EAssign, EConfiguration, EDynamicElement, EStatement, FileMonitorStart, OgreEvent, OgreMatchClause, OgreMatchFilter, OgreMethod, Progress, Read, RTFilterQ, SourceGenerator, StreamMonitor, UriCopy, UriDelete, UriExists, UriList, UriMkDir, UriPattern, UriStageFrom, UriTargetGroup, UriTouch, Write

---

public interface **Stateful**

Defines a class which potentially accumulates runtime additions to a list of child elements that may also contain objects added at build time, or that otherwise is not or cannot be cleared after execution.

Author:
Albert L. Rossi

Method Summary	
int	replace(java.lang.String addName, java.lang.Object value, int index) Method which will set an indexed item in a list-container based on the name of the "add" method and index provided.

FIGURE 10.2
Relationship of interface "Stateful" to implementing classes, including UriList. (Credit: Albert L. Rossi, University of Illinois.)

cover in Sections 10.1.2.1 and 10.1.2.2. We also use it in our group's current development, using web services, in which we develop web service clients to remote servers.

A second model is that of the software integrated circuit. In this case, each component has a set of input and output ports, and the problem of composition of the components

ncsa.ogre.core.ant.taskdefs

Class ErrorHandlerTask

```
java.lang.Object
  |
  +--org.apache.tools.ant.ProjectComponent
        |
        +--org.apache.tools.ant.Task
              |
              +--ncsa.ogre.core.ant.taskdefs.ErrorHandlerTask
```

Direct Known Subclasses:
AddLeadingZeros, Append, CheckSysProps, Deserialize, DeserializeList, EApply, EAssignType, EClone, ECondTask, EContainer, EDumpState, EEval, EExport, EInvoke, ENumericalCast, EnvToSysProps, ERemove, EReturn, ESleep, ESyncExec, ETry, EventQueueStart, EVersion, FileMonitorStart, FindWorkingDir, GetSysProp, GetSysProps, InfixToRPN, MarshalXML, ProxyFetch, Publish, PullEvents, Read, RTExec, Serialize SerializeList, SetSysProp, SourceGenerator, Stop, Tokenize, Uname, UniqueId, UnmarshalXML, UriCopy, UriDelete, UriExists, UriList, UriMkDir, UriStageFrom, UriTouch, WaitForEvent, Write

---

public abstract class **ErrorHandlerTask**
extends org.apache.tools.ant.Task

Wraps task execution in a "try { ... } catch (Throwable)" statement. Options allow various ways of handling the caught throwable.

All OGRE-dependent tasks extend this class rather than Task (exceptions: Print, SvnVersion).

FIGURE 10.3
Leveraging capabilities of one class (ErrorHandlerTask) by many other classes. (Credit: Albert L. Rossi, University of Illinois.)

Constructor Summary
UriList()

Method Summary	
void	addConfiguration(EConfiguration c)
void	addMethodPattern(UriMethodPattern p)
void	addPattern(UriPattern p)
ListResult	getResult()
java.util.List	getSimpleNext()
java.util.List	getSimpleResult()
int	replace(java.lang.String addName, java.lang.Object value, int index) Implementation of Stateful interface method.
void	setDir(java.lang.String path)
void	setFullScan(boolean b)

FIGURE 10.4
Attributes and functions that are a part of UriList. For instance, the attribute setDir will establish the directory one would like to list, and setFullScan is used to indicate whether this is a full (recursive) listing, or a flat listing. (Credit: Albert L. Rossi, University of Illinois.)

into the application needed is simply a matter of connecting the right "type" of ports, whereby type is defined by the port's interface, which advertises the message that the port is capable of addressing. There are plenty examples of software integrated circuits, and are usually depicted as "dataflow" applications. One such example, AVS (Advanced

Visual Systems) [7,8] provides a large set of components that can be harnessed to read and visualize data. In AVS, a user selects components, and interconnects them visually by using ports of the same type to build the application. Once the application is complete, it can be reused by the user to solve the same or similar visualization problems. Another such example is D2K — Data to Knowledge [9], which is a rich dataflow data analysis and mining environment. D2K provides a graphical composition tool, intended as a rapid application development environment, to provide the data application developer the mechanisms to compose D2K components into a D2K application. Once the application is developed, D2K provides many mechanisms to provide simple interfaces to users (e.g., D2K Streamline), and also allows exposing the application developed to other software environments through a web service interface. A third example is another visualization tool, SCIRun [10], developed at the University of Utah. In addition to many advanced visualization capabilities, all composable through a graphical editor, SCIRun allows the incorporation of advanced modeling as software components. Note that in order to incorporate such scientific computing into SCIRun, a developer must adhere to the programming standards in SCIRun. Depending on the application, this may be easy or impossible to achieve.

Note that common to component architectures are the notions of controls and events. For controls, a particular component needs to implement a control interface that allows the component container to watch over component properties and state. Additionally, components may need to send messages to one another. Besides direct connection of ports (as in a software integrated circuit), a container should support an event system that allows any component to broadcast a message, which can then be picked up by other components listening for messages of a particular type. In the distributed environment, this is a particularly important functionality.

Considering the features that a component architecture brings to an object-oriented system, it appears that we have the foundational pieces necessary to build a distributed object system. In Sections 10.1.2.1, we will discuss three such attempts to build distributed object systems, the Common Object Request Broker Architecture (better known as CORBA), Microsoft's COM+, and Java Remote Method Invocation (RMI).

10.1.2.1 CORBA

CORBA is an open, vendor independent, language neutral specification, and related infrastructure that allows multiple CORBA-based applications to interoperate [11]. Applications in CORBA are all objects, as discussed previously. However, CORBA mandates that one separates the interface for the object from the implementation. The interface is described through the Object Management Group (OMG) Interface Definition Language (IDL) [12]. The IDL is the only mechanism other CORBA applications have to access functionality in another application. Note that IDL is language independent, but has mappings to many popular languages such as C++ and Java. Also, IDL requires typing of each CORBA object, and the variables used in the IDL. This works to allow interoperability between the CORBA applications. Practically speaking, after generating IDL, a CORBA developer would compile the IDL into client stubs and object skeletons, which are used in the clients and servers (respectively) to act as proxies to each other. Additionally, the stubs and skeletons work with the Object Request Broker, or ORB, to mediate client operations on object instances to make it appear that the client is directly operating on the object. Additionally, by virtue of every object instance having a unique object reference, a client is able to find the instance that they require. Finally, note that this is a binary standard for object interoperability.

CORBA has been pressed into service in many circumstances. One such effort that has both academic roots and commercial applications is Arjuna transaction system developed by the Distributed Systems Group from the University of Newcastle upon Tyne [13]. This transaction system, implemented in C++ and Java, has been used in a commercial product, and is the foundation for a transactional workflow system.

10.1.2.2 COM+

Microsoft's component architecture, COM+, is an extension from Microsoft's Component Object Model (COM) [14] and Distributed Component Object Model (DCOM) [15] (COM is the core component architecture that is the foundation of Microsoft Windows). The original COM specification allowed for binary function calling between components, strongly-typed groupings of functions into interfaces, a base interface providing dynamic discovery of interfaces from other components and component lifetime management, unique identification provision for components, and a component loader (i.e., container) for establishing and managing component interactions, both cross-process and cross-network. Additionally, COM provided a language-independent component architecture, indeed, one could easily program components in languages as different as C++ and Visual Basic.

A DCOM added a simple twist to COM by allowing the client-component interaction to (managed by the COM runtime in COM) be distributed on a network and by allowing the normal COM runtime connection to be replaced by a DCOM wire-protocol standard, managed by the DCOM runtime. This allows the creation of distributed applications with the same ease as the development of applications on the same system. With this ability to distribute components, in addition to being able to gain remote functionality, a DCOM system could increase its scalability by distributing its workload among many hosts. Indeed, one project at the National Center for Supercomputing Applications (NCSA), NCSA Symera [16] worked to create a virtual supercomputer out of a network of Windows NT systems using DCOM. With this implementation of a high-throughput system much like Condor [17], users with normal COM applications could easily distribute their work among a set of participating Symera hosts, all using DCOM to manage the work.

COM+ [18] aimed to simplify windows component programming by removing much of the code that was necessary to be present in every DCOM component in order to use the system. The critical enhancement for COM+ was that the container now provided all the implementation support that a developer had to provide previously. Essentially, with COM+, a developer could focus on just the component metadata and the methods behind the component interface; the other things such as a class factory, register, query interface, etc., were provided by the framework. Additionally, the interface definition files are now automatically generated by the COM+ runtime, which then allows automatic generation of proxies and stubs for client–server operation.

Though the Microsoft COM architecture evolved in a fashion that improved greatly over time, it was superceded by a web service architecture, .NET. We will cover more on web services in Section 10.2.

10.1.2.3 Java RMI

Sun Microsystem's Java RMI [19] system is different compared with the previous examples of CORBA and COM+. Java RMI leverages Java's object-oriented nature and portability to allow the distribution of objects between clients and servers. Though not language neutral (there is a provision for using Java Native Interface in Java RMI, to connect to non-Java languages), Java RMI exploits the properties of Java to ease the burden of distributing one's computing. For instance, since one can pass full objects across the wire from client to server and vice versa (rather than just the data as arguments, with interface-based approaches), it is possible to take advantage of an object's multiple implementations. Additionally, servers can be quite generic, and take on additional functionality as new objects are delivered to them to be executed. Java RMI also makes load balancing easier, as one can have as many backend resources to deliver the capability to carry out a set of tasks. Finally, updating code and policy to any endpoint is both dynamic and simple. Though Java RMI has many advantages, it has one drawback limiting it: language support besides Java is limited to the Java Native Interface; this substantially limits the flexibility inherent in Java RMI.

10.1.2.4 Common Component Architecture

The Common Component Architecture (CCA) [20] was launched by the U.S. Department of Energy to promote code reuse and component standards to allow complex scientific applications to be composed of reusable software components. The CCA defines a number of requirements to be compliant with the CCA standards. First, each component describes its inputs and outputs using a Scientific IDL (SIDL), and stores the definitions to a repository using the CCA Repository API. The SIDL definitions are used to generate the component stubs; also, components can use services provided by a CCA framework (container). These services include things such as communication, security, and memory management. The component stubs are an element of the CCA standard called Gports, which encompass all component interactions within a framework. Gports have one part that is generic to the framework (and component independent) — (e.g., adding functionality to a component), as well as code implementing the component stub — which by virtue of using standard CCA services, is framework independent.

In some of our work, we used an implementation of a CCA from Indiana University called eXtreme Cluster Administration Toolkit (XCAT) [21]. XCAT was designed to handle applications distributed on a computational grid rather than tightly coupled high performance computing applications. As such, XCAT used an XML-based event system [22], and SOAP (Simple Object Access Protocol) as its communication protocol for remote procedure calls on the ports provided by each component. In the particular collaboration we had with the National Computational Science Alliance, a group of chemical engineers from the University of Illinois and computer scientists from Indiana University worked to solve the problem of electrodeposition of copper into a trench on a silicon wafer. Since the feature size of the trench is approaching a size where continuum codes are having difficulties handling the physics of the deposition, the approach taken was to combine a kinetic Monte Carlo simulation at the trench scale, linked to a finite difference code in the bulk phase (see Figure 10.5) [23,24].

Schematically, we constructed application managers (or proxies) which managed the details of launching, running, and coordinating the two codes, as depicted in Figure 10.6. Note that these were scriptable application managers; we could upload a particular script (in this case cast in jython) that would manage the details peculiar to each application. We would send events to an event channel to signal the other application when a data file was ready to be moved and read in for processing. Note that this was in many ways a proof of concept; this particular implementation, though capable of helping to produce a new science, was not built in a way to work easily with the constraints of production compute resources with their batch queues, startup delays, and so on. Also, it is important to note that this project used grid mechanisms (see Chapter 12) to launch each component. One particularly troublesome aspect of this was that once the components were launched by the portal server (see Figure 10.6), the framework expected to hear back from the components within a certain period of time, or the component would be considered dead, and the computation would fail. This did not work well in the light of our managed resources that computations such as this would typically use, and would provide motivation and requirements for our OGRE workflow system, see Chapter 11.

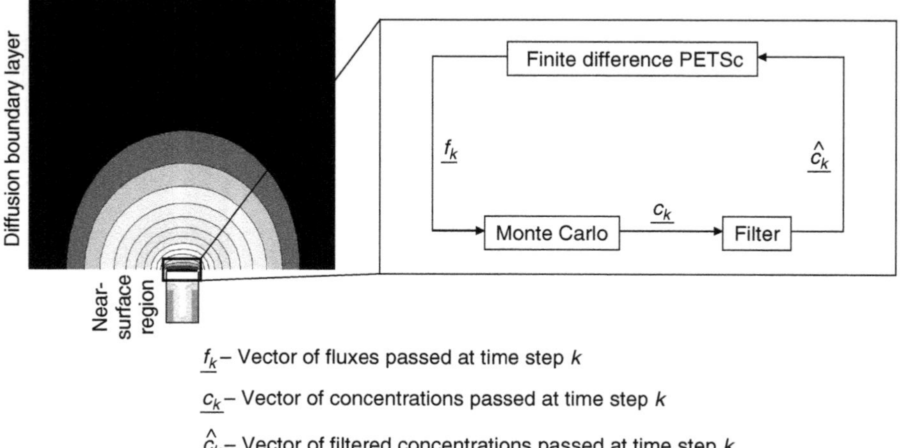

FIGURE 10.5
Physical problem and code layout to solve electrodeposition. Note that the Monte Carlo code provides concentrations to the finite difference code; while the finite difference code provides material fluxes to the Monte Carlo code. (Credit: Richard C. Alkire, University of Illinois.)

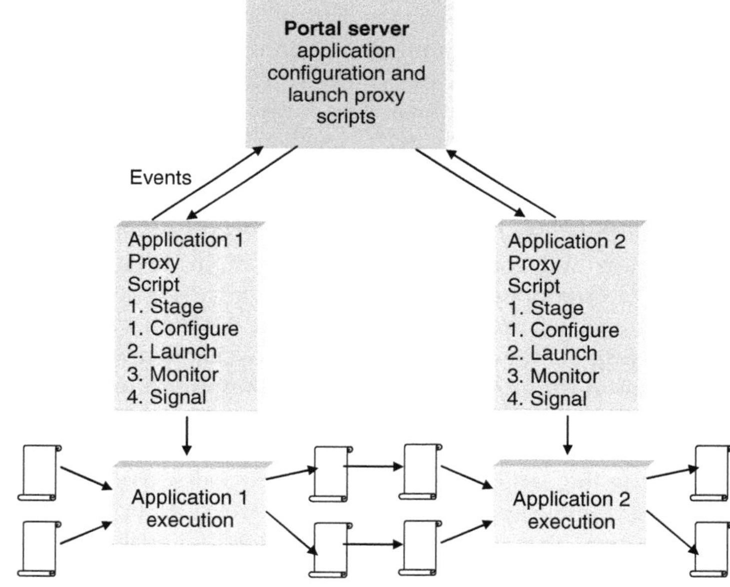

FIGURE 10.6
Configuration of XCAT components to manage execution of a multiscale chemical engineering example. In this case, Application 1 and 2 correspond to the Monte Carlo and finite difference codes. (Credit: Dennis Gannon, Indiana University.)

10.2 Web Services

After considering many of the prevalent component architecture standards in Section 1.1, we will now consider the development of web services. In Section 10.1, one could readily see that there were a multitude of noninteroperating component architecture standards. Although the approaches were technically sound, the lack of interoperability was proving to be a real detriment, especially for some of the business to business applications that distributed computing enabled in the business sector.

Arriving to a solution focused on interoperability, it appeared that separation of the implementation of a component from its interface was critical. A number of information technology companies (IBM, Microsoft, HP, Oracle, etc.) decided to develop a new standard for component interactions — coined web services. A web service focuses on providing an interface to a service for remote clients, to allow you to compose services into the application necessary for the task.

The World Wide Web Consortium (W3C) defines a web services in the following precise way [25]:

> A Web service is a software system designed to support interoperable machine-to-machine interaction over a network. It has an interface described in a machine-processable format (specifically WSDL). Other systems interact with the Web service in a manner prescribed by its description using SOAP messages, typically conveyed using HTTP with an XML serialization in conjunction with other Web-related standards.

Note the emphasis on "machine-to-machine" interaction, web services are not about rendering web pages for users, but rather to enable one code to work with another. Also, the interface specification, WSDL (for Web Service Definition Language), is particularly important. In essence, WSDL represents an interface definition language for web services.

10.2.1 Web Service Standards

WSDL is an XML language used to describe the interface for a particular service [26]. With WSDL, messages going between the client and the server are described abstractly, which then are bound to a network protocol and message format. WSDL is based on the XML [27]. XML describes a class of data objects referred to as XML documents that are a subset of SGML, the Standard Generalized Markup Language [28]. XML is comprised of character data and markup, with the markup describing the documents storage layout and logical structure. XML is described in more depth in Chapter 2. Web services typically use SOAP [29] as a framework for exchanging XML messages. In the context of web services, a SOAP message contains what is necessary to invoke a service or return of the results of such an invocation. Although SOAP can be bound to many different protocols, it is typically bound to http.

As an example, consider our event channel with the relatively simple programming interface [30]. At face value, the channel has a channel implementation class, with three methods — getDatabase — for handling the connection to the database providing persistence to the channel, handleBatch, which handles sending events to the database, and requestEvents, which is responsible for the retrieval of the events. Rendered into WSDL, we have the object described as shown in Figure 10.7. Note that the internal method, getDatabase, which is not relevant to external clients is not described. But, for each of the handleBatch and requestEvents both the request and response messages are described. In the case of our event channel, this WSDL description of the interface is automatically turned into a Java representation by the tooling included with our web service container, Apache Axis [31].

On the client side, to use the service, we have a simple interface, channel, with two methods, handleBatch and requestEvents (see Figure 10.8). Note that this provides a bit of

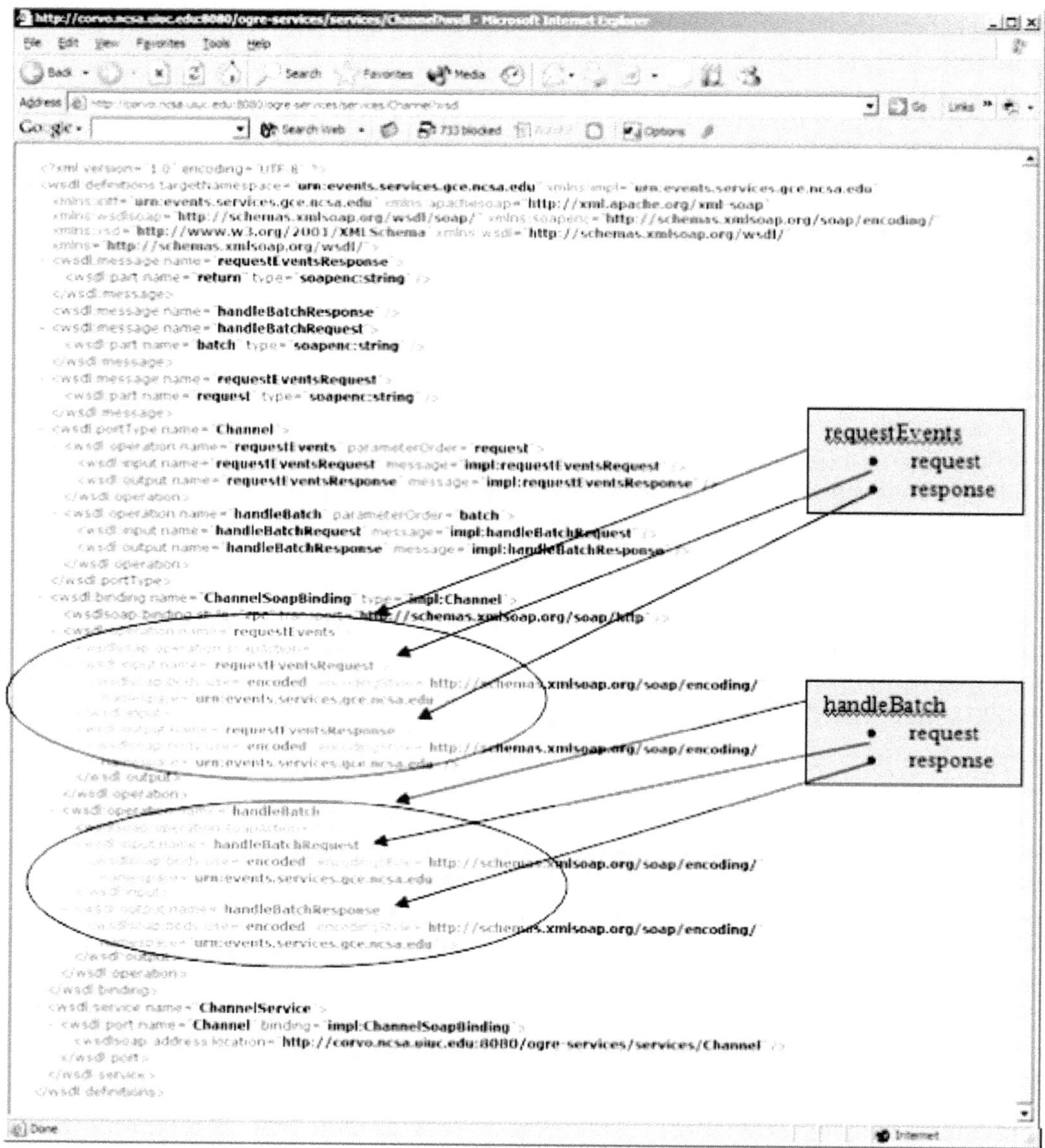

FIGURE 10.7
Rendering of the channelImpl API into WSDL. Note that only the relevant methods for external consumption (i.e., requesting events from the channel and pushing events into the channel) are shown in this public API. (Credit: Albert L. Rossi, University of Illinois.)

transparency as to whether we are using a remote or local instance of the service, as far as the web service client developer is concerned, the two look substantially same.

Although web services were developed with the notion of distributed interoperability, such interoperability is not guaranteed. One group of industry players decided to form a consortium to address the *minimum set of interoperability requirements* across a core set of standards, namely, messaging (SOAP), service description (WSDL), service publication and discovery, and security [32,33]. Note that this profile goes into some depth on areas of web services which have a good deal of implementation

ncsa.ogre.events.xevents.interfaces

Interface Channel

All Superinterfaces:
java.rmi.Remote

All Known Implementing Classes:
ChannelSoapBindingStub

---

public interface **Channel**
extends java.rmi.Remote

Modifies interface originally belonging to the XMessages package designed by Extreme!Laboratory of Indiana University. It has been fitted for use with apache Axis's java2wsdl.

Changes from original: since the Axis wsdl engine does not handle mapping multiple inheritance in interfaces, the relationship between this interface and the Source/Sink interfaces has changed. While their methods are still identical, the web service does not directly implement Source and Sink; it is understood that the remote calls on that service will delegate directly to a Channel object. In our current version, the latter composes Source and Sink objects to which it, in turn, delegates.

Author:
Albert L. Rossi, Aleksander Slominski

Method Summary	
void	handleBatch(java.lang.String batch)
java.lang.String	requestEvents(java.lang.String request)

FIGURE 10.8
Client side interface to access service described by WSDL document in Figure 10.7. (Credit: Albert L. Rossi, University of Illinois.)

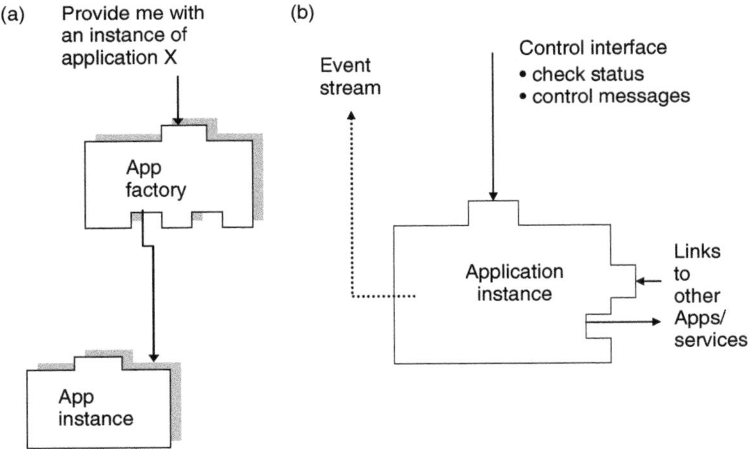

FIGURE 10.9
(a) Schematic of an application factory service creating an instance of the application service. (b) Application instance, with event connections, and connections to other services as well as control port for client-side manipulation of the application service. (Credit: Dennis Gannon, Indiana University.)

experience (messaging and service description) but has minimal requirements for discovery and security. Even with this standard effort ongoing, it is safe to assume that much further development work is needed to have truly interoperable web services.

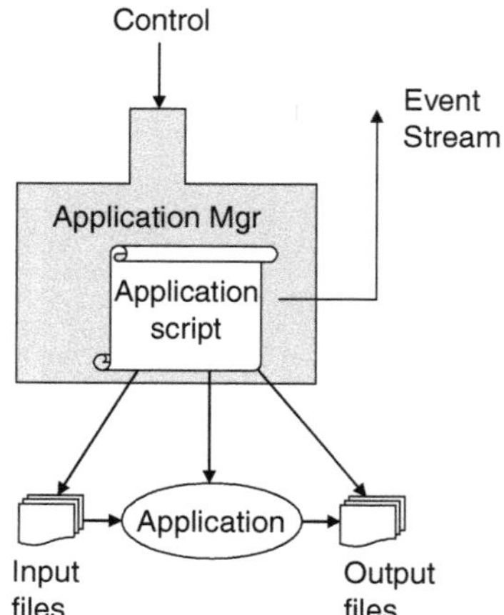

FIGURE 10.10
Scriptable application instance service (application manager) encapsulating legacy application with normal input and output files. (Credit: Dennis Gannon, Indiana University.)

10.2.2 Application Factory

One important pattern of web service behavior is that of an application factory. An application factory is a service that is capable of starting a running instance of an application component. With a factory, the client service contacts the factory and provides the right parameters to pass to the service instance created by the factory. The factory then checks your credentials to ensure you have the right to create the instance of the application, launches the application, and returns a reference to the instance of the application. Note that in this fashion, the running instance is also a service (see Figure 10.9[a]), and one can provide elements of scalability for web services [23].

Typically, application instances connect to an event channel, to handle messages from the application (Figure 10.9[b]). This instance often has a control port, so that clients such as desktop applications or web portals can interact with the application, and may connect to other services such as metadata services.

One may consider wrapping legacy applications (such as FORTRAN Codes) with a web service wrapper. Often, these wrappers are scriptable, and the script manages many of the aspects important to the application code, such as staging files, launching the application, publish event streams etc., (Figure 10.10).

References

1. MODIS home page, http://modis.gsfc.nasa.gov/.
2. CUAHSI hydrologic information system, http://cuahsi.sdsc.edu/HIS/.
3. National operational hydrologic remote sensing center, http://www.nohrsc.nws.gov/.
4. New developments in web services and e-commerce, *IBM Systems Journal*, 2002, http://www.research.ibm.com/journal/sj41-2.html.
5. Object based approaches, Dennis Gannon and Andrew Grimshaw. in Ian Foster and Carl Kesselman, Eds., *The Grid: Blueprint for a New Computing Infrastructure*, chap. 9.
6. *O.G.R.E. User Manual*, http://corvo.ncsa.uiuc.edu/ogre/.
7. http://www.avs.com/index_wf.html.
8. *The AVS Workbook*, http://www.ecu.edu/rave/Visualization/AVSworkbook/TOC.html.

9. *ALG: D2K Overview*, http://alg.ncsa.uiuc.edu/do/tools/d2k.
10. SCIRun — SCI Software — Scientific Computing and Imaging Institute, http://software.sci.utah.edu/scirun.html.
11. Corba FAQ, http://www.omg.org/gettingstarted/corbafaq.htm.
12. OMG IDL, http://www.omg.org/gettingstarted/omg_idl.htm.
13. Distributed systems research group, http://arjuna.ncl.ac.uk/.
14. The component object model: a technical overview, http://msdn.microsoft.com/library/default.asp?url=/library/en-us/dncomg/html/msdn_comppr.asp.
15. DCOM technical overview, http://msdn.microsoft.com/library/default.asp?url=/library/en-us/dncomg/html/msdn_comppr.asp.
16. Access science and technology — NCSA symera: a supercomputer for windows NT, http://access.ncsa.uiuc.edu/Stories/Symbio/Symbio.html.
17. Condor project home page, http://www.cs.wisc.edu/condor/.
18. Object-oriented software development made simple with COM+ runtime services — MSJ, November 1997, http://www.microsoft.com/msj/1197/complus.aspx.
19. Java remote method invocation, http://java.sun.com/products/jdk/rmi/reference/whitepapers/javarmi.html.
20. CCA forum home, http://www.cca-forum.org/index.shtml
21. The XCAT project, http://www.extreme.indiana.edu/xcat/index.html.
22. Grid web services: XMessages/XEvents service, http://www.extreme.indiana.edu/xgws/xmessages/index.html.
23. Programming the grid: distributed software components, P2P and grid web services for scientific applications, in Dennis Gannon et al., *Special Issue of Journal of Cluster Computing*, July 2002, http://www.extreme.indiana.edu/xcat/publications/ProgGrids.pdf.
24. T.O. Drews, E.G. Webb, D.L. Ma, J. Alameda, R.D. Braatz, and R.C. Alkire, Coupled mesoscale-continuum simulations of copper electrodeposition in a trench, *AIChE Journal*, 2004 (in press).
25. Web services architecture, http://www.w3.org/TR/ws-arch/.
26. Web services description language (WSDL) version 2.0 Part 1: core language, http://www.w3.org/TR/2003/WD-wsdl20-20031110/.
27. Extensible markup language (XML) 1.0, 2nd ed., http://www.w3.org/TR/2000/REC-xml-20001006.
28. ISO (International Organization for Standardization). ISO 8879:1986(E). Information processing — text and office systems — standard generalized markup language (SGML). 1st ed. — 1986-10-15. [Geneva]: International Organization for Standardization, 1986.
29. SOAP Version 1.2 Part 1: messaging framework, http://www.w3.org/TR/2003/REC-soap12-part1-20030624/.
30. *O.G.R.E. User Manual*, http://corvo.ncsa.uiuc.edu/ogre-services/.
31. Web services — Axis, http://ws.apache.org/axis/.
32. WS-I > > > WELCOME, http://www.ws-i.org/home.aspx.
33. Basic profile — version 1.0 (Final), http://www.ws-i.org/Profiles/BasicProfile-1.0-2004-04-16.html.
34. UDDI.org, http://www.uddi.org/.

11

Extensible Markup Language

Jay C. Alameda

Contents

In a relatively short period of time, the eXtensible Markup Language (XML) has become a tremendously useful and ubiquitous means of describing data — far outstripping the use of its predecessor, Standard Generalized Markup Language (SGML) [1]. XML has found applications for things as diverse as data description (including use in metadata systems), distributed object descriptions, workflow descriptions as well as message descriptions. In this chapter, we will focus on the basics of data description in XML, including some examples from hydrology, and then look at some specific mechanisms to describe tasks in XML.

11.1 Data Descriptions

Before we seriously consider how to describe data using XML, we should discuss some of the XML fundamentals. First, what exactly is XML? XML, which is short for eXtensible

```
<?xml version="1.0" encoding="UTF-8"?>
<host-list>
    <rm-contacts persistent="false" valid="true">
        <service>jobmanager</service>
        <hostname>cu.ncsa.uiuc.edu</hostname>
        <port></port>
    </rm-contacts>
    <rm-contacts persistent="false" valid="true">
        <service>jobmanager</service>
        <hostname>tg-login.sdsc.teragrid.org</hostname>
        <port></port>
    </rm-contacts>
    <rm-contacts persistent="false" valid="true">
        <service>jobmanager-loadleveler</service>
        <hostname>cu.ncsa.uiuc.edu</hostname>
        <port></port>
    </rm-contacts>
    <rm-contacts persistent="false" valid="true">
        <service>jobmanager-pbs</service>
        <hostname>tg-login.sdsc.teragrid.org</hostname>
        <port></port>
    </rm-contacts>
</host-list>
```

FIGURE 11.1
Sample XML, describing job manager services for remote job submission for prototype job management tool. (Credit: Shawn Hampton, University of Illinois.)

Markup Language, is a document and metadata language proposed by the World Wide Web Consortium (W3C) [2]. As a markup language, XML is easy to use and flexible in the content it can encode. In contrast with HTML, XML is concerned principally with content, not the appearance of the material encoded with XML.

11.1.1 Elements of XML

XML is comprised of a nested sequence of basic elements, each denoted with tags. In Figure 11.1, let us consider a simple sample of XML, pulled from a distributed computing job-launching client, that is used to describe a series of "job managers" that are responsible for accepting computational jobs to be started on compute hosts (we will discuss more regarding job submission for distributed computing in Chapter 12).

Note that we start the XML description with an optional declaration, namely:

```
<?xml version="1.0" encoding="UTF-8" ?>
```

Next, consider the notion of *XML elements*. XML is comprised of a nested sequence of such basic elements. An XML element is denoted by opening tag and closing tag, with intervening content. A simple element in this example is that of <service>, namely:

```
<service>jobmanager</service>
```

In this case, the elements are denoted by the <service> tags, and the content for <service> is jobmanager. It is also possible to enclose no content, for example:

```
<port></port>
```

Note that this can alternatively be more simple with the following:

```
<port />
```

Both representations are functionally equivalent.

XML tags are capable of having attributes, for example, note the attributes we assigned to the <rm-contacts> tag:

```
<rm-contacts persistent="false" valid="true">
```

In this case, persistent and valid are both considered to be attributes of <rm-contacts>. Also note that rm-contacts contains more elements:

```
<rm-contacts persistent="false" valid="true">
    <service>jobmanager</service>
    <hostname>tg-login.sdsc.teragrid.org</hostname>
    <port></port>
</rm-contacts>
```

Finally, observe that the entire XML fragment is contained within the element <host-list>. One thing that is important to realize about XML is that while we may like to manually construct XML to be pleasing to our eyes (with indents, etc.), an equivalent construction is to consider this as an ordered sequence. For example, for the rm-contacts fragment mentioned earlier, one could alternatively represent the XML as the following:

```
<rm-contacts persistent="false"
valid="true"><service>jobmanager</service><hostname>tg-
login.sdsc.teragrid.org</hostname><port></port></rm-
contacts>
```

While we may not enjoy reading such XML fragments, this is one way codes generate XML automatically, as well as how they parse the XML. Both representations function exactly the same way.

We could add comments to our XML file by enclosing the comments with <! Comment hiding here >. For instance, we may want to describe one of the rm-contacts in our example as:

```
<!-- copper uses the loadleveler system for batch submission -->
```

Finally, XML allows arbitrary content addition through the CDATA construct, namely:

```
<![CDATA [arbitrary data for example x=2]]>
```

Note that CDATA hides everything from the parser, so you can hide special characters such as ">" or "& " within a CDATA element to prevent the parser from operating on the characters. XML also provides for the capability to represent special characters such as the following:

```
&amp;   for & use &amp;
&lt;    for < we use &lt;
&gt;    for > we use &gt;
```

and so on. These simple constructs allow the development of rich description paradigms with a large range of applications. Given that, how do we attempt to not degenerate into a large series of noninteroperating descriptions? One tool we can use for this is the provision for XML schemas, discussed in Section 11.1.2.

11.1.2 XML Schema

XML schemas are the foundation for a semantic description for classes of documents [3]. In a schema, we define the *types* and *elements* in a family of documents. An XML Schema Definition (XSD) is defined by the standard XSD namespace. Of course, that is a foundational namespace; we can have one or more XML schema *instance* namespaces, which builds from XSD as a base-level set of definitions. Note that XML schema succeeds the original Document Type Definition (DTD), and has one sizeable advantage over DTD for describing a schema in that a XML schema is comprised of well-formed XML [4].

Let us consider constructing a schema to describe job-submission characteristics of a series of computing hosts as depicted in our example XML in Figure 11.1. We start by constructing a <xs-schema> tag with two attributes, one denoting our own namespace (this is the xmlns attribute, short for XML namespace), and the other indicating that our definitions are built upon the foundational schema in the XS (sometimes referred to as XSD) namespace put forth by the W3C, this through the xmnls:xs attribute. A content-free representation of such a namespace (in this case, "hostlist") declaration can be indicated by the following XML fragment:

```
<xs:schema xmlns="http://gce.ncsa.edu/hostlist"
           xmlns:xs="http://www.w3.org/2001/XMLSchema">
</xs:schema>
```

Note that the strings such as http://gce.ncsa.uiuc.edu/hostlist are not Uniform Resource Locators (URLs) in that they do not describe a physical place, but rather is a generalization of a URL called a Uniform Resource Indicator (URI) [5]. URIs are used to indicate a logical space, in order to be able to address objects.

In our example, we would want to enumerate our elements in the schema. There are many types of elements defined in the W3C Schema [6], also known as the XML schema for XML schemas. A few that we will need to know about for our example are the complexType, sequence, element, and attribute. We will consider one data type for our example, which is relatively simple and generic as it is a string. Considerable types are defined in the XML schema (and documented in [4], but we will not need them for this simple example. Also note that we are building some complexTypes out of simple elements for our example.

Let us start within the <rm-contacts> construction in our simple example, repeated here for one instance:

```
<rm-contacts persistent="false" valid="true">
    <service>jobmanager</service>
```

```
    <hostname>tg-login.sdsc.teragrid.org</hostname>
    <port></port>
</rm-contacts>
```

The simplest place to start is on the inner elements, namely <service>, <hostname>, and <port>. In this case, we could use a simple schema description to handle each one, for example:

```
<xs:element name="service" type="xs:string"/>
<xs:element name="hostname" type="xs:string"/>
<xs:element name="port" type="xs:string"/>
```

Note that, recognizing that the element port denotes something of type integer, we could replace its typing with the following:

```
<xs:element name="port" type="xs:integer"/>
```

We will use the string type for simplicity in the rest of our schema example. A number of simple types and guidelines on constraints for attributes and elements follows in Figure 11.2.

Examining our sample XML again, we notice that the elements <service>, <hostname>, and <port> are part of a sequence of child elements to <rm-contacts>. Armed with this knowledge, we can denote that these are indeed a sequence in our schema:

```
<xs:sequence>
    <xs:element name="service" type="xs:string"/>
    <xs:element name="hostname" type="xs:string"/>
    <xs:element name="port" type="xs:string"/>
</xs:sequence>
```

Looking back at this example, we observe that this particular sequence defines the type for the element <rm-contacts>. Furthermore, we can have anywhere from no such elements to an infinite number of such elements. We could simply denote that with the following construct:

```
<xs:element name="rm-contacts" minOccurs="0" maxOccurs="unbounded">
</xs:element>
```

But, look more closely: <rm-contacts> has two attributes, persistent and valid. We can add these as follows:

```
<xs:element name="rm-contacts" minOccurs="0" maxOccurs="unbounded">
   <xs:attribute name="persistent" type="xs:string"/>
   <xs:attribute name="valid" type="xs:string"/>
</xs:element>
```

Combining the element definition for <rm-contacts> with its constituents, we end up with resulting complex type declaration:

```
<xs:element name="rm-contacts" minOccurs=0 maxOccurs="unbounded">
    <xs:complexType>
```

Element types:	
string	
byte	
unsignedByte	
integer	
positiveInteger	
negativeInteger	
nonNegativeInteger	
nonPositiveInteger	
int	
unsignedInt	
long	
unsignedLog	
short	
unsignedShort	
decimal	
float	
double	
boolean	
time	13:20:00.000
dateTime	2003-09-29T07:07:00.000
date	2003-09-29

Also, for a particular element or attribute, one can attach constraints to the particular element or attribute. These constraints are:

element:	
minOccurs	if zero, then this is an optional element, if one, then this is a required element
maxOccurs	value or unbounded, must be greater than or equal to minOccurs
fixed	set element to a particular constant value
default	default elements apply when elements are empty. An element cannot be both fixed and have a default value.

attribute:	
use	required, optional, or prohibited
fixed	set attribute to a particular constant value
default	default attributes apply when attributes are missing. Default requires a use of "optional"

Figure 11.2
A sampling of data types for elements in XML schemas; as well as constraints for elements and attributes [3].

```
        <xs:sequence>
              <xs:element name="service" type="xs:string"/>
              <xs:element name="hostname" type="xs:string"/>
              <xs:element name="port" type="xs:string"/>
        </xs:sequence>
        <xs:attribute name="persistent" type="xs:string"/>
        <xs:attribute name="valid" type="xs:string"/>
    </xs:complexType>
</xs:element>
```

We are almost finished with our schema. Our outer most element, <host-list>, is also a complex type comprising of a sequence. We can denote the entire ensemble (including the namespace declaration from earlier) as the following:

```
<xs:schema xmlns="http://gce.ncsa.edu/hostlist"
      xmlns:xs="http://www.w3.org/2001/XMLSchema">
   <xs:element name="host-list">
      <xs:complexType>
         <xs:sequence>
            <xs:element name="rm-contacts" minOccurs="0"
                  maxOccurs="unbounded">
               <xs:complexType>
                  <xs:sequence>
                     <xs:element name="service" type="xs:string"/>
                     <xs:element name="hostname" type="xs:string"/>
                     <xs:element name="port" type="xs:string"/>
                  </xs:sequence>
                  <xs:attribute name="persistent" type="xs:string"/>
                  <xs:attribute name="valid" type="xs:string"/>
               </xs:complexType>
            </xs:element>
         </xs:sequence>
      </xs:complexType>
   </xs:element>
</xs:schema>
```

Now that we have gone over some of the basics of XML schemas, we will consider another example; this particular example was developed to support a prototype of a chemical engineering experimentalists data repository. In Figure 11.3, we have a graphical representation of the schema, produced using a popular XML editing tool, XMLSpy [7].

The source XML for the graphical depiction is shown in Figure 11.4. Note that while this is considerably more complex than our introductory example, the majority of the element types are strings, and the main complexity is introduced in the hierarchy of the schema. We will use XML derived from this schema in Section 11.1.3 to illustrate some of the power of the XML transforms.

In close, then although one can easily generate arbitrary XML to describe one's data, an XML schema makes it easier to express what elements and attributes in the XML are supposed to mean, and what they are supposed to contain (through their type). It is also possible to build a hierarchy of complex types. Many tools such as castor and xerces [8] operate on and depend on schemas; XML schemas are also important for mapping data to data objects. A good exercise would be to discover what sorts of publically disclosed schemas are available that are relevant to your area of interest; finding, using, and extending such schemas, and using the schemas as guidelines for encoding your own data help ensure ease of data interoperability.

11.1.3 XML Transforms

The real power of XML lies in our ability to transform it into many useful forms. For instance, one may want to select a portion of XML, and wrap it in a new set of tags. Consider the following fragment of XML:

```
<?xml version="1.0"?>
<purchase id="p001">
    <customer db="cust123"/>
```

```
    <product db="prod345">
           <amount>23.45</amount>
    </product>
</purchase>
```

We will use the eXtensible Stylesheet Language (XSL) along with a transformation tool, XSLT, to perform the desired conversion [9,10]. Without getting into the details (yet) of the transformation, consider this small bit of XSL:

```
<xsl:template match ="customer">
    <buyer><xsl:value-of select="@db"/></buyer>
</xsl:template>
```

In this particular fragment of XSL, which is coded in XML, we use the <xsl: template match="customer"></xsl:template> semantics to indicate that we want to match on a particular element, in this case, the <customer> element. Once we have performed this match, the transform will select value of the db attribute of customer (using <xsl:value-of select="@db"/>), and wrap this value in the new XML element tags, <buyer></buyer>, that is, as

```
<buyer>cust123</buyer>
```

Now let us turn our attention back to the chemical engineering experimentalist's schema, introduced in Section 11.1.2 and displayed in Figure 11.3 and Figure 11.4. One instance of XML encoded using this schema would be the following:

```
<chemical-engineering-experiment id="1">
   <cell id="rotocell">
       <geometry>
          <electrode-diameter>1 micron</electrode-diameter>
```

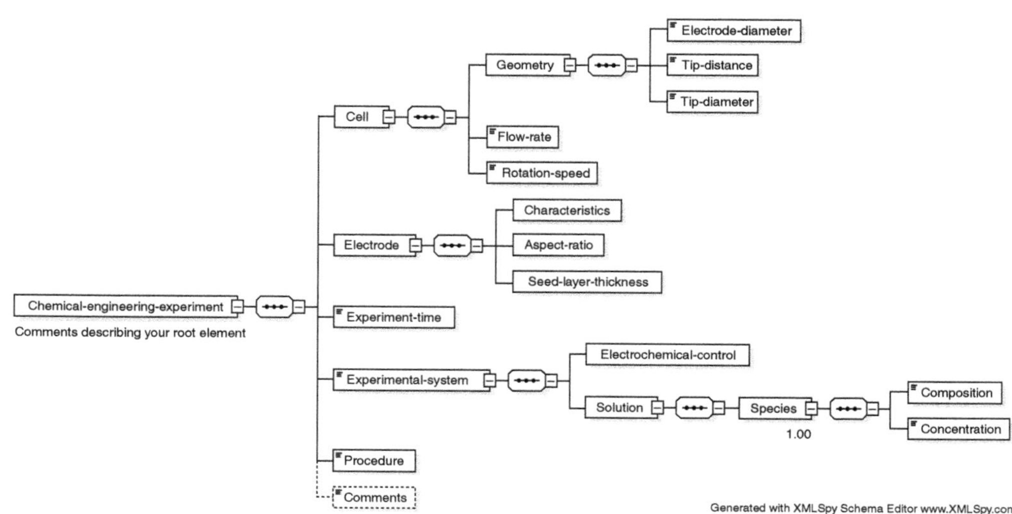

FIGURE 11.3
Schema for prototype data repository for chemical engineering experiments, as rendered by XMLSpy.

```
<?xml version="1.0" encoding="UTF-8"?>
<!-- edited with XML Spy v4.3 U (http://www.xmlspy.com) by Shawn Hampton
     (NCSA University of Illinois) -->
<xs:schema xmlns:xs="http://www.w3.org/2001/XMLSchema" elementFormDefault="qualified"
    attributeFormDefault="unqualified">
   <xs:element name="chemical-engineering-experiment">
     <xs:annotation>
        <xs:documentation>Comment describing your root element</xs:documentation>
     </xs:annotation>
     <xs:complexType>
        <xs:sequence>
           <xs:element name="cell">
              <xs:complexType>
                 <xs:sequence>
                    <xs:element name="geometry">
                       <xs:complexType>
                          <xs:sequence>
                             <xs:element name="electrode-diameter" type="xs:string"/>
                             <xs:element name="tip-distance" type="xs:string"/>
                             <xs:element name="tip-diameter" type="xs:string"/>
                          </xs:sequence>
                       </xs:complexType>
                    </xs:element>
                    <xs:element name="flow-rate" type="xs:string"/>
                    <xs:element name="rotation-speed" type="xs:string"/>
                 </xs:sequence>
              </xs:complexType>
           </xs:element>
           <xs:element name="electrode">
              <xs:complexType>
                 <xs:sequence>
                    <xs:element name="characteristics" type="xs:string"/>
                    <xs:element name="aspect-ratio" type="xs:string"/>
                    <xs:element name="seed-layer-thickness" type="xs:string"/>
                 </xs:sequence>
              </xs:complexType>
           </xs:element>
           <xs:element name="experiment-time" type="xs:string"/>
           <xs:element name="experimental-system">
              <xs:complexType>
                 <xs:sequence>
                    <xs:element name="electrochemical-control" type="xs:string"/>
                    <xs:element name="solution">
                       <xs:complexType>
                          <xs:sequence>
                             <xs:element name="species" maxOccurs="unbounded">
                                <xs:complexType>
                                   <xs:sequence>
                                      <xs:element name="composition" type="xs:string"/>
                                      <xs:element name="concentration" type="xs:string"/>
                                   </xs:sequence>
                                </xs:complexType>
                             </xs:element>
                          </xs:sequence>
                       </xs:complexType>
                    </xs:element>
                 </xs:sequence>
              </xs:complexType>
           </xs:element>
           <xs:element name="procedure" type="xs:string"/>
           <xs:element name="comments" type="xs:string" minOccurs="0"/>
        </xs:sequence>
     </xs:complexType>
   </xs:element>
</xs:schema>
```

FIGURE 11.4
Chemical engineering experimentalist's schema.

```
        <tip-distance>0.1 micron</tip-distance>
        <tip-diameter>0.01 micron</tip-diameter>
      </geometry>
      <flow-rate>1 ml/s</flow-rate>
      <rotation-speed>3 rpm</rotation-speed>
   </cell>
   <electrode>
      <characteristics>silver</characteristics>
      <aspect-ratio>10:1</aspect-ratio>
      <seed-layer-thickness>1 nm</seed-layer-thickness>
   </electrode>
   <experiment-time>3600 s</experiment-time>
   <experimental-system>
      <electrochemical-control>mercury</electrochemical-control>
      <solution>
         <species>
            <composition>CuSO4</composition>
            <concentration>1 M</concentration>
         </species>
      </solution>
   </experimental-system>
   <procedure>Doe 1998 Masters thesis</procedure>
   <comments>We do this while standing on our head. Get much
      better results</comments>
</chemical-engineering-experiment>
```

So, for a particular user, this XML captures many key details of the experiment. Suppose, now, in our example, that we want to use some of our data to talk to a manufacturer of electrochemical cells. However, this manufacturer uses a somewhat different XML schema, for instance:

```
<?xml version="1.0" encoding="UTF-8"?>
<!-- edited with XML Spy v4.3 U (http://www.xmlspy.com) by Shawn
Hampton (NCSA University of Illinois) -->
<xs:schema xmlns:xs="http://www.w3.org/2001/XMLSchema"
elementFormDefault="qualified" attributeFormDefault="unqualified">
  <xs:element name="cell-specification">
      <xs:annotation>
         <xs:documentation>this is used by customers to specify
            their custom electrochemical cell</xs:documentation>
      </xs:annotation>
      <xs:complexType>
         <xs:sequence>
            <xs:element name="echem-cell">
               <xs:complexType>
                  <xs:sequence>
                     <xs:element name="physical characteristics">
                        <xs:complexType>
                          <xs:sequence>
                            <xs:element name="electrode-diameter"
                               type="xs:string"/>
                            <xs:element name="gap"
```

```
                              type="xs:string"/>
                           <xs:element name="span"
                              type="xs:string"/>
                         </xs:sequence>
                       </xs:complexType>
                    </xs:element>
                    <xs:element name="volumerate"
                       type="xs:string"/>
                    <xs:element name="omega" type="xs:string"/>
                 </xs:sequence>
                 <xs:attribute name="id" type="xs:string"/>
              </xs:complexType>
           </xs:element>
           <xs:element name="electrodetype" type="xs:string"/>
        </xs:sequence>
                 <xs:attribute name="id" type="xs:string"/>
     </xs:complexType>
  </xs:element>
</xs:schema>
```

In addition, imagine that we want to place an order with this electrochemical cell manufactuer, with the order being presented to the manufacturer by the following XML fragment.

```
<cell-specification id="1">
   <echem-cell id="rotocell">
      <physical characteristics>
         <electrode-diameter>1 micron</electrode-diameter>
         <gap>0.1 micron</gap>
         <span>0.01 micron</span>
      </physical characteristics>
      <volumerate>1 ml/s</volumerate>
      <omega>3 rpm</omega>
   </echem-cell>
   <electrodetype>silver</electrodetype>
</cell-specification>
```

We will walk through the transformations necessary to handle this case. In this example, we will be using a built-in XSLT available in Apache ANT [11], using the following ANT build script to effect the changes (we will discuss ANT build scripts in Section 11.2.1):

```
<project name="Demo" default="runAll" basedir=".">

    <property name="wrk" value="${user.dir}/examples/tests"/>

  <target name="xslttest">
    <mkdir dir="${wrk}/xslout"/>
    <xslt in="experimentdesign.xml"
          out="${wrk}/xslout/cellbuilder.xml"
          style="expt.to.maker.xsl">
    </xslt>
  </target>
```

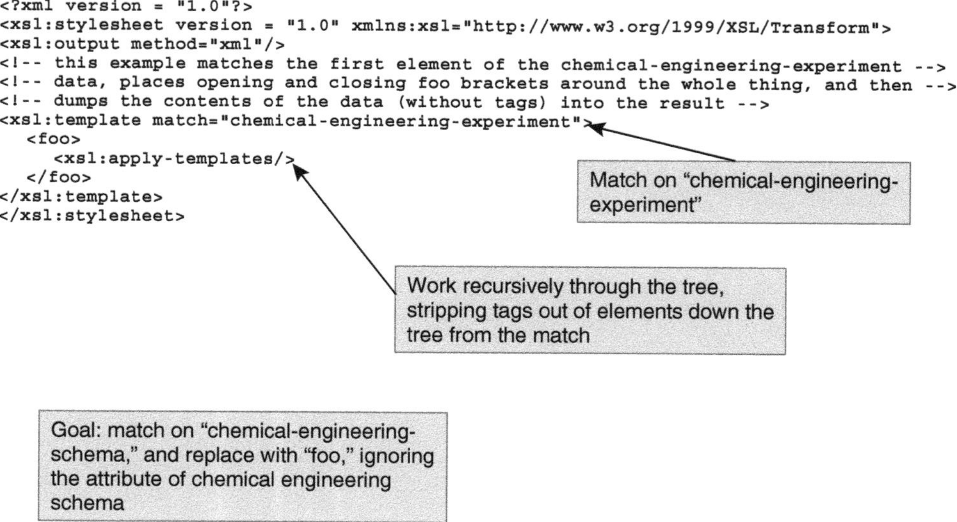

<?xml version = "1.0"?>
<xsl:stylesheet version = "1.0" xmlns:xsl="http://www.w3.org/1999/XSL/Transform">
<xsl:output method="xml"/>
<!-- this example matches the first element of the chemical-engineering-experiment -->
<!-- data, places opening and closing foo brackets around the whole thing, and then -->
<!-- dumps the contents of the data (without tags) into the result -->
<xsl:template match="chemical-engineering-experiment">
<foo>
<xsl:apply-templates/>
</foo>
</xsl:template>
</xsl:stylesheet>
Match on "chemical-engineering-experiment"
Work recursively through the tree, stripping tags out of elements down the tree from the match
Goal: match on "chemical-engineering-schema," and replace with "foo," ignoring the attribute of chemical engineering schema

FIGURE 11.5
Our first transformation, in which we match the outermost elements, and replace the element tags with alternate tags.

Now let us consider how to transform the data from its original schema to the new schema. First, consider the transformation depicted in Figure 11.5. For this first transformation, we will simply match on the <chemical-engineering-experiment> element in our source XML (by use of <xsl:template match="chemical-engineering-experiment">), replace this tag with <foo> and then recur through the rest of the document (using <xsl:apply-templates/>), stripping out all other tags, but leaving the contents of the elements. Finally, note that we are specifying that the output from this transformation is to be XML, by virtue of <xsl:output method="xml"/> in the transformation document. The result, shown in Figure 11.6, is an XML document with only the outermost tags preserved.

Note that we can do this transformation in an alternate fashion, where we select the root element rather than the specific <chemical-engineering-element>, as shown in Figure 11.7. This is preferred to matching on the specific outermost element.

Now that we have performed a very elementary transformation, let us try for some more complexity to get closer to our desired target XML. Consider the transform depicted in Figure 11.8.

With this transformation, we have replaced the outermost element, and then the second level element. Note that we apply the template twice, once for each match. The results of this transformation are shown in Figure 11.9.

Getting closer to our final goal, we will illustrate two more transformations. The next transform is simply a replacement of the third level of elements in the hierarchy, as shown in Figure 11.10.

Finally, in our last illustration of the transformation from the source XML to the desired transformed XML, we will match on the <geometry> element in the source XML, and then select the values of the children nodes, renaming them as appropriate for the desired output XML. We illustrate this transformation in Figure 11.11, and the results in Figure 11.12.

As an exercise for the student, how would one finish this series of transformations to complete the desired output document?

For more details on XML transformations, I recommend consulting some of the references for this chapter [12].

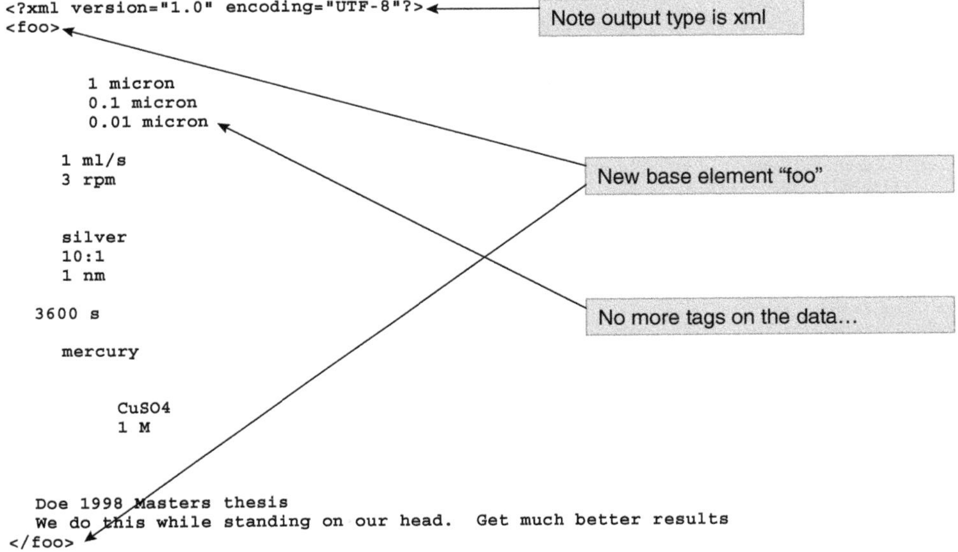

FIGURE 11.6
Results of our first transformation, note that all tags for elements except outermost element corresponding with the source outermost element have been stripped out, leaving the element content only.

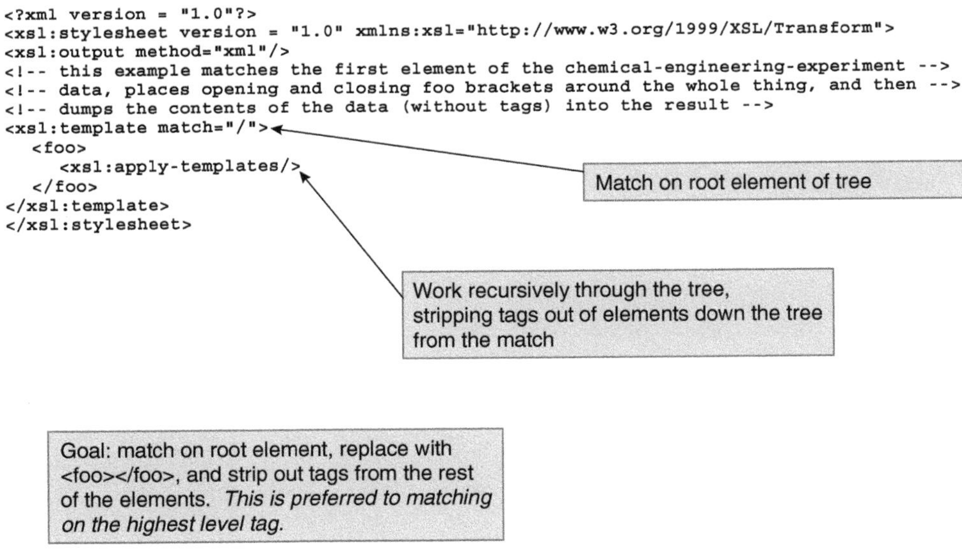

FIGURE 11.7
Replacing outermost elements, but selecting on the root element rather than the specific outermost elements in the source XML.

11.1.4 Data Description and Transformation from Hydrology

We will now present an example hydrologic data description, and transformation, for a particular hydrologic dataset, as developed by Francina Dominguz in our first hydroinformatics course. In particular, note the richness of the schema developed in this

```
<?xml version = "1.0"?>
<xsl:stylesheet version = "1.0" xmlns:xsl="http://www.w3.org/1999/XSL/Transform">
<xsl:output method="xml"/>
<!-- this example matches the first element of the chemical-engineering-experiment -->
<!-- data, places opening and closing foo brackets around the whole thing, and then -->
<!-- dumps the contents of the data (without tags) into the result -->
<!-- now we are trying for the 2nd element, with a translation. -->
<!-- we are ignoring the attributes of the first two elements, as well. -->
<xsl:template match="chemical-engineering-experiment">
   <cell-specification>
      <xsl:apply-templates/>
   </cell-specification>
</xsl:template>
<xsl:template match="cell">
   <echem-cell>
      <xsl:apply-templates/>
   </echem-cell>
</xsl:template>
</xsl:stylesheet>
```

Match on first element of tree, replace with "cell-specification"

Work recursively through the tree, stripping tags out of elements down the tree from the match

Next, match on cell, replace with echem-cell

And, again work recursively through the tree, applying the second template

Goal: Move down the hierarchy of the tree. Start at root, replace with the appropriate tag (still without attribute), move down to cell

FIGURE 11.8
Working through two levels of the hierarchy in our source XML, replacing both levels of tags in the elements selected.

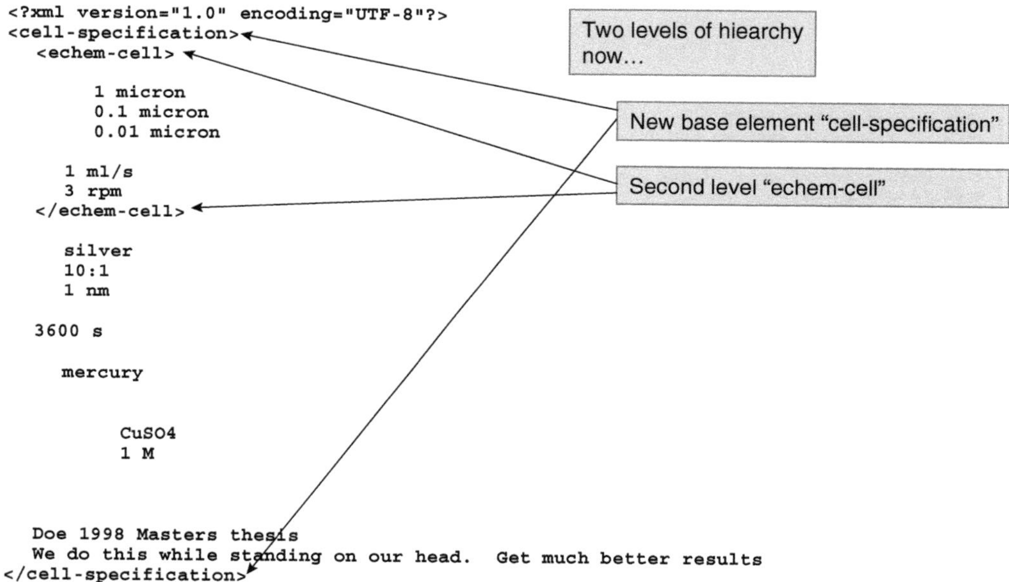

```
<?xml version="1.0" encoding="UTF-8"?>
<cell-specification>
   <echem-cell>

         1 micron
         0.1 micron
         0.01 micron

      1 ml/s
      3 rpm
   </echem-cell>

      silver
      10:1
      1 nm

   3600 s

      mercury

         CuSO4
         1 M

   Doe 1998 Masters thesis
   We do this while standing on our head.  Get much better results
</cell-specification>
```

Two levels of hiearchy now...
New base element "cell-specification"
Second level "echem-cell"

FIGURE 11.9
Results of our two-level transformation specified in Figure 11.8.

example, as well as the complexity of the transformation. The following is the project report and appendices developed by the student to complete the project.

11.1.4.1 Introduction

The objective of this project is to use the XML format, the XSD schemas and the XSLT transformations in the specific context of hydrologic data. A hydrologic dataset in XML

```
<?xml version = "1.0"?>
<xsl:stylesheet version = "1.0" xmlns:xsl="http://www.w3.org/1999/XSL/Transform">
<xsl:output method="xml"/>
<!-- this example matches the first element of the chemical-engineering-experiment -->
<!-- data, places opening and closing foo brackets around the whole thing, and then -->
<!-- dumps the contents of the data (without tags) into the result -->
<!-- now we are trying for the 2nd element, with a translation. -->
<!-- we are ignoring the attributes of the first two elements, as well. -->
<!-- #4  drop into the geometry tree... -->
<xsl:template match="chemical-engineering-experiment">
   <cell-specification>
      <xsl:apply-templates/>
   </cell-specification>
</xsl:template>
<xsl:template match="cell">
   <echem-cell>
      <xsl:apply-templates/>
   </echem-cell>
</xsl:template>
<xsl:template match="geometry">
   <physical-characteristics>
      <xsl:apply-templates/>
   </physical-characteristics>
</xsl:template>
</xsl:stylesheet>
```

Match on third element of tree ("geometry"), replace with "physical-characteristics"
Work recursively through the tree, stripping tags out of elements down the tree from the match

Figure 11.10
Third level of element selection and renaming.

```
<?xml version = "1.0"?>
<xsl:stylesheet version = "1.0" xmlns:xsl="http://www.w3.org/1999/XSL/Transform">
<xsl:output method="xml"/>
<!-- this example matches the first element of the chemical-engineering-experiment -->
<!--  data, places opening and closing foo brackets around the whole thing, and then -->
<!--  dumps the contents of the data (without tags) into the result -->
<!-- now we are trying for the 2nd element, with a translation. -->
<!-- we are ignoring the attributes of the first two elements, as well. -->
<!-- #4  drop into the geometry tree... -->
<!-- #5  extract node values -->
<xsl:template match="chemical-engineering-experiment">
   <cell-specification>
      <xsl:apply-templates/>
   </cell-specification>
</xsl:template>
<xsl:template match="cell">
   <echem-cell>
      <xsl:apply-templates/>
   </echem-cell>
</xsl:template>
<xsl:template match="geometry">
   <physical-characteristics>
         <electrode-diameter>
            <xsl:value-of select="electrode-diameter"/>
         </electrode-diameter>
         <gap>
            <xsl:value-of select="tip-distance"/>
         </gap>
         <span>
            <xsl:value-of select="tip-diameter"/>
         </span>
   </physical-characteristics>
</xsl:template>
</xsl:stylesheet>
```

While matching on third element of tree ("geometry"), replace with "physical-characteristics"
And select values of each of the children nodes, renaming appropriately for the new schema

Figure 11.11
Illustration of renaming nodes, preserving the values for these nodes.

format described by a certain schema is transformed to another dataset in XML described by a different schema. In applied hydrology, this methodology is useful for several reasons. The description of datasets in XML format will allow different platforms to access the data. Furthermore, if the data will be used for hydrologic modeling, the XSLT transformations will allow the modeler to change the XML format to that used in the model.

File Description:
HydroProject.xml: first xml file
HydroProject.xsd: schema describing HydroProject.xml.
HydroProjectDESIGN: design of HydroProject.xsd

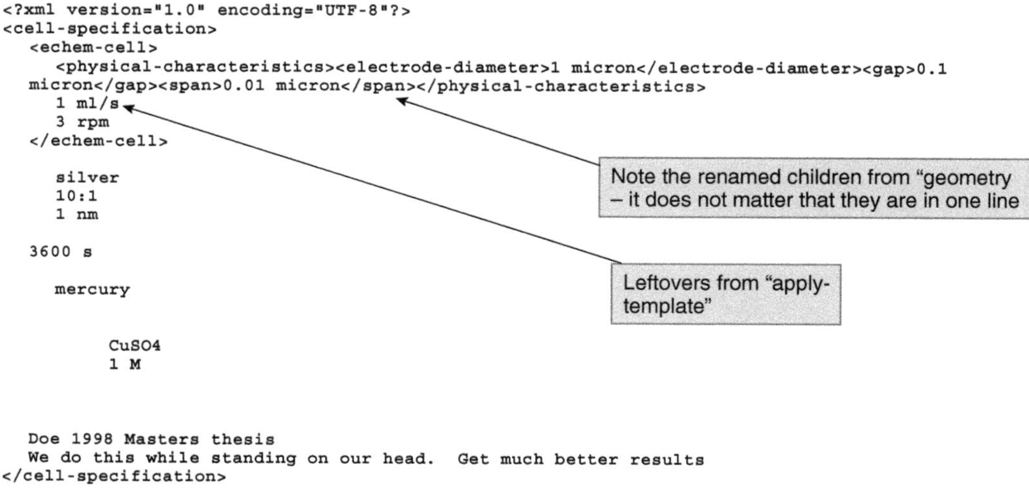

```
<?xml version="1.0" encoding="UTF-8"?>
<cell-specification>
  <echem-cell>
    <physical-characteristics><electrode-diameter>1 micron</electrode-diameter><gap>0.1
  micron</gap><span>0.01 micron</span></physical-characteristics>
    1 ml/s
    3 rpm
  </echem-cell>

    silver
    10:1
    1 nm

  3600 s

    mercury

        CuSO4
        1 M


  Doe 1998 Masters thesis
  We do this while standing on our head.  Get much better results
</cell-specification>
```


FIGURE 11.12
Results of transformation in Figure 11.11. Note the preservation of the node values for the children of the renamed <geometry> node.

Results/xslout/BasinData.xml: second xml file
HydroProjectTransformed.xsd: schema describing BasinData.xml.
HydroProjectTransformedDESIGN: design of HydroProjectTransformed.xsd

Transformation.xsl: transformation

xslthw1.xml code to run the ant program

Data: original data file

***The xml files are not printed out. They would take up too much space and it is easier to view them as a file.

11.1.4.2 Focus Data

The National Oceanic and Atmospheric Administration's Great Lakes Environmental Research Laboratory (NOAA-GLERL) produces several different types of data related to the Great Lakes [13]. Real time, hydrology and hydraulics, and physical properties, biological and chemical data are only some of the datasets available through their Web site http://www.glerl.noaa.gov/data/. The datasets are available in different formats. The historical hydrologic data comes in Microsoft excel files, the forecasted data comes in ASCII files, while the bathymetric and coastal delineation data is also in ASCII format with a header which describes the data and a description of the header in a Microsoft word file. This multiplicity of formats was one of the reasons I chose this dataset. The unification of formats will be useful for hydrologic modeling, and using XML to describe the data will have the advantage of being platform independent.

Some of the available datasets were selected to create the first XML file.

Only data for Lake Superior and Lake Michigan was used.

1. Great Lakes monthly hydrologic data:
   Type of file: Excel document

Table 11.1

Coordinated Great Lakes drainage areas (Coordinating Committee, 1977)

	Superior	Michigan	Huron	St. Clair	Erie	Ontario
Lake Area (sq km)	82,100	57,800	59,800	1,170	25,800	19,000
Land Area (sq km)	128,000	118,000	134,000	15,700	61,000	64,000
Basin Area (sq km)	210,000	176,000	194,000	16,900	86,800	83,000

Source: Great Lakes monthly hydrologic data report.

- Lake and land precipitation for the analyzed period was calculated at GLER using daily station data. The Thiessen polygon method was used.
  Files: *PrecipLake.xls and PrecipLand.xls*
- Lake evaporation is calculated using the great lakes evaporation model. This is a lumped-parameter surface flux and heat-storage model. It uses areal-average daily air temperature, windspeed, humidity, precipitation, and cloudcover.
  File: *Data/ Evaporation.xls*
- Lake levels are calculated as the beginning of month lake levels.
  File: *Levels_BOM(1).xls*

Average annual lake levels from 1990 to 1991 and monthly values from 1990 to 1999 of lake and land precipitation and lake evaporation were selected. The description of each dataset was obtained from the Great Lakes monthly hydrologic data report [13].

2. Lake and land areas:
   Type of file: Word document
3. Bathymetric Data:
   Type of file: ASCII.
   Bathymetric data for Lake Superior comes from the Canada Center for Inland Waters while the data for Lake Michigan was transformed from a 2 min grid obtained from Hughes et al. The data from the Great Lake Environmental Research Laboratory (GLERL) is uniform for all lakes. Each file has a header which describes the data. The information in the header includes: number of east–west grid boxes, number of north–south grid boxes, base latitude, base longitude, grid size, and maximum depth. There are also constants to change from geographic to map coordinates and vise versa.
   Files: *SuperiorBathymetry.txt and MichiganBathymetry.txt.*
4. Forecasted Hydrologic Data:
   Type of file: ASCII
   Three NOAA agencies support estimates of present hydrology and long-term hydrological probabilities for each of the Great Lake Basins: NOAA's GLERL provides hydrological models, near real-time meteorological data reduction procedures, and support for making the outlooks. NOAA's Climate Prediction Center (CPC) produces climatic outlooks that are available each month. The Midwestern Climate Center (MCC) gets climatic data from NOAA's National Climatic Data Center (NCDC) and Environment Canada's Canadian Meteorological Centre and prepares the data in a format usable by the data reduction procedures and models from GLERL.
   Files: *SuperiorForecast.txt and MichiganForecast.txt*

5. Shoreline Data:
   Type of file: ASCII
   The shorelines have been digitized from the national ocean survey charts. The average segment length is 1.24 km. The files describe the geographical coordinates of the endpoints, and each file forms a closed loop.
   Files: *SuperiorShoreline.txt and MichiganShoreline.txt*

11.1.4.3 Initial Schema Description

The diagram of the initial schema, HydroProject.xsd, is shown in Figure 11.13. It was designed thinking from the GLERL point of view. They produce different types of data for all of the Great Lakes Basins. Each child node of *HydroData*: *Geographic-Location*, *Gridded-Data*, *Time series-Data*, and *Forecasting-Data* represents one of the different types of data that they provide. Within these nodes there are instances of the specific type of data for different basins (in this case only Lake Superior and Lake Michigan were included). The basin is just part of the definition of the hydrologic data type.

The *Geographic-Location* node includes the bounding coordinates of the lake and detailed shoreline delineation (only a path to the filesource is included due to the size of this file, and this path leads to a URL thinking of easier data transfer). The *Gridded-Data* node stores the bathymetry of the lakes, and here again, only the path to the file is included due to the amount of data in each file. The *Time series-Data* and *Forecasting-Data* are time series data, and in this case the data is included in the XML document.

11.1.4.4 Transformed Schema Description

The diagram of the transformed schema, HydroProjectTransformed.xsd, is shown in Figure 11.14. It was designed thinking from the modeler's point of view. First of all, the modeler will generally be interested in extracting data from a single basin. In this case, Lake Superior basin was selected, and the hydrologic information regarding only to this basin was extracted. The information has been ordered in a different way thinking of a single basin.

The *Basin-Characterization* node includes the location of the basin, and the area. The *Physical-Data* node stores geodatabases, in this case it has bathymetric data, but in general, DEM, vegetation, soil, and other types of data would be stored in this node. Because of the size of these files, the actual data is not included in the XML document and there is only a reference to the data source (a URL is assigned thinking of data transfer). The grid description is also included.

The *Hydrologic-Data* node stores time series data. It includes historical and forecast data, with very similar formats. Some of the elements in the HydroProject.xml file are transformed into attributes in the transformed XML file. Note that the forecast data has an additional attribute (besides the basin and hydro-data-type) called "quantile" referring to the quantile of the prediction. This is the only difference between the historical and forecast data. The actual data series is included in the XML file with a detailed description of the data.

11.1.4.5 Transformation

The transformation file, Transformation.xsl, is shown in Appendix 2. Note that all the instructions of data extraction from HydroData.xml are only executed if the basin-code = 1 (Lake Superior), this means that it will select only the data related to that basin.

The hierarchy of the templates is very important for the transformation. The root node template is the first one, and then the transform goes through the child nodes (described as different levels). Using the "apply-templates" command ensures all the nodes are tested.

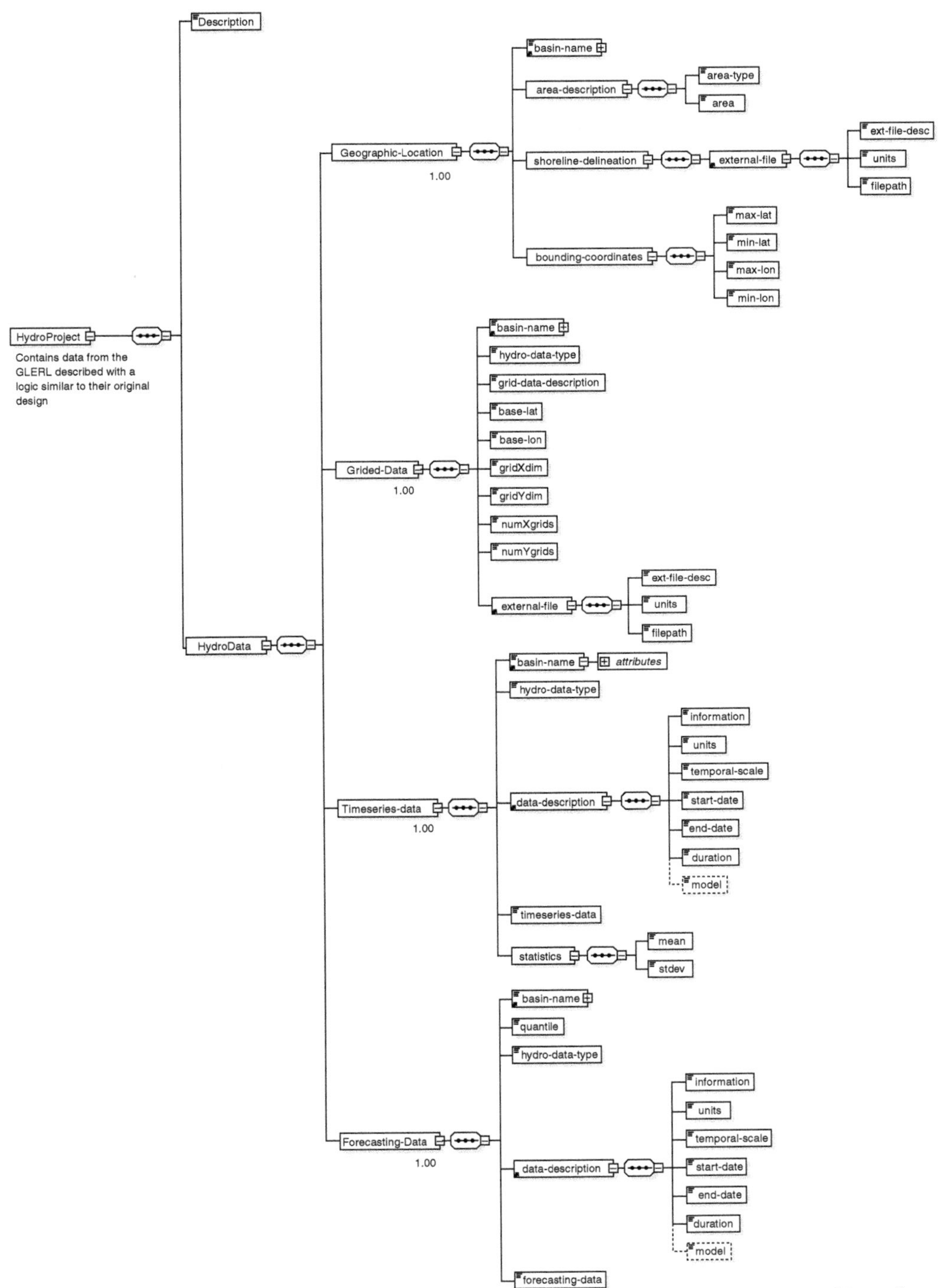

FIGURE 11.13
Design of the initial schema, from the GLERL point of view. (Credit: Francina Dominguz, University of Illinois.)

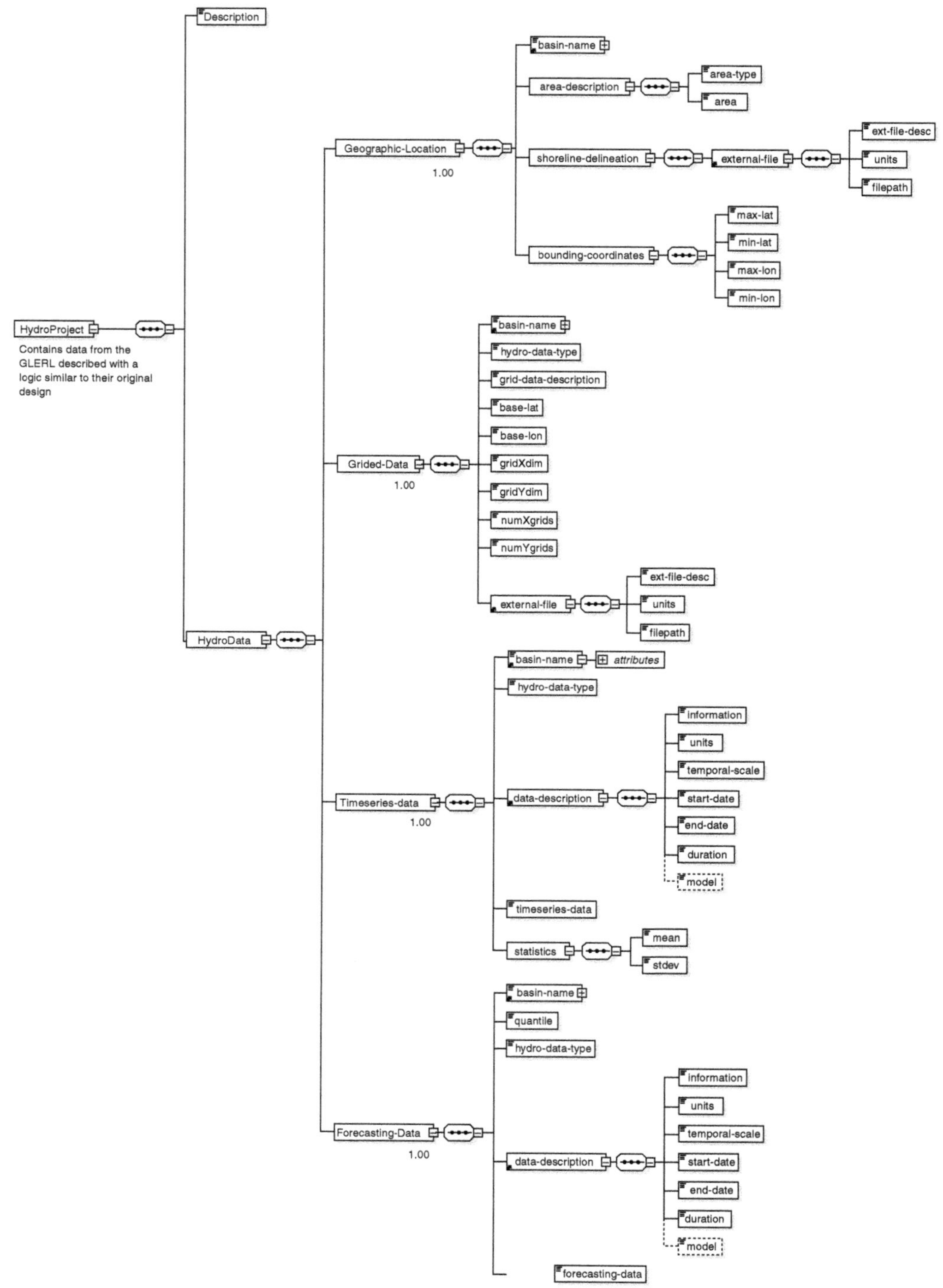

FIGURE 11.14
Design of the transformed schema, from a hydrologic modeler's point of view. (Credit: Francina Dominguz, University of Illinois.)

11.1.4.6 Conclusions

There is a great need in the hydrologic community to unify the formats of existing data. The many different formats for data description create a big problem for a modeler who needs to transform all these different formats into the one required by his or her model. XML is quickly becoming the dominant data description language, as W3Schools says: "XML is a cross-platform, software and hardware independent tool for transmitting information," and this is what makes its incorporation into the hydrologic community so important.

This project provides a simple example of a description of existing hydrologic data in XML format. Furthermore, it also gives an example of a transformation to change the data from a "data provider" design to a "modeler" design. The possibilities of data transformation are endless, and this project only provides one of the many different ways to transform the data.

One of the difficulties in the case of hydrologic modeling is that the size of the data files is very large. Although the data can be described within the XML document, handling of large data files within XML is not so good. The other possibility is to write the path for the data file (which was done in this project), but this is also not an optimum solution. This is one of the drawbacks of the XML language for hydrologic data.

Some of the resources that were useful are listed here:

- XML Spy software (very useful)
- http://www.w3schools.com/
- http://www.whitefrost.com/documents/html/technical/xsl/xslPrimer.html
- http://www.w3.org/TR/xmlschema-0

11.2 Task Descriptions in XML

Now that we have considered some of the basics of XML, including exploring the ability to transform XML as needed, and looked at the application of XML descriptions of scientific data, we would like to turn our attention to using XML to describe scientific tasks, or workflows. We will examine the use of a workflow engine developed at the University of Illinois called the Open Grid computing environments Runtime Engine, or OGRE [14]. There are many other examples of workflow systems using XML to encode their tasks, for example, Chimera [15], or the Business Process Execution Language (BPEL) [16].

11.2.1 Apache Ant

The OGRE is an extension to Apache Ant [17], which is a Java-based build tool. Ant is fairly easy to understand, as it only has three main constructs, namely, projects, targets, and tasks. An Ant project has three attributes, its *name,* the *default* target, and the *basedir,* which is the base directory from which all paths are calculated. An Ant target is a set of tasks which may be dependent on other targets. Finally, tasks are simply pieces of code that can be executed, such as copy files or filesets. Full details on Ant are available in the Ant manual [17].

We will consider the following makefile, used to build our file management tool Trebuchet:

```
<project name="ncsa-middleware-swing-clients" default="compile" basedir=".">

    <!-- THESE PROPERTIES ARE REQUIRED ************ -->
    <property name="project.name" value="ncsa-middleware-swing-clients"/>
```

```
<property name="src.dir" value="src"/>
<property name="build.dir" value="build"/>
<property name="lib.dir" value="lib"/>
<property name="jar.base" value="ncsa"/>
<property name="build.lib" value="${build.dir}/lib"/>
<property name="project.jar" value="${build.lib}/${project.name}.jar"/>
<property name="classes.dir" value="${build.lib}/classes"/>


<!-- CLASSPATH REF ******************************************************** -->
<path id="compileClasspath">
    <fileset dir="lib">
        <include name="compile/*.jar"/>
    </fileset>
    <pathelement path="${java.class.path}"/>
</path>

<path id="runtimeClasspath">
    <pathelement location="${project.jar}"/>

    <fileset dir="lib">
        <include name="compile/*.jar"/>
        <include name="runtime/*.jar"/>
    </fileset>
    <pathelement path="${java.class.path}"/>
</path>

<path id="buildtimeClasspath">
    <pathelement location="${project.jar}"/>

    <fileset dir="lib">
        <include name="compile/*.jar"/>
        <include name="runtime/*.jar"/>
        <include name="buildtime/*.jar"/>
    </fileset>
    <pathelement path="${java.class.path}"/>
</path>


<!-- SET UP ENVIRONMENT *************************************************** -->
<target name="setenv">
    <mkdir dir="${build.dir}"/>
    <mkdir dir="${build.lib}"/>
    <mkdir dir="${classes.dir}"/>
</target>

<!-- BUILD **************************************************************** -->
<target name="build" depends="compile,jar"/>

<!-- COMPILE ************************************************************** -->
<target name="compile" depends="setenv">
    <javac srcdir="${src.dir}"
        destdir="${classes.dir}"
        classpathref="compileClasspath" >
    </javac>
</target>

<!-- JAR ****************************************************************** -->
<target name="jar">
    <mkdir dir="${classes.dir}/mappings"/>
    <copy todir="${classes.dir}">
        <fileset dir="${src.dir}">
```

```
                <include name="**/*" />
            </fileset>
        </copy>

        <jar jarfile="${project.jar}" basedir="${classes.dir}">
            <include name="**/*"/>
        </jar>
    </target>

    <!-- CLEANS EVERYTHING ****************************************************** -->
    <target name="clean">
        <delete quiet="true" dir="dist"/>
        <delete quiet="true" dir="${build.dir}"/>
    </target>

    <!-- RUN ******************************************************************** -->
    <target name="trebuchet" depends="build">
      <java classname="ncsa.swing.SpringFedApplication"
            classpathref="runtimeClasspath"
            fork="true">
        <jvmarg value="-Djava.net.preferIPv4Stack=true" />
        <arg value="spring-trebuchet.xml"/>
      </java>
    </target>

</project>
```

Looking at the first line, one can see that the default target is <compile>:

```
<project name="ncsa-middleware-swing-clients" default="compile" basedir=".">
```

We will also be working out of the current working directory, for all path computations. After this initial definition of our project, we establish the values of some Ant properties (after THESE PROPERTIES ARE REQUIRED comment), as well as compute some paths (after CLASSPATH REF comment) that we will need in our building process:

```
<!-- THESE PROPERTIES ARE REQUIRED ************ -->
<property name="project.name" value="ncsa-middleware-swing-clients"/>

<property name="src.dir" value="src"/>
<property name="build.dir" value="build"/>
<property name="lib.dir" value="lib"/>
<property name="jar.base" value="ncsa"/>
<property name="build.lib" value="${build.dir}/lib"/>
<property name="project.jar" value="${build.lib}/${project.name}.jar"/>
<property name="classes.dir" value="${build.lib}/classes"/>


<!-- CLASSPATH REF ***************************************************** -->
<path id="compileClasspath">
    <fileset dir="lib">
        <include name="compile/*.jar"/>
    </fileset>
    <pathelement path="${java.class.path}"/>
</path>

<path id="runtimeClasspath">
    <pathelement location="${project.jar}"/>
```

```
    <fileset dir="lib">
        <include name="compile/*.jar"/>
        <include name="runtime/*.jar"/>
    </fileset>
    <pathelement path="${java.class.path}"/>
</path>

<path id="buildtimeClasspath">
    <pathelement location="${project.jar}"/>

    <fileset dir="lib">
        <include name="compile/*.jar"/>
        <include name="runtime/*.jar"/>
        <include name="buildtime/*.jar"/>
    </fileset>
    <pathelement path="${java.class.path}"/>
</path>
```

Note that we build the paths using the properties set up in the initial section.
Now consider using the default target of "compile" — if we land in the compiler target,

```
<target name="compile" depends="setenv">
   <javac srcdir="${src.dir}"
          destdir="${classes.dir}"
          classpathref="compileClasspath">
   </javac>
</target>
```

we note a depencency on setenv. So, the target setenv is executed prior to executing compile:

```
<target name="setenv">
    <mkdir dir="${build.dir}"/>
    <mkdir dir="${build.lib}"/>
    <mkdir dir="${classes.dir}"/>
</target>
```

This target establishes the directories that the compile task will populate. Note the specific tasks javac and mkdir in the compile and setenv targets, specifically. These are part of Ant's built-in task set. In order to perform the XML transformations mentioned previously in this chapter (Section 11.1.3), one can use the XSLT task to perform a given transformation, much as in this previously-mentioned Ant script:

```
<project name="Demo" default="runAll" basedir=".">

   <property name="wrk" value="${user.dir}/examples/tests"/>

 <target name="xslttest">
     <mkdir dir="${wrk}/xslout"/>
        <xslt in="experimentdesign.xml"
              out="${wrk}/xslout/cellbuilder.xml"
              style="expt.to.maker.xsl">
        </xslt>
 </target>
```

Note that armed with our knowledge of Ant (as well as knowing that this script was elided from a far larger example script), we could improve it by specifying a valid default, that is

```
<project name="Demo" default="xslttest" basedir=".">

    <property name="wrk" value="${user.dir}/examples/tests"/>

  <target name="xslttest">
    <mkdir dir="${wrk}/xslout"/>
        <xslt in="experimentdesign.xml"
            out="${wrk}/xslout/cellbuilder.xml"
            style="expt.to.maker.xsl">
        </xslt>
  </target>
```

This section is meant to give you a broad overview of the capabilities of Apache Ant as an XML-based scripting language. However, for a more general class of scientific applications, we developed OGRE, discussed in Section 11.2.2.

11.2.2 OGRE

In light of the intrinsic features presented by Apache Ant, our group decided to use this as a foundation for a scientific workflow language execution engine, OGRE. The initial design target for OGRE was to handle workflows on any grid platform, that is, the "microscopic" workflow. These workflow descriptions can vary considerably between problems, but the pattern of "stage input files to temporary scratch space — run executable — store results to persistent storage" is very common in scientific computing. Setting out on this task, we identified a number of requirements for our control layer surrounding a scientific code or codes — specifically, the workflow engine:

- Should not require any alterations to native computational code (especially community codes).
- Should be platform independent.
- Should be capable of making most of the standard system calls associated with a shell script.
- Should be capable of handling repeated/iterated and parallel control structures.
- Should be capable of sending and receiving (asynchronously) messages of various sorts.
- Should be capable of interacting with remote file systems and hosts.
- Should be capable of interacting with web services.
- Should be flexible enough to use with diverse components or services (i.e., a general reusability requirement).
- Should be modular, so that selected pieces can be incorporated into applications, if so desired.
- Should be extensible, allowing developers to add to or modify it.
- Should facilitate translation to and from graphical representation, or composition through a graphical interface.

The most conspicuous element missing from Ant that would make it feasible as a scientific workflow engine is a runtime environment, that is, though one could set properties outside the targets and tasks, one cannot introduce new properties (or variables) as a result of a task's execution, or otherwise affect a running script. We decided to extend Ant's ProjectHelper to accomplish the goal of handling a runtime environment. How this works will be illustrated through some variables; those interested in more details for OGRE should consult the OGRE developer's manual [14].

First, we will examine how the OGRE project helper allows variables to be set in one task, and then used in another task. Consider Figure 11.15, our "hello world" example.

In this particular example — we use the esequence container to create our environment. Within this, we used edeclare to set two variables, $w1$ and $w2$, and then we used the variables in a standard Ant task, echo. We have a second echo outside of the esequence container that does not depend on the environment. Note that performing a global assignment or declaration of a variable can make it span OGRE containers, for example, see Figure 11.16.

OGRE also provides a convenient way to manipulate files, using the URI tasks built into the core tasks for OGRE. Essentially, the URI provides a uniform way to manipulate files or directories, *independent of their location*, through a uniform naming convention of a scheme

Example 1.0

```
<!--////////////////////////////////////////////////////////////////////////////////////////////////////////////////////-->
<!--                              Example 1.0                                        -->
<!--                              "HELLO WORLD"                                      -->
<!--           Demonstrates Simple Assignment & Dereferencing.                       -->
<!--           Demonstrates ability to embed Ant tasks directly {"as is"}            -->
<!--           inside the extended syntax.                                           -->
<!--/////////////////////////////////////////////////////////////////////////////////-->
 <target name="1.0">
   <esequence>                                                         <!-- esequence "container"; creates the initial environment      -->
      <edeclare name="w1" string="HELLOOO"/>                           <!-- simple declaration task                                      -->
      <edeclare globalName="w2" string="WWWW..."/>                     <!-- value goes in the global frame                               -->
      <echo message="A message from Porky Fig: $R{w1} $R{w2}"/>        <!-- global frame is searched if no matching var is in scope      -->
   </esequence>                                                        <!-- 'echo' is an Ant task, but its attribute depends on env;     -->
   <echo message="That's all, folks!">                                 <!-- here, 'echo' has no environment dependencies; does not       -->
 </target>                                                             <!-- require a container                                          -->
```

Figure 11.15
Hello world example script. Note that the variables $w1$ and $w2$ are set in the declare task, but used by the internal Ant task echo, in a fashion similar to normal properties. (Credit: Albert L. Rossi, University of Illinois.)

Example 1.2

```
<!--////////////////////////////////////////////////////////////////////////////////////////////////////////////////////-->
<!--                              Example 1.2                                    -->
<!--           Demonstrates sharing of global environment across targets.        -->
<!--           Note that targets are treated with the customary dependency       -->
<!--           semantics as normal Ant.                                          -->
<!--//////////////////////////////////////////////////////////////////////////// -->
 <target name="1.2.0">
    <esequence>
      <eassign globalName="x" string="assigned by target 1.2.0"/>   <!-- 'assign' and 'declare' are equivalent for globals     -->
    </esequence>
 </target>
 <target name="1.2.1" depends="1.2.0">
     <esequence>
        <echo message="$R(x), dereferenced by target 1.2.1"/>
     </esequence>
 </target>
 <target name="1.2" depends="1.2.1">
     <esequence>
        <edeclare globalName="x" string="assigned by target 1.2"/>  <!-- global can be overwritten using 'declare' or 'assign' -->
        <echo message="$R(x), dereferenced by target 1.2"/>
     </esequence>
 </target>
```

Figure 11.16
Global assignment of variables span specialized OGRE containers. (Credit: Albert L. Rossi, University of Illinois.)

TASK	SCHEME or SCHEME-PAIR
uricopy	*(file,file); (file,ftp); (file,gridftp); (file,mssftp)* *(http,file); (jar,file); (resource,file); (ftp,file); (gridftp,file); (mssftp,file)* *(ftp,ftp); (ftp,gridftp); (ftp,mssftp)* *(gridftp,ftp); (mssftp,ftp)* *(gridftp,gridftp); (gridftp,mssftp)* *(mssftp,gridftp); (mssftp,mssftp)*
uridelete	*file; ftp; gridftp; mssftp*
uriexists	*file; ftp; gridftp; mssftp; resource; classpath; jar*
urilist	*file; http; ftp; gridftp; mssftp; resource; classpath; jar; class*
urimkdir	*file; ftp; gridftp; mssftp*
urimove	*(file,file); (file,ftp); (file,gridftp); (file,mssftp)* *(ftp,file); (gridftp,file); (mssftp,file)* *(ftp,ftp); (ftp,gridftp); (ftp,mssftp)* *(gridftp,ftp); (mssftp,ftp)* *(gridftp,gridftp); (gridftp,mssftp)* *(mssftp,gridftp); (mssftp,mssftp)*
uristagefrom	*file; http; ftp; gridftp; mssftp; resource*
uritouch	*file*

FIGURE 11.17
Supported URI operations and schemes. (Credit: Albert L. Rossi, University of Illinois.)

followed by the specific name that you wish to manipulate. A summary table of supported schemes and URI tasks follows in Figure 11.17.

To illustrate one such use of the URI tasks, consider the following two figures. In the first figure, Figure 11.18, we setup up the global properties for directory manipulation. Note that we use the uripattern task to be able to select the files we wanted to operate on within the subdirectories we chose.

There are many more attributes to OGRE scripts, including the provision for notification, and to integrate web service clients. But rather than trying to replicate the user manual, in Section 11.2.3 let us examine a script for a project called the Modeling Environment for Atmospheric Discovery (MEAD) [18].

11.2.3 Example of Scientific Workflow in XML

In this section, we will consider a workflow script from a slightly dated version of OGRE, to illustrate not the details of the OGRE syntax, but rather how we can accomplish scientific computing on production compute resources through the use of OGRE scripts. The following script was developed for the MEAD project, in order to support the use of the Weather Research and Forecasting (WRF) [19] code on our production resources.

```
<!--                        wrf.xml                             -->

<!--  This is the taskflow script for running preprocessing and the wrf       -->
<!--  executable.                                                             -->
```

GlobalSetup

```
<!--////////////////////////////////////////////////////////////////////////////////////////////////////////////-->
<!--                                Global Setup                        -->
<!--///////////////////////////////////////////////////////////////-->

 <property name="target1" value="file://${user.dir}/etc/copied1"/> <!-- To obtain remote versions of these examples, modify these -->
 <property name="target2" value="file://${user.dir}/etc/copied2"/> <!-- properties, and change the 'sourcePattern' declarations   -->
 <property name="source1" value="file://${user.dir}/docs"/>        <!-- below to conform with the new sources, if any.             -->
 <property name="source2" value="file://${user.dir}/conf"/>        <!-- (examples 5.1 through 5.4 are local by definition)         -->
 <property name="sourcefile" value="file://${user.dir}/build.xml"/>
 <property name="targetfile" value="file://${user.dir}/etc/copied1/build.xml"/>

 <target name="globalSetup">
   <edeclare globalName="sourcePattern1">
      <uripattern baseUri="${source1}">
         <nestedelement include="**/*.jpg"/>
      </uripattern>
   </edeclare>
   <edeclare globalName="sourcePattern2">
      <uripattern baseUri="${source2}">
         <nestedelement exclude="**/*log4j*"/>
      </uripattern>
   </edeclare>
   <edeclare globalName="sourcePattern3">
      <uripattern baseUri="file://${user.dir}">
         <nestedelement include="**/*.xml"/>
      </uripattern>
   </edeclare>
   <edeclare globalName="deletePattern">
      <uripattern baseUri="${target1}">
         <nestedelement include="**/*.jpg"/>
      </uripattern>
   </edeclare>
   <edeclare globalName="resourcePattern">                         <!-- no need to modify this                                      -->
      <uripattern baseUri="classpath://java.class.path">
         <nestedelement include="**ogre.handler.properties"/>
      </uripattern>
   </edeclare>
 </target>
```

FIGURE 11.18
Setup of the patterns to be used in our URI scanning example in Figure 11.19. (Credit: Albert L. Rossi, University of Illinois)

```
<!--  Albert L. Rossi                                                -->
<!--  Grid Computing Environments, NCSA University of Illinois       -->
<!--  revised Ant version 2:        October     2003                 -->

<project name="wrf" default="Exit" basedir=".">

   <property environment="env"/>

   <!-- ////////////////////// SET WORKING DIRECTORY ///////////////////////// -->
   <target name="FindWorkingDirectory">
      <condition property="work.dir" value="${pre.def.dir}">
         <available file="${pre.def.dir}" type="dir"/>
      </condition>
      <condition property="work.dir" value="${env.SCR}">
         <and>
            <not>
               <available file="${work.dir}" type="dir"/>
            </not>
            <available file="${env.SCR}" type="dir"/>
         </and>
      </condition>
      <condition property="work.dir" value="${env.TG_CLUSTER_PFS}">
         <and>
            <not>
               <available file="${work.dir}" type="dir"/>
            </not>
            <available file="${env.TG_CLUSTER_PFS}" type="dir"/>
         </and>
      </condition>
```

```
    <condition property="work.dir" value="${env.TG_CLUSTER_SCRATCH}">
       <and>
          <not>
             <available file="${work.dir}" type="dir"/>
          </not>
          <available file="${env.TG_CLUSTER_SCRATCH}" type="dir"/>
       </and>
    </condition>
    <condition property="work.dir" value="${scr}">
       <and>
          <not>
             <available file="${work.dir}" type="dir"/>
          </not>
          <available file="${scr}" type="dir"/>
       </and>
    </condition>
    <condition property="work.dir" value="${env.PWD}">
       <and>
          <not>
             <available file="${work.dir}" type="dir"/>
          </not>
          <available file="${env.PWD}" type="dir"/>
       </and>
    </condition>
    <condition property="work.dir" value="${user.dir}">
       <and>
          <not>
             <available file="${work.dir}" type="dir"/>
          </not>
          <available file="${user.dir}" type="dir"/>
       </and>
    </condition>
 </target>


 <!-- ///////////////////////// SANITY CHECK //////////////////////////// -->
 <target name="CheckEnvironment" depends="FindWorkingDirectory" if="sanitycheck">
    <esequence>
      <edeclare globalName="env" object="true"/>
      <sysprops envToSysProps="true"/>
      <ereturn name="env" get="environment"/>  <!-- Vector of "name=value" -->
    </esequence>
    <sysprops checkProps="true" errorProperty="failed" rethrow="false">
       <nestedelement name="run.id"/>
       <nestedelement name="fm.timeout"/>
       <nestedelement name="wrf.urls"/>
       <nestedelement name="sounding.file.name"/>
       <nestedelement name="pre.cmd"/>
       <nestedelement name="wrf.cmd"/>
       <nestedelement name="f.ufmtendian"/>
       <nestedelement name="wrf.log.name"/>
       <nestedelement name="t1"/>
       <nestedelement name="t2"/>
       <nestedelement name="x2"/>
       <nestedelement name="y2"/>
       <nestedelement name="tdelay"/>
       <nestedelement name="um"/>
       <nestedelement name="vm"/>
       <nestedelement name="num.steps"/>
       <nestedelement name="save.intvl"/>
       <nestedelement name="nx"/>
       <nestedelement name="ny"/>
       <nestedelement name="x1"/>
       <nestedelement name="y1"/>
       <nestedelement name="rad1"/>
       <nestedelement name="rad2"/>
```

```
      <nestedelement name="storage.url"/>
   </sysprops>
</target>

<!-- ///////////////////////////// ALIVE ////////////////////////////// -->
<target name="Alive" depends="CheckEnvironment" unless="failed">
   <tstamp prefix="alive">
      <format property="TSTAMP" pattern="MM/dd/yyyy hh:mm aa"/>
   </tstamp>
   <progress messageType="text"
             label="Alive"
             message="Alive at host:[${env.HOST}] with path:[${work.dir}]
             at time:[${alive.TSTAMP}]"/>
</target>

<!-- ////////////////////////// STAGE WRF ///////////////////////////// -->
<target name="StageWRF" depends="Alive" unless="failed">
   <esequence>
      <edeclare name="stagedfiles" string=""/>
      <tokenize startTag="[" endTag="]" sequence="${wrf.urls}"/>
      <ereturn name="stagedfiles" get="tokens"/>
      <stage dstDir="${work.dir}"
             untarTo="parent"
             overwrite="true"
             deleteTars="true"
             permissions="ugo+rx"
             chmodType="both">
         <nestedelement srcVector="$R{stagedfiles}"/>
      </stage>
      <move file="${work.dir}/${sounding.file.name}"
            toFile="${work.dir}/input_sounding"/>
   </esequence>
</target>

<!-- ////////////////////// CREATE IDEAL_PARAMS /////////////////////// -->
<target name="CreateIdealParams" depends="StageWRF" unless="failed">
   <write path="${work.dir}/ideal_params" text="true">
      &lt;segment string="${um}" tabs="1"/&gt;
      &lt;segment string="${vm}" lineBreaks="1"/&gt;
      &lt;segment string="${t1}" tabs="1"/&gt;
      &lt;segment string="${x1}" tabs="1"/&gt;
      &lt;segment string="${y1}" tabs="1"/&gt;
      &lt;segment string="1500"  tabs="1"/&gt;
      &lt;segment string="${rad1}" tabs="1"/&gt;
      &lt;segment string="1500" lineBreaks="1"/&gt;
      &lt;segment string="${t2}" tabs="1"/&gt;
      &lt;segment string="${x2}" tabs="1"/&gt;
      &lt;segment string="${y2}" tabs="1"/&gt;
      &lt;segment string="1500" tabs="1"/&gt;
      &lt;segment string="${rad2}" tabs="1"/&gt;
      &lt;segment string="1500" lineBreaks="1"/&gt;
      &lt;segment string="${tdelay}" lineBreaks="1"/&gt;
   </write>
</target>

<!-- ///////////////////// CREATE NAMELIST.INPUT ////////////////////// -->
<target name="CreateNamelist.Input" depends="CreateIdealParams" unless="failed">
   <write path="${work.dir}/namelist.input" text="true">
      &lt;segment string="&amp;amp;namelist_01" lineBreaks="1"/&gt;
      &lt;segment string="time_step_max          = ${num.steps},"   lineBreaks="1"/&gt;
      &lt;segment string="max_dom                = 1,"              lineBreaks="1"/&gt;
      &lt;segment string="dyn_opt                = 2,"              lineBreaks="1"/&gt;
      &lt;segment string="rk_ord                 = 3,"              lineBreaks="1"/&gt;
      &lt;segment string="diff_opt               = 1,"              lineBreaks="1"/&gt;
      &lt;segment string="km_opt                 = 1,"              lineBreaks="1"/&gt;
      &lt;segment string="damp_opt               = 0,"              lineBreaks="1"/&gt;
```

```
&lt;segment string="isfflx                    = 0,"             lineBreaks="1"/&gt;
&lt;segment string="ifsnow                    = 0,"             lineBreaks="1"/&gt;
&lt;segment string="icloud                    = 1,"             lineBreaks="1"/&gt;
&lt;segment string="num_soil_layers           = 5,"             lineBreaks="1"/&gt;
&lt;segment string="spec_bdy_width            = 5,"             lineBreaks="1"/&gt;
&lt;segment string="spec_zone                 = 1,"             lineBreaks="1"/&gt;
&lt;segment string="relax_zone                = 4,"             lineBreaks="1"/&gt;
&lt;segment string="tile_sz_x                 = 0,"             lineBreaks="1"/&gt;
&lt;segment string="tile_sz_y                 = 0,"             lineBreaks="1"/&gt;
&lt;segment string="numtiles                  = 1,"             lineBreaks="1"/&gt;
&lt;segment string="debug_level               = 0 /"            lineBreaks="1"/&gt;
&lt;segment string=" "                                          lineBreaks="1"/&gt;
&lt;segment string="&amp;amp;namelist_02"                       lineBreaks="1"/&gt;
&lt;segment string="grid_id                   = 1,"             lineBreaks="1"/&gt;
&lt;segment string="level                     = 1,"             lineBreaks="1"/&gt;
&lt;segment string="s_we                      = 1,"             lineBreaks="1"/&gt;
&lt;segment string="e_we                      = ${nx},"         lineBreaks="1"/&gt;
&lt;segment string="s_sn                      = 1,"             lineBreaks="1"/&gt;
&lt;segment string="e_sn                      = ${ny},"         lineBreaks="1"/&gt;
&lt;segment string="s_vert                    = 1,"             lineBreaks="1"/&gt;
&lt;segment string="e_vert                    = 60,"            lineBreaks="1"/&gt;
&lt;segment string="time_step_count_output    = ${save.intvl}," lineBreaks="1"/&gt;
&lt;segment string="frames_per_outfile        = 1,"             lineBreaks="1"/&gt;
&lt;segment string="time_step_count_restart   = 9999,"          lineBreaks="1"/&gt;
&lt;segment string="time_step_begin_restart   = 0,"             lineBreaks="1"/&gt;
&lt;segment string="time_step_sound           = 6 /"            lineBreaks="1"/&gt;
&lt;segment string=" "                                          lineBreaks="1"/&gt;
&lt;segment string="&amp;amp;namelist_03"                       lineBreaks="1"/&gt;
&lt;segment string="dx                        = 1000.,"         lineBreaks="1"/&gt;
&lt;segment string="dy                        = 1000.,"         lineBreaks="1"/&gt;
&lt;segment string="dt                        = 6.,"            lineBreaks="1"/&gt;
&lt;segment string="ztop                      = 20000.,"        lineBreaks="1"/&gt;
&lt;segment string="zdamp                     = 5000.,"         lineBreaks="1"/&gt;
&lt;segment string="dampcoef                  = 0.2,"           lineBreaks="1"/&gt;
&lt;segment string="smdiv                     = 0.,"            lineBreaks="1"/&gt;
&lt;segment string="epssm                     = .1,"            lineBreaks="1"/&gt;
&lt;segment string="khdif                     = 300.,"          lineBreaks="1"/&gt;
&lt;segment string="kvdif                     = 500.,"          lineBreaks="1"/&gt;
&lt;segment string="mix_cr_len                = 750.,"          lineBreaks="1"/&gt;
&lt;segment string="radt                      = 0,"             lineBreaks="1"/&gt;
&lt;segment string="bldt                      = 0,"             lineBreaks="1"/&gt;
&lt;segment string="cudt                      = 0,"             lineBreaks="1"/&gt;
&lt;segment string="julyr                     = 0,"             lineBreaks="1"/&gt;
&lt;segment string="julday                    = 1,"             lineBreaks="1"/&gt;
&lt;segment string="gmt                       = 0. /"           lineBreaks="1"/&gt;
&lt;segment string=" "                                          lineBreaks="1"/&gt;
&lt;segment string="&amp;amp;namelist_04"                       lineBreaks="1"/&gt;
&lt;segment string="periodic_x                = .false.,"       lineBreaks="1"/&gt;
&lt;segment string="symmetric_xs              = .false.,"       lineBreaks="1"/&gt;
&lt;segment string="symmetric_xe              = .false.,"       lineBreaks="1"/&gt;
&lt;segment string="open_xs                   = .true.,"        lineBreaks="1"/&gt;
&lt;segment string="open_xe                   = .true.,"        lineBreaks="1"/&gt;
&lt;segment string="periodic_y                = .false.,"       lineBreaks="1"/&gt;
&lt;segment string="symmetric_ys              = .false.,"       lineBreaks="1"/&gt;
&lt;segment string="symmetric_ye              = .false.,"       lineBreaks="1"/&gt;
&lt;segment string="open_ys                   = .true.,"        lineBreaks="1"/&gt;
&lt;segment string="open_ye                   = .true.,"        lineBreaks="1"/&gt;
&lt;segment string="nested                    = .false.,"       lineBreaks="1"/&gt;
&lt;segment string="specified                 = .false.,"       lineBreaks="1"/&gt;
&lt;segment string="top_radiation             = .false.,"       lineBreaks="1"/&gt;
&lt;segment string="chem_opt                  = 0,"             lineBreaks="1"/&gt;
&lt;segment string="mp_physics                = 1,"             lineBreaks="1"/&gt;
&lt;segment string="ra_lw_physics             = 0,"             lineBreaks="1"/&gt;
&lt;segment string="ra_sw_physics             = 0,"             lineBreaks="1"/&gt;
&lt;segment string="bl_sfclay_physics         = 0,"             lineBreaks="1"/&gt;
&lt;segment string="bl_surface_physics        = 0,"             lineBreaks="1"/&gt;
```

```
      &lt;segment string="bl_pbl_physics          = 0,"          lineBreaks="1"/>
      &lt;segment string="cu_physics              = 0,"          lineBreaks="1"/>
      &lt;segment string="h_mom_adv_order         = 5,"          lineBreaks="1"/>
      &lt;segment string="v_mom_adv_order         = 3,"          lineBreaks="1"/>
      &lt;segment string="h_sca_adv_order         = 5,"          lineBreaks="1"/>
      &lt;segment string="v_sca_adv_order         = 3,"          lineBreaks="1"/>
      &lt;segment string="io_form_history         = 2,"          lineBreaks="1"/>
      &lt;segment string="io_form_restart         = 2,"          lineBreaks="1"/>
      &lt;segment string="io_form_input           = 2,"          lineBreaks="1"/>
      &lt;segment string="io_form_boundary        = 2 /"         lineBreaks="1"/>
      &lt;segment string=" "                                     lineBreaks="1"/>
      &lt;segment string="&amp;namelist_05"                     lineBreaks="1"/>
      &lt;segment string="/"                                     lineBreaks="1"/>
      &lt;segment string=" "                                     lineBreaks="1"/>
      &lt;segment string="&amp;namelist_quilt"                  lineBreaks="1"/>
      &lt;segment string="/"                                     lineBreaks="1"/>
   </write>
</target>

<!-- ///////////////////////////// RUN IDEAL ////////////////////////////// -->
<target name="RunIdeal" depends="CreateNamelist.Input" unless="failed">
   <esequence>
      <rtexec execDir="${work.dir}" args="${pre.cmd}" quiet="true">
         <outMonitor file="${work.dir}/ideal.out"/>
         <errMonitor file="${work.dir}/ideal.out"/>
      </rtexec>
      <ereturn globalName="currentExit" get="exitValue"/>
      <econdblock type="else-if">
         <econdtask>
            <eeval>
              <expression infix="$R{currentExit} != 0"/>
            </eeval>
            <esequence>
              <progress messageType="text"
                        label="ideal (preprocessing) failed with exit=$R{currentExit}"/>
              <property name="failed" value="true"/>
            </esequence>
         </econdtask>
         <progress messageType="text"
                label="ideal (preprocessing) completed"/>
      </econdblock>
   </esequence>
</target>

<!-- ///////////////////////// START FILE MONITOR //////////////////////// -->
<target name="StartFileMonitor" depends="RunIdeal" unless="failed">
   <esequence>
      <edeclare globalName="filewait" long="$E{ ${fm.timeout} + 100 }"/>
      <fmstart timeout="${fm.timeout}">
         <fileset dir="${work.dir}">
            <include name="**/wrfout*"/>
         </fileset>
      </fmstart>
      <property name="fmstart" value="true"/>
   </esequence>
</target>

<!-- //////////////////////////// COMPUTATION //////////////////////////// -->
<target name="RunWRF" depends="StartFileMonitor" unless="failed">
   <esequence>
      <progress pid="wrfoutlistener" async="true" label="wrfout"
         messageType="ranged"
         messageMatch="(\p{Alpha}+://)(\p{Graph}+?)(wrfout_(\w{3})_({\}d{6}))"
         messageMatchGroup="3" iterationMatch="\d{6}" rangeMax="${num.steps}"
         newFiles="true"/>
      <rtexec  execDir="${work.dir}" args="${wrf.cmd}" quiet="true">
```

```
          <outMonitor file="${work.dir}/wrf.out"/>
          <errMonitor file="${work.dir}/wrf.out"/>
       </rtexec>
       <ereturn globalName="currentExit" get="exitValue"/>
       <stop reference="wrfoutlistener" delay="2000"/>
       <econdblock type="else-if">
          <econdtask>
             <eeval>
                <expression infix="$R{currentExit} != 0"/>
             </eeval>
             <esequence>
                <progress messageType="text"
                          label="wrf failed with exit = $R{currentExit}"/>
                <property name="failed" value="true"/>
             </esequence>
          </econdtask>
          <progress messageType="text" label="wrf computation completed"/>
       </econdblock>
    </esequence>
</target>

<!-- ////////////////////////// POSTPROCESSING //////////////////////////// -->
<target name="MakeDataDirs" depends="RunWRF" unless="failed">
    <move toDir="${work.dir}/input">
       <fileset dir="${work.dir}">
             <include name="*param*"/>
             <include name="*namelist*"/>
             <include name="*input*"/>
             <include name="*INI*"/>
             <include name="*in*"/>
             <include name="*bdy*"/>
       </fileset>
    </move>

    <move toDir="${work.dir}/output">
       <fileset dir="${work.dir}">
             <include name="*err*"/>
             <include name="*out*"/>
             <include name="*log"/>
             <include name="*.dat"/>
             <include name="*.tar"/>
             <exclude name="*.exe"/>
       </fileset>
    </move>
</target>

<!-- //////////////////////////// RECORD METADATA ///////////////////////// -->
<target name="RecordRunMetadata" depends="MakeDataDirs" unless="failed">
    <runoutput xmlPath="${work.dir}/${run.id}_metadata.xml"
               dstUri="${storage.url}/${run.id}"
               runId="${run.id}"
               useCastor="false">
       <inputFileSet dir="${work.dir}">
          <include name="input/**"/>
       </inputFileSet>
       <outputFileSet dir="${work.dir}">
          <include name="output/**"/>
       </outputFileSet>
    </runoutput>
    <progress messageType="text" label="input/output metadata"
             message="${run.id}_metadata.xml"/>
</target>

<!-- /////////////////////// MOVE EVERYTHING TO MSS ///////////////////// -->
<target name="SaveFailed" depends="RecordRunMetadata" if="failed">
    <tar destfile="${work.dir}/save-failed.tar">
```

```
         <tarfileset dir="${work.dir}">
            <include name="**/*"/>
         </tarfileset>
      </tar>
   </target>

   <target name="MoveToMSS" depends="SaveFailed" >
      <filecopy dstUri="${storage.url}/${run.id}">
         <fileset dir="${work.dir}">
            <include name="input/**"/>
            <include name="output/**"/>
            <include name="*save*"/>
         </fileset>
      </filecopy>

<!--
      <filecopy dstUri="${storage.url}"
                srcUri="file://${work.dir}/${run.id}_metadata.xml"/>
-->

      <progress messageType="text"
                label="files moved to storage"
                message="${storage.url}/${run.id}"/>
   </target>

   <!-- ////////////////////////////// CLEAN UP ////////////////////////////// -->
   <target name="CleanUp" depends="MoveToMSS">
      <delete includeEmptyDirs="true">
         <fileset dir="${work.dir}" includes="**/*"/>
      </delete>
      <progress messageType="boolean" label="wrf run completed"/>
   </target>

   <!-- //////////////////////////////// EXIT //////////////////////////////// -->
   <target name="StopFileMonitor" depends="CleanUp" if="fmstart">
      <esequence>
         <stop reference="filemonitor"/>
      </esequence>
   </target>

   <target name="Exit" depends="StopFileMonitor">
      <waitforpublisher/>
   </target>

</project>
```

The first target is used to establish the current working directory. Note the wide variety of names that our production resources assign to preferred working (or scratch) directories, this long conditional is used to handle the range of semantics in the directories (Figure 11.19):

```
<target name="FindWorkingDirectory">
   <condition property="work.dir" value="${pre.def.dir}">
       <available file="${pre.def.dir}" type="dir"/>
   </condition>
   <condition property="work.dir" value="${env.SCR}">
       <and>
           <not>
             <available file="${work.dir}" type="dir"/>
           </not>
           <available file="${env.SCR}" type="dir"/>
       </and>
   </condition>
```

Example 5.0

```
<!--////////////////////////////////////////////////////////////////////////////////////////////////////////////////////////////-->
<!--                  Example 5.0                               -->
<!--      Unix-Style list print-out of files in two             -->
<!--      directories with pattern-mathcing.                    -->
<!--/////////////////////////////////////////////////////////////-->

  <target name="5.0" depends="globalSetup">
    <esequence>
      <edeclare name="results" null="true"/>
      <urilist fullScan="true">
        <nestedelement pattern="$G{sourcePattern1}"/>
        <nestedelement pattern="$G{sourcePattern2}"/>
        <configuration>                              <!-- used to configure the underlying task and/or datatypes -->
          <attribute name="sortByName" value="true"/>      <!-- settings are specific to the operation; consult the   -->
          <attribute name="fullMetadata" value="true"/>    <!-- task documentation for details                        -->
          <attribute name="getUnixLists" value="true"/>
          <attribute name="useLocalUnixScan" value="true"/>
        </configuration>
      </urilist>
     <ereturn name="result" get="result"/>           <!-- returns a "ListResult" object                           -->
     <print line="$R{result}"/>                      <!-- toString on this object will print out all lists        -->
    </esequence>
  </target>
```

Figure 11.19
Illustration of performing a UNIX style listing of selected members of two different directories. (Credit: Albert L. Rossi, University of Illinois)

```
<condition property="work.dir" value="${env.TG_CLUSTER_PFS}">
    <and>
      <not>
        <available file="${work.dir}" type="dir"/>
      </not>
      <available file="${env.TG_CLUSTER_PFS}" type="dir"/>
    </and>
</condition>
<condition property="work.dir" value="${env.TG_CLUSTER_SCRATCH}">
    <and>
      <not>
         <available file="${work.dir}" type="dir"/>
      </not>
      <available file="${env.TG_CLUSTER_SCRATCH}" type="dir"/>
    </and>
</condition>
<condition property="work.dir" value="${scr}">
    <and>
       <not>
         <available file="${work.dir}" type="dir"/>
       </not>
       <available file="${scr}" type="dir"/>
    </and>
</condition>
<condition property="work.dir" value="${env.PWD}">
    <and>
       <not>
         <available file="${work.dir}" type="dir"/>
       </not>
       <available file="${env.PWD}" type="dir"/>
    </and>
</condition>
<condition property="work.dir" value="${user.dir}">
    <and>
       <not>
         <available file="${work.dir}" type="dir"/>
       </not>
```

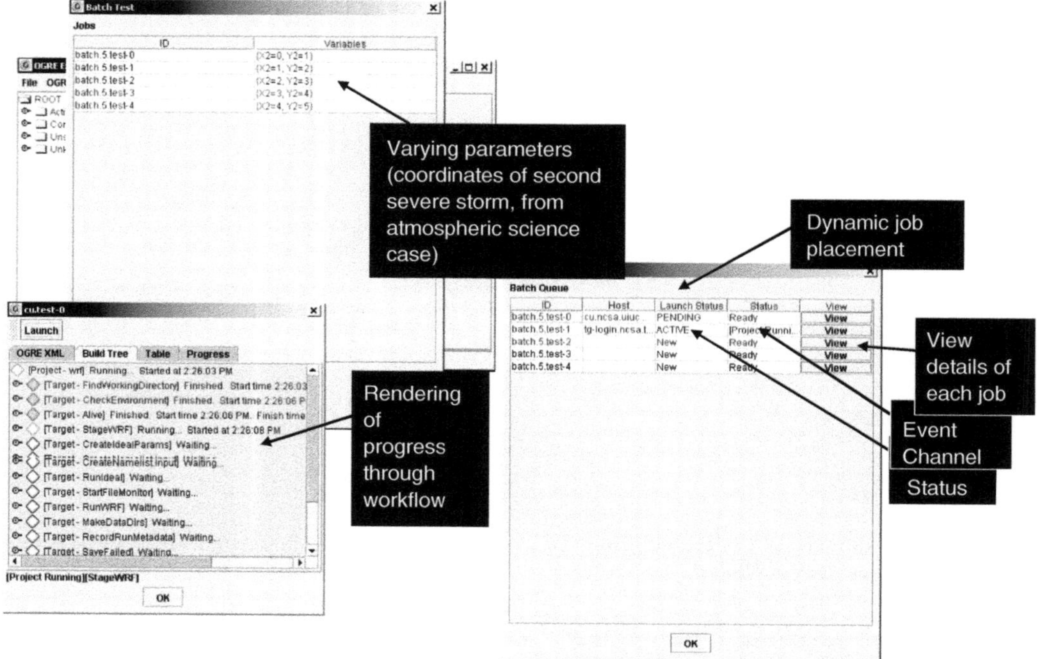

FIGURE 11.20
Rendering of RemoteOGRE interface, used to support launching and monitoring of remote OGRE instances. (Credit: Shawn Hampton, University of Illinois.)

```
            <available file="${user.dir}" type="dir"/>
        </and>
    </condition>
</target>
```

In the next target, we work to pull properties that were assigned at OGRE-launch time, these include a number of important run parameters that ultimately are variable as part of a larger parameter evaluation scheme (Figure 11.20):

```
<!-- //////////////////// SANITY CHECK ///////////////////// -->
<target name="CheckEnvironment" depends="FindWorkingDirectory" if="sanitycheck">
   <esequence>
      <edeclare globalName="env" object="true"/>
      <sysprops envToSysProps="true"/>
      <ereturn name="env" get="environment"/> <!-- Vector of "name=value" -->
   </esequence>
   <sysprops checkProps="true" errorProperty="failed" rethrow="false">
      <nestedelement name="run.id"/>
      <nestedelement name="fm.timeout"/>
      <nestedelement name="wrf.urls"/>
      <nestedelement name="sounding.file.name"/>
      <nestedelement name="pre.cmd"/>
      <nestedelement name="wrf.cmd"/>
      <nestedelement name="f.ufmtendian"/>
      <nestedelement name="wrf.log.name"/>
      <nestedelement name="t1"/>
```

```
      <nestedelement name="t2"/>
      <nestedelement name="x2"/>
      <nestedelement name="y2"/>
      <nestedelement name="tdelay"/>
      <nestedelement name="um"/>
      <nestedelement name="vm"/>
      <nestedelement name="num.steps"/>
      <nestedelement name="save.intvl"/>
      <nestedelement name="nx"/>
      <nestedelement name="ny"/>
      <nestedelement name="x1"/>
      <nestedelement name="y1"/>
      <nestedelement name="rad1"/>
      <nestedelement name="rad2"/>
      <nestedelement name="storage.url"/>
   </sysprops>
</target>
```

The "alive" target is incredibly valuable for debugging on large clusters, as it announces what is node 0 on a large MPI run.

```
<target name="Alive" depends="CheckEnvironment" unless="failed">
   <tstamp prefix="alive">
      <format property="TSTAMP" pattern="MM/dd/yyyy hh:mm aa"/>
   </tstamp>
   <progress messageType="text"
             label="Alive"
             message="Alive at host:[${env.HOST}] with path:[${work.dir}]
             at time:[${alive.TSTAMP}]"/>
</target>
```

Note that we automatically generate "build" events within OGRE, so the completion of each target can be sent to the event channel. Also, the task "progress" issues an event to the event channel with the Alive message; this event can be grabbed and rendered by a user's graphical user interface to notify them where and when the job started.

In the next target, we pull all of our executables from remote storage, to set up our run (we have assumed that we (1) have a library of prebuilt binaries that we can use, and (2) we do not have these codes formally installed on any of our compute platforms).

```
<!-- ////////////////////// STAGE WRF //////////////////////// -->
<target name="StageWRF" depends="Alive" unless="failed">
   <esequence>
       <edeclare name="stagedfiles" string=""/>
       <tokenize startTag="[" endTag="]" sequence="${wrf.urls}"/>
       <ereturn name="stagedfiles" get="tokens"/>
       <stage dstDir="${work.dir}"
              untarTo="parent"
              overwrite="true"
              deleteTars="true"
              permissions="ugo+rx"
              chmodType="both">
         <nestedelement srcVector="$R{stagedfiles}"/>
       </stage>
```

```
        <move file="${work.dir}/${sounding.file.name}"
              toFile="${work.dir}/input_sounding"/>
    </esequence>
</target>
```

Note that we do not stage anything if Alive fails.

Next, we generate an input file for the preprocessor named ideal.exe; note the use of variables established elsewhere:

```
<!-- ////////////////// CREATE IDEAL_PARAMS ////////////////// -->
<target name="CreateIdealParams" depends="StageWRF" unless="failed">
    <write path="${work.dir}/ideal_params" text="true">
        &lt;segment string="${um}" tabs="1"/&gt;
        &lt;segment string="${vm}" lineBreaks="1"/&gt;
        &lt;segment string="${t1}" tabs="1"/&gt;
        &lt;segment string="${x1}" tabs="1"/&gt;
        &lt;segment string="${y1}" tabs="1"/&gt;
        &lt;segment string="1500" tabs="1"/&gt;
        &lt;segment string="${rad1}" tabs="1"/&gt;
        &lt;segment string="1500" lineBreaks="1"/&gt;
        &lt;segment string="${t2}" tabs="1"/&gt;
        &lt;segment string="${x2}" tabs="1"/&gt;
        &lt;segment string="${y2}" tabs="1"/&gt;
        &lt;segment string="1500" tabs="1"/&gt;
        &lt;segment string="${rad2}" tabs="1"/&gt;
        &lt;segment string="1500" lineBreaks="1"/&gt;
        &lt;segment string="${tdelay}" lineBreaks="1"/&gt;
    </write>
</target>
```

Similarly (though not explicitly shown) we perform a similar operation for the namelist input files for WRF.

For the following target, we finally execute the ideal preprocessor. Note that we monitor for success or failure of the preprocessor, and issue an event with the status of the preprocessor run.

```
<!-- /////////////////////// RUN IDEAL /////////////////////// -->
<target name="RunIdeal" depends="CreateNamelist.Input" unless="failed">
   <esequence>
      <rtexec execDir="${work.dir}" args="${pre.cmd}" quiet="true">
         <outMonitor file="${work.dir}/ideal.out"/>
         <errMonitor file="${work.dir}/ideal.out"/>
      </rtexec>
      <ereturn globalName="currentExit" get="exitValue"/>
      <econdblock type="else-if">
         <econdtask>
            <eeval>
               <expression infix="$R{currentExit} != 0"/>
            </eeval>
            <esequence>
               <progress messageType="text"
                         label="ideal (preprocessing) failed with
                         exit = $R{currentExit}"/>
```

```
            <property name="failed" value="true"/>
         </esequence>
      </econdtask>
      <progress messageType="text"
               label="ideal (preprocessing) completed"/>
   </econdblock>
</esequence>
</target>
```

Next, we establish a filemonitor that watches for output files to be issued from the execution of WRF. This allows us to guage progress of the WRF run, for instance, if we know that ten output files will be generated, we can use this knowledge to generate events to render a remote progress bar in a graphical user interface.

```
<!-- //////////////////// START FILE MONITOR //////////////////// -->
<target name="StartFileMonitor" depends="RunIdeal" unless="failed">
   <esequence>
      <edeclare globalName="filewait" long="$E{ ${fm.timeout} + 100 }"/>
      <fmstart timeout="${fm.timeout}">
         <fileset dir="${work.dir}">
            <include name="**/wrfout*"/>
         </fileset>
      </fmstart>
      <property name="fmstart" value="true"/>
   </esequence>
</target>
```

Finally, we can run WRF, building the MPI command line necessary to run WRF dynamically from parameters set outside of the script:

```
<!-- ////////////////////// COMPUTATION ////////////////////// -->
<target name="RunWRF" depends="StartFileMonitor" unless="failed">
   <esequence>
       <progress pid="wrfoutlistener" async="true" label="wrfout"
                 messageType="ranged"
                 messageMatch="(\p{Alpha}+://)
                 (\p{Graph}+?)(wrfout_(\w{3})_(\d{6}))"
                 messageMatchGroup="3" iterationMatch="\d{6}"
                 rangeMax="${num.steps}"
                 newFiles="true"/>
       <rtexec execDir="${work.dir}" args="${wrf.cmd}" quiet="true">
           <outMonitor file="${work.dir}/wrf.out"/>
           <errMonitor file="${work.dir}/wrf.out"/>
       </rtexec>
       <ereturn globalName="currentExit" get="exitValue"/>
       <stop reference="wrfoutlistener" delay="2000"/>
       <econdblock type="else-if">
          <econdtask>
             <eeval>
                <expression infix="$R{currentExit} != 0"/>
             </eeval>
             <esequence>
                <progress messageType="text"
                          label="wrf failed with exit = $R{currentExit}"/>
```

```
                <property name="failed" value="true"/>
            </esequence>
        </econdtask>
        <progress messageType="text" label="wrf computation completed"/>
    </econdblock>
  </esequence>
</target>
```

Note that in this particular target, the rtxec task runs WRF; also, the progress task issues events for each output file that is generated.

There are a number of other targets associated with file cleanup, storing of files, and related tasks — what is important here is to get an idea of what is possible with an XML scripting language such as depicted with OGRE. To close this section, I would like to show Figure 11.20, illustrating one such graphical user interface developed to manage the workflow we just discussed. In this case, the interface managed the execution of a large number of jobs to support a very simple broker scenario. We will discuss broker scenarios some more in the Chapter 12.

References

1. ISO (International Organization for Standardization). ISO 8879:1986(E). Information processing — Text and Office Systems — Standard Generalized Markup Language (SGML), first edition — 1986-10-15. [Geneva]: International Organization for Standardization, 1986.
2. *Extensible Markup Language (XML)* 1.0, second edition. http://www.w3.org/TR/2000/REC-xml-20001006.
3. XML Schema Primer. http://www.w3.org/TR/xmlschema-0/.
4. XML.com: Using W3C XML Schema. http://www.xml.com/pub/a/2000/11/29/schemas/part1.html.
5. http://www.w3.org/Addressing/.
6. http://www.w3.org/2001/XMLSchema.xsd.
7. http://www.xmlspy.com.
8. http://www.w3.org/XML/Schema.
9. http://www.w3.org/TR/xslt.
10. http://www.xml.com/pub/a/2000/08/holman/index.html.
11. Apache Ant — Welcome. http://ant.apache.org.
12. XML Bible, second edition, Chapter 17. http://www.ibiblio.org/xml/books/bible2/chapters/ch17.html.
13. Thomas E. Croley II, Timothy S. Hunter, and S. Keith Martin, Great Lakes Monthly Hydrologic Data, NOAA Data Report ERL GLERL, National Technical Information Service, Springfield, Virginia, 1999.
14. *OGRE User Manual*. http://corvo.ncsa.uiuc.edu/ogre/.
15. *Virtual Data System*. http://www.griphyn.org/chimera/.
16. Business Process Execution Language for Web Services Version 1.1. http://www-128.ibm.com/developerworks/library/ws-bpel/.
17. *Apache Ant User Manual*. http://ant.apache.org/manual/index.html.
18. MEAD. http://www.ncsa.uiuc.edu/expeditions/MEAD/.
19. Weather Research and Forecasting (WRF). http://wrf-model.org/.

12

Grid Computing

Jay C. Alameda

Contents

In the mid-1990s, some enterprising computational and computer scientists wanted to compute and use data resources prevalent in offices, laboratories, even countries and across the world seamlessly and efficiently. They were faced with daunting challenges, namely, each domain required a separate authentication process, and each account and password was different, that is, provided one was able to access the resources in the first place. Additionally, it was tedious to remember where one had stored the data, what that data meant and why that data was there in the first place.

12.1 Grid Genesis

12.1.1 Pregrid Distributed Computing

In the early days, scientists and computer scientists were willing to try a variety of ad hoc schemes to perform distributed computing. For instance, one could use remote shell (rsh) and remote copy (rcp) on unix platforms to implement some rudiments of distributed computing. An rsh could be used to invoke a command on a remote machine, while an rcp was useful for copying data between hosts in a straightforward way.

To implement such schemes, one needed to create a list of trusted machines and users in one's home directory — in a so-called .rhosts file. Provision of such a file allowed access without a password from the trusted hosts to the trusting hosts. To establish a "grid" of computational hosts would involve propogating the .rhosts files among all the hosts. This had some popularity for a while, to the extent that some client–server software packages used this scheme. However, it had the drawback that one compromised account could lead to the demise of all of your trusted machines.

Alternate pre-grid schemes included the use of Kerberos [1] and versions of rsh and rcp that used Kerberos authentication. This proved to be effective within a domain, such as a department, that was willing to establish the Kerberos domain controller, which then controlled all the user accounts for the machines using Kerberos for their authentication. The user burden for such a system was considerably lower than for a careful placement of .rhosts to be certain, but this type of system did not scale very well outside the contained settings, such as within a department.

Less centralized approaches, without the security drawbacks of .rhosts, also used public key authentication for secure shell (ssh) [2]. This would require a user to generate a private/public key pair, and propogate the public key to all the hosts that you want to authenticate to without a password. This does place a burden on the user, in the sense that the user is responsible for propogating his/her public keys. Additionally, symmetric authentication requires placement of public and private key pairs on all hosts in one's computational "grid."

12.1.2 I-WAY 1995

In 1995, twenty computing sites collaborated in a wide-area networking experiment called I-WAY aimed at supporting distributed applications, with the sites interconnected by a 155 mb/sec Asynchronous Transfer Mode (ATM) network [3]. These sites hosted node points consisting of supercomputers and immersive visualization devices [4], that were accessed through a point of presence (I-POP) machine, which presented a consistent interface to the network for a variety of node points "behind" the I-POP. This particular experiment culminated in a set of multisite demonstrations at Supercomputing 95.

In order to achieve the functionality desired for I-WAY, a suite of software called I-Soft, was deployed on the set of I-POP resources. I-Soft had a number of requirements, grouped in terms of user- and site-oriented requirements:

1. User oriented
   - Resource naming and location
   - Uniform programming environment
   - Autoconfiguration and resource characterization
   - Distributed data services
   - Trust management
   - Confidentiality and integrity
2. Site oriented
   - Allow for different use polices
   - Different access mechanism

The I-Soft system that resulted did not satisfy all the requirements stated; in fact, there is still work in progress with today's infrastructure. To make the software task more tractable, the I-Soft software was designed to be deployed on the relatively consistent I-POP systems rather than the core compute or visualization nodes. The I-WAY team used a shared filesystem (Andrew filesystem, AFS) between all I-POP nodes (which was valuable for sharing code and scheduling information between sites), and presented a uniform Kerberos authentication into the I-POP nodes, which allowed access through a modified telnet or ftp client. The path from the I-POP nodes to the target compute or visualization node was through rsh by root (!) over a private network. This allowed for any action to be taken on behalf of an authenticated user into I-POP on the remote compute or visualization resource. With this sort of infrastructure design, it is not surprising that I-POPs were paired consistently with participating sites. Furthermore, scheduling amongst the resources was mostly a cooperative enterprise, with little infrastructure within I-Soft to support more automated scheduling.

From this experiment, it became obvious that to support distributed computing, one needed uniform authentication mechanisms that do not compromise security for participating sites. Furthermore, the applications using the resource demonstrated that it was important to support a variety of ways to use the resources — and that the "uniform programming environment" needed to be addressed. Finally, this experiment was the catalyst necessary to start the Globus Project [5], which has played a central role in bringing the grid — a persistent environment that enables software applications to integrate instruments, displays, computational, and information resources, all physically separated and managed by multiple organizations — to life.

12.2 Protocol-Based Grids

The practical effect of the I-WAY experiment discussed in the previous section was to launch an attempt to define a software infrastructure to solve the distributed resource sharing problem that many had perceived to be necessary to solve problems beyond what is capable by any single organization. This software infrastructure, captured well in the paper "The Anatomy of the Grid" [6] is based on an architecture complete with protocols, services, application programming interfaces, and software development kits became a reality in the Globus Toolkit. With the first versions of the toolkit, though, the focus was on interoperability by well-defined protocols supported by system-level services on computing platforms that needed to interoperate. Hence, I consider these early grids to be "protocol-based grids," in distinct contrast to "service-based grids" that are currently emerging with later versions of the Globus Toolkit, as well as toolkits from the Open Middleware Infrastructure Institute (OMII) [7].

12.2.1 The Globus Toolkit

As the foundation for the computational grid, the Globus Toolkit has shaped much of the early protocol-based grid work. Globus is comprised of three pillars, namely, Resource Management, Data Management, and Information Management. We will focus on the first two pillars as we describe how the grid has played a role in solving certain scientific problems.

The key to success of Globus is its security architecture. In order to help ensure broad adoption by participating sites, the Globus security architects had a goal of better security than is normally implemented at participating sites. The solution derived by Globus, called the Grid Security Infrastructure (GSI) [8], is an extension of the Public Key Infrastructure

(PKI) [9] that allows you to generate a *proxy credential*, which is a short-term credential to allow a process (such as a computational job) to act on your behalf without requiring intervention such as typing a password during the execution of the process.

The public key infrastructure uses digital certificates to assert identity of entities in the computational grid. At the foundation of a digital certificate is the trust implied in the signature chains of the certificate; ultimately, trust in the certificate authorities is used to sign user and host certificates. In the following example of my digital certificate, as revealed by a Globus tool designed to display information about the certificate, one can see that my certificate is signed by the National Center for Supercomputing Applications, which runs a certificate authority for its users:

```
tg-login2 ncsa/jalameda> grid-cert-info -all
Certificate:
    Data:
        Version: 3 (0x2)
        Serial Number: 660 (0x294)
        Signature Algorithm: md5WithRSAEncryption
        Issuer: C=US, O=National Center for Supercomputing
Applications, CN=Certification Authority
        Validity
            Not Before: Sep 13 19:35:03 2004 GMT
            Not After : Sep 13 19:35:03 2006 GMT
        Subject: C=US, O=National Center for Supercomputing
Applications, CN=Jay C. Alameda
        Subject Public Key Info:
            Public Key Algorithm: rsaEncryption
            RSA Public Key: (2048 bit)
                Modulus (2048 bit):
                    00:bd:c2:1b:3b:7e:a0:04:a7:aa:f0:7b:6c:55:1d:
                    10:20:cd:bf:86:4a:08:81:6c:10:18:29:58:53:43:
                    d7:bf:8c:b9:de:d7:26:c3:ce:f4:36:d4:18:e5:6d:
                    a9:45:41:d1:3d:2b:c3:71:30:46:cd:40:8f:32:d0:
                    99:f9:14:48:5a:d5:86:11:4a:91:61:47:22:08:5b:
                    2a:98:ed:b1:43:1a:51:ba:e4:21:19:90:53:2a:07:
                    e1:a8:c4:c4:5d:50:73:15:75:50:36:9f:49:b7:cf:
                    74:c3:62:74:a8:96:ad:12:f1:ad:ff:cc:b0:86:0e:
                    66:4e:d3:8d:d9:c5:ae:9b:72:68:e7:3e:02:37:3d:
                    62:ea:02:b0:7f:be:de:f8:24:c3:e0:9e:8f:af:97:
                    3c:e2:f9:b0:38:11:b0:3d:82:5d:bf:31:e2:c5:00:
                    09:d7:e5:e1:c3:f8:4c:0c:83:d4:bb:76:7e:8d:6f:
                    1a:5b:36:a3:8f:b3:3f:68:70:4b:34:ed:fc:68:a6:
                    cc:22:87:a5:40:f3:00:a9:bc:fd:b5:23:16:30:d5:
                    5a:38:ee:66:44:3f:d3:e5:35:5f:07:2f:54:c0:b2:
                    45:b9:65:5d:ee:57:3c:a7:66:78:df:55:bf:53:d0:
                    93:8f:ca:f9:31:14:87:f8:d8:07:a4:69:2f:95:97:
                    51:97
                Exponent: 65537 (0x10001)
        X509v3 extensions:
            X509v3 Basic Constraints:
                CA:FALSE
            X509v3 Key Usage:
```

```
                Digital Signature, Non Repudiation,
Key Encipherment, Data Encipherment
            X509v3 Subject Key Identifier:
A2:EC:26:F2:FD:D2:DF:61:8E:9F:8F:76:50:8A:FA:11:FE:FA:FB:48
            X509v3 Authority Key Identifier:
keyid:F2:BE:8D:AA:61:49:11:2D:B9:5B:72:24:2A:18:D2:C2:85:C6:98:81
                DirName:/C=US/O=National Center for Supercomputing
Applications/CN=Certification Authority
                serial:00

            X509v3 CRL Distribution Points:
                URI:http://ca.ncsa.uiuc.edu/4a6cd8b1.r0

        Signature Algorithm: md5WithRSAEncryption
            02:25:ae:84:bb:f0:76:7b:f0:3d:15:62:78:71:18:d0:de:69:
            4e:e2:25:09:d6:72:81:8b:35:31:f0:d0:02:be:e7:3f:7a:03:
            49:27:75:36:bb:ca:19:dc:2a:d6:1f:97:3b:88:e1:5c:01:3f:
            7e:e8:e7:a7:bb:12:ae:77:8a:2b:c8:c5:e6:33:d1:e6:45:e9:
            ba:b9:94:2e:8c:04:a4:2a:27:11:ec:a5:de:2d:38:c0:95:75:
            67:a8:f5:70:d9:1e:8a:d4:58:37:d6:6f:a2:3f:fa:b0:d5:50:
            58:5a:ff:99:a2:b9:38:5f:d0:07:f9:91:c3:88:47:4d:8c:1b:
            2f:82:bf:cc:51:c5:b8:4e:2c:4d:41:33:43:3b:5a:6e:53:c9:
            83:9c:66:8f:7e:e3:ef:f7:98:e9:13:0c:2f:55:de:55:19:3a:
            16:da:59:71:6e:fb:2c:51:d1:19:6a:ab:d9:cb:9b:0d:2a:90:
            9c:07:7d:4b:1c:ee:68:01:02:97:c9:f9:c5:e1:36:b0:de:61:
            e3:2b:fe:ad:f7:1f:28:1c:b1:08:f6:a5:50:98:e1:56:56:01:
            6e:4b:06:8d:45:d4:66:69:69:cc:7d:15:69:0a:ab:1b:0e:a6:
            ba:76:f8:a1:5d:1a:af:b2:bd:1a:57:7f:5a:2e:0b:a7:f7:ba:
            d7:50:ef:d9
```

In the case of grid authentication, mutual authentication is necessary to perform any actions on a remote resource; in this case, both the resource and user present each other's signed public keys, which are verified by the appropriate certificate authority's public key at each end of the connection.

12.2.1.1 "Bag of Tools"

In the late 1990s, the Globus Toolkit emerged with a popular set of tools to be able to connect remote resources to form a computational grid. One thing that ensured its popularity was that it was not necessary to adopt the entire toolkit (e.g., by integrating it into your application) in order to derive some benefit from using the toolkit. For instance, one could simply use the command-line tools to create a proxy credential (grid-proxy-init), and then use more command line tools to submit computational jobs (globusrun) and move data (globus-url-copy). In this fashion, the cost of adopting the Globus toolkit was low, as one could perform lots of simple tasks, and also combine the tasks through simple scripts.

12.2.1.2 GRAM

Job submission in the Globus Toolkit is enabled by a low-level protocol, Grid Resource Allocation Manager or GRAM, which is implemented in a system level service called

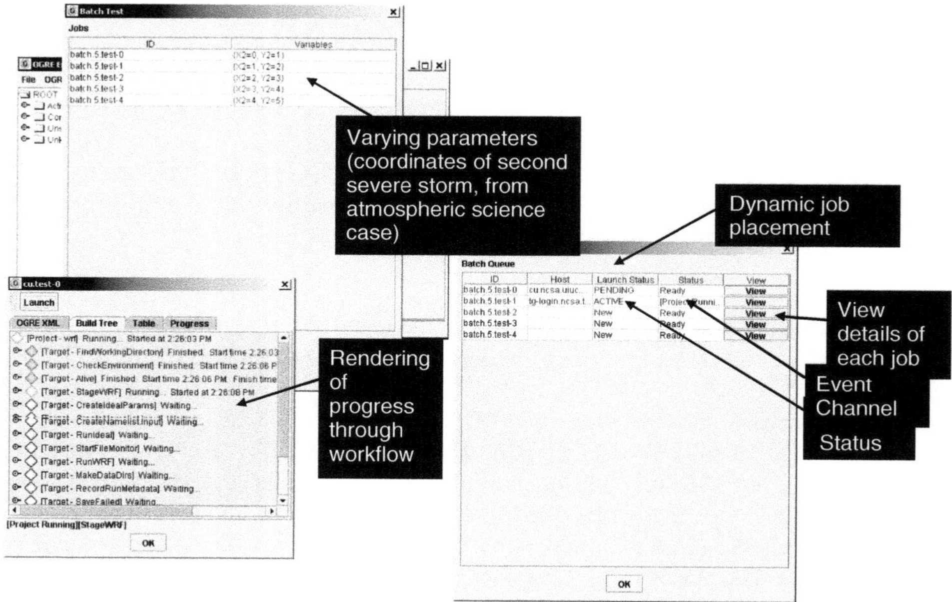

FIGURE 12.1
Graphical user interface "RemoteOGRE," which uses GRAM and GridFTP to launch computational jobs on remote resources. (Credit: Shawn Hampton, University of Illinois.)

a gatekeeper. This gatekeeper interfaces to at least one or more jobmanager; namely, to an interactive jobmanager, who allows execution of commands as a user *without the benefit of a unix shell*, or a batch jobmanager, who automatically generates a batch script from a set of generic job specifications. Interaction with the gatekeeper can be accomplished through command-line tools such as globusrun; alternatively, it is possible to use client libraries such as the Java Commodity Grid [10] to build advanced client tools as depicted in Figure 12.1. The capability to submit jobs through a low-level protocol has been one stable element of the Globus toolkit as it attempts to evolve to a web-service foundation; use of client libraries such as the Java Commodity Grid help to hide some of the changes in job submission that are inevitable as the project evolves.

12.2.1.3 GridFTP

GridFTP refers to a standardized protocol [11] for data transfer on the grid. Servers supporting gridFTP data transfer have been developed by the Globus Project and NCSA; tools such as globus-url-copy (part of the default globus toolset) and uberftp [12] have been developed to interact with the gridftp servers to allow users to move their data as needed on the computational grid. Gridftp has supported performance controls, such as TCP buffer size and the provision for parallel data streams, to allow for the movement of large volumes of scientific data. New versions of the Gridftp server will provide for *striping* of data transfers among a number of gridftp hosts that are connected in parallel to a parallel filesystem. This is necessary to take advantage of some of the emerging high bandwidth transcontinental networks that were deployed to support scientific computing, such as the network dedicated to TeraGrid [13]. Finally, it is possible to build advanced interactive clients such as in Figure 12.2 on top of toolkits such as the Java Commodity Grid to provide for easy data management by end-users.

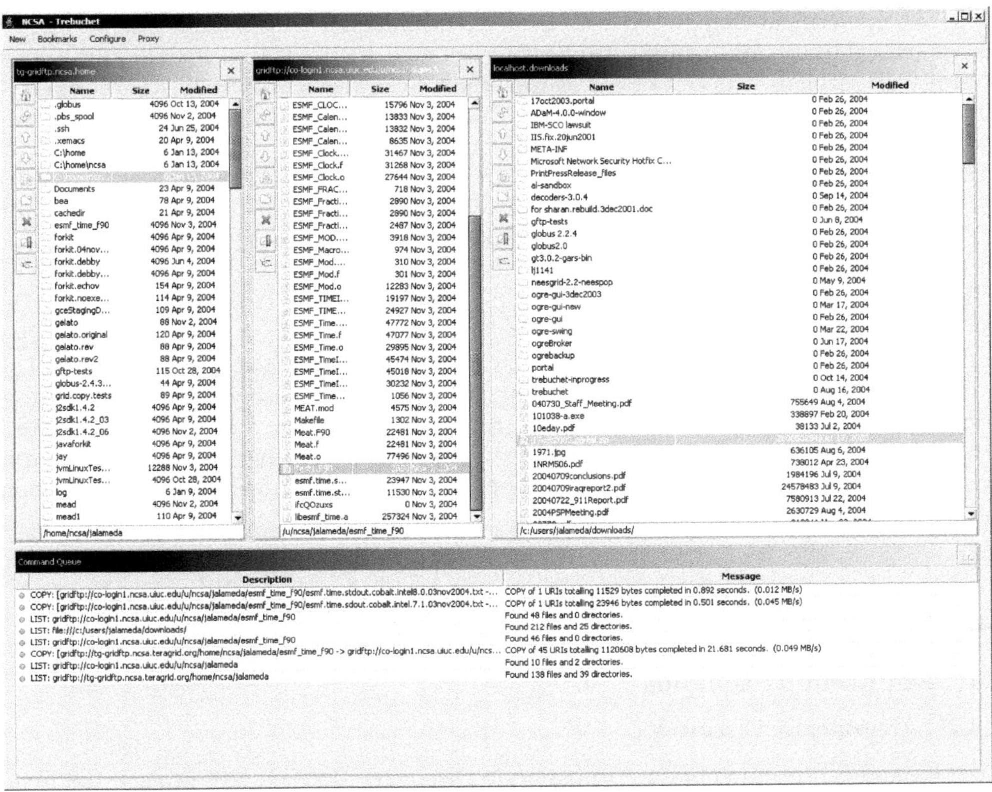

FIGURE 12.2
NCSA Trebuchet, an advanced data management tool which supports gridFTP. (Credit: Shawn Hampton, University of Illinois.)

12.3 Service Grids

From the previous section, one can see that with the protocol-based grids we have a uniform mechanism to submit jobs and move files. What may not be evident, though, is the additional complexity inherent in the approach; for instance, the command-line tools represent user interfaces that are not very easy to comprehend or use. Finally, the protocol-based grids do nothing to address some of the areas where we could conceivably add value to the grid, namely, in the area of scientific data — data collections, descriptions, tracking, aggregations, and so on. In essence, we would like to achieve a level of transparency to the data, including location transparency, and also be able to add some value (from a scientist's point of view) to collections of data. Since we learned that XML can be used to describe both data and service interfaces, it seems that building a grid around web services would be a reasonable enterprise, as one could embed the services with rich functionality (expressed as a programming interface) that would allow for rich services; for example, for building and accessing data collections, for providing location transparency to data, and so on.

12.3.1 Stateful Web Services

Unfortunately, though this is seemingly obvious, the development of such a service environment has had a difficult start. As in the beginning of the development of service-based grids [14], one critical change for web services, the addition of *state* to a web service, was seen as critical to model scientific applications as web services. The first attempt to model stateful web services was embodied as the Open Grid Services Infrastructure (ogsi) [15], with its first implementations in the Globus Toolkit v3.x and OGSI.net [16]. Unfortunately, this development did not meet well with the much larger web service community, so, this particular specification is being deprecated in favor of the Web Services Resource Framework, WS-RF [17] that is being developed by IBM and Argonne National Laboratories and will be implemented in the Globus Toolkit v4 and WSRF.net [18]. However, I would not expect this to be finalized in the near future, as not all the industry partners have weighed in on this specification, not to mention that this is still lacking in experience with real applications. A much more conservative view has been espoused by the U.K. eScience program's OMII, in which their distribution of web-service based middleware will start with the broadly-accepted WS-I specification [19], while the specifications for stateful web services work themselves out. Rather than attempting to capture a moving target here, we recommend following a simple web service approach (as depicted in the earlier chapter), perhaps by following the OMII activity in the United Kingdom, with possible adoption of stateful service paradigms as they mature.

12.4 Application Scenarios

12.4.1 Atmospheric Discovery

In the Linked Environments for Atmospheric Discovery (LEAD) [20] project, the investigators aim to develop information technology to allow people and technologies to interact with the weather. In current practice, a variety of observational tools are used to measure weather parameters independent of the ongoing weather conditions, initializing models on fixed grids that generate forecasts on regular time intervals. In LEAD, we are working to change that, in the sense that the analysis, modeling, data acquisition, and assimilation would all respond to weather conditions. Additionally, the LEAD project will be working to connect experimental weather radars from the CASA project [21] to modify the radar's observational parameters to better measure high impact weather such as tornadoes.

In LEAD, each element of the system is being modeled as a web service. In a full implementation of LEAD, we would use a hierarchy of dynamic orchestration capabilities to be able to connect and steer the web services so that the overall system can dynamically adapt to the changing weather conditions, including modifying observation parameters, modeling parameters, and, even securing necessary compute resources *on demand* in order to be able to derive a forecast before the predicted high-impact weather occurs at a particular location [22]. Figure 12.3 illustrates some of the interactions that LEAD will allow, especially with the arrows between the LEAD tools and the experimental dynamic observations.

12.4.2 Chemical Engineering

Another class of applications is derived from chemical engineering, in the form of multiscale, multiphenomena science, in which the phenomena being examined are modeled in multiple codes and full consideration of the problem requires combinations of the

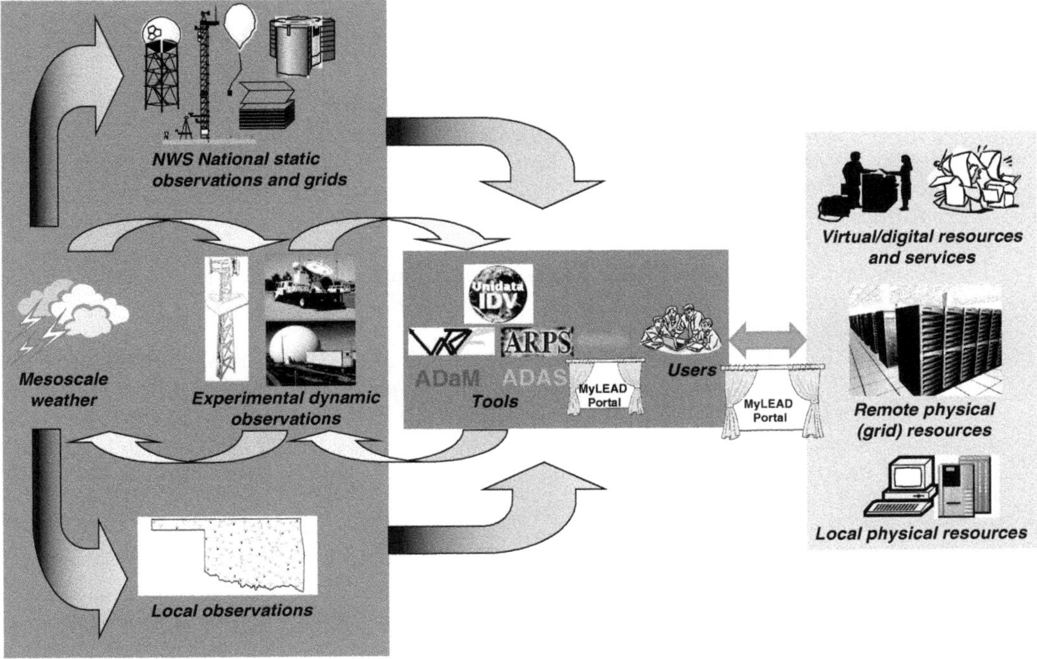

FIGURE 12.3
Many of the interaction mechanisms embodied in the LEAD project. (Credit: Kelvin Droegemeier, University of Oklahoma [22].)

multiple codes, as each of the codes do not handle the full range of phenomena being examined. Examples of this application include copper electrodeposition to form microprocessor conductors, crevice corrosion, and flow through porous media in petroleum reservoirs.

In the copper electrodeposition problem, the important phenomena occur on the surface of the silicon and deposited copper, which is moving and changing shape as the deposit evolves. In order to handle this phenomena we have linked a finite-difference diffusion-migration code to a kinetic Monte Carlo code, in which each code provides a boundary condition to the other code. The original incarnation of the linked codes had the Monte Carlo code evolving over time, only pausing after sending a concentration boundary condition to the continuum code in order to receive the corresponding spatial derivative flux boundary condition from the continuum code, which would provide this estimate after evolving to a pseudo steady state and then terminating.

Our OGRE workflow system (described in Chapter 11) and event system provide the signaling between the codes, triggered by the appearance and disappearance of small files; as well as the mechanism to move the boundary conditions, which were stored as files to the code requiring the respective boundary condition. A subsequent refinement of this problem involved consideration of the resistance difference along the surface of a 300-mm silicon wafer; in this case the model was controlled by a 1D resistive strip model, which was linked to a number of Monte Carlo codes [23]. In this case, the 1D resistance code will initiate Monte Carlo simulations along the surface of the wafer, and then wait for all the Monte Carlo codes to return before continuing with the simulation. In this case, again, the OGRE workflow engine provides the basic infrastructure for signaling between the codes, as well as file transfer services.

References

1. Kerberos: The Network Authentication Protocol. http://web.mit.edu/kerberos/www/.
2. OpenSSH Manual Pages. http://www.openssh.com/manual.html.
3. Software Infrastructure for the I-WAY High Performance Distributed Computing Experiment. I. Foster, J. Geisler, W. Nickless, W. Smith and S. Tuecke Eds. *Proceedings of the 5th IEEE Symposium on High Performance Distributed Computing*, pp. 562–571, 1997.
4. Daves CAVE pages. http://www.evl.uic.edu/pape/CAVE/.
5. I. Foster and C. Kesselman. The Globus Alliance, http://www.globus.org/, and Globus: A Metacomputing Infrastructure Toolkit. *International Journal of Supercomputer Applications*, 11(2): 115–128, 1997.
6. I. Foster, C. Kesselman and S. Tuecke. The Anatomy of the Grid: Enabling Scalable Virtual Organizations. *International Journal of Supercomputer Applications*, 15(3), 2001.
7. OMII: Open Middleware Infrastructure Institute. http://www.omii.ac.uk/.
8. GSI: Key Concepts. http://www-unix.globus.org/toolkit/docs/3.2/gsi/key/index.html.
9. Public-Key Infrastructure (X.509) (pkix) Charter. http://www.ietf.org/html.charters/pkix-charter.html.
10. Cog Kit Wiki. http://www.cogkit.org/.
11. GFD.20 GridFTP: Protocol Extensions to FTP for the Grid. http://www.gridforum.org/documents/GWD-R/GFD-R.020.pdf.
12. NCSA GridFtp Client. http://dims.ncsa.uiuc.edu/set/uberftp/.
13. Welcome to TeraGrid. http://www.teragrid.org.
14. The Physiology of the Grid: An Open Grid Services Architecture for Distributed Systems Integration. I. Foster, C. Kesselman, J. Nick and S. Tuecke Eds. *Open Grid Service Infrastructure WG*, Global Grid Forum, June 22, 2002.
15. Open Grid Services Infrastructure (OGSI) Version 1.0. http://www.ggf.org/documents/GWD-R/GFD-R.015.pdf.
16. OGSI.net Main page. http://www.cs.virginia.edu/~humphrey/GCG/ogsi.net.html.
17. OASIS Web Services Resource Framework (WSRF) TC. http://www.oasis-open.org/committees/tc_home.php?wg_abbrev=wsrf.
18. WSRF.NET. http://www.cs.virginia.edu/~gsw2c/wsrf.net.html.
19. Welcome to the WS-I Organization's Web site. http://www.ws-i.org.
20. Linked Environments for Atmospheric Discovery. http://lead.ou.edu/.
21. CASA Home Page. http://www.casa.umass.edu/.
22. Linked Environments for Atmospheric Discovery (LEAD): Architecture, Technology Road Map and Deployment Strategy. K. K. Droegemeier, V. Chandrasekar, R. Clark, D. Gannon, S. Graves, E. Joseph, M. Ramamurthy, R. Wilhelmson, K. Brewster, B. Domenico, T. Leyton, V. Morris, D. Murray, B. Plale, R. Ramachandran, D. Reed, J. Rushing, D. Weber, A. Wilson, M. Xue and S. Yalda, Preprints, 21st international Conference on Interactive Information Processing Systems for Meteorology, 9–13 January, San Diego, CA, Amer. Meteor. Soc.
23. T. O. Drews, S. Krishnan, J. C. Alameda Jr., D. Gannon, R. D. Braatz and R. C. Alkire. Multiscale Simulations of Copper Electrodeposition onto a Resistive Substrate. *IBM Systems Journal*, 49(1), 2005.

13

Integrated Data Management System

Seongeun Jeong, Yao Liang, and Xu Liang

Contents

13.1 Introduction

Complicated scientific studies such as hydrological modeling involve various datasets from disparate systems and data sources. Development of modeling and analysis techniques to explain complex phenomena on the earth requires scientists to access and maintain the diversity of data at the very heart of Earth science research. With the rapid expansion of the World Wide Web (WWW) such diverse data are obtained from disparate systems and data sources that many agencies or institutions manage. One of the main issues that earth scientists including hydrologists encounter is interoperability between disparate data sources. When scientists use diverse datasets with different formats from various platforms, the datasets need to be converted into interoperable forms in the scientist's platform. This conversion is a time-consuming and sometimes overwhelming process that takes time and energy away from conducting real research. Another challenging issue that earth scientists face is easy access of data analysis and visualization for large datasets with heterogeneous formats. The analysis and visualization tools allow scientists to check the quality of the data to be used and to facilitate new discoveries of hidden properties or laws represented by complex datasets [1,2].

Integrated data management is a systematic approach to effectively address the specific needs discussed earlier for scientific information management. An integrated data management system collects, processes, stores, displays, and disseminates information. The "integration" feature in a system can be defined in two aspects: (1) a system to accommodate all disparate data sources with heterogeneous formats and structures; and (2) a system to perform a variety of tasks from data retrieval, management, aggregation, and merge analysis to data visualization [3]. The rest of this chapter discusses the architecture of the integrated data management system and its implementation focusing on hydrological applications.

13.2 Metadata and Integrated Data Management

13.2.1 Metadata for Data Management

The integrated data management system is developed based on metadata. Here, we will discuss how to use metadata for data management. Metadata are simply information about data. In practice, metadata comprise a structured set of descriptive elements to describe an information/data resource. For example, the data of a newspaper article is composed of the headline and story, whereas metadata describes who wrote it, when and where it was published, and in what section of the newspaper it appears. Although the concept of metadata is simple and clear as seen in this example, the use of metadata in scientific information is not as simple because metadata can be used in various forms. On one hand, metadata can be used to describe elements of actual data such as the name, format, content, and control of the data; on the other hand, metadata can be used to describe information about the quality and condition of the data. For detailed background information on metadata, readers are referred to other relevant chapters.

In relation to data management, however, we find two distinct types of metadata. First, metadata are used to identify the existence of data. Second, metadata documents the content, context, quality, structure, and accessibility of a specific dataset. We use the first type of metadata to query data from various sources and use the second type to provide more detailed information on the data for scientific visualization and analysis [4]. It can be seen that to build an integrated data management system, both types of metadata are necessary.

13.2.2 Architecture of Metadata-Driven Integrated Data Management System

Metadata can be used as an effective integration tool in a data management system because a set of metadata is able to include all the information about the data needed in the data system. An integrated data management system driven by metadata can include several types of metadata that interact with each other. A mechanism that supports the metadata and the interactions among the different types of metadata is called "metadata mechanism." The metadata mechanism performs its functions by interacting with other components of the data management system where it resides. The role of metadata mechanism needs to be better understood in the context of a data system. Thus, we need to understand what constitutes a data system, in particular, an integrated data system. An integrated data management system consists of a database/data storage system coupled with tools to retrieve, integrate, analyze, and visualize data. These components of the system are essential.

An architecture of an integrated data management system is shown in Figure 13.1. The architecture consists of three components: data storage, metadata, and applications. The data storage component includes information regarding data sources that a hydrologist needs to conduct a hydrologic study. The information may include, for example, precipitation, streamflow, air temperature, land surface conditions, etc. Depending on the design, the data storage component can also include other interdisciplinary information such as geomorphology, land use, and ecology. In this architecture, the metadata component contains all information on data and data storage. Also, it may include the definition information about the formats of data prepared by various data centers (e.g., National Climatic Data Center, NCDC), government agencies such as the US Geological Survey (USGS), National Oceanic and Atmospheric Administration (NOAA), and individual data collectors. Whenever there is a new dataset available, the dataset needs to be easily supported by the metadata component. Therefore, it is important to design an extensible metadata component to have new data sources easily incorporated when they become available.

The application component in the architecture implements functionalities of data retrieval, aggregation/merge, visualization, and analysis interacting with the metadata component.

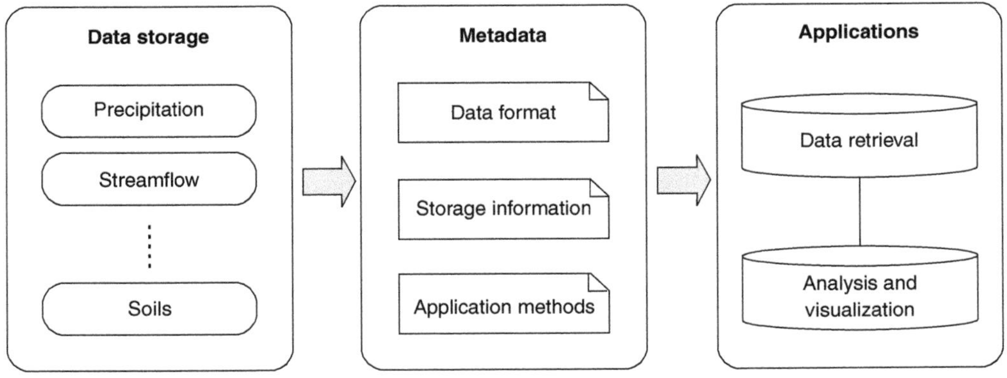

FIGURE 13.1
Architecture of metadata-driven integrated data management system.

To provide a user-friendly environment, the application is usually presented to the user through a Graphical User Interface (GUI). The user provides information to the data management system to retrieve or visualize a specifically requested dataset through the GUI, and the application component displays the requested information through the GUI. A modular approach is desirable for the application component to accommodate a variety of application environments and to be consistent with the flexible and extensible metadata mechanism in the architecture. In other words, when there is a need for a new function, the architecture of the system should allow the adding of a plug-in module for the new function rather than changing the structure of the entire application.

The data storage can take various forms ranging from fully equipped relational databases such as Microsoft (MS) Access and Oracle to simple file systems. An appropriate database should be carefully selected or designed to meet the needs of the system. The data storage does not necessarily have to be a database. If the metadata component includes information on both data and data storage, a simple file system can be used to support the structured organization of massive datasets based on the metadata. We will discuss more about this later.

13.2.3 Components Interaction

Interactions among the components of a data system play an important role in determining its performance. For example, recent development of markup language, especially the eXtensible Markup Language (XML), enables the interactions between metadata written in XML and applications to be more effective. The World Wide Web Consortium (W3C) has recommended the Document Object Model (DOM) to dynamically access and update the content, structure and style of documents. DOM allows the XML document to be interpreted in a standard way by different applications. When metadata are written in XML, applications can access the metadata documents through a hierarchy of objects of type "Node." In an XML document, which is structured as a tree, a node is a point where two or more lines (branches) meet.

One can also think of using a simpler interaction method other than complicated interfaces. For instance, a text file could be an information exchange means to support an application-level interaction. Using temporary swap files as a means of information exchange could be an alternative because it is simple. In Section 13.5.2, we will discuss how to use swap files as an application-level interaction method.

13.3 Metadata Mechanism for Data Management

In Section 13.2, we discussed the concept and overall architecture of an integrated data management system. In this section, we explore the specifics of the metadata mechanism by understanding metadata in the context of an integrated data management system rather than the details of metadata themselves. Detailed implementation processes will be discussed in Section 13.4.

13.3.1 Structure

The structure and role of metadata in a data system largely depend on how they are written. Metadata can be written in many forms using various file types. Before XML emerged, most metadata were written in a text file format. To be serviceable on the Web, many data providers including government agencies present metadata in HyperText Markup Language (HTML) format. Recently, XML has become a standard language for writing metadata. In this section, we use XML to write metadata, allowing information to be self-describing to the application by providing information about the content of a document. Also, XML enables hierarchical structure of documents, thereby providing us with a convenient way to write metadata and interact with applications. To keep abreast of the recent development in metadata design, we focus on using XML as a metadata language in the following.

We identified two distinct types of metadata in Section 13.2.1. The first type may be defined as "retrieval metadata" because it is related to data source identification and retrieval. The second type may be called "application metadata," which is associated with various functions that applications implement including data visualization and analysis. When the two types of metadata interact with each other, a mechanism of metadata is then established to support an integrated data management system.

Retrieval and application metadata need to be divided into more specific "functional types" to perform various tasks of an integrated data system. Because a metadata mechanism includes information about both data and data storage, the functional types of metadata should be articulated to support the tasks. Figure 13.2 illustrates an example of metadata hierarchy employed by a metadata mechanism. The metadata hierarchy consists of five types of metadata: logical category, query, logical directory, format, and application method. The first three types of metadata belong to the retrieval metadata, while the other two belong to the application metadata. Depending on the design of the metadata hierarchy

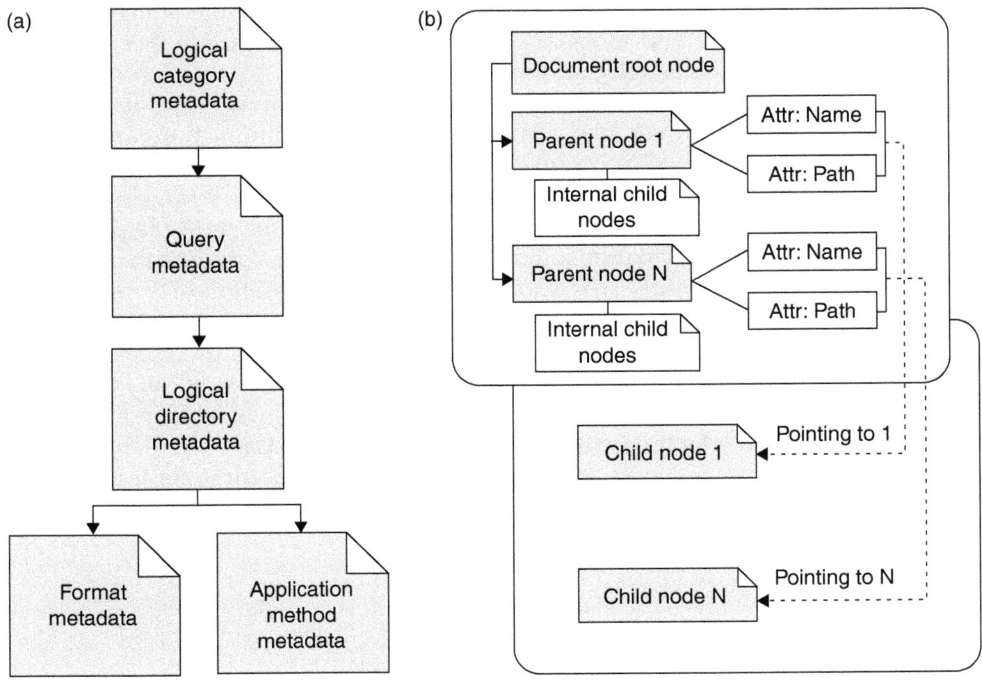

FIGURE 13.2
(a) Metadata interactions within a metadata mechanism. In the metadata for logical directories, the names and paths of metadata for formats and visualization methods are embedded. (b) The basic internal structure of an XML document. Normally a child node pointed by its parent indicates a separate XML metadata document that contains its own child nodes.

of a metadata mechanism, the functional metadata types may vary. This set of five types of metadata is an example of a metadata hierarchy in a metadata mechanism that supports an integrated data management system. In the following sections, we use these five types of metadata as an example to describe an integrated data management system.

The mechanism shown in Figure 13.2 is an interactive process of metadata involved in data retrieval and visualization. The five types of metadata are invoked by application functions in the order shown in Figure 13.2(a). The logical category metadata can be used to form an interface for users to select a category to access data. A data category can be any type used to characterize data. For example, a set of data source categories may include USGS, NOAA, NASA, etc. The query metadata is used to define methods to query data. This type of metadata includes query conditions such as date formats (e.g., YYYYMMDD) and numerical fields of data (e.g., precipitation in millimeters). The logical directory metadata includes hierarchical data storage paths. The logical directory for a dataset is organized in a tree form (see Figure 13.4). The metadata for formats and application methods are used to specify data formats and application methods, respectively. Each type of the metadata is used in the order shown in Figure 13.2(a) to retrieve and visualize data produced by various sources. The details of each metadata type are discussed in Section 13.3.2.

In the mechanism shown in Figure 13.2(a), the previous metadata also provides a pointer to the next metadata to be accessed. The previous metadata includes information such as metadata names and their paths to point to the next metadata. In the structure shown in Figure 13.2(a) and (b), the application is initiated by detecting the root element included in one of the logical category metadata. The application obtains information (e.g., paths to the next step) for the next step from one of the child elements of the root element. Based on the information, the category metadata invokes one of the query metadata. In the same fashion, the query GUI metadata calls the logical directory metadata, which invokes the format and application method metadata. In each transition from one metadata type to another, the user's selection is reflected through GUIs.

Figure 13.2(b) shows the basic structure inside an XML metadata document while Figure 13.2(a) illustrates the overall mechanism (i.e., metadata interactions and hierarchy) of the five types of metadata. The internal structure of an XML document can be understood as a hierarchical collection of nodes, which are equivalent to XML elements (tags). Under the document root nodes in Figure 13.2(b), there could be multiple parent nodes. Each parent node has the information to point to its child. In this conceptualization, a node is more than an element because it is also an abstract representation of a metadata itself by having the name and path to the metadata. When an application perceives a node as an object through DOM, the application is able to access the metadata represented by the node. For example, when a Java-based application accesses parent node 1 in Figure 13.2(b), it actually obtains the values of its attributes. The attribute values include the name and path of the metadata to which parent node 1 points. Thus child node 1, a child metadata represented by a separate XML file, can be accessed through parent node 1. Internal child nodes are used to provide information for special functions such as data query.

13.3.2 Functional Types of Metadata

An XML-based metadata is a well-formed document to define data. However, it cannot perform any functions unless it is interpreted by applications through a metadata mechanism. When each type of the metadata described in Section 13.3.1 is read by applications, the application is then driven by the metadata to conduct a unique function. In this section, we describe the functional types of metadata and discuss the function of each type.

13.3.2.1 Metadata for Logical Categories

The metadata for logical categories is a basic element in a metadata mechanism design. The metadata for logical categories is designed to describe the logical category structure of all datasets stored in a system. "Category" is a collection of data sharing common features. Therefore the "category structure" provides us with information on how the categories of data sources are organized, particularly in a hierarchical way. When we apply the category concept to data sources such as streamflow from USGS and precipitation from NCDC, a category refers to a logical group that is related to the datasets of a data source. The category structure can be used to classify data in any way that the user prefers. For instance, the data may be categorized by regions or by sources.

Categorized datasets comprise a hierarchical collection of nodes, which is composed of parents and their children. Through such a collection, a structural frame of datasets from a data source is established. Figure 13.3 shows a way of categorizing hydrologic datasets to build a metadata hierarchy. As shown in Figure 13.3, hydrological data can be structured through data sources or types. Some users may prefer to retrieve data from data sources such as NOAA and USGS while others may want to access data by data types (e.g., precipitation data). In Figure 13.3, each source or type (i.e., each box) at each level of the tree structure is a node. In a tree structure, a node is a point where two or more lines meet. Traversing the tree structure, one can reach the same data node whether one retrieves according to data source or data type. For example, whether one started from the "NOAA" node or "Precipitation" node, one is able to reach the point precipitation data or radar precipitation data. The point precipitation data are provided by rain gage stations. A category structure such as the one shown in Figure 13.3 determines the node tree of metadata for logical categories. The designer of a metadata mechanism should consider how to construct a node tree. Every node in the tree is represented by metadata through the parent–child relationship. Section 13.4.2 discusses details on how to present the category structure in XML-based metadata.

13.3.2.2 Metadata for Query

The metadata for query is used to generate a query GUI for the user to enter query information. This type of metadata is used to support data search. In conjunction with the metadata

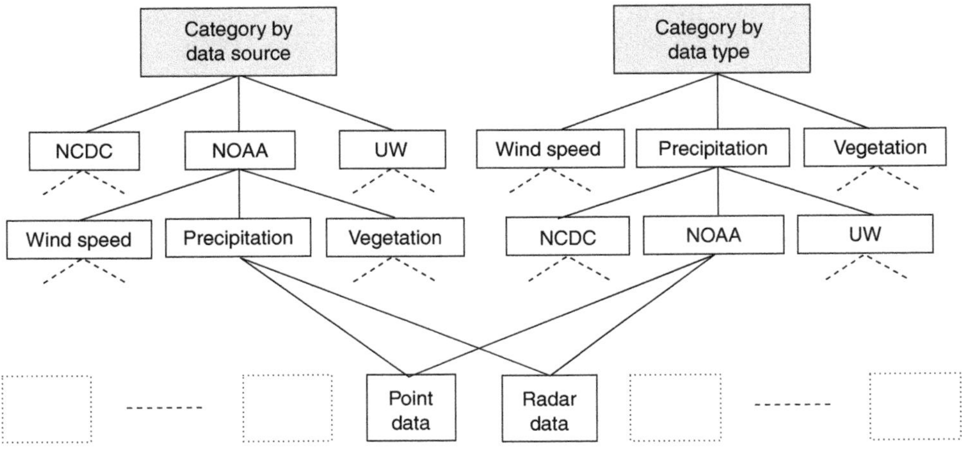

FIGURE 13.3
An example of category structure.

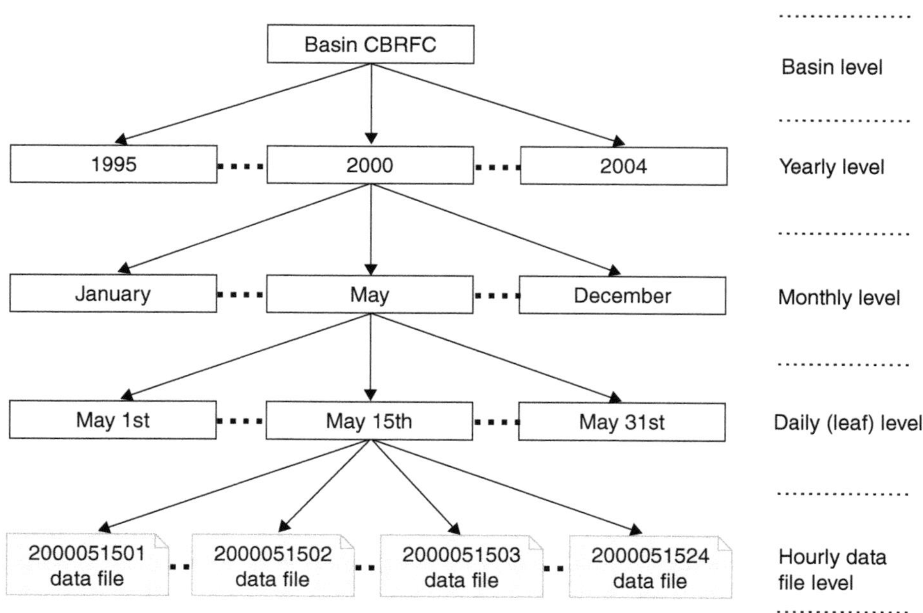

FIGURE 13.4
An example of data structure for data query and logical directories.

for logical directories, this type of metadata allows for establishing a structural data organization. A hydrologic data file is generally composed of several data fields, including dates and measured values. Each field can be used to query the data file. That is, data fields in a data file are extracted to construct a query environment. For instance, streamflow data files from USGS have a data field labeled "measured streamflow." The streamflow field becomes an input parameter for data query on a GUI, when the GUI provides a text box for it. In the text box, one can enter query conditions using keywords such as "from" and "to." By providing these query conditions, one can retrieve all the streamflow data that meet the query conditions. Before we proceed, we suggest that readers consider how to organize data from different data sources to conduct effective data query.

Although datasets produced by a data source can be organized in many ways, we introduce a method to create a tree structure of datasets to provide a better understanding of the query metadata. Figure 13.4 shows a structure of datasets from Hydrologic Data Systems Branch (HDSB) Next Generation Radar (NEXRAD) Stage III. The structure illustrated in Figure 13.4 is mainly associated with metadata for query. Note that this structure is a collection of nodes that forms a tree. The node tree should be designed in such a way that the query functions of an application can traverse each node to reach desired files. For instance, when the user queries HDSB NEXRAD Stage III hourly precipitation data shown in Figure 13.4, the user may enter query conditions such as {*Basin* CBRFC, 2000 05 15 02, 2000 05 15 03}. Based on the query conditions, the application can traverse the related nodes in the tree structure shown in Figure 13.4. For example, the application reads the top node of the tree for the HDSB NEXRAD Stage III precipitation dataset and records the content for each child node. With this information, the application can identify that only Basin CBRFC is qualified for the query condition. Thus all other subdirectories are filtered out. In a similar fashion, the metadata for CBRFC includes nodes for its several child nodes such as the years 2000 and 2004. Since the year 2000 matches the query conditions, the metadata for the year 2000 is accessed and parsed. After two more steps, the application reaches the metadata at the daily level. The daily level includes leaf metadata, which contain the file name and path for each hourly time interval. The term "leaf" indicates the lowest level node in the tree structure. The application records the names and paths of the two files: 2000051502.dat and 2000051503.dat. For information regarding the query function in a GUI environment, see Section 13.5.2.

13.3.2.3 Metadata for Logical Directories

The metadata for logical directories help to find real data files when the application traverses the nodes of a logical directory tree. This type of metadata is used to provide a road map to locate files of interest based on query conditions. The metadata for query define only logical conditions, while the metadata for logical directories provide directions to reach actual data files. A logical directory tree is similar to the directory structure one may have in a local disk drive. This type of metadata records the content of logical directories for all data files within the system. It contains all the metadata names and paths for each node in the logical directory tree, in which a parent node includes its child nodes' names and paths. When a logical directory is supported by the tree structure, a parent node points to its child, thereby enabling data search. Thus, this logical directory structure makes it possible to have the structured organization of disparate datasets in an integrated data system.

As also shown in Figure 13.4, the directory tree is organized according to the dataset structure. Each node represents a metadata file that describes information of the corresponding logical directory. In Figure 13.4, Basin CBRFC is the root of the tree, and its child nodes represent information at the yearly level. In a similar fashion, each yearly node includes monthly nodes, which are usually composed of 12 nodes (i.e., 12 metadata names and their paths). Depending on the query conditions that the user provides, the node invoked by its parent node differs at each level. Note that logical directory metadata are used along with query metadata. Until the application reads the leaf node, which points to actual data files, this process is repeated based on the user's query conditions. In Figure 13.4, the directory structure is made for an hourly dataset. Therefore, the daily nodes are leaf nodes which include location information for the data files.

13.3.2.4 Metadata for Data Formats

Query metadata describes part of the data format for a dataset. However, the format information of a dataset written in the query metadata is not sufficient for the application to interpret the dataset. Through a series of metadata such as the query metadata and logical directory metadata, the application presents a list of data files that the user can select. Based on the user's query conditions, the system reads individual files. However, the system needs to understand a data file in terms of its data format and structure. We use format metadata to provide such information. The format metadata gives instructions on: (1) the data format (e.g., ASCII format); (2) the row or byte from which the real data start; (3) the number of columns and their order; and (4) the physical data type (i.e., integer, long, string, etc.) of each field. In general, if all datasets from a data source follow the same data format and structure, only one metadata of this type is needed for the data source. Thus, all data files belonging to the data source use the same metadata for formatting.

A conceptual scheme of metadata for formats is shown in Figure 13.5, in which "Root_Elem," "Attr," and "Elem" denote the root element, attribute, and element of an XML document, respectively. The *Valid from* attribute is used to describe from which row or byte the real data start. The symbol "+" implies that the *aField* element can be used multiple times. The *aField* element is used to describe each data field in a data file.

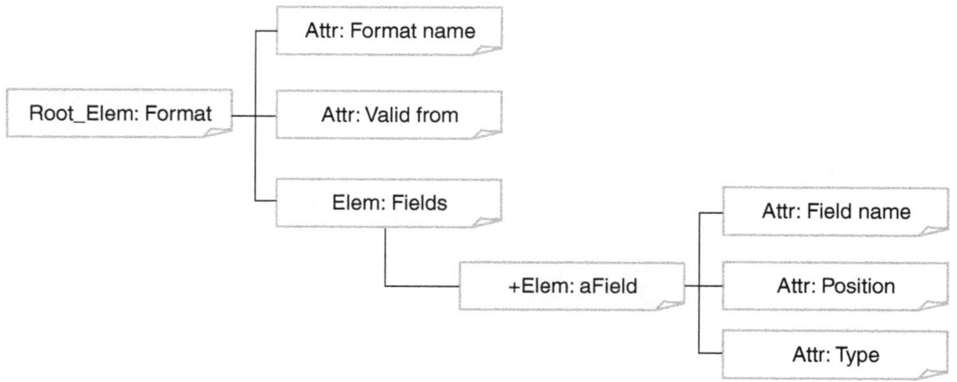

FIGURE 13.5
Schematic representation of the metadata for data formats.

13.3.2.5 Metadata for Application Methods

Using the previous four types of metadata, the user is able to reach the desired data files. The metadata for application methods are designed for additional functionalities such as data visualization. Add-on application toolkits such as data visualization certainly make a data management system more useful and integrative. A data system with this type of metadata is able to perform various operations including visualization and analysis in addition to data retrieval. Depending on the nature of datasets, the applicable methods may differ. However, any dataset from a specific data source usually requires only one metadata for application methods. For instance, all the hourly precipitation data from NCDC require only one metadata file for the application method because the data files are consistent in their format. Details on this type of metadata are explained in Section 13.4.2.

13.4 Data Management System Using Metadata Mechanism

Based on the conceptual framework of the metadata mechanism discussed earlier, this section focuses on the implementation of the metadata mechanism. It also includes discussions on principles of designing a metadata mechanism and implementations of an integrated data management system. Examples of how to write metadata in XML are given, along with practical methods to design metadata. Finally, a system-level application that is supported by the metadata mechanism is presented as a model.

13.4.1 Principles of Implementing a Metadata Mechanism in a Data System

This section provides examples to illustrate how to design metadata, how a plain text-format data file is organized in an XML-based metadata, and how an application reads metadata, using Java to create the application. Note that this chapter focuses on an integrated data management system rather than metadata themselves. However, knowing how to write metadata is important to understand the integrated data management system.

In Section 13.3.2 we discussed five types of metadata for a metadata mechanism. Depending on the designer of metadata, the composition of metadata types may be different. However, those five types would be enough to support an integrated data management system unless there are additional functions for the system to perform. Based on the five types of metadata, designing a metadata mechanism may begin by asking questions about information of datasets such as:

1. What categories (sources) of datasets are available?
2. What is the storage path of each dataset?
3. What is the format of each dataset (ASCII, Big Endian, Little Endian, etc.)?
4. From which row (for ASCII format) or byte (for binary format) does the real data begin for each dataset?
5. How many fields are in each dataset and in what order are they arranged?
6. What is the physical data type (integer, long, etc.) for each field?

To answer question 1, one may identify data sources such as USGS streamflow and NCDC precipitation as different categories. Thus, USGS streamflow datasets may be included in the metadata for logical categories. Storage paths for datasets need to be designated in metadata for logical directories as an answer to question (2). In Section 13.4.2 we illustrate how to write metadata for logical directories. Questions (3) through (6) are associated with metadata for formats, queries, and application methods.

The steps in Figure 13.6 show how to answer these six questions. Figure 13.6(a) shows simple daily precipitation data from NCDC in a text file. By looking inside this file named "ncdc_dppt.txt" we know that this file has four data columns in ASCII format. Thus, we can build a format structure for the file that needs to have elements such as a file format name (e.g., ASCII, Big Endian, etc.) and attributes of data fields (Recall the scheme shown

(a)

Data (Sample NCDC daily precipitation data; File name: ncdc_dppt.txt)

```
2004   1   1   5.1
2004   1   2   0.0
2004   1   3   0.0
2004   1   4   8.2
...
```

(b)

DTD for ncdc_dppt.txt (File name: format.dtd)

```
<!ELEMENT format_mtd (format)>
  <!ELEMENT format (fields)>
    <!ATTLIST format name (ascii|bin_bigEndian|bin_littleEndian)#REQUIRED
              validFrom CDATA #REQUIRED>
    <!ELEMENT fields (aField+)>
      <!ELEMENT aField EMPTY>
       <!ATTLIST aField name ID#REQUIRED
                       position CDATA#REQUIRED
                       unit CDATA#IMPLIED
                       type (String|int8|double64|float32|int32|long64|int16)#REQUIRED>
```

(c)

Metadata for ncdc_dppt.txt (File name: ncdc_dppt_format.xml)

```
<?xml version="1.0" ?>
<!DOCTYPE format_mtd>
<format_mtd>
 <format name="ascii" validFrom="0">
   <fields>
        <aField name="year" position="1" type="int32"/>
        <aField name="month" position="2" type="int32"/>
        <aField name="day" position="3" type="int32"/>
        <aField name="precipitation" position="4" unit="mm" type="float32"/>
   </fields>
 </format>
</format_mtd>
```

FIGURE 13.6
An example of a process that reinterprets a data file using an XML-based metadata.

in Figure 13.5). The attributes of data fields are derived from their intrinsic characteristics including the data type and relative position of each field. Before we write an XML-based metadata for a data file, we need to write a Document Type Definition (DTD) or an XML schema to define the elements and data structure of the XML document. Currently, the new XML schema specification is replacing DTDs. However, since DTDs are easier to understand than XML schema, we introduce and use DTDs here to help readers understand the fundamentals of a document model. A DTD is designed for a type of data files that follow the same definition. The DTD in Figure 13.6(b) defines four elements: *format_mtd, format, fields,* and *aField*. The *format_mtd* element is a root element that controls the entire document. The other three elements define the format and structure of the data file. The *fields* element has the *aField* element as a child. The symbol "+" following the *aField* element implies that "one or more" preceding element is required as mentioned earlier. We use this "one-or-more" frequency operator to ensure that at least one data field (e.g., year, month, streamflow, etc.) exists for the element. To depict the format of each field, the *aField* element has four attributes: *name, position, unit,* and *type*. Often, since some data fields do not have units, the *#IMPLIED* keyword is used to indicate that there is no default value for the optional *unit* attribute. Using those basic elements and attributes, the DTD defines the XML document for the example data file. Furthermore, the simple DTD in Figure 13.6(b) can be used to describe hydrologic data formats.

Figure 13.6(c) shows an XML document whose elements and data structure were defined by the DTD. The value of the *name* attribute of the *format* element indicates that the data file is in ASCII format. The *validFrom* attribute (i.e., 0) indicates that the real data starts from the first row. Each *aField* element and its attributes under the *fields* element are used to define the format and structure of each data column. For more details about writing DTDs, XML schema, or XML documents, readers are referred to other relevant chapters.

The focus of this section is to illustrate how to construct an integrated data management system. Hence, metadata needs to be linked to applications to constitute a complete system. XML currently provides two Application Program Interfaces (APIs) that give instructions on how to read XML documents by applications. These two APIs are the Simple API for XML (SAX) and the DOM mechanism. SAX attempts to define a standard event-based XML API, processing the document in a single pass. On the other hand, the DOM mechanism examines a document and uses it to create a new representation of the data. The DOM mechanism describes how each node in an XML document tree can be represented in the DOM tree as an object. The DOM mechanism is a more organized way to contain data than SAX because SAX pays little attention to the data's tree structure [5]. Thus we use the DOM mechanism to convey information of XML metadata to applications.

In Figure 13.7, using the DOM mechanism and Java, we illustrate how an XML document is interpreted in an application. As shown in Figure 13.7, the application creates a tree

```
...
NodeList format = ((Element)format_mtd.item(0)).getElementsByTagName("format");
NodeList myFields = ((Element)format.item(0)).getElementsByTagName("fields");
String srcFileFmt = ((Element)format.item(0)).getAttribute("name");
String srcFileValidFrom = ((Element)format.item(0)).getAttribute("validFrom");
NodeList fields = ((Element)myFields.item(0)).getElementsByTagName("aField");
int totalFieldsNum = fields.getLength();
for (int i = 0; i < totalFieldsNum; i++)
  {
   String fieldName = ((Element)fields.item(i)).getAttribute("name");
   String fieldPosition = ((Element)fields.item(i)).getAttribute("position");
   String fieldUnit = ((Element)fields.item(i)).getAttribute("unit");
   String fieldType = ((Element)fields.item(i)).getAttribute("type");
   ...
  }
```

FIGURE 13.7
Java (i.e., Java API for XML Processing) code segment for parsing XML metadata shown in Figure 13.6.

representation of "ncdc_dppt_format.xml," which is shown in Figure 13.6(c) using the DOM mechanism. All elements along with their attributes are organized in an invisible tree structure. After parsing ncdc_dppt_format.xml, the application obtains information about the file format. When information such as data storage is provided through other XML metadata, the application then acquires a complete set of information for the data file. The application knows the location and format of the file, and the available application methods. Consequently, it understands and reads data in "ncdc_dppt.txt" that was shown in Figure 13.6(a). Using the ASCII reading function the application reads real data from the beginning of the file since the value of the *validFrom* attribute is "0." In addition, the application creates four array variables labeled "year," "month," "day," and "precipitation" to store each variable's data type. While reading "ncdc_dppt.txt" row by row, the application assigns each field to one of the four array variables based on the *position* attribute.

Figure 13.8 shows a GUI that retrieves a data file from NOAA Distributed Model Intercomparison Project (DMIP) following the procedure mentioned earlier. In the same way that metadata defines a data file, the application retrieves the data file.

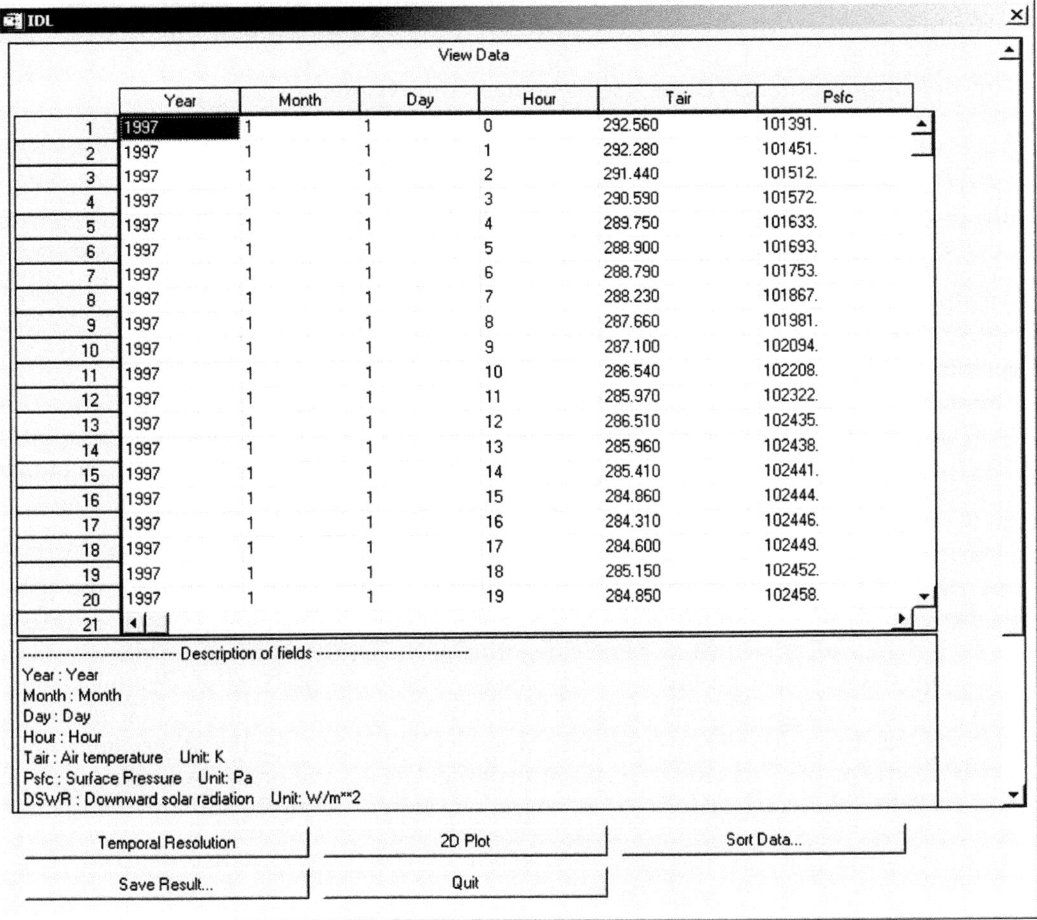

	Year	Month	Day	Hour	Tair	Psfc
1	1997	1	1	0	292.560	101391.
2	1997	1	1	1	292.280	101451.
3	1997	1	1	2	291.440	101512.
4	1997	1	1	3	290.590	101572.
5	1997	1	1	4	289.750	101633.
6	1997	1	1	5	288.900	101693.
7	1997	1	1	6	288.790	101753.
8	1997	1	1	7	288.230	101867.
9	1997	1	1	8	287.660	101981.
10	1997	1	1	9	287.100	102094.
11	1997	1	1	10	286.540	102208.
12	1997	1	1	11	285.970	102322.
13	1997	1	1	12	286.510	102435.
14	1997	1	1	13	285.960	102438.
15	1997	1	1	14	285.410	102441.
16	1997	1	1	15	284.860	102444.
17	1997	1	1	16	284.310	102446.
18	1997	1	1	17	284.600	102449.
19	1997	1	1	18	285.150	102452.
20	1997	1	1	19	284.850	102458.
21						

FIGURE 13.8
Tabular format view of a data file from NOAA DMIP in a GUI that is implemented using the Interactive Data Language (IDL).

```
<?xml version="1.0" ?>
<!DOCTYPE TypesDef>
<TypesDef>
   <aType name="DMIP" display_name="NOAA DMIP" mtd="dts_for_src_DMIP_mtd.xml" mtd_RelaStorepath=".\">
      <description>NOAA Distributed Model Intercomparison Project (DMIP)</description>
   </aType>
   <aType name="HDSB" display_name="HDSB" mtd="dts_for_src_HDSB_mtd.xml" mtd_RelaStorepath=".\">
      <description>Hydrologic Data Systems Branch (HDSB)</description>
   </aType>
   <aType name="NCDC" display_name="NCDC" mtd="dts_for_src_NCDC_mtd.xml" mtd_RelaStorepath=".\">
      <description>National Climatic Data Center</description>
   </aType>
   ...
   ...
</TypesDef>
```

FIGURE 13.9
An example of the metadata for logical categories of which root element is *TypesDef*.

13.4.2 Writing Metadata

This section shows how to write real XML metadata for an integrated data management system. In Section 13.3.2, we discussed five types of metadata. In this section, we demonstrate how each type of the metadata can be written, using example XML documents. The explanation of each XML document is based on the principles discussed in Section 13.4.1. Since we focus on the metadata mechanism and data management system, we do not cover all the details of writing DTDs or XML schema in this chapter.

13.4.2.1 Example on Metadata for Logical Categories

In the previous sections, we discussed the overall concept of metadata for logical categories. Although we can use many different methods to write this type of metadata based on the concept we described, in order to focus on integrated data management, we list all the available sources as a first step in integrating data sources. This individual source becomes a logical category. Each source category is also able to have subcategories, producing a tree structure.

The tree structure for categories is presented in an XML-based metadata in Figure 13.9. The *aType* element has data sources such as DMIP, HDSB, and NCDC as its attributes. If there are more available data sources, those sources can be added in the metadata. The *display_name* attribute indicates the name to be displayed in a GUI. The *mtd* attribute is used to point to a child node (category) of the current node. For instance, one of the current nodes in Figure 13.9 is DMIP. The child nodes of the DMIP node are pointed to by the value of the *mtd* attribute, which is an XML-based metadata with the name "dts_for_src_DMIP_mtd.xml." The XML-based metadata "dts_for_src_DMIP_mtd.xml" may include its children such as basin categories. If the DMIP datasets are classified into basins, the datasets can be accessed by basins, which are used as geographical units for hydrologic studies. The *mtd_RelaStorepath* attribute contains information about the metadata's relative paths. The *aType* element has a child element called *description*, which describes the category defined by the *name* attribute. Depending on one's design logic, the metadata is found in various forms that consist of different elements and attributes.

13.4.2.2 Example on Metadata for Query

Figure 13.10 shows a text file and its corresponding XML query metadata. The data file shown in Figure 13.10(a) is hourly precipitation data from NCDC with six data fields. In Figure 13.10(b), the metadata, which characterizes the data file, provides the user with query information. The values in the columns containing date and precipitation information

(a)

Station_No	Year	Month	Day	Hour	Precipitation
47630	1948	7	1	0	0.5
47630	1948	7	1	1	2.0
47630	1948	7	1	2	1.5
47630	1948	7	1	3	0.9
47630	1948	7	1	4	0.2
47630	1948	7	1	5	0.0

(b)

```
<?xml version="1.0" ?>
<!DOCTYPE OneFileType_mtd (View Source for full doctype...)>
<OneFileType_mtd name="HPCP_asc">
  <allFiles_mtd>
    <oneMtd files_mtd="asc_files_Sacramento_HPCP.xml" files_mtd_relaStorePath=".\" />
  </allFiles_mtd>
  <characterFields>
    <symbolFields>
       <aSymbolField name="sacramento" displayName="Sacramento Precipitation">
         <symbolValue value="047630 SACRAMENTO EX CA" />
       </aSymbolField>
    </symbolFields>
  </characterFields>
  <queryFields>
    <singleDateFields>
       <aSingleDateField name="Year Month Day Hour" displayName="Date"
       fmt="YYYY_MM_DD_HH" />
    </singleDateFields>
    <numericFields>
       <aNumericField name="Station_No" displayName="Cooperative Station Number" />
       <aNumericField name="Year" displayName="Year" />
       <aNumericField name="Month" displayName="Month" />
       <aNumericField name="Day" displayName="Day" />
       <aNumericField name="Hour" displayName="Hour" />
       <aNumericField name="Precipitation" displayName="Precipitation" unit="mm" />
    </numericFields>
  </queryFields>
</OneFileType_mtd>
```

FIGURE 13.10
(a) NCDC hourly precipitation data in ASCII format (b) Metadata for the hourly data file.

can be used to query data files. In Figure 13.10(b), there are three main elements under the root element (i.e., *OneFileType_mtd*). The *allFiles_mtd* and *characterFields* elements are used to convey basic information about data files such as storage paths and file descriptions. The *queryFields* element, which indicates the fields found in the column headings of the data file shown in Figure 13.10(a), is used to represent all query conditions for the data file.

Under the *queryFields* element, there are two child elements, which are *singleDateFields* and *numericFields*. In addition, the child elements of these two include attributes such as *name* and *displayName*. These elements and attributes represent the structure of many scientific data that are widely distributed. Thus the elements and the attributes under the *queryFields* are an interpretation of the NCDC data file in terms of querying. A GUI can be designed in such a way that each leaf element (i.e., *aSingleDateField, aNumericField*) in Figure 13.10(b) constitutes an input menu of the query GUI. The user can provide a query condition in each text box of the menu (e.g., Java TextField class). As long as the data format for NCDC data files is consistent, the user needs only one *queryFields* element as shown in Figure 13.10(b). To alter the format of NCDC precipitation data, the user needs to modify only the metadata without making any changes in the application. For example, NCDC may provide one more data column called "air temperature" in the future. In this case, the user only needs to add the following simple XML syntax under the *numericFields* element: *<aNumericField name="air temperature" displayName="air temperature" unit="K" />*. This simple modification ensures the efficiency and extensibility of metadata mechanism for integrated data management.

```
(a)
<?xmlversion="1.0"?>
<!DOCTYPE OneFileType_mtd2>
<OneFileType_mtd2 name="HDSB_NEXRADIII_Precip_mtd">
 <allFiles_mtd>
   <oneMtd mtd="files_struct_NEXRADIII_Precip.xml"
   relaStorePath=".\NEXRAD_StageIII\" />
 </allFiles_mtd>
 <characterFields>
   <symbolFields>
     <aSymbolField name="Basin_Region" displayName="Basin/Region">
       <symbolValue value="ABRFC" />
       <symbolValue value="CBRFC" />
       <symbolValue value="CNRFC" />
     </aSymbolField>
   </symbolFields>
 </characterFields>
</OneFileType_mtd2>
```

```
(b) (file name: files_struct_NEXRADIII_Precip.xml)
<?xml version="1.0" ?>
<!DOCTYPE Tree_Struct_of_OneFileType>
<Tree_Struct_of_OneFileType>
  <aFile leaf="0" name="CBRFC.xml"
  relaStorePath=".\CBRFC\" mtd="none">
    <characterFields>
      <symbolFieldsRange>
        <aSymbolFieldRange name="Basin_Region">
         <discreteSymbolValue value="CBRFC" />
        </aSymbolFieldRange>
      </symbolFieldsRange>
    </characterFields>
  </aFile> ... ...
</Tree_Struct_of_OneFileType>
```

```
(c) (file name: CBRFC.xml)
<?xml version="1.0" ?>
<!DOCTYPE Tree_Struct_of_OneFileType>
<Tree_Struct_of_OneFileType>
  <aFile leaf="0" name="CBRFC_2000.xml" relaStorePath=".\2000\" mtd="none">
   <queryFields>
    <singleDateFieldsRange>
       <aSingleDateFieldRange name="year month day hour"
       fmt="YYYY_MM_DD_HH">
         <rangeSingleDateValue from="2000 01 01 00" to="2000 12 31 23" />
       </aSingleDateFieldRange>
    </singleDateFieldsRange>
   </queryFields>
  </aFile>
</Tree_Struct_of_OneFileType>
```

FIGURE 13.11
(a) A simple metadata for data query. (b) A metadata for logical directories (files_struct_NEXRADIII_Precip.xml) to correspond to one of the query conditions in the query metadata in Figure 11(a). (c) A metadata for logical directories (CBRFC.xml) invoked by its parent metadata in Figure 13.11(b).

13.4.2.3 Example on Metadata for Logical Directories

Metadata for logical directories show how datasets belonging to a data source are connected to one another. The logical directory is similar to a file directory, and the metadata are used to organize and describe the structure of data files. The metadata for logical directories are hierarchically organized. In the following, a "pointer" will be used to indicate the invocation path of the relevant metadata in a data retrieval process. Figure 13.11(a) shows part of one of the simple query metadata. There are three *symbolValue* elements which means that there are three options (i.e., three basins) available for selecting a dataset. In this design structure, whichever option the user selects, the option is linked with "files_struct_NEXRADIII_Precip.xml," which is a metadata for logical directories. Note that metadata for query invokes metadata for logical directories in the metadata mechanism as shown in Section 13.3.1, where the actual invocation occurs through the application.

Figure 13.11(b) shows an invoked metadata for logical directories, which describes where available data files are and how the application can reach them for basin "CBRFC." Assume that the user selected the "CBRFC" basin among the three basins. "files_struct_NEXRADIII_Precip.xml" in Figure 13.11(b) is written for a basin level, which is an intermediate level to reach a NEXRADIII precipitation data file. The intermediate metadata points to its subordinate metadata shown in Figure 13.11(c), which contains data files for each available year. Note that all years except the year 2000 have been omitted by the user. One may note that there is a difference between the metadata in Figure 13.11(b) and the one in Figure 13.11(c). While the former has the *characterFields* element without the *queryFields* element, the latter has only *queryFields*. We intended these two elements to be optional, by using the following syntax: *<!ELEMENT aFile (description?, characterFields?, queryFields?)>* in which the symbol "?" implies that the preceding element or group can be used or neglected. At the basin level (i.e., Figure 13.11[b]), we do not need a query field because we need all metadata belonging to the specific basin.

In Figure 13.11(c), the attributes of the *aFile* element have three categories of information: (1) whether the metadata is a leaf node; (2) what the metadata is pointing to; and (3) whether it has a related format metadata. The value "0" in the leaf attribute indicates that this metadata in Figure 13.11(b) is not a leaf node. The current metadata points to a file with the value "CBRFC_200005.xml" in the *name* attribute. If the metadata file is a leaf node, the value in the *name* attribute will be a real data file. Since this metadata file is not a leaf node, there is no related format metadata in the *mtd* attribute. Also, if the metadata file is a leaf node, there would be a related format metadata. The format metadata in the leaf node are used to retrieve the real data file specified in the *name* attribute. The attribute values of the *rangeSingleDateValue* element, which is shown in Figure 13.11(c), are used as references for the application to find a desired file. In this example, the application must find all metadata belonging to the year 2000 because the metadata "CBRFC.xml" points to "CBRFC_2000.xml."

13.4.2.4 Example on Metadata for Data Formats

Metadata for data formats are designed to define the physical format of the data. When we design format metadata, we need to consider elements such as the file format (e.g., ASCII), number of fields, and data type of each field. Thus XML-based metadata for formatting would look similar to the one shown in Figure 13.12. In fact, when one opens a USGS streamflow data file, one would find that its format is well described by the metadata in Figure 13.12.

In Figure 13.12, under the *format_mtd* root element there are two child elements: *description* and *format*. The *description* element gives basic information of the data source while the *format* element defines the type of data and the starting row or byte of actual data in the file. The *format* element has a child element labeled *fields*, which also has *field* as its child. The number of the *field* element corresponds to that of the fields in a data file. The *position* attribute indicates the order of the data fields in the file. The *format* attribute has different values such as int32 representing 32 bit signed binary integers. With this information, the data file is formatted into a standard XML document. All of these elements and attributes in Figure 13.12 are a succinct but well-formed representation of a data format.

```
<?xml version="1.0" ?>
<!DOCTYPE format_mtd (View Source for full doctype...)>
<format_mtd>
   <description>USGS Streamflow Format</description>
   <format name="ascii" validFrom="1">
      <fields>
         <field name="year" position="1" type="int32" />
         <field name="month" position="2" type="int32" />
         <field name="day" position="3" type="int32" />
         <field name="streamflow" unit="cfs" position="4" type="int32" />
      </fields>
   </format>
</format_mtd>
```

FIGURE 13.12
An example of metadata for data formats.

```
<?xml version="1.0" ?>
<!DOCTYPE application_mtd>
<application_mtd>
  <description>Application Type for USGS Streamflow</description>
  <applicationTypes>
    <applicationType name="TempReso">
       <method>YYYY_MM_DD Year Month Day Streamflow</method>
    </aplicationType>
    <applicationType name="Plot2D">
       <method>YYYY_MM_DD Year Month Day Streamflow</method>
    </applicationType>
    <applicationType name="Sort">
       <method>YYYY_MM_DD Year Month Day</method>
    </applicationType>
  </applicationTypes>
</application_mtd>
```

FIGURE 13.13
An example of metadata for application methods.

13.4.2.5 Example on Metadata for Application Methods

The metadata for application methods make the system more integrated for data management, performing data visualization, and analysis. This type of metadata is related to application toolkits that support a variety of analysis and visualization methods. In a similar way to the other types of metadata, before we design metadata of this type we need to consider what elements are necessary for the XML metadata. Depending on the application toolkits that will be supported, the content of the metadata varies. However, one may need to identify the types of applications and visualization formats to be implemented. These two elements are basic to any metadata of this type.

Figure 13.13 illustrates a metadata file for application methods that describes USGS streamflow data. In Figure 13.13, the *applicationType* element can have several application methods in its nested attributes such as "2D plots" and "temporal resolution." The values *TempReso*, *Plot2D*, and *Sort* that are contained in the *name* attribute represent temporal resolution, 2D plot, and sort methods, respectively. Each of the string-type values can be uniquely interpreted by the individual applications. The text enclosed by the *method* element defines how the presentation would appear. For example, if the syntax *<method>YYYY_MM_DD Year Month Day Streamflow</method>* is used, the application understands that the x-axis will show dates in YYYY_MM_DD format and the y-axis will represent streamflow. The way the application reads XML metadata depends on the designer. One of the important factors

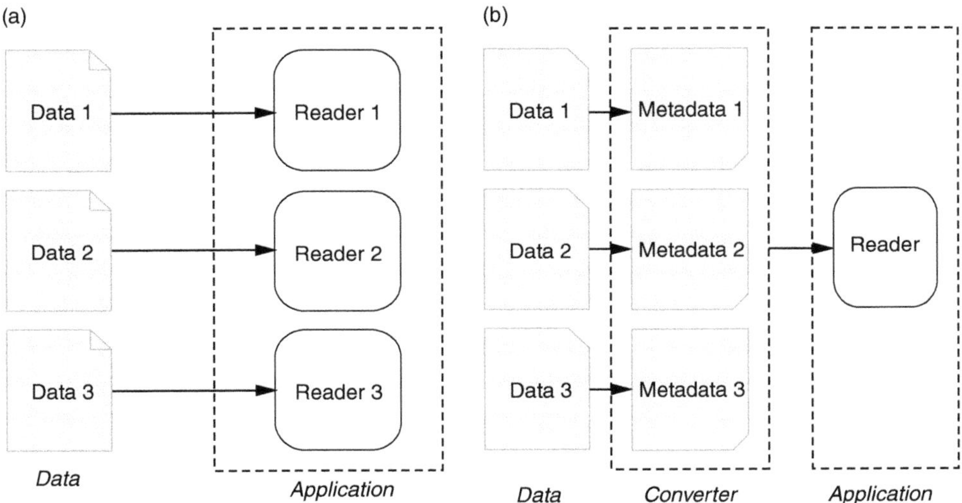

FIGURE 13.14
Traditional approach (a) vs. metadata-driven approach (b) for developing an integrated data management system.

in the implementation of this process is how one organizes the elements and attributes to represent application methods in a tree structure.

13.4.3 Construction of a Metadata-Driven Data Management System

The metadata-driven system whose components have been discussed previously aims at providing an integrated data management environment. This section will discuss a sample system that implements an integrated environment and focuses on the metadata-driven architecture for an integrated data management system. In addition, we present how the basic system can be established and modified flexibly. Since XML is used to write metadata, the DOM mechanism is used for interactions with applications. The following section presents a "complete" prototype system that illustrates how to integrate different data sources and deals with all aspects of data management from data retrieval to visualizations.

13.4.3.1 System Design

In general, there exist two different ways of developing an integrated data management system for disparate datasets. One approach considered is a traditional approach as illustrated in Figure 13.14(a), while the alternative one is based on the metadata-driven data model as shown in Figure 13.14(b). In the traditional approach, the application component needs a special converter for each format. In the metadata-driven approach, a metadata mechanism is used to construct a generic format converter for each data format. When each data format is specified by its corresponding metadata, one generic reader (e.g., embedded in the application) component reads all the formats in a standard way.

The approach outlined in Figure 13.14(b) forms the foundation for metadata-driven integrated data management. A variety of data sources from many disparate systems can be integrated through this approach. In the implementation aspect, the application component could be composed of one or more applications and toolkits, depending on the functionalities that the system needs to perform. In addition, the programming languages used to develop individual applications rely on the designer. Currently, XML parsers that create

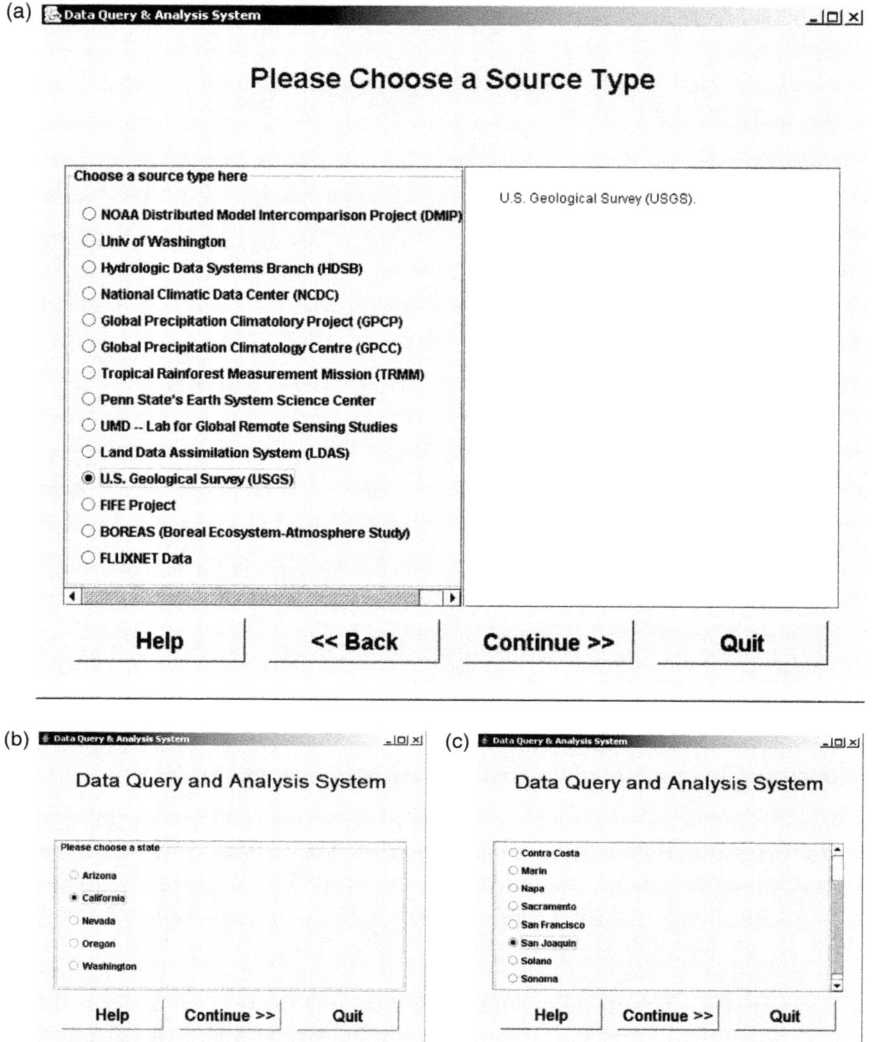

FIGURE 13.15
An example of the implementation of a data management system, where the application is driven by a metadata mechanism through GUIs.

the DOM to facilitate the interactions between XML and applications are available for most programming languages including Java and C++ [6].

13.4.3.2 Implementation

In the metadata-driven approach, applications are driven through GUIs. Figure 13.15 shows a GUI representing various data sources for hydrologic studies. This GUI is a Java-enabled graphical interface represented partially by the metadata shown in Figure 13.9. This graphical version implies that there is an internal interpretation of the XML-based metadata to create a DOM tree. Then the Java application traverses the nodes in the DOM tree and creates the GUI using Java Swing technology [7]. Recall the way XML-based metadata are read by Java through the DOM mechanism shown in Figure 13.7. Using Array or Vector classes

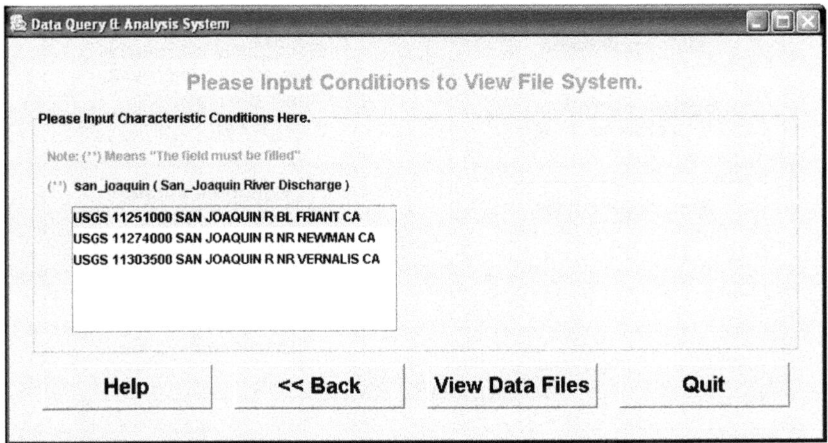

Figure 13.16
A GUI that shows available datasets based on the simple querying process shown in Figure 13.15.

of Java or any equivalents of other languages is efficient to read XML documents and keep the content in memory for graphical representation. Each *aType* element in Figure 13.9 is presented in Figure 13.15 in the form of a button using the Java JRadioButton class interface. Note that the structure of the category metadata in Figure 13.9 is also reflected in the GUI.

Figure 13.15(b) and (c) show how the application interacts with the metadata mechanism in a data management system. This is an example of a query process to reach desired data files in a system. In general, hydrologic data are classified by regions or basins in a tree structure. The GUI in Figure 13.15(b) shows a category of states in the western region of the US whereas the GUI in Figure 13.15(c) shows a category of counties in California. The GUI in Figure 13.15(b) precedes the one in Figure 13.15(c). Therefore, the metadata that is reflected by the previous GUI is a parent node of the one following it. The metadata mechanism is supported by the "pointer" discussed in Section 13.4.2. For example, the node (metadata) for California points to a node that includes all counties belonging to the state. These three interfaces are based on the metadata for logical categories. This structure demonstrates that metadata are hierarchically organized and can be displayed in an application in such a way.

Figure 13.16 shows the query result based on the process presented in Figure 13.15. Let us assume that the user selected San Joaquin at the county level. The next GUI lists the datasets (i.e., USGS streamflow) belonging to San Joaquin. This GUI is enabled by the metadata for query and logical directories consecutively. That is, the metadata for logical categories call the metadata for query, which then invoke the metadata for logical directories. This retrieval process shows that various data sources with different data formats can be interpreted through a single reader.

13.5 Development of an Integrated Data Management System

Up to this point, we have discussed how a metadata mechanism interacts with applications. This section discusses the various aspects of an integrated data management system. There are two main aspects for an integrated data management system: (1) integration of various data sources; and (2) add-on functionalities. In this section, we introduce an example to illustrate various functions used to support the integrated data management.

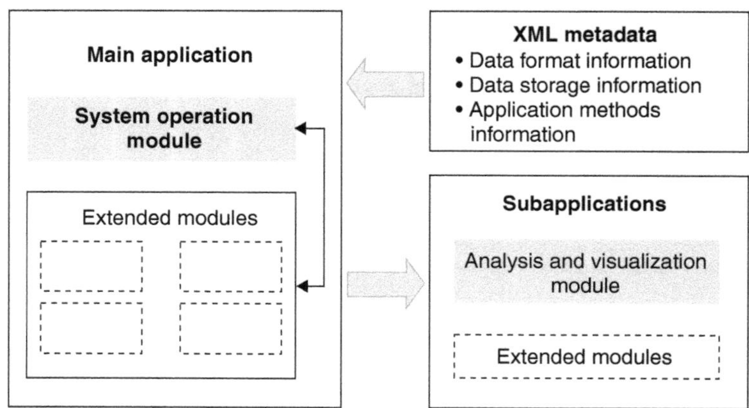

FIGURE 13.17
Design framework for an integrated data management system.

13.5.1 Design Framework

First of all, one needs to decide whether to use only one application for the entire system or more. Second, one needs to consider how to accommodate add-on functionalities in the system to facilitate the additions of new data sources and analysis methods. Third, one must determine what interaction methods to use within the system. Although there are more other considerations one may want to include, the three listed here are the main factors affecting the design of a data management system.

Figure 13.17 shows a basic design framework that considers the three main factors described earlier. The framework takes on the form of modular approach to ensure the extensibility of the system. In the framework, we divided the application component into two parts: a main application and subapplications. In this architecture, the subapplications (e.g., application toolkits) are extended modules with respect to the main application. Because each of these subapplications performs its own specialized functions, this architecture makes the system more flexible and efficient. For example, IDL software, a powerful language and software package in data visualization, may be used to create a visualization application instead of using common programming languages such as Java and C++ [8]. As shown in the main application column of Figure 13.17, the main application is also able to have extended modules. Under this architecture, the system operation module creates GUIs while some extended modules perform functions such as data query and file search. Extended modules may be constructed as a collection of functional libraries.

13.5.2 Implementation of an Integrated Data Management System

Following the design framework shown in Figure 13.17, we introduce a prototype system for integrated data management. We use Java and IDL to implement the application component and XML for the metadata component.

13.5.2.1 System Design

The prototype system architecture shown in Figure 13.18 includes four components: (1) XML-based metadata; (2) a Java-based application; (3) IDL-based visualization and

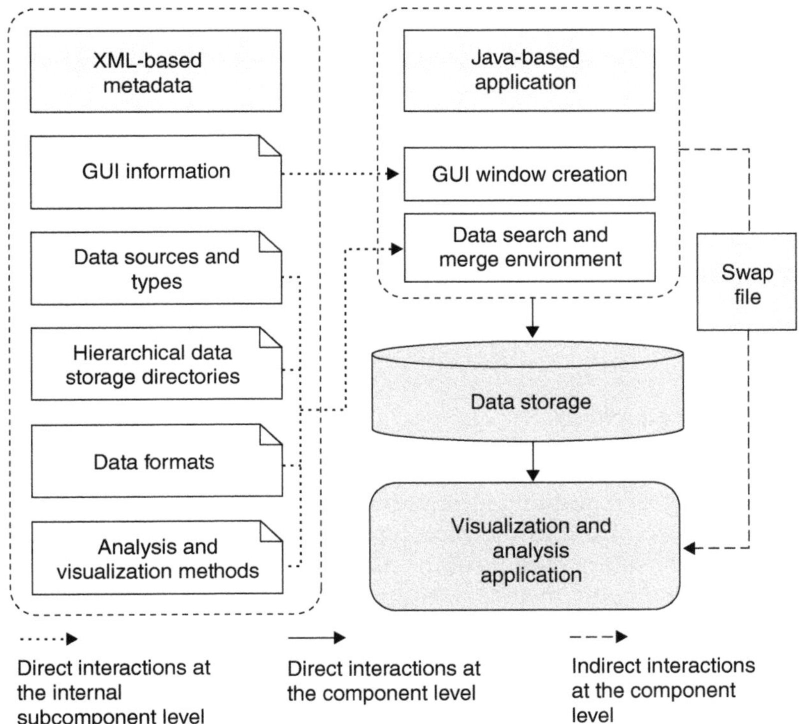

FIGURE 13.18
System architecture of a prototype integrated data management system.

analysis application; and (4) data storage. As shown in Figure 13.18, the Java-based application is at the heart of the system. This application, which functions as a main application, interprets XML-based metadata to obtain all the information for data retrieval, data analysis, and visualization in the system. The visualization and analysis component shown in the right column of Figure 13.18 is implemented by IDL. The IDL-based application visualizes data and provides basic statistical analysis. IDL can access data of various types including the complicated network Common Data Format (netCDF) and Hierarchical Data Format (HDF) [9].

Depending on the availability of a new data source, metadata for the new data source can be added to the existing metadata, due to the extensibility of the metadata mechanism. The application for visualization and analysis is an addition to the main application, and has several functional groups such as statistical analysis and 2D visualization. These functional groups can be extended whenever necessary. The architecture of the four interacting components makes the system an integrated application environment equipped with a specialized reader and format converter for various formats.

As shown in Figure 13.18, there are three types of interactions among the components: (1) direct interactions at the internal subcomponent level; (2) direct interactions at the component level; and (3) indirect interactions at the component level. The direct interactions at the internal subcomponent level occur between XML-based metadata and the main application. Each part of the XML-based metadata directly interacts with a subcomponent in the main application through the DOM mechanism at the subcomponent level (see Figure 13.18). The second type of interaction takes place between the data storage and the applications (i.e., Java and IDL applications). The applications directly access the data storage using file-reading functions. This type of interaction occurs at the component level. For this direct interaction, information about the structure of the data storage is provided by the metadata component. An approach of using swap files is employed for the third type of interaction between the main application and the IDL application. The swap file is used as an intermediate medium for the communication between the two applications. The contents written in a swap file are information collected from XML-based metadata through the main application. The collected information includes data formats, visualization methods, and names of available data files. The IDL application reads the information and performs actions based on it. As shown in Figure 13.18, communication between the Java and IDL applications occurs indirectly at the component level.

13.5.2.2 Individual Functions

Data query

Data query is a process of requesting information from a data system. As discussed earlier, the metadata mechanism used in the prototype system supports data query. The query function is based on the query module of the main application and two types of metadata: the logical directory and query metadata. Figure 13.19 shows a GUI for the query function. The left panel is used to enter query conditions, and the right panel is used to define output formats for data display. In fact, the left panel reflects a representation of the metadata for query, and the right panel reflects that of the metadata for formats. The logical directory metadata is internally used by the query module to find desired data files.

The user may enter query conditions such as {*Basin CBRFC, 2000 05 15 02, 2000 05 15 03*} as shown in Figure 13.19. Recall the discussions about the query metadata in Section 13.4.2. Based on the query conditions, the application reads the root metadata file or top node of the logical directory metadata for the HDSB NEXRAD Stage III precipitation dataset. This

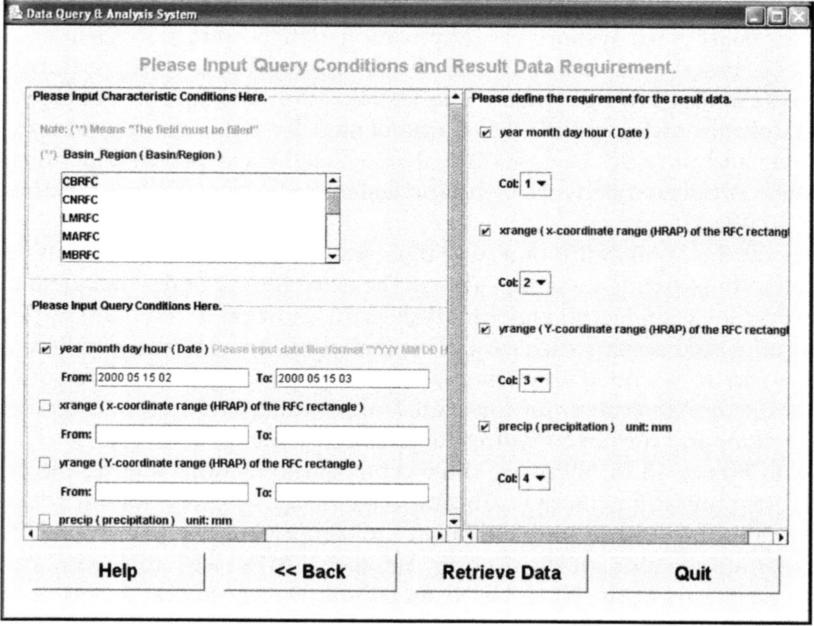

FIGURE 13.19
A GUI to query data and specify output formats.

metadata file describes subdirectories such as CBRFC and CNRFC as shown in the upper left panel of the GUI in Figure 13.19. It also records the content information for each subdirectory. Using this information, the application identifies that only Basin CBRFC is qualified for the query condition. Thus, all other subdirectories are filtered out. Consequently, the application parses the metadata files at the next level, which contain the information on the qualified basin, CBRFC. In a similar fashion, the metadata for CBRFC includes nodes for several subdirectories such as the years 2000 and 2004. Since the year 2000 matches the query condition, the metadata for the year 2000 is accessed and parsed. After more steps, the application reaches the metadata for the daily level. The application records the names and paths of the two files: 2000051502.dat and 2000051503.dat. All these processes occur internally. Since the DOM parser traces only the nodes that are qualified for the query conditions, the performance of the system is efficient.

Data merge

Complicated hydrologic studies such as hydrologic modeling involve various datasets. Furthermore, hydrologic modeling requires comparison among various types of datasets from different sources that cover the same watershed. Such comparison is a useful approach to validate and cross-check the quality of the datasets. Comparing data from different sources can be facilitated by merging the data into a common format. This section discusses how to integrate different data sources using data merge, which is another aspect of the integrated prototype data system.

The data merge function is an advanced feature in the metadata-driven model. Figure 13.20 shows the architecture of the prototype system to which the merge function is added. The four components of the prototype system are presented maintaining the original architecture of the system. The main application component is explained in three extensible parts: (1) a data query module; (2) supporting modules; and (3) a merge module. The merge module is added to the main application of the original system while another query process is added to the query module to support data merging. However, it takes

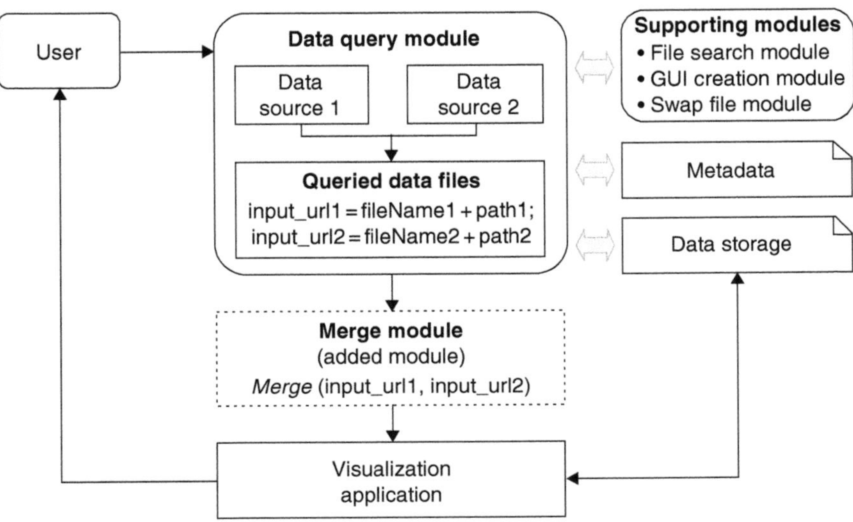

FIGURE 13.20
Design of data merge as an extended function.

advantage of the main application functions to retrieve files by using the same metadata mechanism. There is no change in the supporting modules related to the main application. Thus, this system architecture ensures its extensibility and interoperability.

The merge function is initiated by concurrently parsing two user-selected metadata files. The user may select a pair of two data sources such as precipitation from NCDC and streamflow from USGS. As a result, the metadata and its path for each source at each step or node are internally stored in Java Stack classes. In addition to the architecture of the merge function, Figure 13.20 also shows the process of the merge function. For example, the data query module parses two sources named *data source 1* and *data source 2*. Each source is composed of a tree of nodes, and the nodes are parsed through the original mechanism of the main application, except that in the merge case the mechanism parses each source individually. When the application parses nodes, it uses the same metadata as those used for a normal retrieval process.

The action of merging two files is performed by the Java application. After a series of parsing processes, the two files of the leaf node, which are *input_url1* and *input_url2*, are identified. The merge function of the Java application binds the two selected files and writes them into a single file. At the same time, a swap file is created to convey the format information of the merged file to the visualization application for further analysis. This information transfer involves the metadata mechanism and supporting modules. The same format metadata as the ones for the normal retrieval process are used for the merge function. The format metadata for the two sources are used to create new format information for the merged data file, and the newly created format information for the merged file is written in a swap file. Then the visualization application reads the merged file.

The GUI for merging two data files is shown in Figure 13.21. The GUI consists of two panels that contain: (1) two list boxes for viewing data lists; and (2) a menu for viewing

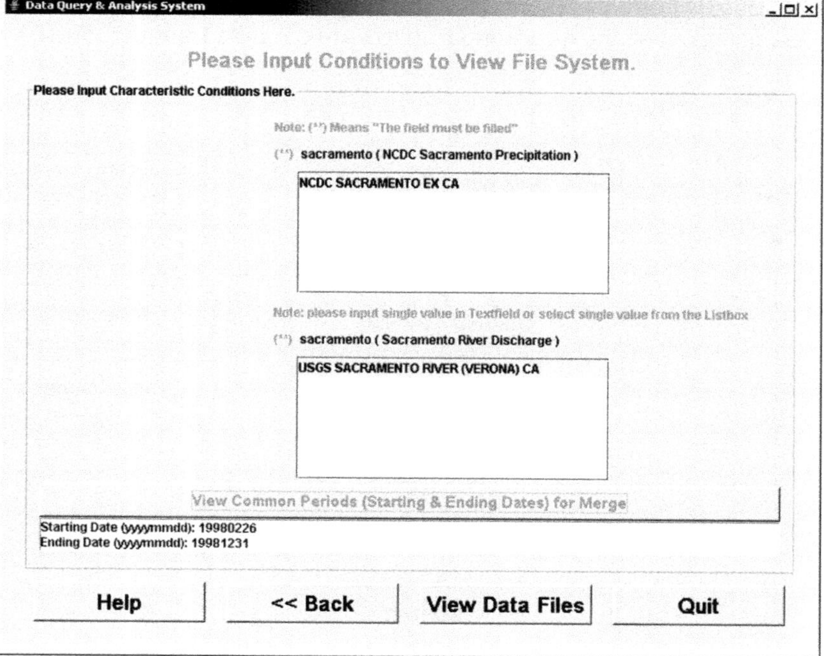

FIGURE 13.21
Graphical user interface for data merging.

common periods between the two files. Each of the two list boxes on the GUI represents a list of files that are available for a data merge process. In a hydrological context, these two sets of data need to belong to the same watershed for comparison in general. Another necessary task for conducting the data merge is to identify common temporal periods that the two data files cover, because the merge function is meaningful only for two data files that have a common period. Thus, the second panel provides the user with information on common periods that the two data files have before the user retrieves the merged data file. If there is no common period, an error message will appear on the GUI.

Data analysis and visualization

Data visualization greatly enhances the identification of trends and patterns inherent in massive scientific data. Data visualization has become an essential part of hydrologic research where comparative analyses among many data types and datasets are performed. Under the prototype system architecture, data analysis and visualization functions are performed at the final step following the data retrieval process. These functions are additional modules (i.e., add-on toolkits) to the main application, and more functions can be added as needed in the future. These functions make the system more complete to be an integrated system. We introduce 2D and 3D analysis and visualization examples in the following. The application for analysis and visualization in the system is implemented by IDL.

Figure 13.22 illustrates a plot of 2D time series data for the Blue River basin in Oklahoma. The left panel shows a time series for each field (i.e., measured precipitation, simulated evapotranspiration and soil moisture, and observed and simulated streamflow). The two vertical lines in the left panel show the range whose zoom-in display is shown on the right panel. The statistics of each field within the chosen range of the two vertical lines are listed

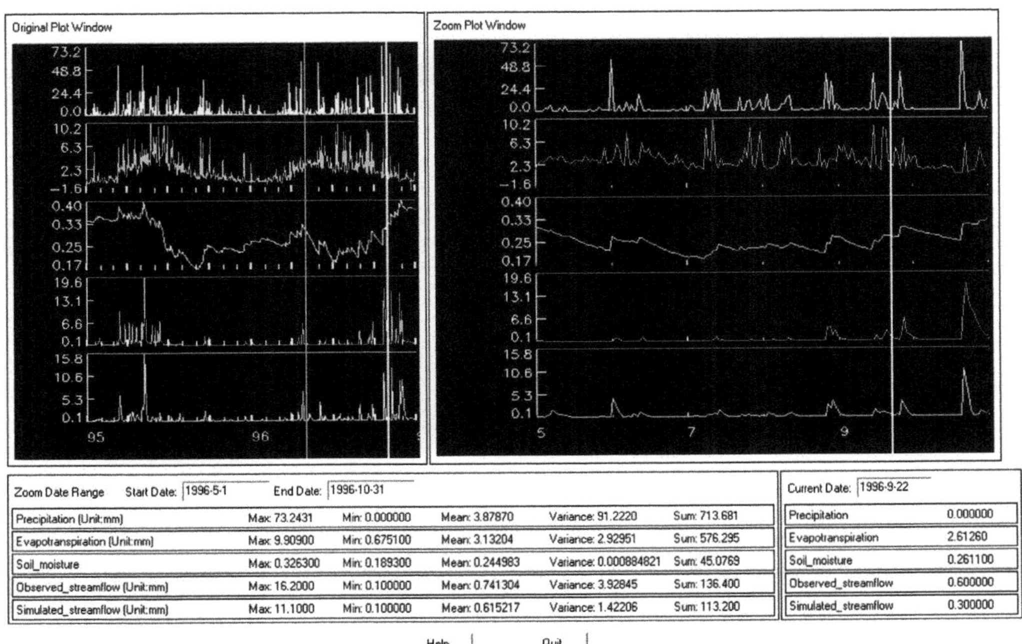

FIGURE 13.22
A snap shot of a 2D time series display for the data read in from a data source for the Blue River basin in Oklahoma.

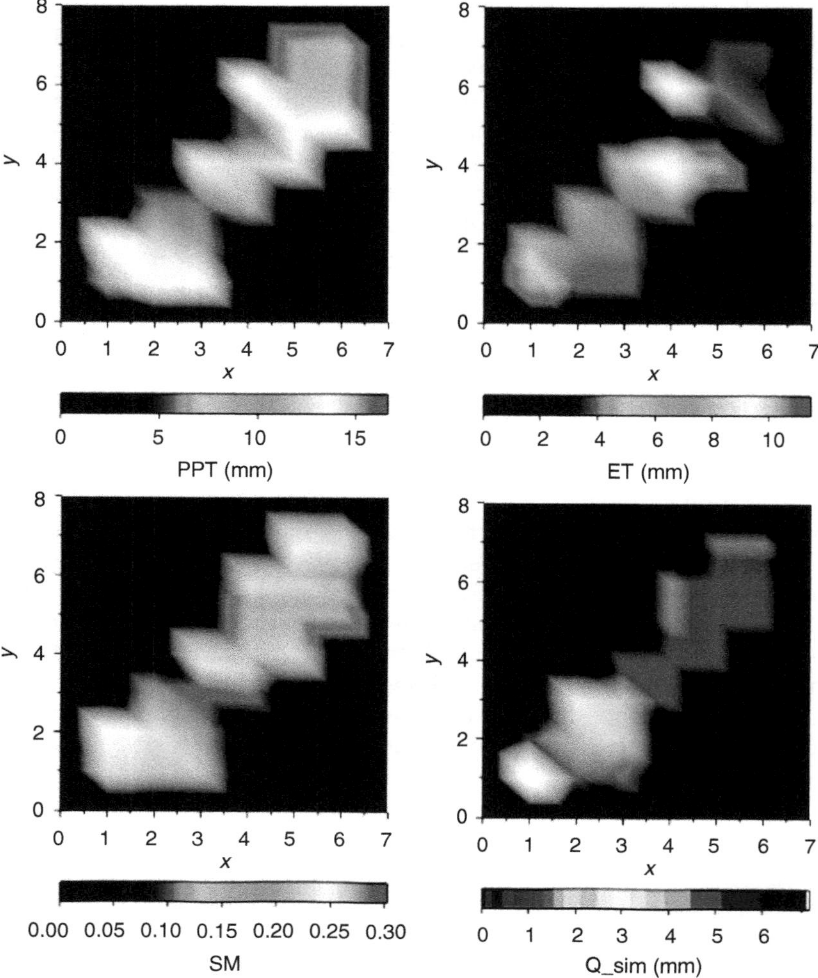

FIGURE 13.23
A 2D color map view of the spatially distributed daily precipitation from NEXRAD, and evapotranspiration, soil moisture, and runoff from a model simulation on October 2, 1995 over the Blue River watershed corresponding to the precipitation.

under the plots on the left. The single vertical line shown in the right panel of the time series indicates the probing location of the time series, and its corresponding values for each field are shown on the right.

Figure 13.23 is an example of a 2D color map that displays the data in a spatial context. The 2D color map is an areal representation of 2D scalar fields with the same value. The 3D surface rendering is used to depict the characteristics of data over a spatial area as a function of spatial variables. Figure 13.24 shows an example of a 3D surface rendering using the NEXRAD Stage III dataset in the Blue River basin. The precipitation dataset for the watershed is represented in the form of $z = f(x, y)$, where z is the precipitation at each grid over the watershed, and x and y are the coordinates for each grid. As shown in Figure 13.24, the z values are rendered into the map where the same z value is represented by the same color. 3D animation is based on the 3D surface rendering, but it is more dynamic than the surface rendering. By selecting the animation option, the 3D surface rendering is animated

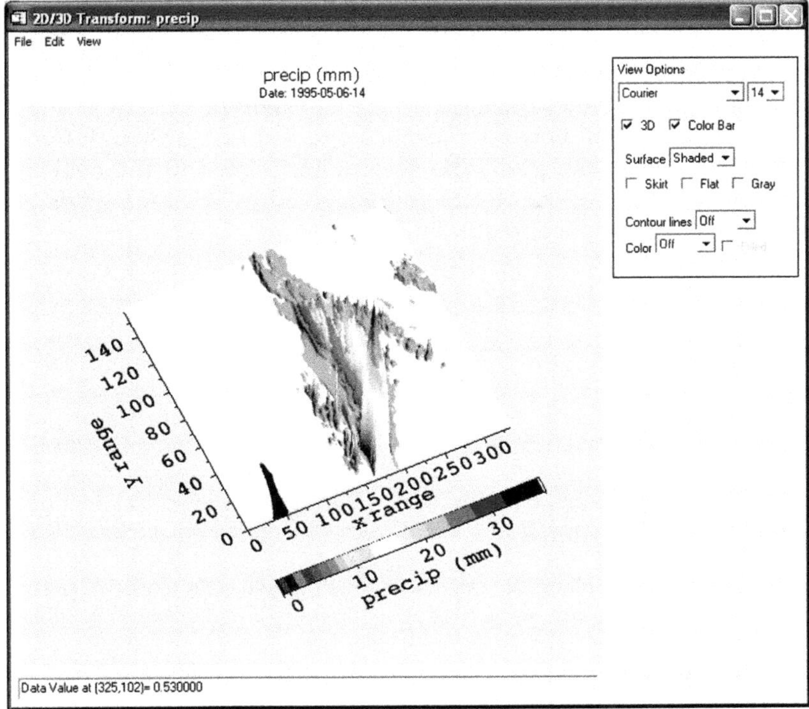

FIGURE 13.24
A 3D view of the NEXRAD Stage III hourly precipitation (in millimeters) over the Blue River region on May 6, 1995, 2 P.M. Options for files and viewing methods are available on the top menu and right panel.

with respect to time. The prototype system supports animation functions for each of the spatial data visualization features.

13.6 Conclusions

In this chapter we cover a variety of topics related to the development of an integrated data management system and provide relevant examples. We have focused on two main aspects of the integrated data management system: integration of various data sources and add-on functionalities. A metadata-driven approach is employed to develop the integrated data management system. We present a design framework and procedures of a metadata mechanism to manage data in an integrated data system, and provide examples to show the implementation of the mechanism. This chapter also provides readers with a comprehensive and fundamental understanding of the architecture of an integrated data management system. A sample architecture for an integrated data management system is introduced. This system architecture provides the reader with an appreciation of the complexity of an integrated data management system's structure.

The system architecture is designed by using an XML-based metadata-driven data model and is applied to develop a prototype integrated data management system. The metadata-driven prototype data management system helps readers understand how datasets with different formats from different sources can be effectively retrieved and visualized in a single system by using a generic format converter, the metadata mechanism. Additionally, the interactions among the components that constitute the prototype system are depicted in detail so that readers are provided with the design knowledge of the interaction methods as well as the components of an integrated data management system. We also illustrate the extensible and integrative characteristics of the prototype system to allow for future modifications in dealing with new data formats and sources. Although the data sources and applications shown in the prototype system are related to the field of hydrology, the ideas, approaches, and system architecture involved in developing this prototype are general, and can be used/applied to other fields.

References

1. Liu, Z., Liang, Y., and Liang, X., Integrated data management, retrieval and visualization system for earth science datasets, in *Proceedings of the 17th Conference on Hydrology (CD version)*, Long Beach, CA, 2003.
2. Liu, Z., and Liang, Y., Design of metadata in a hydrological integrated scientific data management system, in *Proceedings of the 7th Joint Conference on Information Science* (JCIS), Cary, 2003, pp. 418–421.
3. Jeong, S., Liang, Y., and Liang, X., Design of an integrated data retrieval, analysis, and visualization system: application in the hydrology domain, *Environmental Modelling and Software* (in review).
4. Helly, J., Staudigel, H., and Koppers, A., Scalable models of data sharing in earth sciences, *Geochemical Geophysical Geosystem*, 4(1), 2003, 1010.
5. McLaughlin, B., *Java and XML*, O'Reilly, Sebastopol, 2000, pp. 341–348.
6. W3C, Document object model (DOM) level 3 core specification version 1.0, W3C Recommendation 07 April 2004.
7. Geary, D.M., *Graphic Java Mastering the JFC Vol. II: Swing*, 3rd ed., Sun Microsystems Press, Palo Alto, CA, 1999.
8. Fanning, D.W., *IDL Programming Techniques*, 2nd ed., Fanning Software Consulting, Fort Collins, 2000, p. 129.
9. Gumley, L.E., *Practical IDL Programming*, Morgan Kaufmann Publishers, San Francisco, CA, 2002, p. 172.

Section IV

Data Processing and Analysis

14

Introduction to Data Processing

Peter Bajcsy

CONTENTS

14.1 Introduction to Section IV

Let us assume that we are given a real life problem in one of the application areas where geographic information and its analysis would be an inherent part of the most optimal solution. One can encounter such problems in a very large number of application domains, for example, hydrology, atmospheric science, agriculture, environmental engineering, water quality, ecology, watershed management, insurance, earthquake engineering, geology, plant biology, land use planning, and military. We named only those application domains that the authors have been involved in during the past years but there are many more to list if we would want to create an exhaustive list.

It is very hard to describe the whole gamut of real life problems that application engineers are working on. One could classify the majority of these problems based on a very typical sequence of data processing and analysis steps that lead to a final problem solution. This typical sequence of data processing and analysis steps includes (1) analysis of data sources (sensing, data acquisition, data transmission), (2) data representation (raster, boundary, point data), (3) registration of spatial coordinate systems (spatial registration, georeferencing) and temporal synchronization, (4) data integration (integration and fusion of related or unrelated features over common spatial and temporal regions), (5) feature extraction, (6) feature selection, and (7) feature analysis and decision support. The goal of this chapter is to overview the above steps as illustrated in Figure 14.1. The organization of Section IV (Chapters 14 to 21) follows the above sequence of processing steps as well. Miscelaneous chapter materials can also be found in Appendices C, D, E and F.

Figure 14.1 also shows a reference to additional three components, such as data storage, data communication, and visualization. These three components are clearly present in the design of a complete end-to-end solution but are not our major focus in Section IV. We will

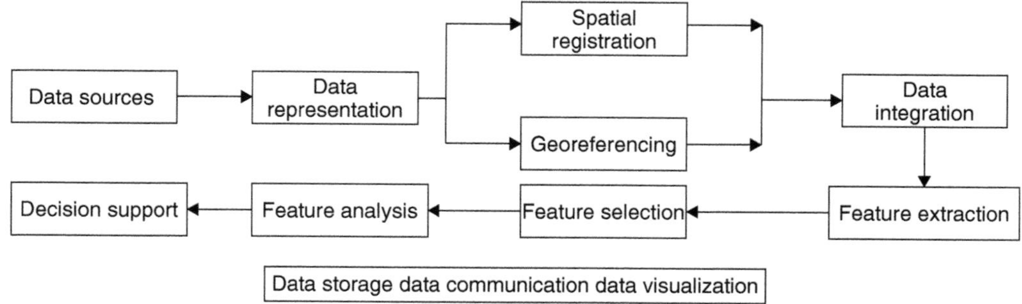

FIGURE 14.1
An overview of data processing and analysis chapter.

present examples of data storage, communication, and visualization without any detailed description.

Our objectives are (1) to improve the reader's data processing and analysis understanding, (2) to encourage seeking optimal solutions, and (3) to help plan efforts (and resources) for solving problems according to their specific challenges. We believe that these objectives can be achieved by (1) clearly defining terms (especially important for multi-disciplinary teams), (2) stating problems and their constraints, (3) providing multiple approaches based on the past work, (4) elaborating on at least one solution, and (5) demonstrating a selected solution on example problems, either real or contrived. We adopted this structure while preparing each chapter of Section IV.

Before reading the Section IV, many readers would probably ask at least two questions. First, why would the presented material be useful to them? Second, what makes the authors believe in the importance of the presented material? These questions and their answers are equally important to novices as well to experts. We devote the next two subsections to answer those two questions by presenting a motivation example and an overview of research funds.

14.2 Motivation Example

You are asked to prepare a plan for managing geographically distributed resources, for example, pricing of customer's repairs or deployment of a police force. Your goal is to incorporate (1) geographical information, for example, county or zip-code boundaries, terrain elevation or forest cover, and (2) historical information, for instance, records about customer's repairs or crime rate reports, into the final solution. Given the geographically distributed resources, the problem is about finding the best partition of a given geographical area that is (1) based on all available information, (2) formed by aggregations of known boundaries, (3) constrained by spatial locations of known boundaries, and (4) obtained as a maximum or a minimum of a certain objective function (or an error metric), such as the result leads to fair pricing of customer's repairs or adequate deployment of police for crime prevention.

The input data consist of raster, boundary, and point data types. The boundary and point data types are also referred to as vector data. Raster data represent grid-based information collected by primarily camera-based sensors, for example, satellite or aerial images.

Boundary data are used for representing contours of man-made or naturally formed territories. Point data are usually tables with information associated with a certain geographical location or region (e.g., a customer's permanent residence). The heterogeneity of all input data types creates a challenge for anyone who seeks an optimal solution or is about to make an important decision. One has to overcome issues related to (1) data representation (e.g., efficiency tradeoffs between vector data information retrieval and data storage), (2) georeferencing raster, boundary, and point data, (3) statistical feature extraction from raster data over georeferenced boundaries, (4) aggregation of boundaries into territorial partitions based on similarity of features, (5) error evaluation of multiple territorial partitions, and (6) visualization of input raster and vector data, as well as extracted features and obtained territorial partitions. The material presented in Section IV addresses the above issues in detail and it should be useful to readers designing solutions to similar problems.

14.3 NSF Funded Applications

The National Science Foundation (NSF) has been known to fund research projects identified by the scientific community as the top priority problems. We present a selected subset of four large NSF Information Technology Research (ITR) awards given in 2003 to demonstrate the relevance of the material in this book to the work to be conducted under current NSF funded projects. All NSF ITR awards are planned for five years and there might be a continuation of many of these projects after the end dates. The outlines of these projects are provided below:

1. *Simplifying the development of grid applications* (*Amount: $8.25 million*): While grid computing promises to help solve problems in science and other fields by connecting computers, databases, and people, the current difficulty of writing efficient programs to take advantage of such diverse resources limits its use. This project will create software tools to simplify and accelerate the development of grid applications and services. It will also look at novel scheduling techniques based on abstract "virtual grids" to deliver high efficiency on grids of real machines. Improved usability and efficiency should greatly expand the community of grid users and developers.

2. *100 Megabits per second to 100 million households* (*Amount: $7.5 million*): The growing demand for communications combined with newly emerging technologies have created a once-in-a-century opportunity to upgrade the country's network infrastructure, bringing 100 megabit-per-second Internet access to millions of homes in America within the next few years. Such a capability will have a dramatic effect on daily life, but only if the network is much more reliable, anticipates applications not yet envisioned, is economically sustainable for the long run, is easier to use and operate and is more secure than the Internet is today. The project will design blueprints for this near-future network by applying principles from security, economics and network research.

3. *Mobile robotic sensor networks* (*Amount: $7.5 million*): This project will deploy a new class of aerial, suspended robotic sensors that can monitor their own sensing performance and move themselves along a light, easily deployed network of cable systems to make the most of their monitoring abilities. Test networks will explore and monitor a mountain stream ecosystem from the ground to the treetops for global change indicators and monitor coastal wetlands and urban rivers for biological pathogens. By configuring itself autonomously, the sensor network can select the best sensing tools and arrangements for each scientific task. The same technologies could one day be applied to securing and monitoring public infrastructure.

4. *Predicting high-impact local weather (Amount: $11.25 million)*: Today's weather forecast models run on fixed schedules over fixed regions, independent of the type of weather that may be occurring. This project will develop grid-computing environments for on-demand detection, simulation, and prediction of high-impact local weather, such as thunderstorms and lake-effect snows. With new tools for orchestrating complex flows of information, hazardous weather detection systems and forecast models will be able to reconfigure themselves in response to evolving weather and also guide the operation of observing systems such as Doppler radars to provide optimal input for the models. This project will hasten the transition of research results to operational use and bring the teaching benefits of sophisticated atmospheric science tools into high school classrooms for the first time. Ultimately, this project will help reduce the hundreds of lives lost and $13 billion of economic damage that hazardous weather causes each year in the United States.

One should also mention the Consortium of Universities for the Advancement of Hydrologic Science, Inc. (CUAHSI) [1] as another effort to integrate hydrological research and development. Our material would be directly related to the CUAHSI effort and to analysis presented in Arc Hydro [2].

14.4 Overview of Section IV

Section IV presents data processing and analysis steps according to Figure 14.1. In this chapter, we introduced a set of basic processing and data analysis steps, followed by a motivation example, and research and development (R&D) projects as funded by NSF. Chapter 15 describes multiple data sources including measurement devices (sensors), storage media and databases, and data generators. We outline two examples (data generation for scene modeling purposes and data acquisition using wireless sensor networks for monitoring purposes) to describe what type of understanding of data sources is essential for successful data analysis and data interpretation. Chapter 16 presents data representations and their tradeoffs with respect to information retrieval efficiency and data representation size in a computer memory. We explain the tradeoffs between raster and vector data representations followed by comparing (1) two types of vector data (U.S. Census Bureau TIGER/Line and ESRI Shapefile formats) and (2) three standard types of raster data representation, BIL, BIP, and BSQ. Chapter 17 summarizes registration techniques and includes a general registration of multiple spatial coordinate systems (possibly temporally changing). Specific registration techniques using georeferencing information are presented in Chapter 18. Mathematical formulas and details about geographic projection types are provided in Appendices 18.A and 18.B. Chapter 19 presents approaches to data integration problems for (1) multiple raster image with heterogeneous geographic projections, data accuracy, spatial, and spectral resolution (denoted as a shallow integration problem) and (2) raster and vector data types for spatially and temporally mismatched data of related measurements (denoted as a deep integration problem). Chapter 20 describes why feature extraction is an important step in analyses and the challenges of automating raster and point data feature extraction. In Chapter 21, we present approaches to feature selection, where features can be either spectral measurements or any generic features. Furthermore, we illustrate how all data processing steps described in previous chapters are utilized for feature analysis and decision support motivated by the example Section 14.2. Finally, Appendices C and D provides review questions of Section IV, and Appendices E and F outlines a term project that exercises knowledge learnt in Section IV combined with the HDF and XML technologies described in the previous sections of this book.

14.5 Terminology

As part of the introduction to this chapter, one needs to clearly define the terminology used. It is known that terms might have several meanings depending on application domains and our goal is to avoid any terminology confusion.

The most common term used in data analysis is feature or attribute. Feature or attribute is usually a physical measurement or category associated with spatial location and temporal instance. Features can be divided into continuous, for example, terrain elevation, or categorical, for example, forest label.

Territory or region refers to a geographic area delineated by a boundary and characterized by attributes. In general, territories can be man-made, as in County, Census Tract, Census Block Group, Zip Code (ZCTA), Police District territories, or they may be "naturally formed," such as ecoregions or forest type regions.

Vector data are understood as data referring to boundary or point information, while raster data denote grid information (also called images). An example of vector data would be the U.S. Census bureau boundary information stored in TIGER/Line files and an example of raster data would be a digital elevation map.

Additional terms related to the motivation example described in Section 14.2 would include territorial neighbors and geographical territorial partition. Territorial neighbors are defined as a list of spatially neighboring territories for a given territory. Geographical territorial partition is formed by aggregating multiple territories that divide any geographical area.

References

1. The Consortium of Universities for the Advancement of Hydrologic Science, Inc. (CUAHSI), accessible at URL http://www.cuahsi.org/.
2. Maidment, D.R., *Arc Hydro, GIS for Water Resources*, ESRI Press, Redlands, CA, 2002. http://www.posc.org/ Epicentre.2_2/DataModel/ExamplesofUsage/eu_cs.html.

15

Understanding Data Sources

Peter Bajcsy

Contents

15.1 Introduction

In this chapter we focus on data sources. It is necessary to emphasize that in order to perform data analysis one needs to fully understand sensor measurements, data acquisition processes, data representation, and preprocessing steps that lead to the data to be analyzed. The frequent ignorance about these issues is a motivation factor for us to present a few fundamental considerations about data sources and what they imply about data for future data analyses and simulations.

It is very hard to enumerate all possible data sources since data for analyses can come from measurement devices, storage media, single or distributed sets of databases, or data generators. A short list of data sources could include small sensors, large instruments, data archives, flat files, databases, or simulation programs. Furthermore, data producers serving different end use applications add extra dimensionality to any data provided for analysis. This dimensionality is related to data organization, datum accuracy, preservation of application relevant information and finally, data format.

Based on the type of end applications (data consumers), data producers could be classified as government agencies, research institutes, and commercial companies. A combination of data sources and data producers leads to data heterogeneity. It poses quite often a formidable task for a researcher to integrate and understand data before any analysis takes place. The types of heterogeneity include (1) various data types, such as, raster, vector, point, (2) multiple file formats and storage media, (3) different datum accuracy and uncertainty including spatial, temporal, or spectral information, (4) unknown sensor characteristics, for example, dependency of accuracy on frequency or temperature, and (5) disparate methods for measuring the same physical phenomenon, for instance, temperature measurements with Mercury- or thermistor-based thermometer.

Next, we present a short overview of data sources from data producers that are divided into three categories: (1) data measurement devices (sensors), (2) storage media and databases, and (3) simulation programs. We describe then an example of data generation for modeling bidirectional reflectance functions (BRDFs) in order to illustrate the importance of fundamental (first-principle, physics-based) data understanding. We also outline another example of data acquisition using wireless ad hoc sensor networks to demonstrate the need for data source understanding.

15.2 Data Sources from Data Producers

15.2.1 Data Measurement Instruments and Sensors

Instruments and sensors used for research in Geographic Information Systems (GIS) can be classified based on their placement into (1) satellite, (2) airborne, (3) laboratory, (4) ground, and (5) underground. Most common devices generating raster data are denoted as camera-based sensors. These sensors operate at different spectral wavelengths to explore material surface and subsurface properties. Outdoor measurements, as needed in hydrology, atmospheric science, or environmental science, frequently utilize sun illumination. The measurements are collected as reflected, refracted, or transmitted spectral components of the interactions between materials and incident light. Examples of such sensors would include photographic sensors (visible and near infrared [NIR] spectrum), thermal infrared (IR) sensors, synthetic aperture radar (SAR) sensors, or multispectral (MS) and hyperspectral (HS) sensors.

Most common sensors generating point data provide information about the air (wind speed and direction, temperature), water (water discharge, nitrate–nitrogen level), or the earth both above ground (precipitation, soil type, corn yield), and below (geological boring logs, earthquake faults). The underlying physical principle of individual sensors varies greatly as a function of measured variable. It would be difficult to describe the whole gamut of point sensors starting from a simple Mercury-based thermometer to a complex Nitrate–Nitrogen chemical sensor. There are several commercial companies specializing in manufacturing sensors. Some examples include hydrology sensors by MBT GmbH [1] and Vaisala Inc. [2], machinery health monitoring sensors by Oceana Sensors [3], MEMS sensors by Analog Devices [4], Advanced Micro Devices (AMD) Inc. [5], and "smart" sensors by Crossbow Inc. [6]. Many companies deliver only specialized sensing equipment for certain applications domains, for example, equipment for measuring soil properties by Veris Technologies [7], or corn yield in agricultural combines made by John Deere [8]. The current trend in sensor development is not only to miniaturize the sensors and make them "smarter" by adding communication and computational capabilities, but also to embed them into various systems and materials. For example, material defect monitoring wireless sensors can be embedded into heavy machinery and indicate fatigue before any failure occurs as in the advanced Structural Health Integrated Electronic Life Determination (SHIELD) systems [9].

15.2.2 Storage Media and Databases

Perhaps, to extract and analyze the data from multiple storage media and databases poses a much bigger challenge for researchers than to generate the data [10]. In order to extract and analyze information from distributed databases, distributed heterogeneous data must be gathered, characterized, and cleaned. These processing steps might be very time-consuming because they might require multiple scans of large distributed databases to ensure the data quality defined by application domain experts and computer scientists. From a semantic integration viewpoint, there are quite often challenges due to the heterogeneous and distributed nature of data. The preprocessing steps might require the data to be transformed (e.g., the Normalized Difference Vegetation Index [NDVI] computation), linked with distributed annotation or metadata files (e.g., geographic locations and attribute descriptions), or more exactly specified using auxiliary programs running on a remote server (e.g., making subsets and identifying a spatial match or any temporal changes).

Based on the aforementioned data quality and integration issues, the need for automated preprocessing techniques becomes eminent, and understanding of data sources is critical to any successful data analysis. The strategies for addressing these issues involve data cleaning using exploratory data mining (EDM), data preprocessing, and semantic integration techniques [11–13]. Data quality and integration issues are very complex and require extra considerations before data from archives on multiple storage media, as well as data stored in flat files and databases, can be analyzed. The National Archives and Records Administration (NARA) investigate many of these problems (see http://www.archives.gov/) in collaboration with research institutions and supercomputer centers, for instance, with NCSA in Illinois and SDSC in California.

15.2.3 Data Generators

Due to the constantly increasing computational power of modern computers, there have been several advancements in the development of accurate physics-based simulation programs. Simulation programs have been developed for multiple application domains, such as, weather predictions to improve the forecast accuracy of significant weather features across multiple scales [14]; finite element analysis in structural engineering [15]; atmospheric spectral modeling [16]; synthetic aperture radar prediction of realistic far-field and near-field radar signatures for 3D target modeling [17,18] and scene clutter [19,20]; radiometric modeling of various multispectral/hyperspectral imaging sensors [21]; or thermal prediction and modeling military applications and the automobile industry [22]. These simulation programs have served as a means for weather prediction, robust structural design, spectral sensor modeling, target prediction, and scene prediction. In general, simulation programs are extremely useful in new sensor design and algorithm development (by providing baseline datasets), as well as a means for tasking aid (e.g., implementations of autonomous robotic applications guided with the help of simulated data), human operator training, and for analyst aid (e.g., interpretation of measured data with the use of simulated data).

While simulation programs are extremely useful for researchers and developers, they also operate under a certain set of theoretical assumptions regardless of their model complexity. It is imperative that a user be familiar with these assumptions in order to benefit from the data generated by complex simulations. Another aspect of simulation programs is the understanding of input and output parameters. It is recommended to pay attention not only to the meaning of modeled parameters but also to the uncertainty of measured values that become inputs to a simulation. Finally, experimental measurements and simulations usually generate metadata about the data acquisition and preprocessing that might use inconsistent keywords for denoting variables and assumptions. Once again, following standard terms and organization of metadata entries are critical for correct data processing [23].

15.3 Example of Data Generation for Modeling BRDFs

Let us assume that our objective is to establish a mapping between pixel intensities of a camera-based sensor and scene/object characteristics as a function of viewing and illumination angles. This mapping is frequently denoted as a Bidirectional Reflectance Distribution Function (BRDF) and the scenario for modeling BRDFs is illustrated in Figure 15.1.

The BRDF modeling problem poses several challenges due to (1) complex geometric and photometric properties of scene/object materials, (2) outdoor environment, and (3) seasonal changes. The complexity of geometric and photometric properties of scene/object materials can be demonstrated by inspecting specular surfaces (artificial grass), diffuse surfaces (concrete), isotropic surfaces (leather), anisotropic surfaces (straw), surfaces with large height variations (pebbles), surfaces with small height variations (sandpaper), pastel surfaces (cotton), colored surfaces (velvet), natural surfaces (lettuce), and man-made surfaces (sponge). These example surfaces are presented in Figure 15.2. One has to understand all variations and formulate adequate models to predict variations of these kinds. Additional modeling challenges are related to the dynamic outdoor environment with a complex 3D structure. For example, researchers at the University of Nebraska are modeling solar radiation interactions at the earth surface (see URL: http://snrs.unl.edu/agmet/908/wheat_canopy.htm). The appearances of wheat canopy (1) on one day (e.g., see the images of May 8, 1999, 55° Solar Zenith Angle, [36.77 N; 97.13 W] near Ponca City, Oklahoma at the URL) and (2) on three different days during the wheat growing season clearly illustrate the complexity of scene modeling.

15.3.1 Approaches to BRDF Modeling

Approaches to BRDF modeling rely on theoretical models derived from Maxwell's equations and dictate data generation for modeling. Light and its interaction with materials

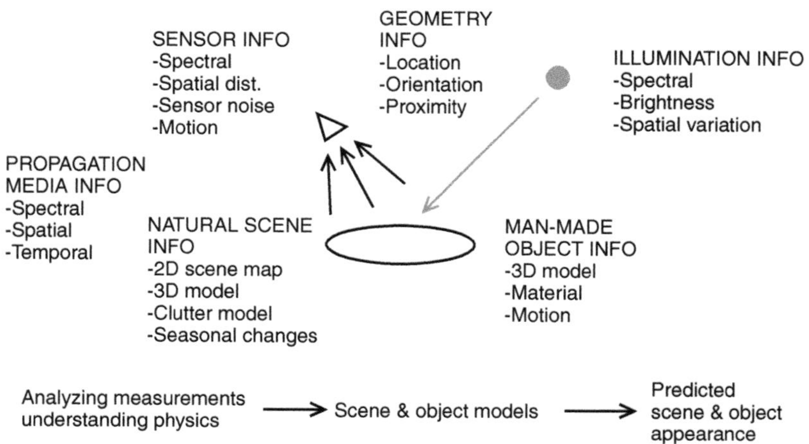

FIGURE 15.1
Modeling spectral scene/object characteristics as a function of viewing and illumination angles.

is viewed either using wave (Electromagnetics) or particle (Quantum Physics) light models [24]. Electromagnetics light model is based on the assumption that light propagates as a spatio-temporal wave with a certain wavelength. The theoretical wave model explains the interaction of light and materials in terms of interference and diffraction and with respect to the light wavelength and the material characteristics. A simplification of the wave light model is derived from Maxwell's equations under the assumptions of geometric optics, that is, the light propagates as a set of rays with a wavelength much smaller than the obstacle size on material surface. This simplified theoretical model limits the interaction of light and materials to reflection and refraction. Particle light models are much more complex than wave models since the light is represented by a set of discrete particles having energy (also called photons). Particle models are based on Bohr's theory of the hydrogen atom, cavity radiation (energy quantization), the photoelectric effect (saturation of kinetic energy), and the Compton effect (wavelength change). To our knowledge, particle light models have not been used for modeling spectral scene/object characteristics.

The most general physics-based surface appearance modeling of spectral scene/object characteristics according to wave light models are the BRDF at a coarse scale and Bidirectional Texture Functions (BTF) at a fine scale [25,26]. The difference between BRDF and BTF is illustrated in Figure 15.3 and Table 15.1. BRDF and BTF depend on incident illumination angles, viewing angles, incident wavelengths, and the interaction of light with each material patch (surface and body properties) that contributes to one value measured by a sensor, for example, one pixel value of a camera sensor. In general, the goal of BTF modeling is to predict a spatial distribution of pixel values given incident illumination angles, viewing angles, wavelength, and the viewed material class description. In a BTF model, we either assume that the light propagation is invariant to the sensors and the

FIGURE 15.2
Complexity of geometric and photometric properties. Materials shown from left to right, and top to down: artificial grass, concrete, leather, straw, pebbles, sandpaper, cotton, velvet, lettuce, and sponge. (Pictures were obtained from the CURET database at http://www1.cs.columbia.edu/CAVE/curet/.)

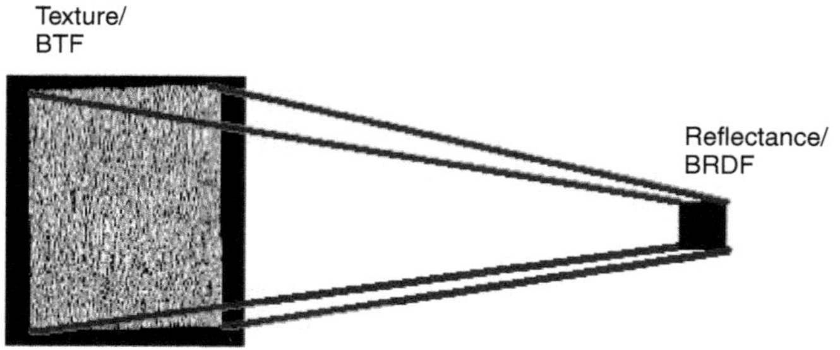

Figure 15.3
The relationship between BRDF and BTF in terms of surface modeling scale.

Table 15.1
Surface appearance models.

Surface appearance model	Pair (θ_i, θ_r)	Range of (θ_i, θ_r)
Coarse-scale	Reflectance	BRDF
Fine-scale	Texture	BTF

media through which the light propagates, or we compensate for any changes due to sensor and light propagation media characteristics in the model. BRDF/BTF have been used in computer vision for object recognition (texture model-based estimation) [27,28], robotics for depth and surface roughness estimation (photometric stereo and depth from shading) [25,26], and in computer graphics for virtual world rendering, training, and simulations (texture model-based generation) [29–32]. Applications of BRDF/BTF in GIS would include material classification (land use maps) or appearance prediction based on satellite and aerial photography [33].

15.3.2 Physics-Based Understanding of BRDFs

Data generation for BRDF/BTF modeling requires understanding of (1) the basic physics [24], (2) currently developed models [25,26,29,34], and (3) the underlying assumptions of multiple models. We will only introduce the top level BRDF/BTF concepts based on the interaction between a planar surface material patch dA and incident light rays with radiance L_i as illustrated in Figure 15.4. Variables referring to incident and reflected rays are denoted with i and r subscripts, respectively. The amount of reflected light impinging on a camera sensor is calculated according to the mathematical definition of reflectance coefficient provided in Equation 15.1. The reflectance coefficient can be defined theoretically in nine different ways depending on the combination of incident and reflected angles. The notation $\rho(.\text{->}.)$ in Equation 15.1 refers to bidirectional, directional-conical, directional-hemispherical, conical-directional, biconical, conical-hemispherical, hemispherical-directional, hemispherical-conical, and bihemispherical models.

$$\begin{aligned}
&\text{Reflectance } \rho = \Phi_{\text{reflected}}/\Phi_{\text{incident}} \text{ (dB, \%)}\\
&\rho(\vec{\omega}_i \to \vec{\omega}_r); \rho(\vec{\omega}_i \to \Delta\vec{\omega}_r); \rho(\vec{\omega}_i \to 2\pi); \rho(\Delta\vec{\omega}_i \to \vec{\omega}_r);\\
&\rho(\Delta\vec{\omega}_i \to \Delta\vec{\omega}_r); \rho(\Delta\vec{\omega}_i \to 2\pi); \rho(2\pi \to \vec{\omega}_r); \rho(2\pi \to \Delta\vec{\omega}_r);\\
&\rho(2\pi \to 2\pi),
\end{aligned} \tag{15.1}$$

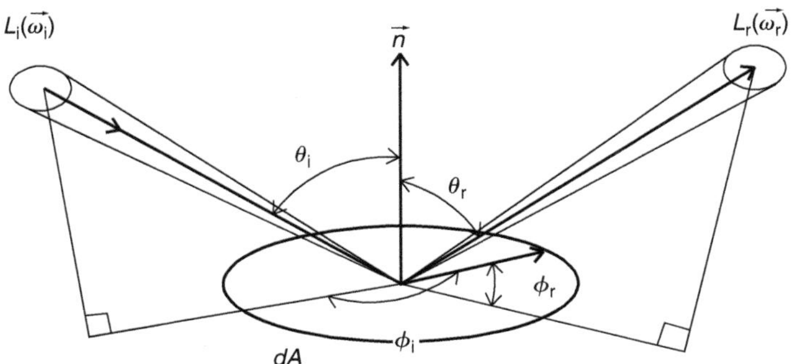

FIGURE 15.4
Reflectance modeling.

where $\Phi_{\text{reflected}}$ is the reflected radiant flux, Φ_{incident} is the incident radiant flux, $\vec{\omega}_r$ and $\vec{\omega}_i$ are the reflected and incident unit solid angles, and ρ is the fraction of the total radiant flux incident upon a surface that is reflected. Reflectance ρ varies according to the wavelength distribution of the incident radiation. Radiant flux is the rate of emission or transmission of radiant energy.

The actual BRDF is then defined as a ratio of reflected radiance over effective incident radiance (see equation [15.2]) and it is a function of incident and reflected angles. Radiance is the flux density of radiant energy per unit solid angle and per unit projected area of radiating surface (Photometric Radiance = Luminance in nit = $[\text{Wm}^{-2}/\text{sr}]$. Irradiance is the density of radiation incident on a given surface (Photometric Irradiance = Illuminance in lux = $[\text{Wm}^{-2}]$). We also provide definitions of radiosity and radiant intensity since they are also frequently used when photometric properties are described.

Radiosity is the density of radiation departing a given surface (Photometric Radiosity = luminosity in lux = $[\text{Wm}^{-2}]$). Radiant Intensity is the radiant energy per unit solid angle (Photometric Radiant Intensity = luminous intensity in Candela = [W/sr])

$$\text{BRDF: } f_r(\theta_i, \phi_i; \theta_r, \phi_r) = \frac{L_r(\theta_r, \phi_r)}{L_i(\theta_i, \phi_i) \cos\theta_i \, d\omega_i} \quad (\text{sr}^{-1}), \tag{15.2}$$

where L_i and L_r are the incident and reflected radiance values, the term $L_i(\theta_i, \phi_i)\cos\theta_i d\omega$ is the irradiance, and f_r is the BRDF value for a set of incident and reflected angular values $(\theta_i, \phi_i; \theta_r, \phi_r)$ in 1/sr.

BRDF satisfies Helmholtz reciprocity principle in Equation 15.3 that can be easily interpreted as the invariance to the order of viewing and illumination angles. Many surfaces can be classified as isotropic, that is, their reflectance is independent of the azimuth angle (see Equation 15.3), which simplifies modeling computation.

Helmholtz Reciprocity Principle

$$f_r(\vec{\omega}_i \to \vec{\omega}_r) = f_r(\vec{\omega}_r \to \vec{\omega}_i), \tag{15.3}$$

Isotropic Surface

$$f_r(\theta_i, \phi_i + \phi; \theta_r, \phi_r + \phi) = f_r(\theta_i, \phi_i; \theta_r, \phi_r).$$

While generating data for modeling, one has to be aware of all assumptions and constraints. For example, additional λ-wavelength dependencies would be required for

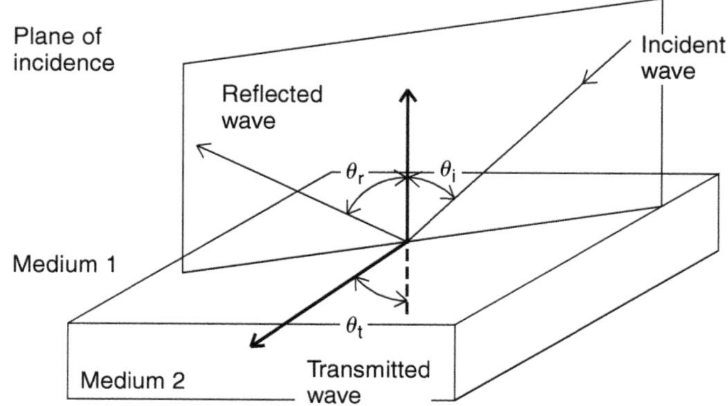

FIGURE 15.5
Modeling BRDFs using geometrical optics. An incident light is decomposed into reflected and refracted components.

modeling absorbent surfaces (see Equation 15.4) or fluorescent surfaces (see Equation 15.5).

$$\text{BRDF}: f_r(\theta_i, \phi_i; \theta_r, \phi_r; \lambda) = \frac{L_r(\theta_r, \phi_r; \lambda)}{L_i(\theta_i, \phi_i; \lambda) \cos\theta_i \, d\omega_i} \quad (\text{sr}^{-1}), \tag{15.4}$$

$$\text{BRDF}: f_r(\theta_i, \phi_i; \theta_r, \phi_r; \lambda_i, \lambda_r) = \frac{L_r(\theta_r, \phi_r; \lambda_i, \lambda_r)}{L_i(\theta_i, \phi_i; \lambda_i, \lambda_r) \cos\theta_i \, d\omega_i} \quad (\text{sr}^{-1}). \tag{15.5}$$

The BRDF/BTF models currently used are based on the geometric optics assumptions described before. Any interaction of light and a material surface is governed by the laws of reflection (Equation 15.6) and refraction (or Snell's law in Equation 15.7). The interaction under geometric optics assumptions is depicted in Figure 15.5.

Law of Reflection

$$\theta_i = \theta_r, \tag{15.6}$$

Snell's Law of Refraction

$$\frac{\sin\theta_r}{\sin\theta_i} = \frac{n_{\text{medium of incident wave}}}{n_{\text{medium of transmitted wave}}}; \tag{15.7}$$

Refractive Index

$$n_{\text{medium}} = \frac{c_{\text{vacuum}}}{c_{\text{medium}}} = f(\lambda), \tag{15.8}$$

where λ is the incident wavelength, θ_i is the angle subtended between the incident ray and the normal to the interface, θ_r is the angle subtended between the refracted ray and the normal to the interface according to Figure 15.5, n is the refractive index of medium, and c is the speed of light in medium.

15.3.3 Data Generation Challenges

In order to generate data for spectral BRDF/BTF modeling, one has to select a wavelength range and a spectral resolution. It might be necessary to use multiple sensors to cover a desired wavelength range, for example, thermal infrared (IR) and color visible spectrum cameras, or to use very specialized cameras to achieve a desired spectral resolution, such as hyperspectral cameras [35] made by Opto-Knowledge Systems, Inc. and using liquid crystal tunable filters [36]. Given an illumination source and a scene/object of interest, one has to measure all input parameters for a selected theoretically derived BRDF/BTF model. However, scenes might be very complex or have a random change of 3D geometry (wind blowing on a corn field) and multiple modeling approaches have to be considered.

Many times, the high complexity of BTF cannot be easily modeled with analytical models due to the randomness of outdoor scene surfaces (geometrical surface randomness) or due to the difficulties of deriving analytical models for real materials (complex interaction of light with materials). For example, it is very hard to model every blade of grass with accurate 3D geometry. Likewise, it is hard to model materials with wavelength dependencies, for example, wavelength dependent absorbent surfaces and fluorescing materials, or the light and material interactions with quantum light behavior, for example, energy saturation and saturation of kinetic energy. Furthermore, any outdoor BTF modeling requires considering temporally changing spectral properties of solar illumination, seasonal vegetation property changes, and atmospheric effects on the illumination. From the BTF model-building viewpoint, very complex analytical models usually require large computational resources that might make the models impractical for any real application.

The aforementioned challenges in BTF modeling have to be well understood and addressed before data generation. Data generation, sensor setup, postprocessing, data analysis, and data synthesis components have to be fully coordinated with mutual understanding of the common goal.

15.4 Example of Data Acquisitions Using Wireless Sensor Networks

Let us assume that our objective is to acquire data from multiple spatially (geographically) distributed locations using wireless and wired sensors. The data could represent any physical quantity of interest, perhaps temperature [37] or sound [38]. Our goal is to monitor these measured entities over time, evaluate them and possibly take actions based on their values. This type of data acquisition problem poses several system design challenges. In order to meet our goal by deploying sensors, sensing physical variables, calibrating received measurements, analyzing data, and triggering actions, one has to become intimately familiar with the data sources, for example, wireless point sensors.

We will devote this section to describing issues that arise while using one of the emerging sensor technologies, called "smart" sensors as the point sensors (also known as sensor *"motes"* or sensor *"nodes"*). The term "mote" was coined in the area of micro electromechanical systems (MEMS) and low power technology sensor literature [39–41]. Created with integrated circuit (IC) technology and combined with computational logic, these "smart" sensors have the benefit of small size, low cost, and power consumption, and, the capability to perform on-board computation. The built-in support for wireless communication makes it possible to deploy them in remote, humanly inaccessible locations. We chose to describe a particular case of motes, called MICA [42], that are manufactured by Crossbow Inc. [6]. The associated software is developed by the University of California at Berkeley [43].

In order to design the most optimal data acquisition system with a given set of wireless MICA sensors, one has to understand the hardware and software while asking questions about (1) optimal spatial arrangement of sensors, (2) synchronization of data acquisitions at multiple sensor sites, (3) effective communication mechanism for receiving information from many sensors, (4) the impact of other wireless sensors on the designed system, (5) calibration of collected measurements, and (6) data fusion from multiple sensors. These questions can be viewed as a small subset of all considerations about a data acquisition system using wireless sensor networks, and serve the purpose of illustrating why and how understanding of data sources is closely related to data analysis. Next, we describe the MICA hardware and software components, and then explain the tradeoffs between information loss and information content evaluated while designing a data acquisition system based on our above considerations.

15.4.1 Understanding Hardware and Software Components

The MICA hardware provides (1) computational capabilities (4 MHz Atmega 128L processor), (2) memory for program and data storage (128 Kbytes Flash, 4 Kbytes Static Random Access Memory [SRAM], and 4 Kbytes of Electrically Erasable Programmable Read-Only Memory [EEPROM]), and (3) communication devices (916 MHz radio transceiver with a maximum data rate of 40 Kbits/sec) for transferring information between motes and a base station (see Figure 15.6). Each MICA board contains also a connector for attaching a sensor board (e.g., the plug-in MTS101CA sensor board attached through a 51-pin expansion connector as it can be seen in Figure 15.7) and a battery pack (2 AA batteries) to power the MICA and its attached sensor board. There are currently three versions of MICA (MICA1, MICA2, and MICA2dot), which differ in their hardware design and

FIGURE 15.6
MICA1 (left) and MICA2 (right) mother-boards.

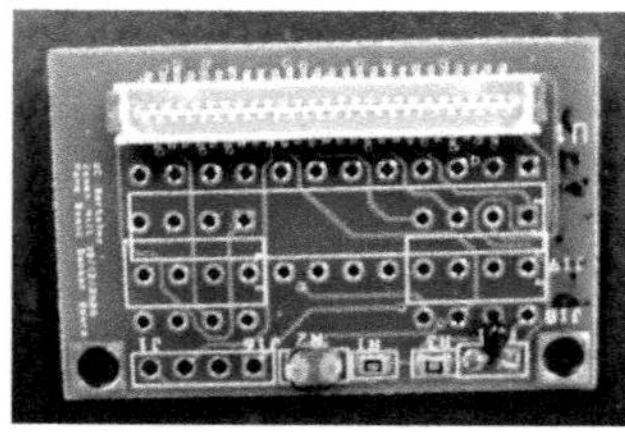

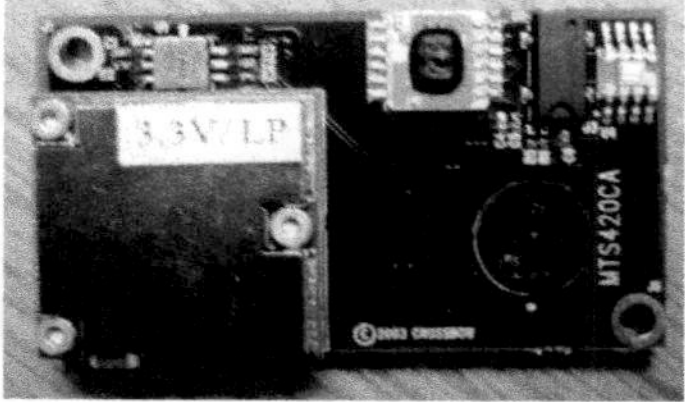

FIGURE 15.7
Multiple sensor boards that can be attached to the MICA mother-boards and can have a precision thermistor and a light/photocell sensor (left), buzzer, microphone, accelerometer, magnetometer (middle), or processing for Global Positioning System (GPS) sensor (right — metal case).

represent a progressive line of motes. Each MICA is operated by a tiny operating system called tinyOS (or TOS) that is uploaded to the EEPROM memory via serial port (MICA2 and MICA2dot) or parallel port (MICA1). Multiple MICAs form a sensor network (i.e., sensors can exchange information). The sensors in the network can send their measurements to another sensor or to a base station either directly (single-hop networks) or via other sensors by propagating information from sensor to sensor (multihop networks). The fact that the sensor locations (1) are chosen arbitrarily *a priori* and (2) can change over time, led to the term "ad hoc" sensor networks to reflect the randomness of network spatial configuration.

One writes programs in tinyOS that can turn on individual sensors, digitize measured values, store the values in a buffer and either perform preprogrammed computation with data or transmit data to a base station. TinyOS is an open source software platform and allows networking, power management, and sensor measurement details to be abstracted from an application development. The key to tinyOSs functionality is the NesC (network-embedded-systems-C) compiler, which compiles tinyOS programs. NesC has a C like structure and provides several advantages, such as, interfaces, wire error detection, automatic document generation, and facilitation of significant code optimizations. TinyOS also consists of a tiny scheduler and a graph of components (or a set of *"command"* and *"event"* handlers known in C programming language).

While our goal is to monitor and process measured physical entities over time, the hardware and software capabilities described above have severe limitations. First, sensor motes have limited processing and data storage capabilities [44]. While writing an application code, one should avoid writing highly complex programs or expect to accumulate a large quantity of data on the sensor. Second, power is an expensive commodity. The two AA batteries supply power to sensors over only a finite time and the power consumption is proportional to mote's activities (i.e., the amount of computation and communication). Keeping energy conservation in mind, the processor has three sleep modes: (1) "idle," which just shuts the processor off; (2) "power down," which shuts everything off except the watch-dog (interrupt monitor); and (3) "power save," which is similar to power-down, but leaves an asynchronous timer running. The processor, radio, and a typical sensor load consume a power of about 100 mW. This figure can be compared with the 30 μW drawn when all electrical components are inactive (in sleep modes). The overall system must embrace the philosophy of getting the work done as quickly as possible and then going into sleep mode [45].

15.4.2 Maximizing Information Content

From a data analysis viewpoint, we are looking for maximum information content and a minimum information loss in a data acquisition system. Information content represents primarily (1) spatial locations of measurements, (2) time of data acquisitions, and (3) calibrated values. The word calibrated refers to the value in physical units, for example, [−20°C, 50°C] rather than to the raw value coming out of an analog to digital converter (ADC). Spatial and temporal information about each measurement can be obtained by adding an appropriate sensor. For example, spatial locations of MICAs can be measured by attaching Global Position System (GPS) sensors in outdoor environments or by measuring a time delay between acoustic and electro-magnetic waves (acoustic time-of-flight ranging) [46–48]. In order to save (1) the cost of a time sensor and (2) the energy consumed by time data acquisition, a relative time can be derived from a processor's counter since there is no internal clock on a MICA board. Although one eliminates the need for a time sensor, the problem of synchronization of data acquisitions arises. Regardless of a chosen synchronization technique [49,50], for instance, by broadcasting a RESET signal to all sensors and adding the offset due to a distance [51,52], it is important to incorporate the uncertainty of spatial and temporal measurements into the data analysis. One should also be aware that measuring and transmitting spatial and temporal information would consume not only bandwidth but also energy. Lastly, a conversion of raw values to calibrated values in engineering units is a critical step to provide information content. There are sensors that (1) are calibrated, (2) come with calibration equations, or (3) have to be calibrated with a calibration gauge. For instance, the raw values from the thermistor on MTS101CA [53] can be converted to degrees Kelvin with accuracy of ±0.2 K by using the following approximation over the 0°C to 50°C temperature range

$$1/T(K) = a + b^{*}\mathrm{Ln}(\mathrm{Rthr}) + c^{*}[\mathrm{Ln}(\mathrm{Rthr})]^{3}, \qquad (15.9)$$

where Rthr = $R1$(ADC_FS-ADC)/ADC, a = 0.001010024, b = 0.000242127, c = 0.000000146, $R1$ = 10 K, ADC_FS = 1024, and ADC = output value from sensor's ADC measurement. Calibrated values with spatial and temporal information are valuable not only for data analysis but also for calibration of other sensors, for example, calibrating spectral cameras or depth maps obtained from stereo vision [37,48].

In order to quantify the information lost, let us define information loss (IL) as a percentage of the difference between the number of values measured by a sensor and the number of values received at the base station over the total number of expected MICA readings from all the active MICA sensors according to Equation 15.9. Let n represent the expected number of values (readings) in each packet being sent with its counter value (time estimate).

$$\mathrm{IL} = \frac{\sum_{i}^{\mathrm{numSensors}} \sum_{j}^{\mathrm{numPackets}} \mathrm{counter}(\mathrm{packet}(i, j+1)) - (\mathrm{counter}(i, \mathrm{packet}(j)) + n)}{\sum_{i}^{\mathrm{numSensors}} \mathrm{counter}(\mathrm{packet}(i, \mathrm{numPackets})) - \mathrm{counter}(\mathrm{packet}(i, 0))}. \qquad (15.10)$$

One could enumerate a few factors that impact the information loss including (1) communication mechanism [54–56], (2) the number of operating motes, (3) distance between a sensor and a base station (or signal strength), (4) spatial arrangement of sensors [57], and (5) other wireless devices operating at a similar frequency. Since a wireless environment with ad hoc multihop sensor networks forms a very complex system, most of these factors are currently under research in order to be fully understood. One can obtain experimental estimates of information loss for simple scenarios, such as a single-hop MICA sensor network in a laboratory environment with increasing numbers of independent sensors. Figure 15.8 shows experimental data for such a case [51], where a set of independent sensors is broadcasting meaningless data with respect to a sensor of interest after every 150 msec. This very simple experiment demonstrates a nonlinear dependency of the information loss on the number of similarly powered devices using the same 916 MHz frequency but working independently (asynchronously).

The description of the aforementioned issues in a particular data acquisition system using MICA sensors highlights the importance of understanding data sources. It is hard to conceive a set of confident data-driven conclusions based on data analysis without understanding of data acquisition system limitations, such as, (1) sensor limitations (power/CPU/memory constraints), (2) data communication limitations (limited spatial range, delays, and packet loses), (3) system reliability (sensor failure, MICA board failure), and (4) data accuracy and uncertainty (sensor accuracy, synchronization accuracy, spatial registration accuracy, temperature dependency).

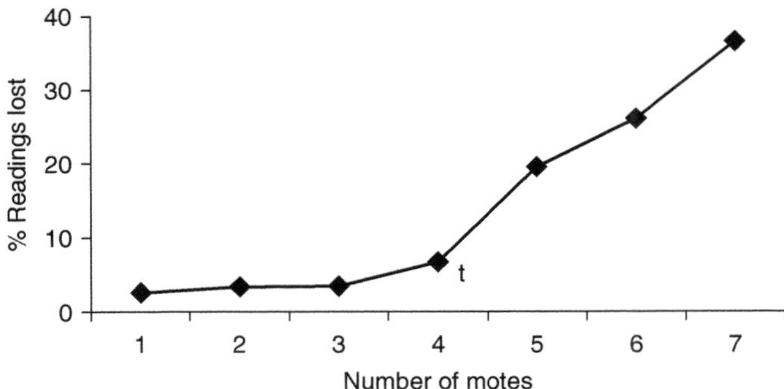

FIGURE 15.8
Percentage of readings lost as a function of number of motes in "other" network.

15.5 Summary

In summary, data sources dictate format, feature quality, spatial coverage, and temporal accuracy. It is essential to understand the issues related to data sources, data producers, storage media, data quality, data transfer (communication), data access and retrieval, and semantic meaning. We illustrated the importance of understanding data sources in the two examples about (1) data generation for BRDF modeling purposes and (2) data acquisition using wireless sensor networks for monitoring purposes. Although any selection of data sources is driven primarily by (1) application needs, (2) data quality, and (3) the associated cost, one should understand the importance of all above issues for a successful and meaningful data analysis.

References

1. MBT GmbH, Germany, Manufacturer of hydrology sensors, URL: http://www.m-b-t.com/english/Products.html.
2. Vaisala Inc., Finland, Manufacturer of hydrology sensors, URL: http://www.vaisala.com/.
3. Oceana Sensors, Virginia, USA, Manufacturer of machinery health monitoring sensors, URL: http://www.oceanasensor.com/.
4. Analog Devices, MA, USA, Manufacturer of MEMS sensors, URL: http://www.analog.com/index.html.
5. Advanced Micro Devices (AMD), Inc., CA, USA, Manufacturer of MEMS sensors for EPA, URL: http://yosemite.epa.gov/r9/sfund/overview.nsf/.
6. Crossbow Inc., URL: http://www.xbow.com.
7. Veris Technologies, 2003, 601 N. Broadway, Salina, KS 67401, http://www.veristech.com.
8. John Deere, IL, USA, Manufacturer of equipment for corn yield in agricultural combines, URL: http://www.deere.com/en_US/deerecom/usa_canada.html.
9. The SHIELD project, the advanced Structural Health Integrated Electronic Life Determination (SHIELD) systems, URL: http://jazz.nist.gov/atpcf/prjbriefs/prjbrief.cfm?ProjectNumber=00-00-5306 and the URL of the main lead (Caterpillar, Inc., IL, USA): http://www.cat.com/.
10. Holloway, A., R.K. van Laar, R.W. Tothill, and D. Bowtell. Options available from start to finish for obtaining data from {DNA} microarrays {II}. *Nature Genetics Supplement*, 2002, 32: 481–489.
11. Dasu, T. and T. Johnson. *Exploratory Data Mining and Data Cleaning*, John Wiley & Sons, NJ, 2003.

12. Han, J. and M. Kamber, *Data Mining: Concepts and Techniques*, Morgan Kaufmann Publishers, San Francisco, CA, p. 550, 2001.
13. Bajcsy, P., J. Han, L. Liu, and J. Young, Survey of bio-data analysis from data mining perspective, in Jason T.L. Wang, Mohammed J. Zaki, Hannu T.T. Toivonen, and Dennis Shasha, Eds., *Data Mining in Bioinformatics*, Springer Verlag, New York, 2004, pp. 9–39, chap. 2.
14. Weather predictions using the Weather Research Forecast (WRF) model, effort led by the National Center for Atmospheric Research (NCAR) Mesoscale and Microscale Meteorology Division (MMM), URL: http://wrf-model.org/.
15. ABAQUS, developed by ABAQUS Inc., finite element analysis in structural engineering, URL: http://www.abaqus.com/.
16. MODTRAN, developed by AFRL/VSBT in collaboration with Spectral Sciences, Inc., MODerate spectral resolution atmospheric TRANSsmittance algorithm and computer model for atmospheric spectral modeling, URL: http://www.vs.afrl.af.mil/Division/VSBYB/modtran4.html.
17. Xpatch, developed by SAIC/Demaco, IL, synthetic aperture radar prediction of realistic far-field and near-field radar signatures for 3D target modeling, URL: http:www.saic.com/products/software/xpatch/.
18. Bajcsy, P. and A. Chaudhuri, Benefits of high resolution SAR for ATR of targets in proximity, in *Proceedings of IEEE Radar Conference*, Long Beach, CA, April 22–25, 2002.
19. Ulaby, F.T. and M.C. Dobson, *Handbook of Radar Scattering Statistics for Terrain*, Norwood, MA: Artech House, 1989.
20. Bajcsy, P., D. Sullivan, and P. Ryan, Estimating Weibull distribution parameters, in C. Dagli, et al., Eds., *Intelligent Engineering Systems Through Artificial Neural Networks*, ASME Press, New York, Vol. 10, pp. 677–682, 2000.
21. DIRSIG, developed by Digital Imaging and Remote Sensing Laboratory at the Rochester Institute of Technology, radiometric modeling of various multispectral and hyperspectral imaging sensors, URL: http://www.cis.rit.edu/research/dirs/research/dirsig.html.
22. MuSES, RadTherm, and WinTherm, developed by ThermoAnalytics, Inc., thermal prediction, URL: http://www.thermoanalytics.com/products/.
23. Bajcsy, P., Metadata information generation, in *Proceedings of NIMA/NSF Motion Imagery Workshop*, Herndon, VA, November 29–30, 2001.
24. Balanis, C.A., *Advanced Engineering Electromagnetics*, John Wiley & Sons, New York, p. 981, 1989.
25. Dana, K.J. et al., Reflectance and texture of real-world surfaces. *ACM Transactions on Graphics*, 1999, 18: 1–34.
26. Dana, K.J. and S.K. Nayar, Correlation Model for 3D Texture. *Proceedings of the International Conference Computer Vision*. Vol. 2, pp. 602–608, Corfu, Greece, September 20–25, 1999.
27. Leung, T.K. and J. Malik, Recognizing Surfaces using Three-Dimensional Textons. *Proceedings of the International Conference Computer Vision*. Vol. 2, pp. 1010, Corfu, Greece, September 20–25, 1999.
28. Panjwani, D.K. and G. Healey, Markov random field models for unsupervised segmentation of textured color images. *IEEE Transactions on Pattern Analysis and Machine Intelligence*, 1995, 17: 939–954.
29. Bajcsy, P., R. Kooper, and D. Andersh, Multisensor scene modeling using statistical models for bidirectional texture functions, SPIE 2004, Algorithms for SAR Imagery XI, April 12–15, 2004.
30. Lu, S.Y. and K.S. Fu, Stochastic tree grammar inference for texture synthesis and discrimination. *Computer Graphics and Image Processing*, 1979, 9: 234–245.
31. Portilla, J. and E.P. Simoncelli, A parametric texture model based on joint statistics of complex wavelet coefficients. *International Journal of Computer Vision*, 2000, 40: 49–70.
32. Zhu, S.C., Y. Wu, and D. Mumford, Filters random fields and maximum entropy (FRAME): to a unified theory for texture modeling. *International Journal of Computer Vision*, 1998, 27: 1–20.
33. Varma, M. and A. Zisserman, Classifying Images of Materials: Achieving Viewpoint and Illumination Independence. *Proceedings of the 7th European Conference on Computer Vision*. Vol. 3, pp. 255–271, Copenhagen, Denmark, 2002.

34. CUReT, Columbia-Utrecht reflectance and texture database, http://www1.cs.columbia.edu/CAVE/curet/html/about.html.
35. Official Opto-Knowledge Systems Inc. Web site, Available at http://www.techexpo.com/WWW/opto-knowledge/index.html.
36. Chrien, T., C. Chovit, and P. Miller, Imaging spectrometry using liquid crystal tunable filters, Jet Propulsion Laboratory and Cambridge Research and Instrumentation Inc., April, 1993, http://www.cri-inc.com/instruments/data/pdf/LCTF_JPL_0300.pdf.
37. Bajcsy, P. and S. Saha, A new thermal infrared camera calibration approach using wireless MEMS sensors, in *Communication Networks and Distributed Systems Modeling and Simulation Conference (CNDS 2004)*, January 19–22, 2004, San Diego, CA.
38. Wang, H., D. Estrin, and L. Girod, Preprocessing in a tiered sensor network for habitat monitoring, *EURASIP Journal on Applied Signal Processing*, 2003, 4: 392–401.
39. MICA: The commercialization of microsensor motes, URL: http://www.sensorsmag.com/articles/0402/40/.
40. A. Mainwaing, R. Szewczyk, D. Culler, J. Anderson. "Wireless Sensor Networks for Habitat Monitoring". *In ACM International Workshop on Wireless Sensor Networks and Applications (WSNA)*, 2002.
41. Gupta, P. and P.R. Kumar, The capacity of wireless networks. *IEEE Transactions on Information Theory*, 2000, 46: 388–404.
42. MICA product specifications, URL: http://www.xbow.com/Products/Product_pdf_files/Wireless_pdf/MICA.pdf.
43. TinyOS specifications, URL: http://webs.cs.berkeley.edu/tos.
44. J. Heidemann, F. Silva, C. Intanagonwiwat, R. Govindan, D. Estrin and D. Ganesan. "Building efficient wireless sensor networks with low-level naming". *In Proceedings of the Symposium on Operating Systems Principles (SOSP)*, pp. 146–159, Chateau Lake Louise, Banff, Alberta, Canada, October 2001, ACM Press.
45. Raghavendra, C.S. and Suresh Singh, PAMAS — power aware multi-access protocol with signaling for ad hoc networks. *ACM Computer Communication Review*, 1998, 28: 5–26.
46. Whitehouse, K. and D. Culler, Macro-calibration in sensor/actuator networks, *Mobile Networks and Applications Journal (MONET), Special Issue on Wireless Sensor Networks*, June, 2003, ACM Press.
47. Whitehouse K., Culler D. "Macro-calibration in Sensor/Actuator Networks." *Mobile Networks and Applications Journal (MONET), Special Issue on Wireless Sensor Networks*. Volume 8 , Issue 4 (August, 2003). pp. 463–472, ACM Press.
48. Scherba, D. and P. Bajcsy, Depth estimation by fusing stereo and wireless sensor locations, Technical report NCSA-ALG-04-0008, October, 2004.
49. J. Elson and D. Estrin. Time Synchronization for Wireless Sensor Networks. Parallel and Distributed Processing Symposium., *Proceedings 15th International*, April 23–27, 2001, pp. 1965–1970 (IEEE Conference Proceedings).
50. J. Elson, L. Girod, and D. Estrin. Fine-Grained Network Time Synchronization using Reference Broadcasts. *In Proceedings of the Fifth Symposium on Operating Systems Design and Implementation (OSDI 2002)*, pp. 147–163, Boston, MA, December 2002, ACM Press.
51. Saha, S. and P. Bajcsy, System design issues in a single-hop wireless sensor network, in *Proceedings of the 2nd IASTED International Conference on Communications, Internet and Information Technology (CIIT)*, pp. 743–748, November, 2003.
52. Saha, S. and P. Bajcsy, System design issues for applications using wireless sensor networks, Technical Report NCSA-ALG-03-0001, August, 2003.
53. Specification of the sensor board MTS MDA series, URL: http://www.xbow.com/Support/Support_pdf_files/MTS-MDA_Series_User_Manual_RevB.pdf.
54. Woo, A. and D. Culler, A transmission control scheme for media access in sensor networks, in *Proceedings of the 7th ACM/IEEE International Conference on Mobile Computing and Networking*, Rome, Italy, pp. 221–235, 2001.
55. Intanagonwiwat, C., R. Govindan, and D. Estrin, Directed diffusion: a scalable and robust communication paradigm for sensor networks, in *ACM/IEEE International Conference on Mobile Computing and Networks (MobiCom 2000)*, August 2000, Boston, MA.

56. Lam, S.S., A carrier sense multiple access protocol for local networks, *Computer Networks*, 1980, 4: 21–32.
57. Krishnamachari, B., D. Estrin, and S. Wicker, Modelling data-centric routing in wireless sensor networks, IEEE Infocom 2002, (also Technical report CENG 02-14), http://www2.parc.com/spl/members/zhao/stanford-cs428/readings/Networking/Krishnamachari_infocom02.pdf.

16

Data Representation

Peter Bajcsy

CONTENTS

16.1 Introduction

In this chapter, we would like to elaborate on the issue of data representation that was introduced in one of the previous chapters about "Managing and Accessing Large Datasets." Our overview of computer data representation presents issues related to the tradeoffs between information retrieval efficiency and data representation size in computer memory as a function of data organization.

There are two basic data structures used in Geographic Information System (GIS). They are raster (or cellular) data and vector data structures [1, Chapter 15]. The need for memory efficient representations of a variety of types of GIS information motivates the use of the raster and vector data structures in GIS applications. Examples of raster-based information would include terrain maps, land use maps, land cover maps, and weather maps. Boundaries (also called contours or outlines) of parcels, ecoregions, Census tracts or U.S. postal zip codes would be examples of vector data. A visualization of raster, vector and tabular information is in Figure 16.1.

As mentioned earlier, (1) raster measurements (images) can be acquired by satellite or airborne sensors, (2) point spatial measurements can come from associating global positioning system (GPS) values and physical attributes, or from address databases, and (3) boundary spatial measurements can be formed based on man-made attributes, such as, zip code, or based on environmental characteristics, such as ecoregions. These three data types can

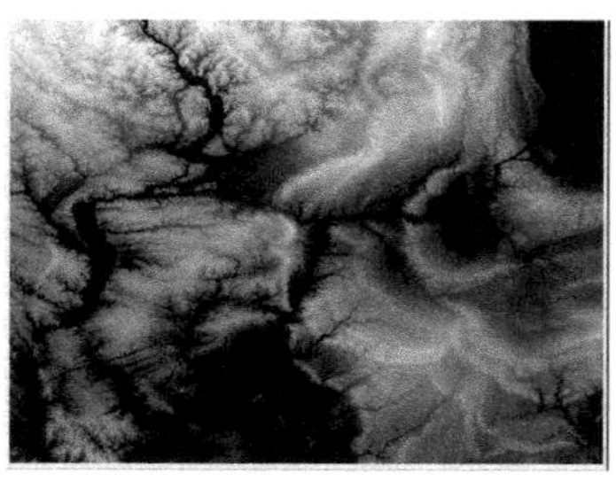

1	lat	lng	value
2	41.32110	-89.5531	194.100
3	41.07110	-89.4697	197.830
4	40.94611	-89.4697	138.479
5	40.94611	-91.2197	217.650
6	42.42111	-91.2197	279.200
7	41.07111	-91.2197	228.070
8	41.07111	-90.9697	161.900

FIGURE 16.1
Examples of raster data (left), vector structures (boundaries of Illinois counties — middle), and point information (right) also classified as vector data structure.

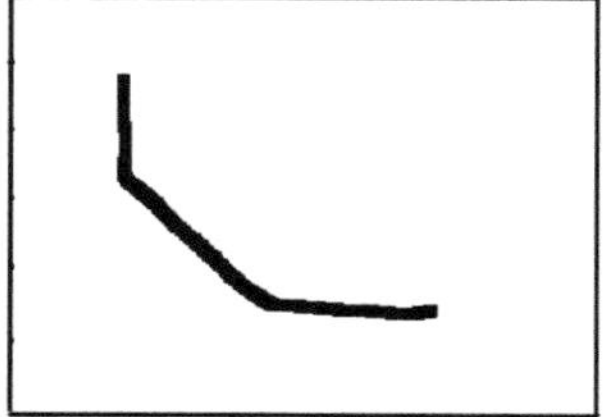

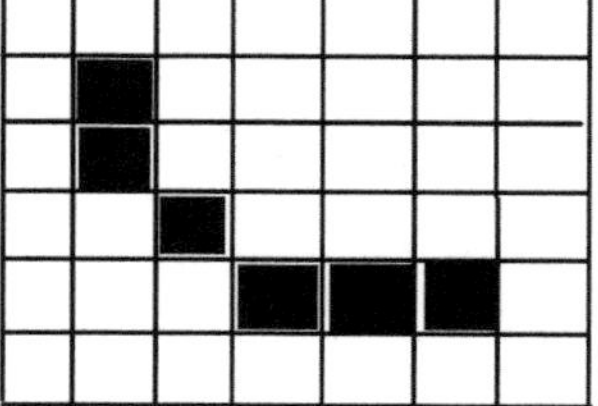

FIGURE 16.2
Raster and vector data representation tradeoffs. Examples of data representation of a piece-wise linear line in a vector format (left) with four points or in a raster format (right) with at least 4 × 5 points enclosing the line.

represent the same or related information. There is a need to understand the tradeoffs of any data type and its computer representation with respect to information retrieval and memory storage efficiency. The primary topic of this chapter is the overview of these tradeoffs in order to achieve efficient information access and minimize computational requirements during software development.

Given the limited capacity of a computer memory, we are looking for a solution that stores spatial information with the fewest number of bits. For example, storing a set of points individually in a table would require storing two double precision values for latitude and longitude in addition to all associated attributes per point. If the same set of points formed a partial boundary of multiple regions then one could refer to the boundary segment multiple times with a boundary index rather than storing the same points multiple times. If the same set of points formed a grid, then one could simply store a single location, the grid spacing parameters, and all associated spatially ordered values to achieve very efficient data storage. One can immediately see why there are many data representation schemes in GIS. An illustration of data representation tradeoffs for raster and vector data types is shown in Figure 16.2.

Given the finite speed of data exchanges between computing hardware devices, we are looking for an optimal data organization or representation that would minimize data retrieval. For example, accessing a point in a table with a specified geographic location might require searching two or six columns with latitude and longitude in decimal or deg/min/sec formats. Retrieving one boundary from a set of boundaries might be as simple as finding the first point and accessing the next N points or as complicated as searching for each boundary point in a large pool of indexed points. In the case of raster data, accessing a point requires conversion from a geographic coordinate system to an image (row/column) coordinate system followed by accessing the pixel. We will devote the rest of this chapter in describing the efficiency tradeoffs between memory size of data representation and information retrieval.

16.2 Vector Data Types

Vector data types (point and boundary information) are more memory efficient than raster data types for storing boundary information. Vector data contain points, lines, arcs, polygons, or any combinations of these elements. Any vector data element can be represented in a reference coordinate system defined by a latitude/longitude, Universal Transverse Mercator (UTM), or pixel coordinate system. The challenge in representing vector data is to organize the data such that the positions and geographic meanings of vector data elements are efficiently stored in memory and easily extracted.

There exist already numerous file formats for storing vector data [1]. Among all vector data organizations in files, the following data structures have been used frequently: location list data structure (LLS), point dictionary structure (PDS), dual independent map encoding structure (DIME), chain file structure (CFS), digital line graphs (DLGs), and topologically integrated geographic encoding and referencing (TIGER) files. For detailed description of each data structure we refer a reader to [1]. Here, we overview boundary information extraction from (1) Shapefiles defined by the Environmental Systems Research Institute (ESRI) [2] and stored in an LLS data structure and (2) Census 2000 TIGER/Line files defined by the U.S. Census Bureau [3] and saved in TIGER file data structures [4]. We show how internal vector data organizations of Shapefiles and TIGER/Line files impact information retrieval and storage requirements [5]. Additional tradeoff comparisons of DLG, TIGER, and LLS (ESRI Shapefile) can be found in Reference [6].

16.2.1 ESRI Shapefile Vector File Format

A Shapefile is a special data file format that stores nontopological geometry and attribute information for the spatial features in a dataset. The geometry for a feature is stored as a shape comprising a set of vector coordinates in a location list data structure (LLS). Shapefiles can support point, line, arc, and area features. Area features are represented as closed loop polygons.

A Shapefile must strictly conform to the ESRI (Environmental Systems Research Institute) specifications [7]. It consists of a main file, an index file, and a dBASE table.

The **main file** is a direct access, variable-record-length file in which each record describes a shape with a list of its vertices. In the **index file**, each record contains the offset of the corresponding main file record from the beginning of the main file. The **dBASE table** contains feature attributes with one record per feature. The one-to-one relationship between geometry and attributes is based on a record number. Attribute records in the dBASE file must be in the same order as records in the main file.

All file names adhere to the ESRI Shapefile naming convention. Each file name consists of the prefix name, followed by a three-letter file extension separated by a comma. The main file, the index file, and the dBASE file have the same prefix. The suffix for the main file is ".shp." The suffix for the index file is ".shx." The suffix for the dBASE table is ".dbf." A naming example of a Shapefile file would be counties.shp (main file), counties.shx (index file), and counties.dbf (DBASE table).

Shapefiles do not have the processing overhead when extracting boundary information (as in a TIGER file), which leads to faster drawing speed and easier boundary information editing. Shapefiles handle single features that overlap or are noncontiguous. However, the memory representation efficiency of Shapefiles is worse than TIGER files because replicas of multiple points are stored several times to provide fast access to each boundary (see Section 16.2.2).

To illustrate data representation of Shapefiles, we describe an example of an array-based data structure, called a ShapeObject, in the I2K software package [8]. This data structure was designed for efficient storage and easy retrieval of boundary data according to ESRI Shapefile format specification. The ShapeObject array structure is composed of two types of arrays: data-holding arrays and index arrays. There are separate data-holding arrays to store the number of parts in each boundary, the number of points in each boundary, the number of points in each part of a boundary and the bounding boxes, and type of each boundary. There is one single array to store the actual points of all boundaries. Some data-holding arrays have a corresponding index array. This index array stores the starting index of each boundary's information in the corresponding data-holding array. The data-holding arrays can be accessed directly according to the boundary one is interested in. The dBASE table information is loaded into a tabular data structure (TableObject) that consists of multiple columns to accommodate categorical and continuous tabular variables. The TableObject has to be associated with the ShapeObject.

Let us assume that we would like to retrieve the second boundary from an instance of ShapeObject. First, we obtain the starting index of that list of boundary points from its corresponding index array containing starting indices of all boundaries. Let this index be i. Second, we retrieve the number of points, n, in the second boundary, from the array storing the number of points in each boundary. Third, to obtain the points of the second boundary, we traverse the common array storing points for all boundaries, starting from the ith point and accessing n subsequent points. Fourth, we might retrieve all associated attributes of the second boundary by extracting attributes from the second row of each column in the TableObject data structure. The purpose of this simple explanation is to give an insight into the ShapeObject and TableObject design considerations for interested software developers.

One can also consider additional data structure design issues for vector data fusion. For example, fusing crime reports associated with Illinois and Indiana counties as shown in Figure 16.3 requires loading multiple Shapefiles, registering points into the same coordinate

FIGURE 16.3
A visualization of spatial boundary information obtained from two Shapefiles of Illinois and Indiana counties loaded together.

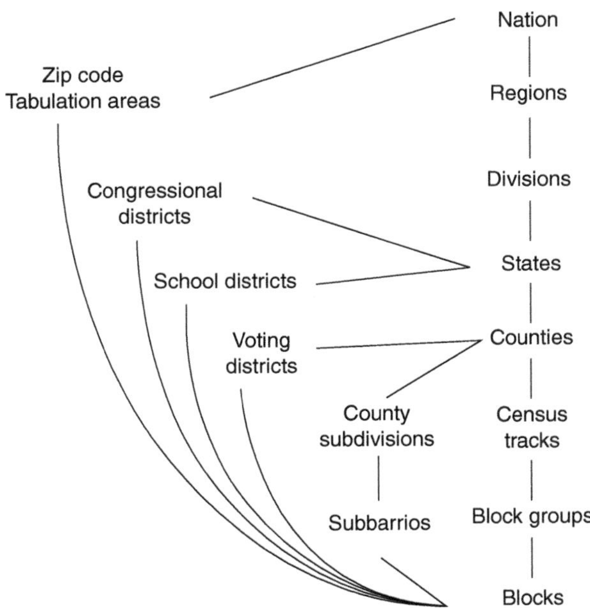

FIGURE 16.4
Hierarchical structure of the U.S. Census Bureau territories.

system, integrating related information, and viewing spatial boundary information and associated attributes. These steps are achieved in I2K software by converting a java linked-list (Vector) to a ShapeObject and inserting the shape records of all Shapefiles into a single linked list structure.

16.2.2 Census 200 TIGER/Line Vector File Format

The Census 2000 TIGER/Line Files provide geographical information on the boundaries of counties, zip codes, voting districts, and a geographic hierarchy of census relevant territories. For example, the census relevant territories are census tracts that are composed of block groups, which are in turn composed of blocks as is shown in Figure 16.4. Census 2000 TIGER/Line files also contain information on roads, rivers, landmarks, airports, etc, including both latitude/longitude coordinates and corresponding addresses [3]. A detailed digital map of the United States, including the ability to look up addresses, could therefore be created through processing of the TIGER/Line files.

16.2.2.1 File Format Description

Because the density of data in the TIGER/Line files comes at the price of a complex encoding, extracting all available information from TIGER/Line files is a major task. Our description next is primarily focused on extracting boundary information of regions while extracting other available information in TIGER/Line files is left out for reasons of brevity.

TIGER/Line files are based on an elaboration of the chain file structure (CFS) [1], where the primary element of information is an edge. Each edge has a unique identification (ID) number (Tiger/Line ID or TLID) and is defined by two end points. In addition, each edge then has polygons associated with its left and right sides, which in turn are associated with

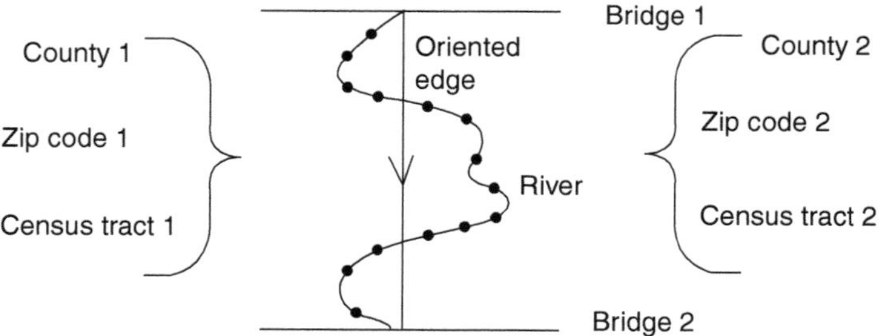

FIGURE 16.5
Illustration of the role of shape points.

a county, zip code, census tract, etc. The edge is also associated with a set of shape points, which provide the actual form an edge takes. The use of shape points allows for fewer polygons to be stored.

To illustrate the role of shape points in Figure 16.5, imagine a winding river that is crossed by two bridges a mile apart, and that the river is a county boundary and therefore of interest to the user. The erratic path of the river requires many points to define it, but the regions on either side of it do not change from one point to the next, only when the next bridge is reached. In this case, the two-bridge/river intersections would be the end points of an edge and the exact path of the river would be represented as shape points. As a result, only one set of polygons (one on either side of the river) is necessary to represent the boundary information of many small, shape-defining edges of a boundary.

This kind of vector representation has significant advantages over other methods in terms of storage space and data representation. To emphasize this point, consider that many boundaries will share the same border edges. These boundaries belong to not only neighboring regions of the same type, but also to different kinds of regions in the geographic hierarchy. As a result, storing the data contained in the TIGER/Line files in a basic location list data structure (LLS), such as ESRI Shapefiles, where every boundary stores its own latitude/longitude points, would introduce a significant amount of redundancy to an already restrictively large data set.

In contrast to its apparent storage efficiency, the TIGER/Line vector data representation is inefficient for boundary information retrieval and requires extensive processing. From a retrieval standpoint, an efficient representation would enable direct recovery of the entire boundary of a region as a list of consecutive points. The conversion between the memory efficient (concise) and retrieval efficient forms of the data is quite laborious in terms of both software development and computation time. We address some of the conversion issues in Section 16.2.3.

Another advantage of the TIGER/Line file representation is that each type of GIS information is self-contained in a subset of files. As a result, users can process only the desired information by loading a selected subset of relevant files. For example, each primary region (county) is fully represented by a maximum of 17 files. Therefore, the landmark information is separate from the county boundary definition information, which is separate from the street address information, etc. Those files that are relevant to the boundary point extraction, and the attributes of those files that are of interest, are the following:

Record Type 1: Edge ID (TLID), Lat/Long of End Points

Record Type 2: TLID, Shape Points

Record Type I: TLID, Polygon ID Left, Polygon ID Right

Record Type S: Polygon ID, Zip Code, County, Census Tract, Block Group, etc.

Record Type P: Polygon ID, Internal Point (Lat/Long).

As part of our description of the Census 2000 data, we would like to mention the difference between the U.S. postal service code areas (or zip codes) and the Census 2000 zip code analog. The zip code information provided in the TIGER/Line files does not give the actual boundaries of zip codes. They are instead Zip Code Tabulation Areas (ZCTAs), which are the Census Bureau's best attempt at representing zip code boundaries. This approximation is necessary because zip codes are not based on geographic regions, but rather on postal routes. Zip codes are therefore collections of points, not boundaries, and can be overlapping. For example, a large office building may have its own zip code while all the buildings around it share another zip code. The U.S. census bureau also made coverage of ZCTA boundaries contiguous, meaning that all of the United States is assigned to a ZCTA. Bodies of water therefore have their own ZCTAs that are designated by a five-digit code ending in "HH." Furthermore, some regions could not be appropriately defined as distinct ZCTAs. They are designated by the first three digits of the zip codes that the region's zip codes have in common and by the suffix "XX."

16.2.3 Vector Format Conversions

As mentioned above, a Shapefile is an information retrieval efficient file format and a TIGER/Line file is data storage efficient format. We present an example of how to convert TIGER/Line files to Shapefiles for those end users that are more concerned about the speed of interactive analysis (e.g., georeferencing, clustering, visualization) than about memory limitations. Our goal is to generate a data structure that holds region names, latitude/longitude points defining a region boundary, neighboring regions to each boundary, and an internal point of the region. From an information retrieval standpoint, this represents a much more efficient organization of data than the TIGER/Line data, and can be exported to Shapefiles. Another advantage of Shapefiles is that many commercial GIS packages support loading and visualization of this format, for example, ArcExplorer (available from the ESRI web site [2,9]).

The underlying principle of the conversion process from TIGER/Line files to ESRI Shapefiles could be compared to sorting points according to the order of boundary edges. This is illustrated in Figure 16.6. In reality, the conversion process begins by loading the raw TIGER/Line files into 2D table-like data structures by making use of manually developed meta data files. Since the TIGER/Line files are fixed-width encoded flat files, meta data is necessary to define the indices of the first and last characters for each attribute in the lines of the flat file. This information, the attributes' names, and their type (integer, floating point number, string, etc.) come from meta data files provided by the Census Bureau. The final piece of information contained in the meta data file is a "Remove Column" field, which dictates whether or not the attribute will be dropped from the table as it is read in. Attributes that are not used during the processing are removed early on for the sake of memory efficiency. The meta information for each Record Type is stored in a comma-separated-value (csv) file, which can easily be parsed into a tabular data structure (TableObject), then accessed in that form by the routine that parses the main data file.

Once the TIGER/Line data are in the form of tables, they are streamed through a complex system of procedures, including conversion to several intermediate data structures, before being inserted into Hierarchical Boundary Objects (HBoundary). Each HBoundary represents one type of region (county, census track, etc.) for a single state. It can be also viewed as one master list of boundary points that all boundaries reference by pointers. The memory

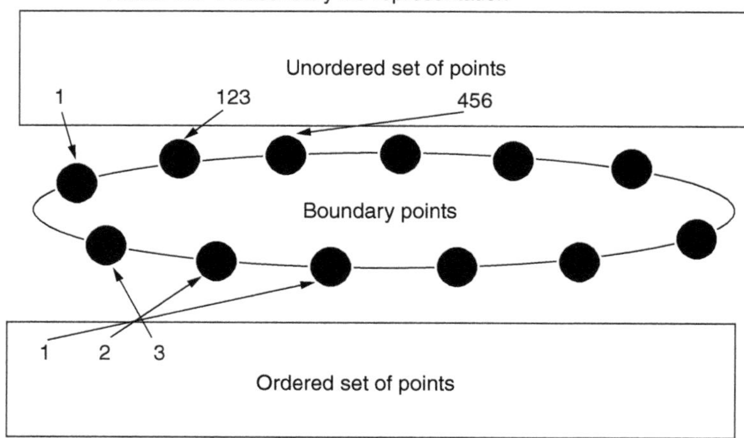

Figure 16.6
The TIGER/Line to ESRI Shapefiles conversion of boundary representation can be viewed as a transformation from an unordered set of points to a clockwise ordered set of points.

savings for each point that is shared by two counties, two census tracts, and two block group boundaries is 30 bytes. For the state of Illinois, this optimization translated into a 38% reduction in memory usage (16.45 MB vs. 26.64 MB). A high level view of the TIGER to Shapefile conversion is given in Figure 16.7.

While writing all information into a Shapefile format, one has to prepare information about the file length, the number of records, the type of records and the global bounding box that encloses all the records, as well as the information required for the index file. Every Shapefile must have a corresponding DBF file. The DBF file contains per shape-record attributes and must have as many rows as there are records in the corresponding main Shapefile. If the DBF file is not written out together with SHP and SHX files, then commercial GIS tools will not be able to open the Shapefile. Thus, even if there is no information for the DBF file, one has to create a DBF file with dummy fields. While this was not an issue in the case of TIGER/Line to Shapefile conversion, one could encounter other GIS processing operations that store only boundary points without any boundary information, for example, the case of automatic extraction of iso-contours from historical maps into ESRI Shapefiles [10].

16.3 Raster Data Types

Raster data files have been deployed in many diverse file formats originating from (1) various communities (e.g., JPG, TIFF, GIF, BMP, TGA, PICT, and PNG — photography community; MPEG, WMV, and AVI — video processing; GeoTIFF, LAN, ENVI IMG, DEM, HDF EOS, and GRID — GIS community; NITF and MSTAR PHX — military), (2) manufacturers of various camera-based instruments (e.g., MCID — confocal microscopes, FITS — astronomy telescopes), (3) image processing programs (e.g., PSD or PDD — Adobe Photoshop), and (4) simulation programs (e.g., ABAQUS files — 2D finite element analysis prediction). Thus, it has been very difficult to compare analytical software tools that operate only on a subset of file formats, and many times only on proprietary file formats. On the

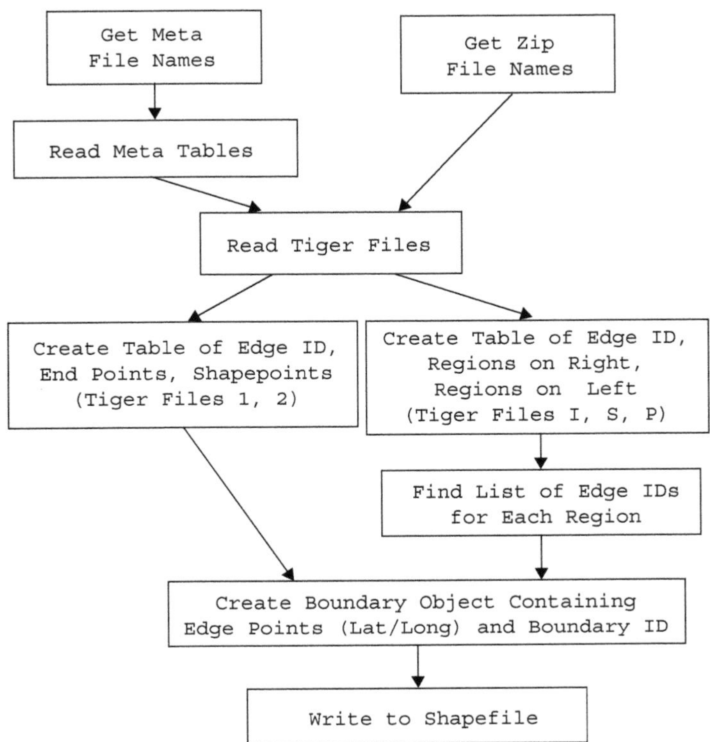

FIGURE 16.7
A flow chart of boundary information extraction from Census 2000 TIGER/Line files (memory efficient format) into the ESRI shape file format with location list data structures (retrieval efficient format).

other side, if a diversity of external file formats could be resolved then an internal raster data representation would likely map into only one out of three data organization schemes. In this section, we briefly describe the three different raster data organization schemes (1) to illustrate that there is no need to have many raster file formats, and (2) to explain the tradeoffs when choosing a raster data representation.

In order to access a value in a raster dataset, one has to define an origin of the coordinate system and the organization schema of multiple values at each row/column location. The row/column location is also denoted as line/pixel location. The origin in most image analyses is defined as the image upper left corner, and the positive horizontal axis corresponds to increasing columns and the negative vertical axis corresponds to increasing rows. The three basic raster data organization schemes include Band Interleaved by Pixel (BIP), Band Interleaved by Line (BIL), and Band Sequential (BSQ) formats. The schemes for storing values are illustrated in Figure 16.8.

16.3.1 BIP, BIL, and BSQ Schemas

The BIP schema provides data organization that keeps all pixel values together and the pixel attributes are ordered from left upper-corner to the right bottom-corner line by line. The values are ordered as follows: value1(row = 0, column = 0), value2(row = 0, column = 0),. . .,value$N$(row = 0, column = 0), value1(row = 0, column = 1), value2(row = 0, column = 1),. . .,value$N$(row = 0, column = 1), and so on. This type of

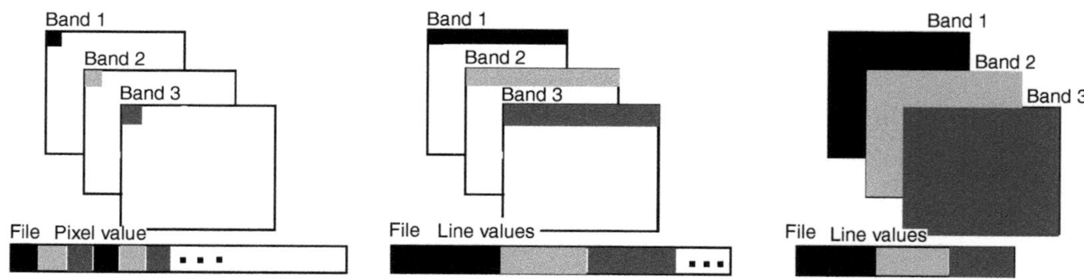

FIGURE 16.8
Three organization schemas for raster data: BIP (left), BIL (middle), and BSQ (right).

data organization is very efficient for performing spatially local operations but less efficient for displaying single band information.

The BIL schema is probably the oldest one. It keeps all row data of one band together and orders row information by band index and then by row index. The order of values can be described as: value1(row = 0, column = 0), value1(row = 0, column = 1),...,value1(row = 0, column = max), value2(row = 0, column = 0), value2(row = 0, column = 1),...,value2(row = 0, column = max),..., value*N*(row = 0, column = 0), value*N*(row = 0, column = 1),...,value*N*(row = 0, column = max), value1(row = 1, column = 0), value1(row = 1, column = 1),...,value1(row = 1, column = max), and so on. This arrangement was convenient for early spectral sensors (e.g., push broom cameras with a line of sensors). However, it is not the most efficient organization for easy access to pixel information, as needed for processing and visualization purposes.

The BSQ schema organizes data by band index from first to last. This particular scheme is convenient for spectral sensors that acquire one band of data in a sequence and hence can directly store the data onto a storage media. Furthermore, it is probably the most efficient organization for visualization since the bands can be retrieved as blocks of memory and displayed as visual layers. However, when spatial subareas have to be either extracted or processed then this schema becomes less efficient than the other data arrangements.

16.3.2 Selection of Raster Data Representation

The overall scrutiny of raster data organizations includes considerations of storage/representation size and data retrieval efficiency with respect to required spatial data processing and data visualization. To illustrate the need for understanding the tradeoffs of raster data representations, we list a few very common questions one can ask. Should we store 100 colors as a single band image with a look-up table or a three band image with byte values? How do we represent data efficiently if we are processing (1) subsets of bands, or (2) sets of lines or (3) subareas of images? How do we convert images efficiently from one to another data organization schema? Should we represent compressed or uncompressed data? How do we display 120 bands of one hyperspectral image? How do we display and process images larger than the size of a computer's RAM memory? What color space should we use for displaying images (e.g., red, green, and blue or hue, saturation and value)?

Answering the above questions requires analysis prior to data representation. Figure 16.9 shows how one would decide whether to choose three band image (three bytes per pixel)

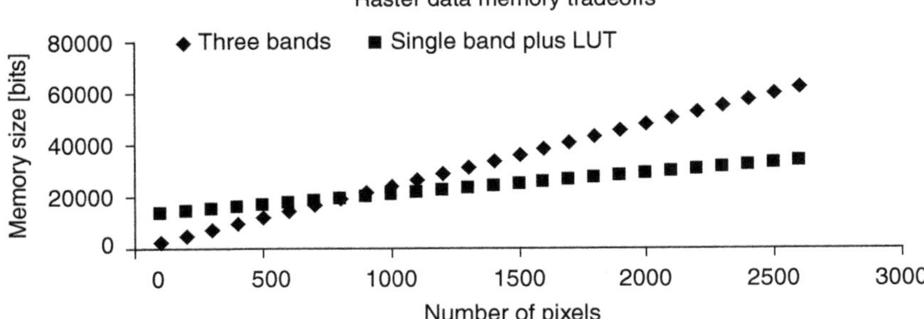

FIGURE 16.9
The tradeoff between raster data representation as three bands or one band plus color look up table (LUT) and memory size, for 200 unique colors. The threshold image size (number of pixels) for choosing a storage efficient representation is 800 pixels.

or single band image (one byte per pixel) plus a color look-up table (one double precision value per color). The values of memory size in Figure 16.9 are computed using the equation below for 200 color values as a function of image size (number of pixels).

$$\begin{aligned} MemSize &= NumPixel * sampTypePixel * sampPerPixel \\ &\quad + NumLUTColors * sampTypeLUTColor, \end{aligned} \tag{16.1}$$

where the second term is zero in the case of three bands, sampTypePixel = 8 bits (1 byte), sampTypeLUTColor = 64 bits, NumLUTColors = 200 color values and sampPerPixel = 1 or 3 samples.

To answer the above question about compressed vs. uncompressed data representations requires studying lossless and lossy compression techniques. For example, lossy Lempel-Ziv Compression [11] or Discrete Cosine Transform (DCT) [12] might be unacceptable for accurate analysis. Furthermore, it is apparent that any lossless compression, for example, Huffman coding or Run-Length coding [12], would increase memory efficiency but decrease information retrieval efficiency. These types of considerations led to designs of sophisticated file formats, such as, the National Imagery Transmission Format (NITF) [13] or Hierarchical Definition Format (HDF) [14] that can contain both compressed data for fast access and uncompressed data for accurate analysis.

In many cases, end users do not have the choice of data representation. It is only the choice of a developer to optimally select raster data representation according to (1) the specifications of instruments generating the raster data, (2) expected information content, (3) required preprocessing, and (4) the specifications of end-user's applications, such as, visual devices and interactive interfaces. In general, one has to be prepared for data representation conversions, and the fundamental knowledge of the aforementioned tradeoff issues is essential.

16.4 Summary

We presented several data representation tradeoffs for vector and raster data types. There are multiple tradeoffs between (1) vector and raster data representation, (2) multiple vector representations, and (3) multiple raster representations. We illustrated the information retrieval and memory size tradeoffs for a particular case of TIGER/Line and ESRI Shapefile vector files and the challenges in data representation conversion. Interested readers can find detailed theoretical and experimental comparisons for ESRI Shapefile, TIGER/Line and DLG-3 files in [6]. In addition, we overviewed raster data representation schemas and listed a set of standard questions one has to resolve in order to optimize raster data representation.

References

1. Campbell, James B., *Introduction to Remote Sensing*, 2nd ed. The Guilford Press, New York, 1996.
2. ESRI Web site: http://www.esri.com/.
3. U.S. Census Bureau, Census 2000. URL: http://www.census.gov/main/www/cen2000.html.
4. Miller, Catherine L., Tiger/Line files technical documentation. UA 2000. U.S. Department of Commerce, Geography Division, U.S. Census Bureau. http://www.census.gov/geo/www/tiger/tigerua/ua2ktgr.pdf.
5. Groves, P., Saha, S., and Bajcsy, P., Boundary information storage, retrieval, georeferencing and visualization, Technical report NCSA-ALG-03-0001, February 2003.
6. Clutter, D. and Bajcsy, P. Storage and retrieval efficiency evaluations of boundary data representations for LLS, TIGER and DLG Data structures, Technical report NCSA-ALG-04-0007, October 2004.
7. ESRI Shape file, file format specification. http://www.esri.com/library/whitepapers/pdfs/shapefile.pdf.
8. Bajcsy, P. et al., Image to knowledge, documentation at URL: http://alg.ncsa.uiuc.edu/tools/docs/i2k/manual/index.html.
9. ArcExplorer product description at ESRI Web site: http://www.esri.com/.
10. Bajcsy, P., Automatic extraction of isocontours from historical maps, in *Proceedings of the 7th World Multiconference on Systemics, Cybernetics and Informatics (SCI 2003)*, Orlando, FL, July 27–30, 2003.
11. Ziv, J. and Lempel, A., A universal algorithm for sequential data compression, *IEEE Transactions on Information Theory*, IT-23, 1977, 337–343.
12. Miano, J. *Compressed Image File Formats: JPEG, PNG, GIF, XBM, BMP*, ACM Press, Addison-Wesley, 1999.
13. The National Image Transmission Format (NITF) Technical Board (NTB), description at URL: http://164.214.2.51/ntb/intro.html.
14. Hierarchical Definition Format (HDF), home web page and specifications, URL: http://hdf.ncsa.uiuc.edu/.

17

Spatial Registration

Peter Bajcsy

Contents

17.1 Introduction

The problem of image registration occurs in many application domains, such as medicine, plant biology, microscopy, structural engineering, agriculture, geographic information systems, and military applications [1–4]. The registration problem arises due to (1) multiple image sensors viewing the same spatial area, for example, visible spectrum and thermal infrared images, (2) one sensor viewing the same spatial area under multiple view angles in time, for example, images from a robot-operated camera or airborne cameras, or (3) one sensor viewing multiple depth-adjacent spatial areas, for example, cross-sections of three-dimensional (3D) medical tissues. Figure 17.1 illustrates these three registration problem scenarios. The last scenario is also frequently denoted as image alignment.

In general, registration is understood as a coordinate system transformation of one or many datasets into a reference coordinate system. Registration can be applied not only to image (raster) data types but also to vector data types. For instance, for calibration purposes, a set of temperature point measurements (vector data) could be registered to thermal infrared images (raster data) as long as registered datasets share a common field of view and were taken at the same time [5]. One can find registration problems with any dimensional data (e.g., 1D electroencephalography signals, 2D aerial images, and 3D confocal microscopy volumes).

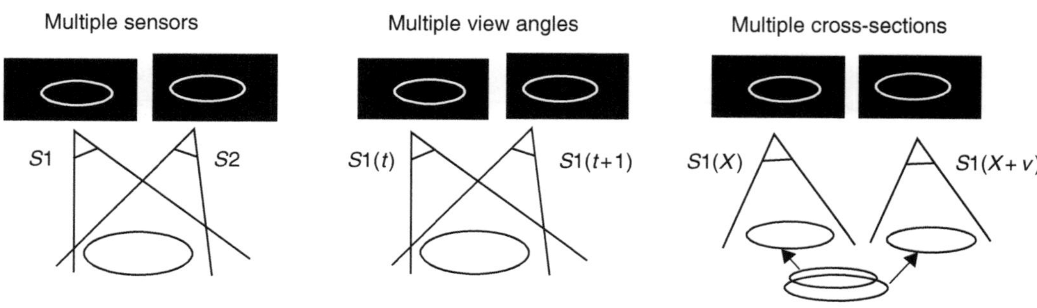

Figure 17.1
Registration problem scenarios.

We will describe in the rest of this chapter how to spatially register datasets based on intrinsic properties of data (e.g., landmark points or salient geometrical features) rather than extrinsic data properties (e.g., artificially introduced fiduciary markers). Our primary focus is on image (raster) registration techniques.

17.2 Spatial Registration Steps

Spatial registration requires performing the following steps. First, select a reference coordinate system. Second, determine locations of matching salient features in multiple datasets. This step is also denoted as finding spatial correspondences. Third, select a registration transformation model that will compensate for geometric distortions. Fourth, estimate the transformation parameters that correspond to the registration transformation model. Fifth, evaluate registration accuracy with a selected metric. One should be aware that in many cases it is very hard to assess objectively the registration accuracy due to the lack of *a priori* information about datasets. Nevertheless, these registration steps would still be commonly used with the understanding that all registration uncertainties have to be considered as well. Next, we will overview spatial registration steps and describe computational issues by giving an example.

17.2.1 Reference Coordinate System

A reference coordinate system is usually defined by a user and depends on a particular application. For example, it would be preferable to choose a reference coordinate system for a sequence of aerial images (see Figure 17.2) that conforms to a map projection of other datasets. This type of registration is also called image rectification. Imagery that was transformed by the orthogonal projection (perpendicular to the ground plane) and was freed from geometric distortions is called orthoimagery. In many other applications, such as, video registration or 3D medical volume reconstruction, any image in a set of processed images can be chosen as the reference coordinate system.

17.2.2 Feature Selection and Finding Spatial Correspondences

Finding spatial correspondences in sets of images can be approached by (1) manual selection of matching points based on visual inspection, (2) automatic salient feature

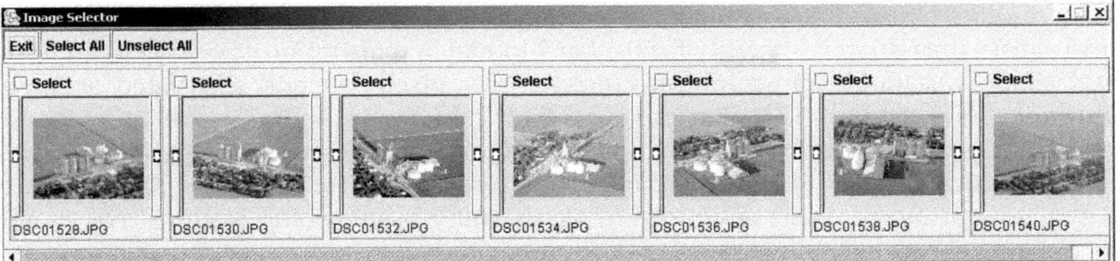

FIGURE 17.2
A sequence of aerial photographs of the same spatial location.

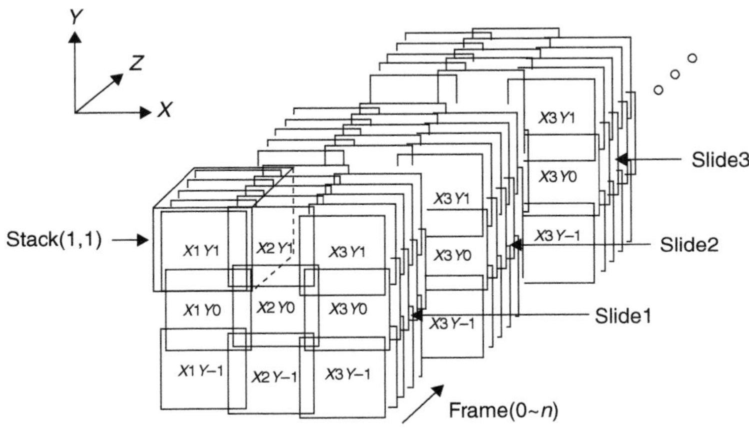

FIGURE 17.3
Illustration of 3D volume reconstruction consisting of mosaicking and cross-section alignment.

detection (e.g., edge detection) and manual assignment of feature correspondences (semi-automated selection), and (3) fully automatic process for finding matching points or features, for example, template-based matching or automated feature matching. While a fully automated process is highly desirable, the use of each approach should be based on its performance robustness. For instance, 3D volume reconstruction from confocal microscope images (acquired from multiple cross-sections at several spatial views) consists of stitching together images of overlapping adjacent spatial views of the same cross-section (also called mosaicking) and aligning multiple cross-sections to form a 3D volume. An illustration of the 3D volume reconstruction task is in Figure 17.3. The problem of image mosaicking (or registration in x–y plane) can be fully automated since the acquisition process and the corresponding transformations are well constrained. However, the problem of image alignment (or registration along z-axis) is much harder to automate since the process of cross-section preparation (sample warping due to slicing or structural changes due to bifurcating structures) before image acquisition are characterized with a large number of degrees of freedom. If possible then intrinsic image properties (salient features of the cross-sections) are combined with the extrinsic image properties induced by artificial objects (e.g., fiduciary markers like needless or skin markers). Similarly, image registration of objects under different illumination in visible spectrum imagery or in different spectral imagery can be described by large intensity variations of spatially common features. Predicting the intensity variations requires very complex models (see Chapter 15 on modeling BRDFs).

The problem of finding spatial correspondences can be alleviated by searching for an invariant feature space, especially for the hard-to-match registration cases due to intensity and shape variations across images. Instead of a pixel pair and associated acquired attributes, one can use (1) higher order geometrical features, such as lines, contours or other shapes, (2) statistical features of geometrical shapes, or (3) features derived after input data transformation. Although there might not exist a "perfect" invariant feature space, registration accuracy could be improved significantly by matching features that should be almost geometrically invariant (road boundaries or distributions of contours [6]), statistically invariant (sample mean of a parking lot) or invariant in a selected sub-subspace (average hue of grass patch in the hue-saturation-value (HSV) color space). Under the assumption of statistical pixel invariance, spatial filtering and noise removal techniques are quite frequently used in practice to achieve better registration.

17.2.3 Registration Transformation Model

Given a set of corresponding features, one has to select a registration transformation model that will compensate for geometric distortions. The choice of a registration transformation model is usually based on assumptions about data acquisition. In general, the transformation models can be categorized as linear or nonlinear. Linear transformations are functions between two vector spaces that respect the arithmetical operations addition and scalar multiplication defined on vector spaces. Examples of linear transformation models would include rigid body, affine, perspective, or projective mappings. A *rigid body transformation* takes into account translational and rotational deformations and it preserves all distances. An *affine transformation* maps parallel lines onto parallel lines and can compensate for any translation, rotation, scale, and shear distortions. A *perspective transformation* maps an image from the 3D world coordinate system to the 2D graphics display with projection lines converging at the eye point. For example, objects appear smaller if they are further away from the eye point. A *projective transformation* is the composition of a pair of perspective projections and corresponds to the perceived positions of observed objects when the viewpoint of the observer changes. This type of transformation preserves incidence (the relationships describing how objects meet each other) and cross-ratio (the ratio of ratios of point distances) but does not preserve sizes and angles. If the data acquisition process distorts images by nonlinear warping then nonlinear transformation models are appropriate. An example of a nonlinear transformation function would be a rational polynomial (cubic) model that is used for mapping 3D terrain datasets to 2D images.

17.2.4 Finding Transformation Model Parameters

The fourth step in the spatial registration process is as follows: given a registration transformation model and a set of corresponding features, we have to find the most optimal transformation parameters for a selected metric. All model parameters can be estimated by direct or search-oriented approaches. An algorithm using the direct approach computes transformation parameters from the corresponding features. An algorithm with a search-oriented approach tries to find the most optimal parameters by starting from an initial guess, and then searching in the space of possible parameters by evaluating each parameter configuration using a goodness-of-match metric. We list a few registration methods as examples of various techniques for model parameter estimation, for example, point mapping registration methods [7–10], the Fourier based registration method [11], elastic model based registration [12,13], or correlation and sequential registration methods [14,15].

17.2.5 Transformation Accuracy Evaluation

As an evaluation feedback, the resulting transformation accuracy has to be computed by using ground truth measurements and a chosen similarity metric. Establishing ground truth registration measurements is usually very hard, expensive, and frequently based on human visual inspection. In terms of similarity metrics, many metrics have been investigated in the past and can range from correlation-based to high-level-based metrics. Examples of correlation-based metrics include correlation coefficients [7], normalized cross-correlation [8], phase-correlation (Fourier invariance) [11], statistical correlation and matched filters [14], and feature difference-based (sum of absolute differences of intensity [15], sum of absolute differences of contours [16], and contour/surface differences [17]). One could also use maximum likelihood estimator (MLE) as a metric to match distributions of image contours [6], or maximize mutual information of a model and image by stochastic search (analog to gradient descent) [18,19]. Examples of high-level-based metrics could be tree and graph distance [20], and syntactic matching: automata [21]. While there does not exist any theory for choosing a particular metric, users have opted many times to choose metrics based on their minimal computational cost.

17.3 Computational Issues Related to Spatial Registration

Let us assume that we would like to design a system for automatic image registration. Our task is to estimate parameters of a registration transformation model U (also called a deformation field), such that a warped image I after transformation matches a reference image T (also denoted as a template). The warped image I is defined on a regular lattice $X = (x, y, z)^T$ and U is defined as $U(u, v, w)^T = U(u(X), v(X), w(X))^T$. We can obtain estimation of U parameters by maximizing the cost function E with a similarity metric according to the Equation 17.1.

$$E = \int \text{similarity}(I(X), T(X - U(X)))\, dX. \tag{17.1}$$

17.3.1 Complexity of Registration Transformation Models

In order to understand the computational issues, we describe the computational complexity of each item in Equation 17.1. First, registration transformation models U can be written in a form of a matrix. For instance, the affine transformation model can be represented as

$$\begin{bmatrix} x' \\ y' \end{bmatrix} = \begin{bmatrix} a_{00} & a_{01} \\ a_{10} & a_{11} \end{bmatrix} \begin{bmatrix} x \\ y \end{bmatrix} + \begin{bmatrix} t_x \\ t_y \end{bmatrix}. \tag{17.2}$$

The $(x', y') = U(u(X), v(X))$ values are the transformed coordinates of $X = (x, y)$. The four parameters, a_{00}, a_{10}, a_{01}, and a_{11}, represent a 2×2 matrix compensating for scale, rotation, and shear distortions in the final image. The two parameters, t_x and t_y, represent a 2D vector of $X = (x, y)$ translation. Similarly, we can represent the transformation models U (1) for the perspective model as $(x', y') = (-fx/(z - f), -fy/(z - f))$, where x, y, and z are world coordinates and f is the focal length of the camera, (2) for the projective model as $(x', y') = ((a_{11}x + a_{12}y + a_{13})/(a_{31}x + a_{32}y + a_{33}), (a_{21}x + a_{22}y + a_{23})/(a_{31}x + a_{32}y + a_{33}))$, or (3) for a nonlinear rational polynomial (cubic) model transforming 3D coordinates

$X = (x, y, z)$ as

$$(x', y') = (\{a_{1,1}x^3 + a_{1,2}x^2z + a_{1,3}x^2 + a_{1,4}x^2 + a_{1,5}xy^2 + a_{1,6}xyz + a_{1,7}xy + a_{1,8}xz^2 + a_{1,9}xz + a_{1,10}x + a_{1,11}y^3 + a_{1,12}y^2z + a_{1,13}y^2 + a_{1,14}yz^2 + a_{1,15}yz + a_{1,16}y + a_{1,17}z^3 + a_{1,18}z^2 + a_{1,19}z + a_{1,20}\}\{a_{2,1}x^3 + a_{2,2}x^2z + a_{2,3}x^2 + a_{2,4}x^2 + a_{2,5}xy^2 + a_{2,6}xyz + a_{2,7}xy + a_{2,8}xz^2 + a_{2,9}xz + a_{2,10}x + a_{2,11}y^3 + a_{2,12}y^2z + a_{2,13}y^2 + a_{2,14}yz^2 + a_{2,15}yz + a_{2,16}y + a_{2,17}z^3 + a_{2,18}z^2 + a_{2,19}z + a_{2,20}\}^{-1}, \{a_{3,1}x^3 + a_{3,2}x^2z + a_{3,3}x^2 + a_{3,4}x^2 + a_{3,5}xy^2 + a_{3,6}xyz + a_{3,7}xy + a_{3,8}xz^2 + a_{3,9}xz + a_{3,10}x + a_{3,11}y^3 + a_{3,12}y^2z + a_{3,13}y^2 + a_{3,14}yz^2 + a_{3,15}yz + a_{3,16}y + a_{3,17}z^3 + a_{3,18}z^2 + a_{3,19}z + a_{3,20}\}\{a_{4,1}x^3 + a_{4,2}x^2z + a_{4,3}x^2 + a_{4,4}x^2 + a_{4,5}xy^2 + a_{4,6}xyz + a_{4,7}xy + a_{4,8}xz^2 + a_{4,9}xz + a_{4,10}x + a_{4,11}y^3 + a_{4,12}y^2z + a_{4,13}y^2 + a_{4,14}yz^2 + a_{4,15}yz + a_{4,16}y + a_{4,17}z^3 + a_{4,18}z^2 + a_{4,19}z + a_{4,20}\}^{-1})$$

17.3.2 Estimating Transformation Model Parameters

Estimating model parameters is usually less computationally expensive using a direct approach than using a search-oriented approach. For example, using a direct approach with an affine transformation model and at least three pairs of corresponding points, one would compute model parameters by solving an over-determined set of linear equations. The equations are constructed by inserting values of $X = (x, y)$ and $X' = (x', y')$ (matching points) into the Equation 17.2, and solving for a and t parameters.

If a search-oriented approach using correlation is chosen, then an automated registration procedure has to compute a similarity measure of each possible match of a template (reference) image T and a warped image I according to the transformation model U. This is a very computationally expensive step since it includes (1) many templates satisfying a selected transformation model, (2) a large search space for finding the best match in a warped image I, and (3) the cost of computing the similarity metric. The set of templates, for instance using an affine transformation model, would include all possible rotations, translations, scales, and shears of a template region. The search space would depend on the size of the warped image and the particular search scenario depicted in Figure 17.4. The computation of a similarity measure would depend on the complexity of the similarity metric — for example, the normalized product of T and I defined as

$$c(u, v) = \frac{\sum_x \sum_y T(x, y) I(x - u, y - v)}{\left[\sum_x \sum_y T^2(x - u, y - v)\right]^{1/2}},$$

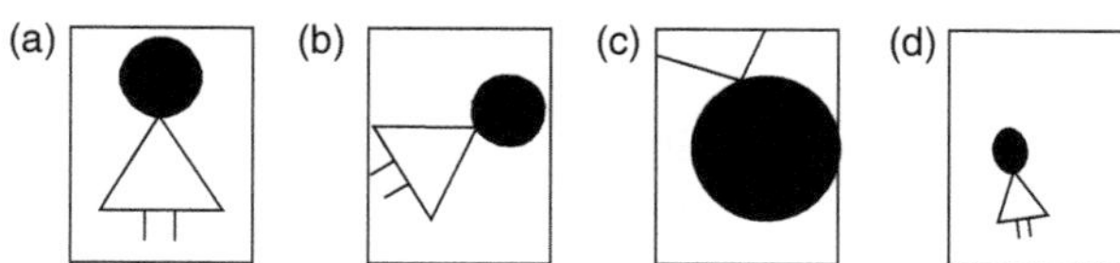

FIGURE 17.4
Three search space scenarios: (a) the reference image, (b) a target image with approximately equal size, (c) a target image with larger size, and (d) a target image with smaller size in comparison with the reference image.

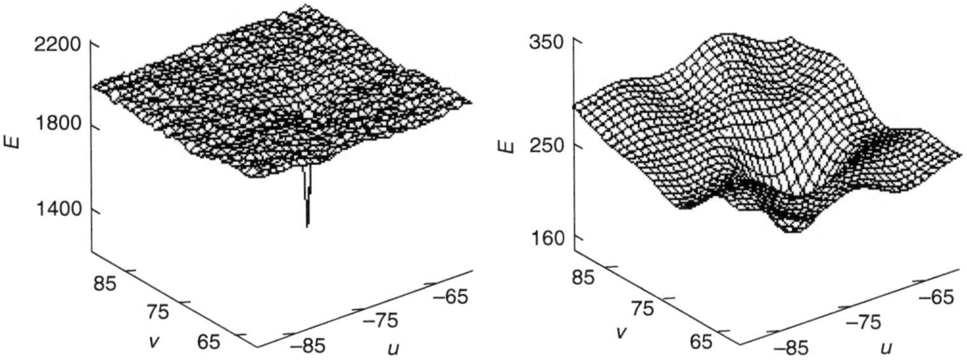

FIGURE 17.5
Detecting a match between a template T and a warped image I based on a similarity measure surface with narrow (left) or wide (right) peaks.

or normalized correlation defined as

$$c(u, v) = \frac{\text{covariance}(I, T)}{\sigma_I \sigma_T}$$

$$= \frac{\sum_x \sum_y (T(x, y) - \mu_T)(I(x - u, y - v) - \mu_I)}{\left[\sum_x \sum_y (I(x - u, y - u) - \mu_I)^2 \sum_x \sum_y (T(x, y) - \mu_T)^2\right]^{1/2}},$$

where μ_I, μ_T and σ_I, σ_T are sample means and standard deviations of warped image and template image.

Finally, one has to detect a match between a template T and a warped image I by selecting the maximum similarity point in a search space. This seemingly straightforward step can be treacherous in the presence of noise. If the peaks in the similarity map are too narrow then the registration result is spatially very accurate but is less robust to noise. On the other hand, if the peaks are wide then the spatial registration accuracy is worse but the noise robustness is better (see Figure 17.5). The above consideration leads us to the explanation why spatial filtering is often used as a pre-processing step before image registration.

As in many other applications, one would strive to achieve a tradeoff between registration accuracy and registration computational requirements. The registration accuracy depends on a choice of matching points, their spatial distribution, a selected registration transformation model, and the interpolation model used during image transformation (e.g., image rotation leading to noninteger pixel locations). On the other side, computational requirements are proportional to the complexity of the registration model, the similarity metric, and the search space (or data size) for finding correspondence points. To minimize the computational requirements, many registration approaches abandon the exhaustive "brute force" search and engage other search strategies [1,2,7,8], such as multi-resolution, dynamic programming, partial differential equations, or multi-dimensional optimization methods. In terms of registration accuracy, one should be aware that global registration methods that estimate a single optimal transformation using matched points could not account for local distortions, such as sensor nonlinearity or a nonlinearity from atmospheric conditions. It is possible to use a combination of multiple local transformations, where each local model affects only a sufficiently small neighborhood with its weighted transformation, but the search complexity (and hence the computational cost) comes at the price of the registration accuracy.

17.4 Summary

We would like to conclude this chapter by a few practical considerations. The choice of manual, semi-automated, and fully automated registration methods should be driven by an assessment of the image acquisition process, the amount of data, the processing time constraints, and the available human and computer resources. Other choices related to the registration model should be made based on a good understanding of the image acquisition process and application requirements.

It should also be mentioned that while finding correspondence points is a challenging task, it is not true that the more points one selects the more accurate registration result he/she obtains. For instance, given an affine registration model and a set of matching points, the optimal parameters can be obtained by solving an over-determined set of linear equations with the unknown transformation parameters. A new pair of matching points might lead to a new set of optimal parameters with a larger global registration error (residual) than the set of parameters obtained without adding a new point. Furthermore, one should

FIGURE 17.6
Illustrations of manual point selection (top), transformation of the right image to the coordinate system of the left image using an affine registration model, and a visual overlay of the registered images (bottom). The visual registration quality assessment in I2K software is enabled by changing (a) transparency (horizontal slider bar) and (b) intensity amplification (vertical slider bar) of the images.

be aware that a larger number of registration model parameters (complexity) means that more correspondence points have to be found, which in turn causes the optimal registration parameters to become more susceptible to noise. It is quite frequent to use visual inspection as the final registration quality assessment, for example, by overlaying images and changing their transparency as it is illustrated in Figure 17.6.

References

1. Hill, D., P. Batchelor, M. Holden, and D. Hawkes, Medical image registration, *Physics in Medicine and Biology*, 46, 2001, R1–R45.
2. Sawhney H., S. Harpreet, J., Asmuth, and C. Jane, Geospatial Registration, *Proceedings of Image Understanding Workshop, DARPA97*, pp. 849–856, 1997.
3. Maintz, J.B.A. and M.A. Viergever, A survey of medical image registration, *Medical Image Analysis*, 2, 1998, 1–36.
4. Brown, L.G., A survey of image registration techniques, *ACM Computing Surveys*, 24(4), pp. 325–376, December, 1992.
5. Bajcsy, P. and S. Saha, A new thermal infrared camera calibration approach using wireless MEMS sensors, *Communication Networks and Distributed Systems Modeling and Simulation Conference (CNDS 2004)*, January 19–22, San Diego, CA.
6. Govindu, V. and C. Shekhar, Alignment using distributions of local geometric properties, *IEEE Transactions on Pattern Analysis and Machine Intelligence*, 21, 1999, 1031–1043.
7. Goshtasby, A., Template matching in rotated images, *IEEE Transactions on Pattern Analysis and Machine Intelligence*, 7, 1985, 338–344.
8. Ranade, S. and A. Rosenfeld, Point pattern matching by relaxation, *Pattern Recognition* 12, pp. 269–275, 1980.
9. Stockman, G., S. Kopstein, and S. Benett, Matching images to models for registration and object detection via clustering, *IEEE Transactions on Pattern Analysis and Machine Intelligence*, 4, 1982, 229–241.
10. Lee, S.-C. and P. Bajcsy, Feature based registration of fluorescent LSCM imagery using region centroids, *SPIE Conference on Medical Imaging*, February 12–17, 2005, San Diego, CA.
11. Castro, E. and C. Morandi, Registration of translated and rotated images using finite Fourier transforms, *IEEE Transactions on Pattern Analysis and Machine Intelligence*, 5, 1987, 700–703.
12. Tom, B.C.S., S.N. Efstratiadis, and A.K. Katsaggelos, Motion estimation of skeletonized angiographic images using elastic registration, *IEEE Transactions on Medical Imaging*, 13, 1994, 450–460.
13. Kybic, J., Elastic image registration using parametric deformation models, *PhD thesis*, no. 2439, at EPFL, Lausanne, Switzerland, 2001.
14. Pratt, W., *Digital Image Processing*, John Wiley & Sons, Inc., New York, 1978.
15. Barnea, D. and H. Silverman, A class of algorithms for fast digital registration, *IEEE Transactions on Computers*, C-21, 1972, 179–186.
16. Barrow, H.G., J.M. Tenenbaum, R.C. Bolles, and H.C. Wolf, Parametric correspondence and chamfer matching: two new techniques for image matching, *Proceedings of the Fifth International Joint Conference on Artificial Intelligence*, Cambridge, MA, 1977, pp. 659–663.
17. Brown, L.G., A survey of image registration techniques, *ACM Computing Surveys*, 1992.
18. Viola, P. and W.M. Wells III, Alignment by maximization of mutual information, *International Journal of Computer Vision*, 24, 1997, 137–154.
19. Wells III, W.M., P. Viola, H. Atsumi, S. Nakajima, and R. Kikinis, Multi-modal volume registration by maximization of mutual information, *Medical Image Analysis*, 1, 1996, 35–51.
20. Mohr, R., T. Pavlidis, and A. Sanfeliu, *Structural Pattern Analysis*, World Scientific, Teaneck, NJ, 1990.
21. Bunke, H. and A. Sanfeliu, Eds., *Syntactic and Structural Pattern Recognition, Theory and Applications*, World Scientific, Teaneck, NJ, 1990.

18

Georeferencing

Peter Bajcsy

CONTENTS

18.1 Introduction

This chapter is devoted to the problem of spatial registration with georeferencing information, also called georegistration. Similar to the general case of spatial registration, georegistration can also be applied to raster and vector data types. For instance, boundaries of counties (vector data) can be registered to elevation maps (raster data) or to boundaries of watersheds (vector data), as long as the registered datasets are accompanied with sufficient georeferencing information.

Georeferencing, or geographic referencing, is the name given to the process of assigning a real-world location to a location on a map. Often, this is done using the familiar latitude and longitude system as illustrated in Figure 18.1. Latitude (lat) and longitude (long) describe points in three-dimensional (3D) space (see Section 18.2), while maps are inherently two-dimensional (2D) representations. Modern maps are usually stored as digital images since they represent raster information similar to 2D image information. Thus, the end result

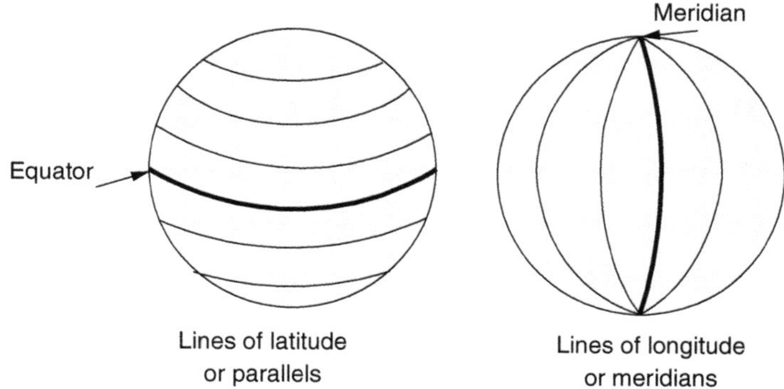

FIGURE 18.1
Illustration of lines of latitude (left) and longitude (longitude). Latitude — the angular distance north or south of the equator measured in degrees along a meridian as on a map or globe. Longitude — the angular distance on the earth or on a globe or map, east or west of the prime meridian at Greenwich, England, to the point on the earth's surface for which the longitude is being ascertained, expressed either in degrees or in hours, minutes, and seconds. Meridian — a great circle on the earth's surface passing through both geophysical poles.

of georeferencing is the ability to retrieve the lat/long coordinates for any point on the georeferenced digital image.

Why would one introduce georeferencing? Lat/long coordinates define a point on a 3D model of the Earth, while map coordinates represent a pixel — a row and column location on a 2D grid obtained from projecting some 3D model of the Earth onto a plane. 2D maps are easy to display and facilitate distance measurements, while 3D coordinates are accurate but cumbersome for computing distance, and have no standard length for different degrees of latitude and longitude. The best way to define the position of an object on a map is with relative horizontal (column) and vertical (row) distances. For example, "Three kilometers north of object A and two kilometers west of object B." In contrast, the best way to specify the position on a sphere is with relative angular offsets like "Five degrees north and six degrees west of A." Thus, the reason for introducing georeferencing is the fact that georeferencing transformations reflect the geometric properties of the earth model and projection plane, and support a mechanism for expressing distances and relative positional information across different coordinate systems.

Given 2D and 3D coordinate systems, there are three possible classes of transformations. First, those that convert from 2D-to-2D, bringing into alignment points that represent identical real-world locations. An example of this is multimodal raster data fusion would be, for instance, the fusion of synthetic aperture radar (SAR) images with electro-optical (EO) or infrared (IR) images. The second class of transformations, 3D-to-3D, is used for information integration, such as the overlay of road network and river information. The third class of transformations operates between 2D and 3D coordinates, for example, row/column and lat/long coordinates. Identifying the location of water wells defined in 3D lat/long coordinates on a 2D digital terrain map is an application example of the 3D-to-2D transformations. The transformations across 2D and 3D coordinate systems, where one of the coordinate systems represents the Earth, are also referred to as georegistration transformations. The third class of transformations is also concerned with georeferencing maps (2D raster data) with contours or boundaries of regions (3D vector data). Examples of these three transformation classes are shown in Figure 18.2. In general, one can register measurements obtained from the Earth's surface (raster and vector data types) by using these three classes of georeferencing transformations.

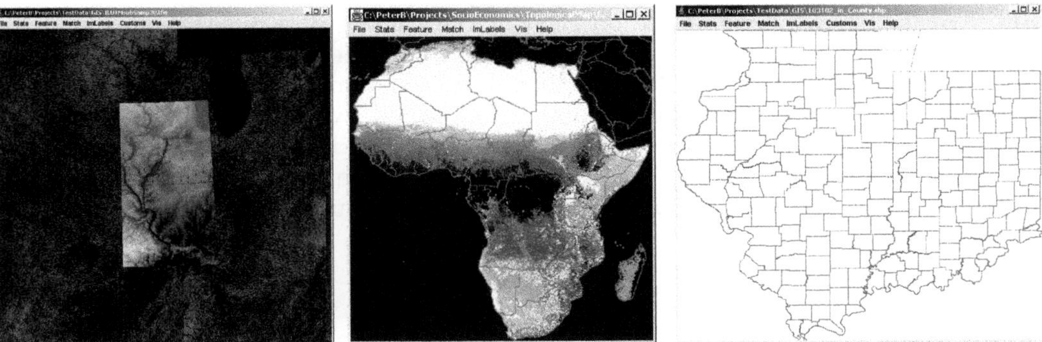

FIGURE 18.2
Examples of three classes of georeferencing transformations generated by using the I2K software [1]. Left: 2D-to-2D transformation of two digital elevation raster files and one land use raster file. Middle: 3D-to-2D transformation of a vector file with country boundaries and a raster file with land use labels. Right: 3D-to-3D transformation of two vector files with county boundaries.

Fields that apply georeferencing include geology, hydrology, atmospheric science, archeology, earthquake engineering, forestry, environmental engineering, water quality research, ecological research, military operations, and many others. For example, the petroleum industry depends upon precise drilling locations, so accurate georeferenced maps are very important. The U.S. Geological Survey (USGS) and the National Geospatial-Intelligence Agency (NGA) both create and make extensive use of georeferenced images, many of which are available to the private sector.

Given the broad spectrum of users, one would like to design software that performs georeferencing automatically so that the end users can easily integrate datasets, extract interesting features, and analyze them. Users of georeferencing information frequently exchange data and therefore have adopted some georeferencing terminology and Earth modeling standards. However, not all georeferenced maps follow the same standards, so identical information content can be represented using different data formats or map types. Furthermore, the existence of multiple Earth models and several geographic projections requires different sets of equations for transformations between 2D and 3D coordinate systems depending on the particular model and map projection, for example, Molodensky equations [2] (see also Appendix 4), Lambert equations [3], etc. Disparate methods of packaging geographic information with digital maps create difficulties for automatically locating each piece of georeferencing information in a file. Thus, when designing an automatic georeferencing system, one faces the challenges of (1) format dependent search for georeferencing information, (2) uniqueness of each found piece of georeferencing information (e.g., multiple conflicting values for the same georeferencing parameter), and (3) the complexity inherent in modeling the Earth and all possible geographic projections.

The current solutions to the automatic georeferencing problem include the ESRI suite of tools [4] (ArcGIS, ArcView, ArcExplorer [5], supporting around 63 coordinate transformations [6]), the Environment for Visualizing Images (ENVI) from Research Systems, Inc. [7], a suite of applications known as the USGS Mapping Science Software [8] from the USGS, or Image To Knowledge (I2K) [1] from NCSA UIUC, just to name a few. The countless hours that have been spent on building solutions by multiple commercial, academic, and government institutions are a testament to the fact that the solution to the automatic georeferencing problem is very much in need. Thus, it is indeed important to understand the theory, basic assumptions, real-world issues, and limitations of any georeferencing software in order to derive reasonable conclusions based on any processed data.

In this chapter, we describe the problem of automatic georeferencing given multiple datasets (raster and vector) with georeferencing information in 2D and 3D real-world coordinate systems. First, we outline the theory and standards used for proper geographic referencing of digital images (see Section 18.2). Second, we present the most common geographic transformations in Section 18.3. Finally, we describe details of finding georeferencing information in TIFF and ERDAS IMG file formats in the "Finding Georeferencing Information" Section 18.4 and summarize the problem of automated map georeferencing in Section 18.5.

18.2 Georeferencing Models

We present an overview of Earth models and geographic projections. Details about georeferencing models can be found in [9] or [6]. In our overview, we start with a map that contains image data and the associated georeferencing map information. We are looking for information about the map projection that includes the Earth shape approximation, projection center and the tie point. Our goal is to resolve the coordinate system of the given image to allow for a mapping between a location on the image and a location on the Earth.

It is common to use some type of model for the Earth in developing a coordinate system. Ordinarily, a sphere or spheroid is used to approximate the shape of the Earth. One should be aware of spatially varying model accuracy (as illustrated in Figure 18.3) before selecting the most appropriate Earth model for georeferencing.

To develop a projected coordinate system, we must provide some mechanism for applying our image plane to the Earth model. Projected coordinate systems can be selected from a range of established projections, for example, conformal, equal area, equidistant, true-direction, conic, cylindrical, or planar [9]. Examples of planar–polar and cylindrical–normal projection coordinate systems are shown in Figure 18.4. Each projected coordinate system comes with a set of linear and angular projection parameters. Linear parameters include "false easting and northing" (a linear value applied to the origin of the x- and y-coordinate

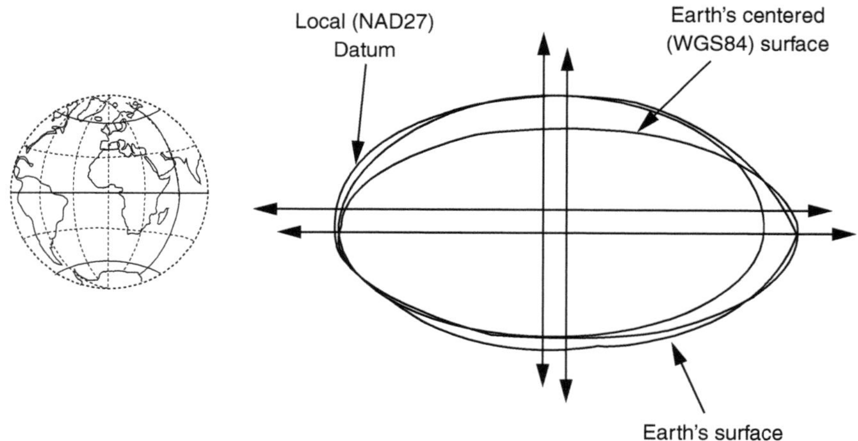

FIGURE 18.3
At left, we see an example of a projection coordinate system, based on a particular Earth model. The illustration on the right shows the spatially varying accuracy of two datums (NAD27 and WGS84) that are used to model the Earth.

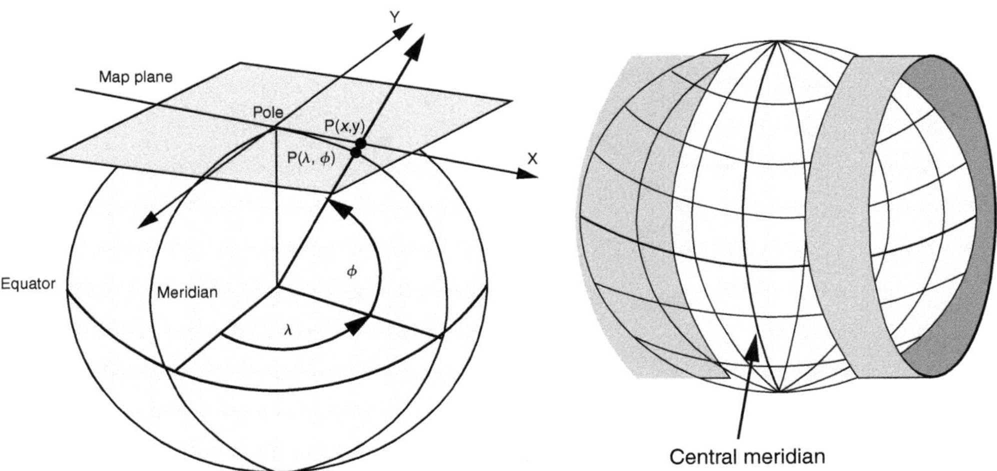

FIGURE 18.4
Examples of planar–polar (left) and cylindrical–normal (right) projection coordinate systems.

system) and "scale factor" (a unit-less value applied to the center point or line of a map projection). Angular parameters include (1) azimuth (the center line of a projection), (2) the central meridian (also called the longitude of origin or longitude of center), which is the origin of the x-coordinates, and (3) the central parallel (also called the latitude of origin or the latitude of center), which is the origin of the y-coordinates.

A geographic projection usually conveys the manner in which the 3D features of the Earth have been distorted onto a 2D image. The types of metadata relevant to a projection include the *projection center* (where the distortion of the map is minimal) and the *insertion point* (a point on the image used to extrapolate data about other points on the image). However, the formulas one would use for automatic georeferencing depend upon additional map parameters. There are hundreds of map projections in use, and the way the distortion occurs influences the usefulness of the map in different domains. A list of nearly all map projections in use today can be found in Reference 9. Fortunately, organizations like the USGS and the Defense Mapping Agency use only a small subset of these.

18.3 Geographic Transformations

We will focus on six geographic coordinate transformations given three types of location representations: 2D map pixels (column and row), 3D lat/long coordinates, and 2D Universal Transverse Mercator (UTM) coordinates (zone, northing, and easting). These six transformations are shown in Figure 18.5, and will be explained next.

The motivation for the transformations between 2D map and 3D lat/long coordinate systems is fairly obvious, because one would like to know the lat/long values of any pixel in a georeferenced image. However, the third location representation, UTM, might not be familiar to most people. The UTM coordinate system is a Cartesian coordinate system developed by the oil industry and explained further in Appendix 3. It is useful for specifying a number of points on a map without having to refer to lat and long. Furthermore, UTM values facilitate easy metric distance calculations, which are difficult in lat/long coordinates, where the distance between two adjacent degrees is dependent upon their location relative to the

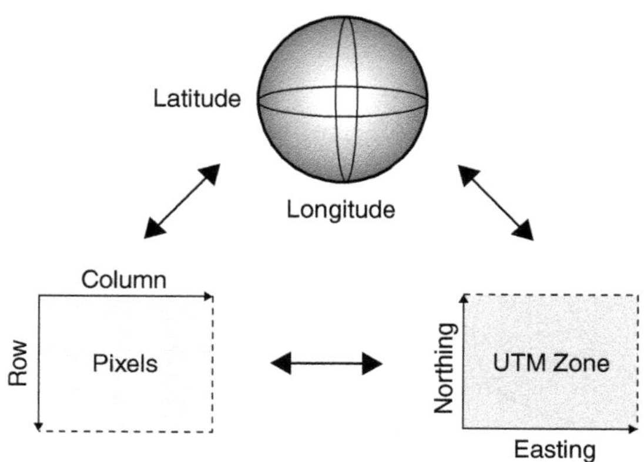

FIGURE 18.5
Most common geographic coordinate transformations.

equator and the prime meridian. In UTM terminology, the horizontal (x) value is called *easting* and the vertical (y) coordinate is called *northing*. A UTM coordinate pair is called a *northing/easting* value.

In order to perform the desired georeferencing transformations, a *model type* (also called datum in the literature) must be set based on the map parameters that are loaded. A *datum*, when used in its strictest sense, refers to a complete specification of a mapping system including the ellipsoid, the map projection, and the coordinate system used. This term seems to work well when referring to something like the Ordnance Survey of Great Britain (OSGB) 1936 datum that has an implied, official ellipsoid (Airy 1830), among other standard parameters. However, the term seems to have slipped into much weaker usage, sometimes only referring to a subset of the necessary attributes. Thus, we avoid the issue entirely and refer to *model type*, which encompasses all of the necessary attributes.

The UTM coordinates are used in maps with various types of projections. Each projection type, along with the various parameters needed to perform the transformations, make up a model type. Each map projection requires a separate set of transformation procedures or algorithms. The complexity of transformation methods is significantly less for a sphere model, such as the Lambert Azimuthal Equal Area projection, than for an ellipsoid model, but the sphere model is appropriate as a model for the Earth only on large-scale projections. For any UTM projection, a standard ellipsoid should always be specified.

18.3.1 Lat/Long to and from UTM Northing/Easting

The UTM system is 2D and the lat/long system is 3D, making transformations between them quite complex. The two transformation methods between UTM and lat/long are based on equations in [2,3,6]. The transformation from UTM to lat/long is referred to as *Transformation 1*, and the transformation from lat/long to UTM is referred to as *Transformation 2*.

A model type must specify a number of parameters in order to perform ***Transformation 1*** and ***Transformation 2***. We present what must be known about the 3D object and about the 2D plane next.

Ellipsoid Parameters to Define Lat/Long Coordinates: As mentioned above, a geometric model is often used to approximate the Earth. The geodetic ellipsoids used as models should always be referred to as *oblate ellipsoids of revolution*, but are referred to simply as *ellipsoids* in all georeferencing literature. An oblate ellipsoid of revolution is an ellipsoid formed by

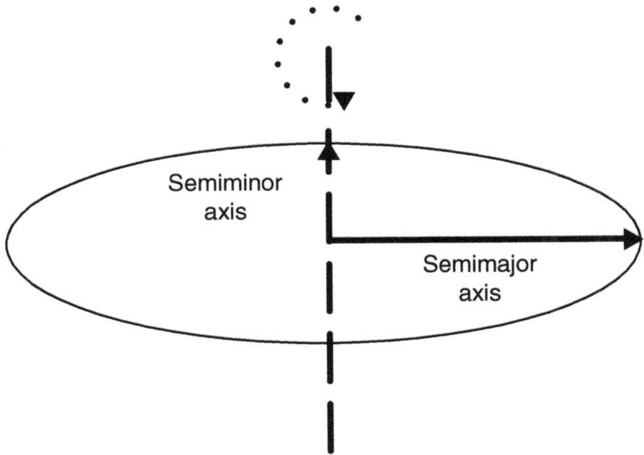

FIGURE 18.6
Ellipsoid parameters.

rotating an ellipse out of its plane around its minor (shorter) axis, as shown in Figure 18.6. All geodetic ellipsoids can be specified by two quantities: a semimajor axis and a semiminor axis. If these two quantities are the same, the ellipsoid is a sphere.

Over approximately the past 200 years, many different measurements have been made of the Earth in an attempt to find the values for the semimajor and semiminor axes that either provide the most accurate representation of the Earth overall, or are simply most accurate for the purposes of whomever is carrying out the measurement. Whenever a particular group made an official measurement of these values, they gave a name to the resulting ellipsoid. Dozens of these ellipsoids have been calculated and then discarded as newer ellipsoid models were developed. The names of these official ellipsoids are usually some combination of letters and numbers, such as WGS84 or GRS80.

The equations used in Transformations 1 and 2 make use of two values to define an ellipsoid: the semimajor axis and the square of the eccentricity of the ellipsoid. The *eccentricity* value, denoted as e, is a measure of the amount by which an ellipse varies from a circle. The square of the eccentricity is calculated with this equation:

$$e^2 = 2f - f^2. \tag{18.1}$$

The *flattening* of the ellipse, f, is defined by:

$$f = (a - b)/a. \tag{18.2}$$

In the flattening equation, a is the ellipsoid semimajor axis and b is the ellipsoid semiminor axis. Often $(a - b)$ is quite small compared to a, and it is common to provide the flattening quantity as $1/f$, or the *inverse flattening*. Because of these relationships, the required ellipsoid parameters can be derived from any one of the three sets of information: (1) the radius of the sphere for the cases where a sphere is used for a model, (2) the semimajor axis and the semiminor axis, or (3) the semimajor axis and inverse flattening value.

18.3.1.1 The 2D UTM Map Parameters

In general, UTM map parameters should include information pertinent to a map's 2D coordinate system and information about the 3D to 2D projection used to obtain the map. Ideally, the map parameters will provide at least the following information: projection type, projection center, false easting, false northing, UTM zone, vertical resolution, horizontal resolution, and the tie point.

Of all map parameters, the map projection is perhaps the most important because it determines the equations used to perform the transformations. Most of the other map parameters are used as inputs to that particular set of equations. Multiple model types might have the same map projection, but other map parameters could be different.

The map resulting from a projection will not have the same amount of distortion at each point. Different parts of the map plane are oriented to the Earth model differently, resulting in varying degrees of distortion. Generally, the distortion on a map will be minimal at its *projection center*. It is important to note that the projection center may or may not be in the viewable area of an image.

Two other quantities that are specific to a projection are the *false easting* and *false northing*. In a UTM map projection, points to the west or south of the projection center can have negative UTM values. In order to avoid the inconvenience introduced by negative values, false easting and false northing values are often added to computed northings and eastings to keep all values positive.

18.3.2 The UTM Northing/Easting to and from Pixel Column/Row

The UTM northing/easting to pixel-valued column/row transformations are denoted as *Transformations 3 and 4*. Since these are both horizontal/vertical 2D coordinate systems, the transformations are linear equations mapping from one Cartesian system to another. We present what must be known about the 2D coordinate systems next.

18.3.2.1 Tie Point/Insertion Point

A *tie point*, also called an *insertion point*, is a pair of coordinates $[(i, j), (x, y)]$ that ties together a pixel value (i, j) and its UTM equivalent (x, y). A tie point is actually specified as $[(i, j, k), (x, y, z)]$ for use with 3D digital maps. All 2D images simply specify 0 for k and z.

The (i, j) coordinate is taken from pixel space with the i value specifying the column number of the pixel, and the j value specifying the row number of the pixel. The (x, y) value is taken from the UTM coordinate plane, with x specifying the easting value (horizontal measurement), and y specifying the northing value (vertical measurement).

To illustrate, a tie point might be [(0,0), (400,000, 600,000)], indicating that the pixel (0,0) has a UTM easting value of 400,000 m and a UTM northing value of 600,000 m. Note that the pixel space starts with (0,0) in the upper left corner of the image, with increasing column values to the right and increasing row values moving down, as indicated by the arrows in Figure 18.5.

18.3.2.2 Resolution/Scale

The other information required for Transformations 3 and 4 is the *resolution* or *scale* of the image. By knowing the number of meters represented by one pixel, a simple calculation can be performed to find the UTM values for any pixel, or the nearest pixel for any specified UTM value. To make the resolution as general as possible, horizontal and vertical values are usually specified separately. The horizontal resolution specifies the number of meters traversed when moving one column in pixel space. The vertical resolution specifies the number of meters traversed when moving one row in pixel space. Example resolution values would be 30 meters/pixel in the x direction (the horizontal resolution), and 40 meters/pixel in the y direction (the vertical resolution).

18.3.2.3 Parameters and Units of Measure

The requisite tie point and resolution parameters can be specified in meters for the tie point and meters/pixel for the resolution, or in terms of lat/long degrees for the tie point and degrees/meter for the resolution. Examples 1 and 2 demonstrate these different units of measure:

Example 1:

Tie point:	[(0,0), (400,000,600,000)]	*UTM values are in meters,*
Resolution:	x: 30.0000 y: 40.0000	*Both values are in meters/pixel.*

Example 2:

Tie point:	[(0,0), (−87,45)]	*UTM values are in degrees,*
Resolution:	x: .00027778 y: .00027778	*Both values are in degrees/meter.*

In Example 1, pixel (0,0) has a UTM easting value of 400,000 m and a UTM northing value of 600,000 m. One pixel in the x direction corresponds to 30 m, while one pixel in the y direction corresponds to 40 m. In Example 2, the upper left corner of the image has lat/long value 87 degrees west longitude and 45 degrees north latitude. The resolution values have been nicely chosen because both x and y resolutions are one second of arc measurement. Recall that a second is 1/60th of a minute, which is 1/60th of a degree. Thus 1/3600th (.00027778) of a degree is a second in arc measurement. Taking the tie point and resolutions together, the pixel at column 1, row 1 has lat/long value 86 degrees, 59 min, 59 sec west longitude and 44 degrees, 59 min, 59 sec north lat.

18.3.3 Lat/Long to and from Pixel Column/Row

The most frequent application of georeferencing is to transform from a pixel value, that is, a column and row of an image, to a lat/long pair. Transformation 5 is the term we will use for this task, while Transformation 6 does the reverse procedure — going from a lat/long pair to an image column and row.

These lat/long ↔ column/row transformations are accomplished via calls to Transformations 1 through 4 discussed above. Specifically, Transformation 5 is carried out by a call to Transformation 4 followed by a call to Transformation 1. Calling Transformation 2 and then calling Transformation 3 accomplishes Transformation 6.

18.4 Finding Georeferencing Information

How should one find the necessary georeferencing information in a broad spectrum of GIS file formats? While some file formats were designed to contain georeferencing information, for example the ERDAS IMG file format [10], the file types of most digital maps were not designed to accommodate georeferencing metadata. For these file formats, of which the Tagged Image File Format (TIFF) [11–14] is one example, the metadata must be provided in some manner for the map to be of any use. Although a number of approaches have been taken, the methods for including the metadata have not yet reached a point of standardization that enables a novice to quickly find the necessary information.

In this section we describe how geographic referencing information can be extracted from TIFF and ERDAS IMG files. A detailed summary can be found in References 15 and 16.

18.4.1 TIFF Files

The presented description about georeferencing information extraction from TIFF world files (tfw) and private tags is based on the literature available in References 13 and 14. The current solution for the TIFF format uses a combination of information extraction approaches that are based on three variants of TIFF files containing georeferencing information. Assuming that sufficient data is provided in some combination of the three sources, we can use the georeferencing interface described earlier. The three TIFF file variants are: (1) one or more standardized files are distributed along with TIFF image data as .tfw or .txt files, (2) the metadata is encoded in the image file using private TIFF tags, and (3) an extension of the TIFF format called *GeoTIFF* is used.

Given multiple information sources, it is possible to read conflicting values for the same geographic field. For example, tie points and resolution values (Section 18.3.2) can be specified in both the TIFF world file and the private TIFF tags. Furthermore, some georeferencing information may be specified either in deg/min/sec or UTM meters. In the absence of necessary information, some transformations can be approximated, but some transformations may not be possible. For example, if both a tfw file and private tags 33550 (ModelPixelScale) and 33922 (ModelTiePoint) are missing, Transformations 3 and 4 cannot be performed. To understand the intricacies of all these cases, we present a specific description of each TIFF file variants.

18.4.1.1 TIFF World Files

Geospecific information for a map in the TIFF format is often included in separate file with the same root file name, but with the extension '.tfw' [13]. A tfw file is a short ASCII file that contains six values as shown in Table 18.1, four of those being the values discussed in Section 18.3.2.

Table 18.1 shows typical values and descriptions for a TIFF world file. In looking at the tfw values, observe that the easting and northing values are available directly from the file. The horizontal and vertical resolution factors are also in the file, and in this example the vertical resolution is a negative value. This negative value, while at first confusing, has a simple explanation. The presence of a tfw file implies that the (i,j) value of the tie point is either the upper left corner or the lower left corner of the image. If the vertical resolution is given as a negative value, then the tie point of the image is implied to be the upper left, that is (0,0). If the vertical resolution is given as a positive value, then the tie point of the image is implied to be the lower left, that is (0, *number_of_rows*), where *number_of_rows* is the number of rows in the image.

The rotation factors of an image are useful if the image is somehow misaligned. It is also important to note that it is not possible to specify the additional z and k values of a tie point in a tfw file.

TABLE 18.1

Sample TIFF world file values.

Value	Description
30.000000	Column (horizontal) resolution
0.000000	Rotation in the horizontal direction
0.000000	Rotation in the vertical direction
−30.000000	Row (vertical) resolution
250000.00	Easting value of tie point
650000.00	Northing value of tie point

18.4.1.2 Private TIFF Tags

The TIFF file format uses a data structure known as a *tag* to supply information about an image. Every tag within the TIFF specification has a particular number associated with it, called its *value*. Any software that reads a TIFF file reads a tag structure by first examining the tag value, and then interpreting the meaning of the tag based on the tag value.

The TIFF specification [11] defines the values and meanings of every valid tag, so no single entity can create a TIFF tag of arbitrary value for its exclusive use. However, it is possible for a company to register one or more TIFF tags for private use, and a number of these so-called private tags are quite useful for geographic referencing [14]. We will describe next the following four selected private tags: 33550, 33922, 34735, and 34737.

Private tag 33550 is referred to as the *ModelPixelScaleTag* and consists of three IEEE double precision floating-point numbers as shown in Table 18.2. Each value gives the resolution for a given direction. For simple 2D maps, the 'z' resolution is specified as 0. This structure has built-in support for 3D digital maps.

Private tag 33922 is referred to as the *ModelTiePointTag*. It usually consists of six IEEE double precision floating-point numbers organized as shown in Table 18.3. Tag 33922 may contain a sequence of [*(I,J,K), (X,Y,Z)*] values, specifying multiple tie points. Third, the private tag 34735 is called the *GeoKeyDirectoryTag*. This tag is used to implement the GeoKey data structure for GeoTIFF files, an extension of TIFF. See Section 18.4.1.3 for details.

Private tag 34737 is referred to as the *GeoAsciiParamsTag*. This tag holds the ASCII values of any GeoKeys that contain ASCII characters. An example is GeoKey 3073 that is explained in Section 18.4.1.3.

18.4.1.3 GeoTIFF Keys

Private TIFF tag solutions worked well for the individual companies who registered the tags. However, for third-party GIS companies, the solution was not ideal because

TABLE 18.2
Private tag 33550 — *"ModelPixelScaleTag"*.

Private tag 33550	
Standard name	**Description**
ScaleX	Horizontal (x) resolution
ScaleY	Vertical (y) resolution
ScaleZ	Elevation (z) resolution

TABLE 18.3
Private tag 33922 — *"ModelTiePointTag"*.

Private tag 33922	
Standard name	**Description**
I	Column number of tie point in pixel space
J	Row number of tie point in pixel space
K	Elevation value of pixel (I, J)
X	Easting value in model space
Y	Northing value in model space
Z	Elevation of (X, Y) in model space

TABLE 18.4

GeoKey 1024 — *"GTModelTypeGeoKey"*.

GeoKey 1024	
Possible value	**Description**
1	A projection coordinate system is being used
2	A geographic coordinate system is being used
3	A geocentric coordinate system is being used

their software had to interoperate with each company's standards. The third-party companies wanted a public standard that would be platform independent and available to anyone. The GeoTIFF format is an open standard built from the TIFF 6.0 specification. Its purpose is to include geographic information in a TIFF document in the form of GeoTIFF keys.

GeoTIFF keys, commonly known as *GeoKeys*, can be described as data structures built on top of the tag structure. A TIFF reader that does not support the GeoKey specification reads all GeoKeys simply as TIFF tags it does not understand. Following the example of the TIFF tag, each GeoKey has a unique value called an *ID number*. By efficiently using the TIFF tag structure, the GeoTIFF standard allows for GeoKeys with ID numbers ranging from 0 to 65,535 while using a maximum of six registered tags. Two of these registered tags, GeoKey DirectoryTag and GeoAscii ParamTag, were listed in the previous section. We will describe next the following four GeoKeys: 1024, 1026, 3072, and 3073.

GeoKey 1024 is called the *GTModelTypeGeoKey*. It consists of one short (2-byte unsigned integer) value that specifies the general type of coordinate system used. The possible values will indicate if the image uses a geographic, projected, or geocentric coordinate system (Table 18.4).

GeoKey 1026 is the *GTRasterTypeGeoKey*. The purpose of this key is to remove the ambiguity inherent in representing a coordinate system with raster data. The key contains one short (2-byte unsigned integer) with two possible values that indicate how to interpret a raster coordinate (a pixel). As an example, consider a raster coordinate of (1,1). If the value of GeoKey 1026 is 1, it indicates that the raster coordinate (1,1) refers to the entire square grid with boundaries (0,0) in the upper left and (1,1) in the lower right. More generally, in this system a coordinate (i, j) refers to the one unit square area with (i, j) as its lower right corner. In contrast, if the value of GeoKey 1026 is two, a pixel value refers only to point indicated by the intersection in pixel space of a vertical line i units to the right of the origin with a horizontal line j units down from the origin. In our example for raster coordinate (1,1), this would be the point one unit right from the origin and one unit down. For details see Section 2.5.2.2 in Reference 12.

GeoKey 3072 is the *ProjectedCSTypeGeoKey*. If GeoKey 1024 has value one, then this key indicates the specific projection that was used. There are 12,761 allowable values for this key, ranging from 20,000 to 32,760, with each one a well-defined projection on a particular geographic location (Table 18.5). See Section 6.3.3.1 in Reference 12 for details.

GeoKey 3073, the *PCSCitationGeoKey*, exists to compensate for the fact that GeoKey 3072 is not a user-friendly way to specify a map projection. The value of GeoKey 3073 is an ASCII string intended to give the user easy to understand information about the map projection. Often it consists of just a few words providing information such as the UTM zone (see Appendix 3), the ellipsoid used, or other geographic metadata about the image. An example of extracted geographic information from a TIFF file format is shown in Figure 18.7.

Table 18.5

GeoKey 1026 — *"GTRasterTypeGeoKey"*.

GeoKey 1026	
Possible value	**Description**
1	Raster pixel represents area
2	Raster pixel represents point

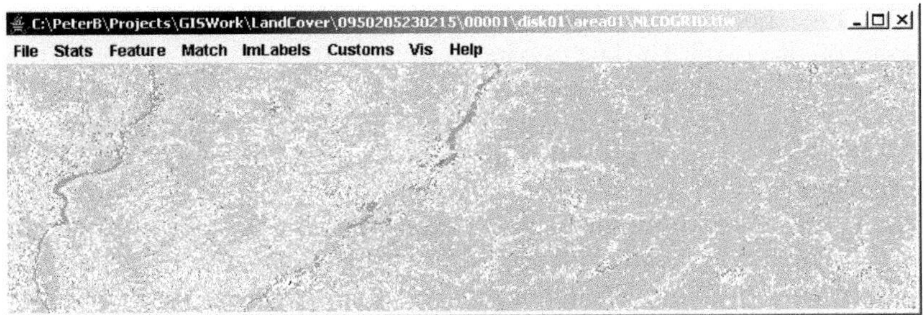

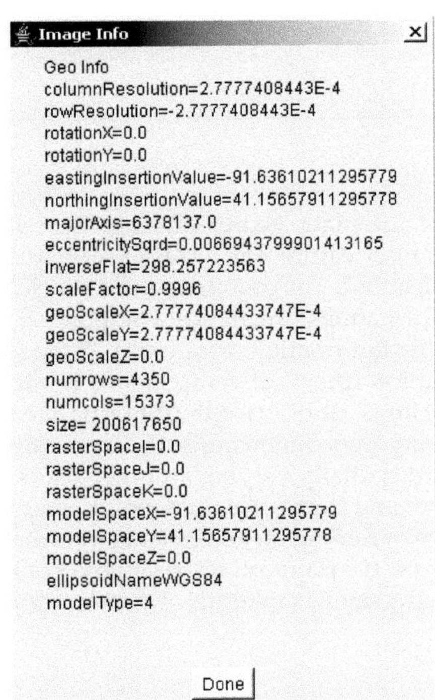

Figure 18.7
A land cover map image (3-band categorical RGB byte data) and its georeferencing information (private TIFF tags, TIFF world file, and USGS Standard Metadata txt) extracted from the TIFF file format by the I2K software.

18.4.2 ERDAS IMG Files

ERDAS IMG format stores georeferencing map information in a separate header file. The header file has the same root file name as the image, but with '.hdr' as its extension. Many types of metadata about the img file are included in the header file in a "keyword = value" format, with the important keyword for geographic referencing being *map info*.

The map info entry in the header file has the format:

$$\text{map info} = \{projection\_name, i, j, x, y, horizontal\_resolution, vertical\_resolution$$
$$UTM\_zone, \text{"}North\text{"} or \text{"}South\text{"}\}$$

The individual values in the map info entry are: (1) *projection_name* — the type of projection used in the image, often UTM (See Appendix 3), (2) $[(i, j),(x, y)]$ — the tie point, (3) *horizontal_resolution* — the horizontal resolution of a pixel, and (4) *vertical_resolution* — the vertical resolution of a pixel. The last two individual values are only present if the projection specified is the UTM Northern Hemisphere or the UTM Southern Hemisphere projection: (5) *UTM_zone* — the UTM zone in which the region is located (See Appendix 3), and (6) *"North"* or *"South"* — indicates if the image is in the Northern or Southern Hemisphere of the specified zone.

It is apparent that the extraction of georeferencing information is much simpler from the ERDAS IMG files than from TIFF files. Although the ERDAS IMG header files might not encapsulate all possible georeferencing parameters, file formats of this type could become simple and eventually a standard representation of georeferencing information in GIS domain.

18.5 Summary

This chapter has focused on the problem of automatic georeferencing. Georeferencing is about finding "real-world" locations (3D) for map features (or 2D raster set features) and the need for it is in a broad spectrum of GIS applications. The problem solution includes 3D and 2D geographic models and mapping transformations (theoretical component), as well as strategies for finding necessary georeferencing information in various GIS file formats and dealing with missing or conflicting information (practical component). While designing automated georeferencing software is often an issue of breadth (differing file types, projections, metadata standards, etc.), using any georeferencing software requires a good understanding of registration limitations and spatially varying accuracy. Thus, one could analyze this problem from a software designer and from a user viewpoint. These two viewpoints coincide when it comes to the problem of finding georeferencing information in GIS file formats, as it was illustrated in Section 18.4. If a standard representation of georeferencing information in all GIS files existed then the savings in terms of development time and data usability would be tremendous.

References

1. Bajcsy, P. et al., Image To Knowledge, documentation at URL: http://alg.ncsa.uiuc.edu/tools/docs/i2k/manual/index.html.

2. Molodensky equations reference. http://www.posc.org/Epicentre.2_2/DataModel/ExamplesofUsage/eu_cs34h.html.
3. Lambert projection equations http://mathworld.wolfram.com/LambertAzimuthalEqual-AreaProjection.html.
4. ESRI Web site: http://www.esri.com/.
5. ArcExplorer product description at ESRI Web site: http://www.esri.com/.
6. *Understanding Map Projections*. Kennedy, Melita. Kopp, Steve. Environmental Systems Research Institute, Inc.
7. Research Systems, Inc., The Environment of Visualizing Images (ENVI), URL: http://www.rsinc.com/envi/index.cfm.
8. USGS mapping science software, URL: http://mapping.usgs.gov/www/products/software.html.
9. In *Map Projections — A Reference Manual*. L.M. Bugayevsky and J.P. Snyder, Taylor and Francis, London, 1995.
10. ERDAS IMG file format description, URL: http://gdal.velocet.ca/projects/imagine/iau_docu0.pdf.
11. TIFF Revision 6.0. Aldus Corporation. Copyright 1986–88, 1992.
12. GeoTIFF Revision 1.0. Ritter, Niles. Ruth, Michael, 2000.
13. Information on TFW files: http://www.genaware.com/html/support/faqs/imagis/imagis15.htm.
14. Information on private TIFF tags: http://remotesensing.org/geotiff/spec/geotiff2.6.html.
15. Alumbaugh, T.J. and P. Bajcsy, Georeferencing maps with contours in I2K, ALG NCSA Technical report, alg02-001, October 11, 2002.
16. Bajcsy, P. and T.J. Alumbaugh, Georeferencing maps with contours, in *Proceedings of the 7th World Multiconference on Systemics, Cybernetics and Informatics (SCI 2003)*, Orlando, FL, July 27–30, 2003.

19

Data Integration

Peter Bajcsy

Contents

19.1 Introduction

Let us assume that we must collect data (or buy already acquired data from a vendor) and analyze spatial and temporal characteristics of a piece of land on the Earth with a limited budget. The cost of data collection is usually proportional to the accuracy of acquired data (instrument selection), spatial and temporal sampling rate (number and resolution of measurements), number of attributes (e.g., spectral bands), spatial coverage, and temporal duration. The cost vs. data quality trade-offs or the availability of data sources from multiple instruments [1–4] are encountered in many applications, and lead us to the problem of data integration and data fusion. This chapter is devoted to the problem of data integration and data fusion for raster and vector data types with the focus on datasets with a geographic component.

In general, data integration could be defined as an overall process that resolves data heterogeneity and datum selection. Data heterogeneity is understood as differences in spatial and temporal resolution, data accuracy (number of bits per value), data type (raster and vector), and geographic projection. Datum selection corresponds to the problem of choosing a value over spatially and temporally overlapping multi-instrument datasets. Based on this definition of data integration, we differentiate data integration as shallow and deep in the following way. Shallow data integration refers to resolving data heterogeneity while deep data integration refers to both, data heterogeneity and datum selection. The terms "shallow" and "deep" could also be related to syntactic and semantic integration occurring in data management problems [5,6]. In this chapter, we would use the term data fusion for deep data integration to discriminate these two cases. Other usages of the terms fusion and integration can also be found in the literature [7].

According to Department of Defense (DOD), the Joint Directors of Laboratories [8], data fusion is a multilevel, multifaceted process dealing with the automatic detection, association, correlation, estimation, and combination of data and information from single and multiple sources. According to Reference 9, "data fusion techniques combine data from multiple sensors, and related information from associated databases, to achieve improved accuracy and more specific inferences that could be achieved by the use of a single sensor alone." A web search for "data fusion" will return many application specific definitions (see URL: http://www.data-fusion.org). The most general definition of data fusion would be as the seamless integration of data from disparate sources. Other definitions are based on relating data fission and data fusion by viewing them as forward and reverse data transformations. For example, data fission could be considered the result of developing separate datasets from a single source of data, such as the National Oceanic and Atmospheric Administration (NOAA) Polar Orbiting Environmental Satellites. The data are "separated at birth" for storage in different locations. An example of data fission describes legacy data processing from satellite imagery. From this perspective, data fusion would be the process of rejoining or integrating these data. Ideally, the ultimate process would reevaluate the original separation of the data, and revise the processing technique to keep the data integrated, or to facilitate the reintegration of the data.

Why would anyone need data integration? It is apparent that multi-instrument measurements of the same phenomenon (the same time and space) contain (1) complementary or redundant information, (2) varying accuracy of spatial and temporal data, and (3) instrument-dependent uncertainty of measurements. Furthermore, differences among multi-instrument datasets arise from multiple types of camera-based aerial or satellite sensors and their associated measurement accuracy, as well as from several data representations preferred by sensor-operating agencies. Other factors contributing to data heterogeneity include (1) temporally changing data gathering and storage techniques, and (2) the application and sponsor specific requirements for every particular data collection. Thus, our motivation for data integration is (1) to overcome data heterogeneities, (2) to add complementary information and reduce redundant information from different data sources, (3) to increase the accuracy of analyzed data, and (4) to reduce uncertainty of acquired measurements. Our ultimate goal is to increase information gain by utilizing multiple sources.

Geographical Information Systems (GIS) applications of data integration could include combining raster maps [10], map fusion with a point set (e.g., elevation map updates with building heights) or integration of multiple point sets (e.g., nests of eagles and hawks). The purpose of GIS data integration could be visualization, geographic planning, or

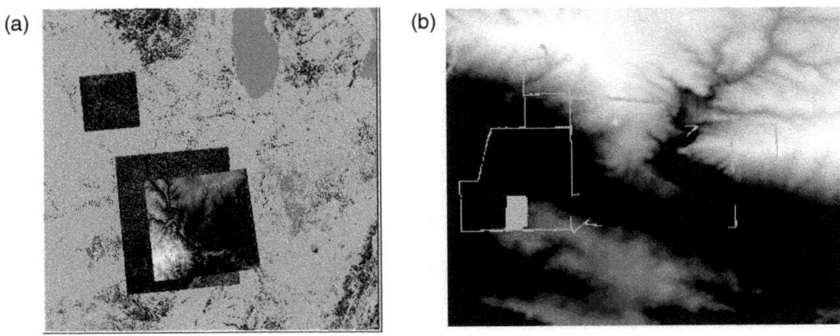

FIGURE 19.1
Examples of raster and raster data integration, and of raster data fusion with a dense set of accurate point measurements. (a) Integration of elevation map, forest label map, and land use maps. (b) Integration of elevation map with a sequence of accurate elevation measurements (green overlay.)

statistical processing and modeling of spatial regions in environmental restoration, battlefield preparation, or precision farming applications. Examples of map integration and map updates with point data are illustrated in Figure 19.1. Besides GIS applications, there are several other application areas for data integration, for example, in aeronautics, structural engineering, bioinformatics and medicine, or robotics. In aircraft navigation, a commercial pilot or an automatic landing system would combine elevation data with a weather map and Landsat imagery to make decisions about landing [37]. In structural health monitoring applications, civil engineers would determine material characteristics by combining temporal point displacement measurements and raster stress measurements, followed by computing material stress during loading [11,12]. In bioinformatics, biologists would combine microarray images with tabular gene descriptions and DNA sequencing information [13], while in medicine, researches and practitioners would integrate multi-instrument data obtained from Magnetic Resonance Imaging (MRI), Computer Tomography (CT), and x-ray instruments. In robotics, a robot director could use both gestures and voice commands to control a robot in a hazardous environment [14,15], or the robot could become autonomous using multisensor fusion [16].

The current commercial solutions for GIS data integration include primarily the Economic and Social Research Institute (ESRI) ArcGIS tools [17]. Other solutions have been developed for data integration of some combination of raster and vector data, such as, ERDAS [18], MCIDAS [19], ENVI [20], NASA EOS Data and Information System (EOSDIS) software, and NCSA I2K [21]. The ESRI ArcGIS tools combine geographic information with attributes to represent complex spatial relationships for easier decision making. These tools reformat the input spatial data to the area of coverage, combine it with the attribute data, and project the combined data into a geographic database to form new data relationships. The most frequently used tool in the ArcGIS system would be ArcMap (raster and vector data display and overlay), spatial analyst (geographic projection and feature extraction), and ArcToolBox (data type transformations). Similarly, NASA EOSDIS integrates user-defined data products from NASA's Earth science research satellites and field measurements programs, and facilitates the manipulation of data from all sources.

The rest of this chapter will focus on (1) spatial interpolation with kriging in Section 19.2, (2) shallow integration of geospatial raster data (map integration problem) in Section 19.3, and (3) deep integration of raster and vector data (fusion of images and point sets) in Section 19.4. Our objective is to present theoretical and practical issues that one has to consider while designing a fully automated data integration system. By "fully automated" we mean that a user specifies N input files with georeferencing information, and the software outputs a dataset with (1) maximum spatial and temporal information content, and (2) maximum accuracy and minimum uncertainty per output value.

19.2 Spatial Interpolation with Kriging

We describe a spatial kriging method for interpolating values from a set of points. A data integration process frequently involves interpolation or extrapolation of vector or low-resolution raster data. The purpose of interpolation and extrapolation is to match the dense spatial resolution of raster data with the sparse spatial resolution of vector data. Although there are many methods for interpolation and extrapolation prediction, such as the least-square polynomial, bilinear, or B-spline models, the kriging technique is commonly used in GIS applications. Examples of GIS usages include (1) hydrologists modeling underground water sources [22], or (2) mining companies predicting concentrations of a particular mineral when given only a small number of sample points [23].

A kriging technique can be described as a method for modeling spatial variation of a continuous variable from a given set of Two-Dimensional (2D) points inside a geographic domain. The goal is to use these data points to predict the values of a variable throughout the relevant spatial domain. Kriging is a topic covered in many textbooks [24], and our description of kriging gives only a brief summary of the concepts relevant for this work. The most important tool is the variogram, or more correctly the semivariogram, denoted as γ. This function takes a vector value representing the difference between two locations in the data, usually called a "lag," as its argument. Its value for a given lag represents how much the feature can be expected to vary over that lag distance and direction. This is a tremendously powerful tool, as a dataset may vary much in one direction (say the north–south direction) and hardly at all in another (say the east–west direction). A semivariogram that takes into account these differences in direction is called an anisotropic semivariogram. Another type of semivariogram, called an isotropic semivariogram, is a function of only the magnitude of the lag distance. As one would imagine, this is significantly easier to implement.

In order to perform ordinary kriging, one must decide on the type of variogram to use, and how to calculate this variogram for the prediction purposes. Most of the section is devoted to a description of the prediction process, and we will use a variogram that is isotropic. That is, the argument to the variogram is only the magnitude of the lag vector, but not its direction. This means our variogram is a scalar function of a scalar variable. It is common in variogram construction to find some way to estimate the variogram first, and an isotropic variogram makes this process easier.

19.2.1 Variogram Estimation

The notation for this section follows closely with Reference 24. The variable we are trying to predict is collectively denoted as Z. Our set of locations is $\{s_1, \ldots, s_n\}$. The value of our variable at a location s_i is $Z(s_i)$. The prediction of our variable at a location s_0 is denoted as $p(Z, s_0)$.

19.2.1.1 Method of Moments Estimator

To understand the characteristics of the variogram associated with a dataset, it is a general practice to estimate it by some means. We will demonstrate the standard method-of-moments variogram estimator, also known as the "classical" variogram estimator. The formula for the "classical" variogram estimator is:

$$2\hat{\gamma}(h_k) = \frac{1}{|N(h_k)|} \sum_{N(h_k)} (Z(s_i) - Z(s_j))^2, \qquad h \in R^d, \tag{19.1}$$

where the set of pairs, $N(h_k)$, is defined as:

$$N(h_k) \equiv \{(s_i, s_j) : \text{Dist}(s_i, s_j) = h_k : i, j = 1, \ldots, n\}. \tag{19.2}$$

Throughout this section, we will usually denote $\hat{\gamma}_k = \hat{\gamma}(h_k)$ as the estimator at a distance h_k. From a mathematical viewpoint, the set $N(h_k)$ is well defined. However, from a computational viewpoint, the size of the set $N(h_k)$ for a given h_k may be quite small or even zero unless processing extremely large datasets. Thus, we must use reasonable values of h_k and a small tolerance Δ_t before computing parameters of the classical estimator. The issue related to populating the set $N(h_k)$ for a chosen h_k is described in Section 19.2.1.2.

19.2.1.2 Populating $N(h_k)$

To find reasonable values of h_k, one can iterate through a dataset (or some portion of a dataset) and record the distances between a particular point and several of its neighbors. If the distances are computed for all point pairs then this computation requires $(n \cdot (n-1))/2$ distance calculations. This may persuade some researches to use only a sample of their full dataset for any distance calculations. These calculations are ordinarily performed using the simple Euclidean distance (even if the points are specified in latitude and in longitude). It would also be possible to approximate the distances between any two points in meters using a geodetic ellipsoid; however the distances involved for most datasets are sufficiently small that this step is unnecessary. One can begin by finding the minimum and the maximum distance between points in the sample of data and using those values as the minimum and the maximum values of h for the estimator. By deciding on a rather arbitrary number of total points for the estimator, one can choose the rest of the values of h_k for the estimator and then calculate each $\hat{\gamma}_k$.

To populate $N(h_k)$ for each of our values of h_k, we cannot insist that the distances between two points be exactly h_k. The distance between any two points would normally be computed as a single or double precision floating point number, so it is unlikely that this distance would agree with any value of distance h_k up to its highest precision. Thus, a tolerance value should be chosen and refined as needed. There is no concrete rule for the size of each set $N(h_k)$, which we denote as P. Following the advice of Cressie [24], P should be larger or equal to 30 ($P \geq 30$) pairs of points such that the distance between each pair falls within the tolerance for each value of h_k.

$$\|\text{Dist}(s_i - s_j) - h_k\| < \Delta_t. \tag{19.3}$$

Therefore, the first step is to compute distances for a small subset of the points in the data (perhaps 10 to 25%) and then compute the minimum and the maximum distance values. Second, one can partition the interval of [min, max] into subintervals as illustrated in Figure 19.2. The partition should, in general, be evenly spaced throughout [min, max]; however, fine-tuning may be necessary to achieve the desired number of pairs for each lag distance.

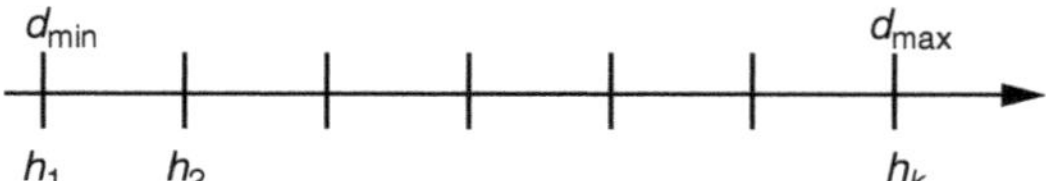

FIGURE 19.2
Finding K distance values.

19.2.2 Selecting and Fitting the Model

Given a set of estimated values $\hat{\gamma}_k = \hat{\gamma}(h_k)$, $k = 1, \ldots, K$, $K = 10$ (i.e., the classical variogram estimator), the next step is to choose a continuous model variogram and fit the model's parameters to the estimated values. There is a great freedom of choice for this step, and the success of the kriging process is mostly determined by the choice of semivariogram model. A semivariogram must have certain properties to be statistically valid (see Reference 24 for details). For example, a valid semivariogram must always give nonnegative answers for positive values of a lag distance h. A simplified power model for the semivariogram in Equation 19.4 is one of the models that would guarantee the earlier property.

$$\gamma(h) = a^2 \|h\|^{\alpha}. \tag{19.4}$$

19.2.3 "Linear" Least Squared Fitting to Power Model

Let us assume that we wish to use the semivariogram model given in equation (19.4). To fit data to our model, we seek values for parameters a and α that predict $\hat{\gamma}_k = \hat{\gamma}(h_k)$, $k = 1, \ldots, K$, $K = 10$ as close as possible. We choose to fit our model in a linear least squares fashion, which we can do with some manipulation of our original formula. Note that for some h, we would like our model to be such that

$$\begin{aligned}
\hat{\gamma}(h) &= a^2 \|h\|^{\alpha}, \\
\ln(\hat{\gamma}(h)) &= \ln(a^2 \|h\|^{\alpha}), \\
\ln(\hat{\gamma}(h)) &= 2\ln(a) + \alpha \ln(\|h\|).
\end{aligned}$$

Writing this equation in vector form yields:

$$\begin{bmatrix} \ln(\|h\|) & 1 \end{bmatrix} \cdot \begin{bmatrix} \alpha \\ 2\ln(a) \end{bmatrix} = \ln(\hat{\gamma}(h)).$$

Now, we find values for α and $2\ln(a)$ that minimize the two norm of the residual of the following linear system:

$$\begin{bmatrix} \ln(\|h_1\|), & 1 \\ \vdots & \\ \ln(\|h_{10}\|), & 1 \end{bmatrix} \cdot \begin{bmatrix} \alpha \\ 2\ln(a) \end{bmatrix} = \begin{bmatrix} \ln(\hat{\gamma}(h_1)) \\ \vdots \\ \ln(\hat{\gamma}(h_{10})) \end{bmatrix}.$$

Using QR factorization, we can solve α and $2\ln(a)$. This solution gives us the valid semivariogram used to perform predictions. As one would expect, there are numerous ways to construct a semivariogram, so the steps taken here may vary. We use the variogram for a prediction in Section 19.2.4.

19.2.4 Predicting a Value

Setting up Linear System: Following closely with Cressie ([24], Section 3.2), we seek to solve the following linear system:

$$\Gamma\lambda = \gamma_0,$$

where Γ, γ_0, and λ are as follows (recalling that s_0 is our value to predict):

Γ: an $n \times n$ matrix whose (ith, jth) entry is $\gamma(s_i - s_j)$

γ_0: an $n \times 1$ column vector whose ith entry is $\gamma(s_0 - s_i)$

λ: an $n \times 1$ column vector where each entry is a weight to be used in the final prediction

Solving Linear System: For some datasets, it may be too costly to use the entire dataset in the prediction process (i.e., $n = |Z(s)|$). For such cases, it may be more computationally feasible to sample the data throughout the space and use this sample to form the matrix Γ (and corresponding vectors γ_0, and λ). After solving the system, we end up with a column vector of weights, λ. We can multiply the transpose of this vector by a column vector of our n points (however they are chosen) as follows:

$$p(Z, s_0) = \lambda' \cdot Z, \qquad Z = \begin{bmatrix} Z(s_1) \\ \vdots \\ Z(s_n) \end{bmatrix}.$$

This yields the prediction, $p(Z, s_0)$.

19.2.5 Practical Considerations

While kriging is used as one step in integration of raster and vector data, it can also be used for spatial prediction purposes. We present two examples related to crop yield prediction and image object removal, to illustrate (1) other kriging applications than data fusion and (2) some practical considerations about kriging.

First, one can use kriging for crop yield prediction over a dense grid of spatial locations. An example dataset is shown in Figure 19.1(b). The input data contained about 18,000 latitude/longitude points collected by harvesting machines equipped with Global Positing System (GPS) in August 2003 in central Illinois. As the machines drove through the fields, they were able to record, among other things, the volume of crop harvested at that location (hereafter referred to as "yield").

As described in Sections 19.2.1 to 19.2.4, kriging requires selection of several parameters including the estimator type, M, P, Δ_t, and h_k. The choice of these parameters should be optimized with respect to any particular input dataset and is usually constrained by the available computational resources. Further, the density of this particular yield data set greatly affected the conditioning of the linear system, making it nearly singular when nearly all of the data was used. This issue can be resolved by selecting a subset of the input points n distributed throughout the geographic domain so that the resulting linear system is well conditioned and fairly cheap to solve. To illustrate the accuracy of such a prediction, we chose semivariogram estimator with isotropic model, and the parameters $M = 1000$, $P = 30$, $\Delta_t = 5 * 10^{-6}$, $n = 600$, and $h_k = 0.002 + k * 0.0006$; $k = 0, 1, 2, 3$; $h_k = 0.02 + (k - 4) * 0.005$; $k = 4, 5, 6, 7, 8, 9$. We predicted the yield values using I2K [21] for a few arbitrarily chosen

Table 19.1

Illustration of kriging accuracy for crop yield prediction.

Latitude	Longitude	Actual yield	Predicted yield
40.408634	89.074655	135	150.01
40.413923	89.074691	147	128.85
40.411642	89.074255	180	158.48
40.410427	89.075334	172	155.83
40.411624	89.074609	180	161.18

(a) 
(b) 
(c) 

Figure 19.3
Kriging technique used for image object removal. (a) Original image and the original image after removing, (b) the pedestrian in the image center, and (c) the car in the lower right corner.

points already in our input dataset. Table 19.1 shows a comparison of actual yield values and the predicted yield values. Although these yield predictions are sensible values, it is unclear how to adjust model parameters to improve the prediction accuracy and what the ultimate accuracy would be with a different set of parameters.

The second interesting application of kriging is image object removal. The use of kriging is slightly different from the previous application scenarios because the input data is a dense set of grid points (an image) with a region of missing values (any removed object). The missing values have to be predicted in order to simulate previously occluded background by any removed object. An example of an image object removal is shown in Figure 19.3. In this case, in addition to kriging model and parameter selections, one has to (1) partition an original image into color homogeneous regions to find the sample subsets for building prediction models, (2) estimate segments of the occluded object background (e.g., a car to be removed at the sky-ground horizon), and (3) develop the mapping between sample subsets and the segments of the predicted background. Thus, automated or semiautomated image object removal becomes a very difficult problem since predicting occluded values involves many decisions based on higher-level image understanding, for instance, implying boundaries of partially occluded objects.

19.3 Shallow Integration of Geospatial Raster Data

Shallow integration of geospatial raster data could be one part of map integration. Maps are viewed as one example of geospatial raster data, and they are frequently mosaicked from multiple map tiles. According to Reference 25, a photographic mosaic can be defined as "an assemblage of photographs, each of which shows part of a region..., put together in such a way that each point in the individual photographs appears once and only once... and variation of scale from part to part... is minimized." Mosaicking is then defined as "the process of constructing a mosaic from [these] photographs." Often, mosaicking is used in GIS to "stitch together" maps of similar scale and projection type.

However, shallow integration of any geospatial raster data can be a difficult task and may involve combining images of differing spatial resolution (spatial scale), spectral resolution (radiometric resolution), number of features per map location, and geographic projection. Furthermore, there are map differences resulting from the digital nature of the map pieces, such as the datum representation and data structures used to store map information. It is desirable to form a resulting integrated image that maintains a consistent geographic projection and resolution throughout the image. Thus, the task of map data shallow integration is about resolving all map differences automatically to produce one consistent map.

In general, shallow integration can be performed with images that come with or without georeferencing information. A general description of map integration without georeferencing information can be found in Reference 10. The solutions in literature follow the process described in Chapter 17, "Spatial Registration."

Let us focus on shallow integration of maps with georeferencing information. Our task is to integrate maps automatically by resolving their dissimilar geographic projections, horizontal and vertical spatial resolutions, data types (byte, short, integer, long, float, or double precision), and number of bands, while assuming that map information is stored in the same data structures. We refer to this problem as shallow integration of raster data because various maps are adjusted without fusing their values. While the number of map bands (also called attributes, features, or channels) vary, the physical meanings of bands in multiple maps do not necessarily correlate and, hence, we do not have to perform datum value selection. This assumption would not hold in the case of maps containing spectral information of variable spectral resolution. For example, this would be the case if the task is to integrate hyperspectral images with 3 nm and 10 nm spectral resolution, and partially overlapping wavelength ranges of (400 nm, 720 nm) and (650 nm, 1100 nm).

We rely heavily on the ability to calculate the latitude and the longitude of any point on any of the map pieces during the integration process. In general, it would not be effective to simply tile the images together based on the tie points. Each image may have its own projection type and thus we would simply produce an image where we georeference with one set of equations in one map area and with a different set of equations in another part of the map. In addition, we would be unable to follow consistent lines of latitude and longitude with such a system.

Why would anyone need a theoretical modeling framework for data integration? The need for a theoretical modeling framework arises when automation and consistency are our goals. Without a theoretical framework, it is hard to design an automated data integration system or perform data integration consistently since the integration would always rely on human operators. In the past, multisensor data integration problem has been often posed as an extension to signal processing, detection estimation and identification, and multiple modeling approaches have been proposed [26–28].

Next, we elaborate on three fundamental challenges of data integrations. The three challenges are related to (1) theoretical modeling of shallow map integration, (2) modeling of integration constraints and data processing operators, and (3) defining metrics for meeting integration constraints. First, we present two theoretical modeling frameworks for describing the shallow map integration problem, such as the framework based on modeling theory and the framework based on optimization of indexed sets. Then, we illustrate shallow integration of two raster files using the framework based on optimization of indexed sets.

19.3.1 Framework Based on Model Theory

The first framework comes from an area of mathematics known as model theory [29] and consists of (1) formal languages to characterize the data acquisition (the world and the sensing process), (2) models to represent input data, operations on data, and relations among the data, and (3) theories to capture symbolic knowledge about sensing [26]. Model theory concerns itself with expressing logical statements in a formal language, and then analyzing mathematical structures (known as models) that interpret these languages.

This framework has been applied to the remote sensing field in an attempt to give mathematical precision to the term "data fusion" [26]. As an example, we apply the ideas of the framework to a fire detection system. The system has the ability to sense temperature variations in a space and to detect fire-like imagery from video cameras placed within the space. We define as follows:

G — A logical statement for our goals. We wish to detect fire and/or the precursor to fire (excessive smoke, sparks) within a room.

$L1, L2$ — Two languages associated with the temperature sensor and the video sensor, respectively. The language consists of variables, constants, and relation symbols. $L1$ could contain, initially, a set of variables for the temperature at discrete points throughout the space. It could also contain logical relations about how the temperature can change throughout the space, given the atmosphere within the space and what is known about the airflow in the area. For example, $L1$ could state that the temperature at two locations x meters apart that cannot vary by more than some tolerance value or else we assume the sensors are broken and give the "error" state. $L2$ could contain a set of variables for each pixel (or a subset of all pixels) within the video image. $L2$ could then have relations declaring what is and what is not detected as a fire. For example, a patch of all black pixels would be defined as "not fire" in the language.

$M1, M2$ — Two models for $L1$ and $L2$, respectively. Let us say that $M1$ is a description of a particular thermodynamic theory detailing how temperatures should change from one discrete location to another within the space. $M2$ could be a set of template images (or a set of rules) that describe what fire should look like in a video image system. Some of the pictures may have smoky fires; some would be fires with no smoke. It is important to note that in general, as in this example, a language can have many models.

$T1, T2$ — Two formal theories that describe knowledge about the observed space and the sensing equipment. $T1$ contains our rules for taking temperature measurements and deciding if there is a fire. $T2$ contains our rules analyzing video images and deciding if there is a fire.

With the following definitions out of the way, we can ask the all-important question: given these two systems, when should we signal that there is a fire in the space? When $T1$ indicates there is a fire? When $T2$ indicates there is a fire? Only when both detect a fire? Do $T1$ or $T2$ have certain confidence ratings so we can threshold the signal (more than 80% sure, signal fire)? The answer, in this framework, is satisfied by a so-called "fused theory," T, and a fused model, M. To give the description of T and M, we make use of the so-called "satisfaction" operator, $\models$. M and T then have the following properties:

$$M \models G$$
$$M \models T$$

$M \models T \Rightarrow M1 \models T1$ and $M2 \models T2$ or
$M1 \models T1$ and $M2 \models T2 \Rightarrow M \models T$

As is the case with much of the model theory, our answer is declarative in nature and does not provide an algorithm for its computation. The properties of the answer are known, but a process deriving it from known values $M1$, $M2$, $T1$, and $T2$ is not. This is an area of on-going research. Typically, remote sensing systems by themselves would have certain software components, so it may seem odd to write a program *about* a program. While the benefits of this kind of formal verification are numerous, they are clearly outside the scope of this text. We refer the reader for rigorous definitions of this theoretical framework to Reference [26].

19.3.2 Framework Based on Optimization of Indexed Sets

The second framework is based on (1) indexing images according to their integration parameters (resolution, projection, etc.), (2) introducing image integration constraints and optimization functions, and (3) searching for optimal map integration solution by considering all admissible map operators [30]. The overview of this approach is presented in Figure 19.4, where ω represents the set of map specific integration parameters. We describe this framework in more detail in Section 19.3.2.1. To begin, we formally introduce an indexed set A, to hold our universe of possible image sets, a set of map integration constraints, and a formal problem statement. We then show one theoretical solution for resolving map dissimilarities.

19.3.2.1 Indexed Set of Images

We are given a set of n images, each of which we denote as $i_j, j = 1, \ldots, n$. We call the set of all n images I. Every image in I has a set of parameters: resolution (r), number of bands (b), geographic projection (p), and sample type (s) defined as the number of bits per sample. For all the images, we collect all of the unique parameters into sets R, B, P, and S.

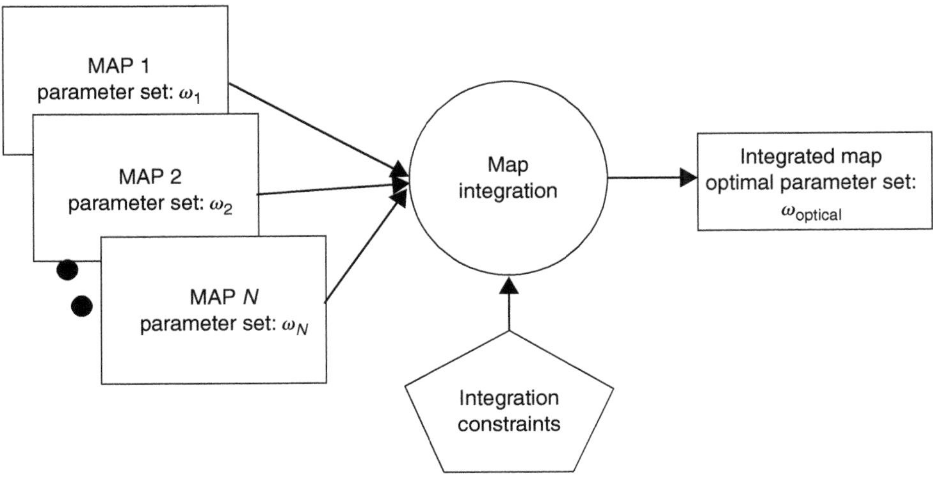

FIGURE 19.4
An overview of the optimization approach using indexed sets. The value ω represents the set of map specific integration parameters.

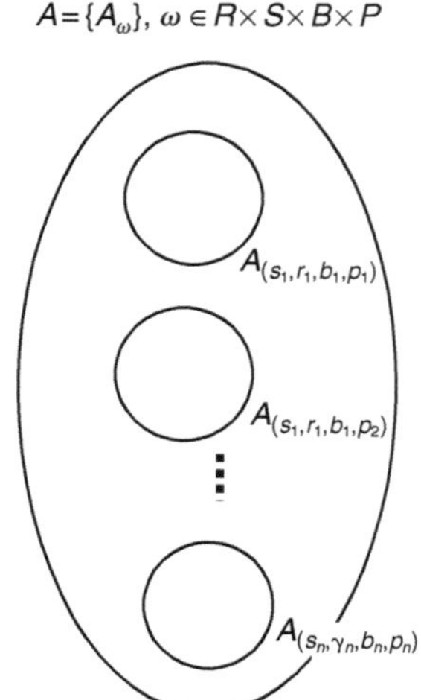

FIGURE 19.5
A set of all transformed images based on all combinations of parameters from $R \times S \times B \times P$.

(e.g., $S = \{s_1, s_2, \ldots, s_m\}$). Every image can be transformed into another image with any collection of values from any of the sets. We then consider the indexed family of sets A, which contains sets of images indexed by elements from the Cartesian product $R \times S \times B \times P$ (see Figure 19.5). Formally, we write $A = \{A_\omega\}$, $\omega \in R \times S \times B \times P$, $j = 1, \ldots, n$. For example, for $\omega = (r_1, s_2, b_1, p_3)$, A_ω is a possible set in A that contains all the images transformed to that set of parameters. We then define $f = \{f_\omega : I \rightarrow A_\omega\}$, $\omega \in R \times S \times B \times P$ to be the indexed family of functions from our original set of images to an element in A.

19.3.2.2 Map (Image) Integration Constraints

In general, any sequence of data preprocessing operations, such as data integration, would seek objectives including (1) preservation of information content, (2) minimization of computational requirements for processing, and (3) minimization of information uncertainty (or maximize data accuracy) of the final data product. While many integration constraints would strive to achieve one of these objectives, the specific choice of integration constraints is application dependent.

For illustration purposes, we could map these general objectives of any data processing into a set of constraints imposed on (1) information entropy (H) that represents information content, (2) data size in memory (M) that corresponds to computational requirements, and (3) spatial error from geographic reprojection (E) that maps to one type of information uncertainty. In this case, the shallow integration of raster data would correspond to an optimization problem defined as follows. Find the "best" f_ω that will maximize the information entropy (H) and minimize both the size in memory (M) of an image and any spatial error from geographic reprojection (E). Thus, for an indexed set of images I, and a function f_ω, we would find the optimum choice of ω_{optimal} as:

$$\omega_{\text{optimal}} = \max_{\omega} \sum_{j=1}^{n} \frac{H(i_{j,\omega})}{M(i_{j,\omega})E(i_{j,\omega})}, \tag{19.5}$$

where $i_{j,\omega}$ in A_ω indicates the image i_j and the parameters ω that are used for transforming the image by using the function $f_\omega(i_j)$. The three variables H, M, and E can be viewed as separate dimensions of our data with always nonnegative quantities.

19.3.2.3 Finding Optimal Integration Solutions

Given the theoretical framework based on optimization of indexed sets, how would one find the optimal set of parameters including resolution (r), number of bands (b), geographic projection (p), and sample type (s) (or ω_{optimal})? One approach is to analyze the models for variables introduced by integration constraints. For example, the changes of entropy (H), memory (M), and spatial error (E) as a function of $\omega \in R \times S \times B \times P$. Every operation for resolving map dissimilarities might change one or more of the optimized variables. We now introduce the list of all considered image operators in Table 19.2 and show the effect of image operators on the values of $h = H(i_{j,\omega})$, $e = E(i_{j,\omega})$, and $m = M(i_{j,\omega})$ in Table 19.3.

To clarify Table 19.3, the term "constant" refers to no changes due to an operator. For instance, we cannot affect the amount of information in an image by upsampling the image, thus $\Delta_{R\uparrow}h = h$. The term "linearly increasing or decreasing" refers to a linear model with a multiplicative factor larger or smaller than one. For example, the effect of $\Delta_{S\uparrow}$ (or $\Delta_{S\downarrow}$) on m is modeled as $\Delta_{S\uparrow}M$ = #bits old sample type/#bits new sample type$\cdot M$ with the ratio greater than one for $\Delta_{S\uparrow}$ and smaller than one for $\Delta_{S\downarrow}$. The term "approximately constant" refers to slight changes that should not have a significant impact on a variable. It is also

Table 19.2

Definition of operators that changes values of entropy, spatial error, and memory size.

Symbols for image operators	Denoted image operations
$\Delta_{R\uparrow}$	Spatial upsampling
$\Delta_{R\downarrow}$	Spatial subsampling
$\Delta_{S\uparrow}$	Changing the sample type up (more bits per pixel)
$\Delta_{S\downarrow}$	Changing the sample type down (fewer bits per pixel)
$\Delta_{P\downarrow}$	Changing the projection type
$\Delta_{B\uparrow}$	Spectral upsampling (larger number of bands)
$\Delta_{B\downarrow}$	Spectral subsampling (smaller number of bands)

Table 19.3

Models for changes of entropy "H," spatial error "E," and memory size "M" due to individual image operators described in Table 19.2. The g entries denote a model that has to be obtained either experimentally or analytically using known image content.

	$\Delta_{R\uparrow}$	$\Delta_{R\downarrow}$	$\Delta_{S\uparrow}$	$\Delta_{S\downarrow}$	$\Delta_{P\downarrow}$	$\Delta_{B\uparrow}$	$\Delta_{B\downarrow}$
M	Linearly increasing	Linearly decreasing	Linearly increasing	Linearly decreasing	Approximately constant	Linearly increasing	Linearly decreasing
H	constant	g_1	Constant	g_2	Approximately constant	Constant	g_3
E	Constant	g_4	Constant	Constant	g_5	Constant	Constant

assumed that any reprojection changes can only increase spatial error with respect to the original image projection and therefore it is sufficient to consider only $\Delta_{P\downarrow}$. The g entries in Table 19.3 can be modeled analytically by assuming a known image content, for example, a checkerboard pattern image (deterministic image content) or constant intensity image with superimposed Gaussian noise (statistical image content).

19.3.2.4 Quality Metrics for Map Integration

How do we decide which map integration method is "best?" In order to find the optimal set of parameters, the variables defined by integration constraints have to be evaluated for every considered combination of operators applied to the original images. For instance, entropy and computer memory variables can be defined using straightforward formulas. The entropy measure H is defined as $H = -\sum_{k=1}^{\text{numColors}} p_k \ln(p_k)$, where p_k is the probability of color k. The memory size M is directly proportional to the number of rows, numbers of columns, samples per pixel, and the sample type, $M = \text{numRows} * \text{numCols} * \text{sampPerPixel} * \text{sampType}$.

It is a much harder problem to define a general metric for spatial error introduced due to geographic reprojections. Multiple spatial error metrics could be defined for specific map integration applications. For instance, if we refer to the motivation example in Chapter 14, and our goal is to extract statistics from integrated images over closed area boundaries, then we would incorporate the accuracy of extracted statistics into a spatial error metric. In this case, we could define a spatial error metric E of an image i_j by computing the difference between any set of continuous or categorical features defined over closed area boundaries before and after reprojection. Equation 19.6 shows an error metric for raster data representing continuous variables. This metric would measure, for instance, a change in average elevation per closed boundary due to a geographic reprojection.

$$E_{\text{NewProjection}}^{\text{ContinuousVariable}}(i_j) = \frac{\sum_{k=1}^{\#\text{boundaries}} |F_{\text{NewProjection}}(i_j, k) - F_{\text{OldProjection}}(i_j, k)|}{\#\ \text{boundaries}}, \tag{19.6}$$

where E is the error, i_j is the jth integrated image i, and F is the statistical feature computed from continuous raster variable over a boundary k.

Similarly, Equation 19.7 illustrates an error metric for raster data representing categorical variables. The error values obtained using this metric would correspond to a change in frequencies of occurrence of categorical labels due to a geographic reprojection, for example, deciduous and coniferous forest labels.

$$E_{\text{NewProjection}}^{\text{CategoricalVariable}}(i_j) = \left\{ \sum_{k=1}^{\#\text{boundaries}} \left| \sum_{m=1}^{\#\text{CategoricalLabels}} h_{\text{NewProjection}}(i_j, k, m) - h_{\text{OldProjection}}(i_j, k, m) \right| \right\} \{\#\text{boundaries}\}^{-1}, \tag{19.7}$$

where E is the error, i_j is the jth integrated image i, and h is the frequency of occurrence of a categorical label m over a boundary k.

19.3.3 Illustration Example

The purpose of this section is to illustrate the use of shallow map integration framework based on optimization of indexed sets. Let us consider two images, i_1 and i_2, which are of the same geographic projection, and have the same number of bands and the same size. Image i_1 has a resolution of 30 m/pixel, an initial entropy measurement of $h_1 = 76.00$, and a size of $m_1 = 152\,000$ bytes. Image i_2 has a geographic resolution of 1000 m/pixel, an initial entropy measurement of 85.00, and a size of $m_2 = 425\,000$ bytes. Also, i_1 has a sample type of float (IEEE floating point numbers represented with 32 bits), while i_2 has a sample type of double (IEEE double precision numbers represented with 64 bits). Since the band number and projection of the two images are both equal, we will ignore their parameters in our set A and corresponding family of functions f_ω. Therefore, $A = \{A_\omega\}, \omega \in \{r_1, r_2\} \times \{s_1, s_2\}$, where $r_1 = 30$, $r_2 = 1000$, $s_1 =$ float (or "32"), $s_2 =$ double (or "64"). Throughout, we denote $A_\omega = A_{r_1,s_1}$ as $A_{1,1}$ and similarly for the rest of the A_ω. The description of each set within A is in Table 19.4.

We are searching for the optimum choice of f_ω as:

$$\omega_{\text{optimal}} = \max_{\omega} \sum_{j=1}^{2} \frac{H(i_{j,w})}{M(i_{j,\omega})}; \qquad \omega \in (30, 1000)x(\text{"32"}, \text{"64"}). \tag{19.8}$$

Geometrically, we want to maximize the slope (see Figure 19.6) of the line determined by a particular choice of f_ω.

The calculations here evaluate changes of h and m due to all possible image operators to match spatial resolution and sample type variables.

TABLE 19.4

Description of the input parameter set A.

Spatial resolution\sample type (m/pixel)	Float ("32")	Double ("64")
30	$A_{1,1}$	$A_{1,2}$
1000	$A_{2,1}$	$A_{2,2}$

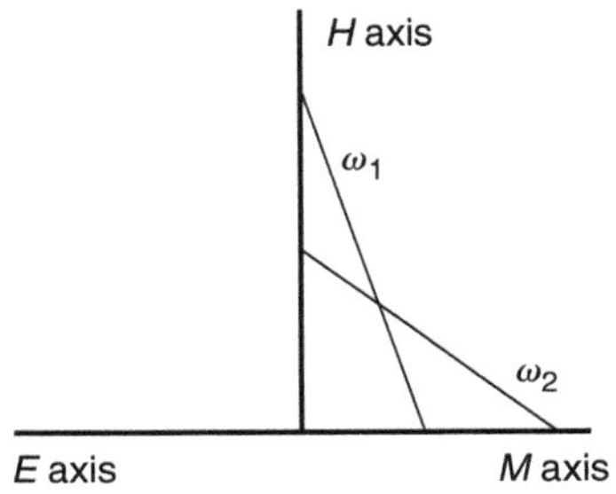

FIGURE 19.6
Optimization along information energy H and data size in memory M axes.

For i_1, we have:

$$A_{11}\colon \frac{h_1}{m_1} \Rightarrow \frac{h_1}{m_1}, \qquad A_{12}\colon \frac{h_1}{m_1} \Rightarrow \frac{h_1}{2\cdot m_1}, \tag{19.9}$$

$$A_{21}\colon \frac{h_1}{m_1} \Rightarrow \frac{\Delta_{R\downarrow}h_1}{\Delta_{R\downarrow}m_1} = \frac{g_1(h_1)}{0.03m_1}, \qquad A_{22}\colon \frac{h_1}{m_1} \Rightarrow \frac{\Delta_{R\downarrow}h_1}{\Delta_{R\downarrow}(2m_1)} = \frac{g_1(h_1)}{0.06m_1}. \tag{19.10}$$

For i_2, we have:

$$A_{11}\colon \frac{h_2}{m_2} \Rightarrow \frac{\Delta_{s\downarrow}h_2}{\Delta_{R\uparrow}(0.5m_2)} = \frac{g_2(h_2)}{16.66m_2}, \qquad A_{12}\colon \frac{h_2}{m_2} \Rightarrow \frac{h_2}{\Delta_{R\uparrow}m_2} = \frac{h_2}{33.33m_2}, \tag{19.11}$$

$$A_{21}\colon \frac{h_2}{m_2} \Rightarrow \frac{\Delta_{s\downarrow}h_2}{0.5m_2} = \frac{g_2(h_2)}{0.5m_2}, \qquad A_{22}\colon \frac{h_2}{m_2} \Rightarrow \frac{h_2}{m_2}. \tag{19.12}$$

For our choice of f_ω, we must determine the functions g_1 and g_2 in order to find the optimal $\omega_{\text{optimal}} = \max_\omega\{(A^{i_1}_{1,1} + A^{i_2}_{1,1}); (A^{i_1}_{1,2} + A^{i_2}_{1,2}); (A^{i_1}_{2,1} + A^{i_2}_{2,1}); (A^{i_1}_{2,2} + A^{i_2}_{2,2})\}$.

Analytical Modeling of $g_1(h) = \Delta_{R\downarrow}h$ function: The effect of subsampling on the entropy of an image is difficult to model given its dependence on the image itself. One could experiment with various real images and some simulated images to develop a model. For instance, an example model can be derived by using a least square fit to values derived from a simulated image with a standard white and black checkerboard and with added Gaussian noise. Table 19.5 and Equation 19.20 show such results such results according to [30].

$$g_1(h_{\text{orig}}) = \left(-0.2467 \cdot \left(\frac{r_{\text{orig}}}{r_{\text{new}}}\right) + .481468\right) + h_{\text{orig}}, \tag{19.13}$$

where r_{orig} is the original spatial resolution of the image and r_{new} is the desired spatial resolution.

Analytical Modeling of $g_2(\lg(s)) = \Delta_{S\downarrow}h$ function: The impact of sample type on the entropy of an image has to be modeled again by making observations and by choosing appropriate analytical models. For example, a model with the Lagrange-type polynomials

TABLE 19.5

Entropy as a function of subsampling ratio. The diffrence entropy value is obtained by subtracting the entropy of subsampled data from the entropy of original data (subsampling = 1).

Subsampling ratio	Entropy value	Difference entropy value
1	84.147194	0
2	84.17162	−0.024426
4	84.27841	−0.131216
6	83.55234	0.594854
8	83.813095	0.334099

Table 19.6

Numerical values for each choice of map parameters.

	$A_{1,1}$	$A_{1,2}$	$A_{2,1}$	$A_{2,2}$
$\frac{H(i_1)}{M(i_1)} + \frac{H(i_2)}{M(i_2)}$	0.0005697	0.00028497	0.01729	0.008

was reported in [30] and the mathematical model of $g_2(n)$ is presented in Equation 19.14.

$$g_2(n = \lg_2(s_{\text{new}})) = h_{\text{orig}} \left(\frac{2^{n-1}}{2^{b_{\text{orig}}}} + T_1 + T_2 + T_3 \right) + m \cdot (2^{b_{\text{orig}}} - 2^n), \tag{19.14}$$

where $b_{\text{orig}} = \lg_2(s_{\text{orig}})$ is the logarithm base 2 of the original number of bits per pixel for the image, $n = \lg_2(s_{\text{new}})$ is the logarithm base 2 of the number of bits per pixel for the new sample type, m is a constant (chosen to be −0.05), and T_1, T_2, and T_3 are defined as:

$$T_1 = \frac{(n - (b_{\text{orig}} - 2)) \cdot (n - (b_{\text{orig}} - 3))}{12}, \tag{19.15}$$

$$T_2 = \frac{(n - b_{\text{orig}}) \cdot (n - (b_{\text{orig}} - 2)) * (n - (b_{\text{orig}} - 3))}{-24/7}, \tag{19.16}$$

$$T_3 = \frac{(n - b_{\text{orig}}) \cdot (n - (b_{\text{orig}} - 1)) \cdot (n - (b_{\text{orig}} - 2))}{-96}. \tag{19.17}$$

Selection of Optimal Values: Given the analytical formulas for $g_1(h) = \Delta_{R\downarrow}h$ and $g_2(\lg(s)) = \Delta_{S\downarrow}h$, we determine the optimal set of parameters:

$$\omega_{\text{optimal}} = \max_{\omega}\{(A^{i_1}_{1,1} + A^{i_2}_{1,1}); (A^{i_1}_{1,2} + A^{i_2}_{1,2}); (A^{i_1}_{2,1} + A^{i_2}_{2,1}); (A^{i_1}_{2,2} + A^{i_2}_{2,2})\}.$$

Working out our numerical examples yields the results shown in Table 19.6. Thus, we select $\omega = (r_2, s_1)$ equal to 1000 m/pixel resolution and floating point sample type for ω_{optimal}. These parameters are considered as "optimal" given the theoretical model.

19.4 Deep Integration of Raster and Vector Data

In this section, we focus on deep integration of raster and vector data, and specifically, on fusion of images and point sets. This particular data fusion problem occurs in many scientific measurements utilizing multiple instruments that collect spatially overlapped data simultaneously. For example, in GIS [31,32], geographically distributed point measurements have to be fused with satellite raster measurements. In hazard monitoring systems [33], fire hazard monitoring requires a data fusion of point temperature measurements and thermal infrared images. In structural engineering domain, material structural health analysis combines point displacement measurements [34] and photoelastic-based strain raster measurements [35].

Let us assume that we are given a digital elevation map with spatial resolution 30 m/pixel and the accuracy of elevation values is around ±50 m. Given a set of points with associated elevation values that are more accurate than ±50 m, the task is to fuse the map and point data to obtain more accurate map elevation values. This problem requires careful considerations of the uncertainty distribution around the spatial set of accurate elevation values so that new values are either map values or the interpolated values from the point set. For instance, if an accurate point coincides with a map pixel location then the choice of a new elevation value is clearly the point value. However, if one has to choose a new value in between two accurate point values then the choice depends on the uncertainty of the map value in comparison with the uncertainty of the interpolated value from the point set.

19.4.1 Framework Based on Uncertainty Analysis

In general, it is not trivial to theoretically model raster and vector data fusion because the number of variables is very large. The fusion variables include specific characteristics of instruments and sensors, the types of integration processing steps and the execution order of fusion steps. Furthermore, the information needed for modeling fusion might not be known, for instance, temporal and spatial variations of the physical phenomenon observed by sensors and instruments. Nevertheless, the importance of theoretical modeling of raster and vector data fusion (deep data integration) is significant for designing reliable and robust data acquisition systems, as well as for conducting one time, very expensive experiments. For example, during very expensive destructive testing of structural materials, one would use theoretical models for optimal experimental setup including sensor selection, sensor spacing, temporal sampling, camera orientation, or camera distance (raster data resolution).

The two fundamental issues of data fusion are about resolving data heterogeneities and about selecting the "best" measurement in terms of accuracy and uncertainty. Thus, a theoretical model of data fusion processes incorporates models of all fused variables from multiple input data sources, and an uncertainty/accuracy model for selecting the final value fused from all input measurements with minimum uncertainty and maximum accuracy. It is usually more difficult to model data uncertainty throughout a fusion process than to model the individual variables. We outline one of the approaches to modeling data fusion by estimating sensor data uncertainty [11,12].

19.4.1.1 Data Fusion Model

Let us consider a fusion scenario illustrated in Figure 19.7. We assume that an uncertainty model is composed of error contributions due to (1) each sensor by itself, for example, sensor noise, (2) transformations of measured values to obtain comparable physical entities for data fusion and to calibrate sensor measurements, (3) spatial interpolation that is needed to match different spatial resolutions of multisensor data or to match geometric disagreements of multiple sensors, and (4) temporal interpolation that has to take place if multisensor acquisitions are not accurately synchronized.

The mathematical description of the proposed global uncertainty model ε is presented in Equation 19.18.

$$\begin{aligned}\varepsilon(x,t,\lambda^{\text{input}}) = f(&\varepsilon^{\text{sensor}}(x,t,\lambda^{\text{input}}), \varepsilon^{\text{transformation}}(\lambda^{\text{input}}, \Psi(\lambda^{\text{input}})),\\ &\varepsilon^{\text{spatial adjustment}}(x), \varepsilon^{\text{temporal adjustment}}(t))\end{aligned} \tag{19.18}$$

where λ^{input} is a set of the measured physical entities, x is a spatial location in the local coordinate system, t is a time frame, ψ is a set of variable transformation functions, and f is a function of the order of the data fusion steps. The variables ε^X correspond to the uncertainty contributions due to sensor device, transformation of input variable to output variable, spatial adjustment, and temporal adjustment. The fusion of raster and vector data

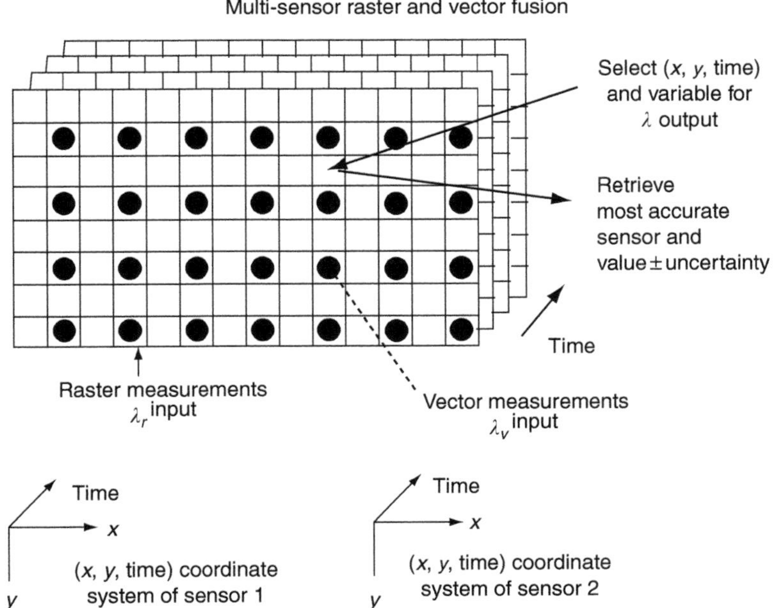

FIGURE 19.7
An example of multisensor raster and vector data fusion. Vector measurements are acquired by a set of point sensors and raster measurements come from a camera.

is performed by minimizing the uncertainty of measurements λ_i^{input} at any spatial location x and time instance t after performing all necessary transformations $\psi_i(\lambda^{\text{input}})$ to obtain comparable physical entities λ^{output}. Equation 19.19 describes the fusion criterion.

$$\lambda^{\text{output}}(x,t) = \psi_k(x,t,\lambda_k^{\text{input}}), \qquad k = \arg\min_i(\varepsilon_i(x,t,\lambda_i^{\text{input}})), \tag{19.19}$$

where $\lambda_i^{\text{input}} \in \lambda^{\text{input}}$, $0 \le i \le$ number of sensors, λ_i^{input} are different input physical entities and each linear function ψ_i ($\psi_i \in \psi$) associated with λ_i^{input} converts the input entity to a comparable (output) physical entity which is to be fused into λ^{output}.

19.4.2 Example of Raster and Vector Data Fusion

Let us analyze a specific case of raster λ_r^{input} and vector λ_v^{input} data fusion by considering one dataset of each type as illustrated in Figure 19.8. In this case, the heterogeneity of data occurs due to different (1) sensor coordinate systems, (2) temporal sampling (synchronization), (3) measured variables, and (4) data types (raster vs. vector). These data heterogeneities can be resolved by (1) finding a coordinate system transformation to register datasets, (2) identifying a common time instance and applying temporal resampling, (3) performing analytical transformations of measured variables, and (4) fitting spatial interpolation/extrapolation models to vector data and applying spatial resampling to match raster data resolution.

Although resolving data heterogeneities is necessary for data fusion, each data operation changes uncertainty associated with a spatial and temporal sensor measurement and hence uncertainty modeling has to include all data processing steps. Figure 19.8 also shows the uncertainty model components associated with each processing step of the data fusion

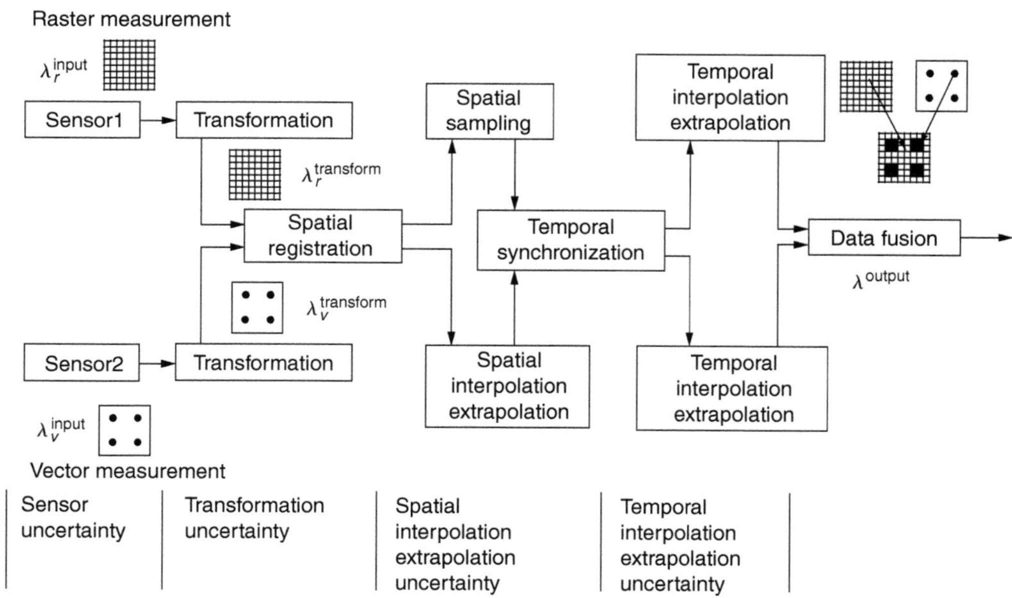

FIGURE 19.8
An overview of raster and vector data fusion process and associated uncertainty model components.

process. While some of the processing steps can occur multiple times during data fusion, for instance, vector data transformation before and after spatial registration, other processing steps might occur only with raster or vector data but not with both, for example, spatial interpolation of vector data or spatial sampling of raster data.

Let us assume that uncertainty of raster or vector measurements at a measured spatial grid (raster data) or point locations (vector data) is constant and always smaller than uncertainty at any other location, for example, locations between two measured points. According to the fusion criterion in Equation 19.18, the fusion rule for raster and vector input data is described in Equations 19.20 and 19.21.

$$\lambda^{\text{output}}(x,t) = \begin{cases} \lambda_v^{\text{output}}(x,t): & \text{if } \varepsilon_r(x,t,\lambda_r^{\text{input}}) > \varepsilon_v(x,t,\lambda_v^{\text{input}}), \\ \lambda_r^{\text{output}}(x,t): & \text{otherwise} \end{cases} \tag{19.20}$$

$$\varepsilon(x,t,\lambda^{\text{input}}) = \min(\varepsilon_v(x,t,\lambda_v^{\text{input}}), \varepsilon_r(x,t,\lambda_r^{\text{input}})) \tag{19.21}$$

where $\lambda_r^{\text{output}}$ and $\lambda_v^{\text{output}}$ are the *processed* raster and vector data values in the registered coordinate system, ε_r and ε_v represent the uncertainty of raster and vector data respectively after processing (e.g., variable or coordinate transformations, spatial and temporal interpolations and etc.). By developing mathematical uncertainty models for ε_v and ε_r, and combining the models with the fusion rule, we can create an uncertainty mask shown in Figure 19.9. This image mask shows regions that contain measurements with the lowest uncertainty from vector and raster sensors (dark circular shapes in the image mask denote smaller uncertainty in raster data). Given the uncertainty mask, the raster and vector data fusion can be achieved by value selection according to the mask labels.

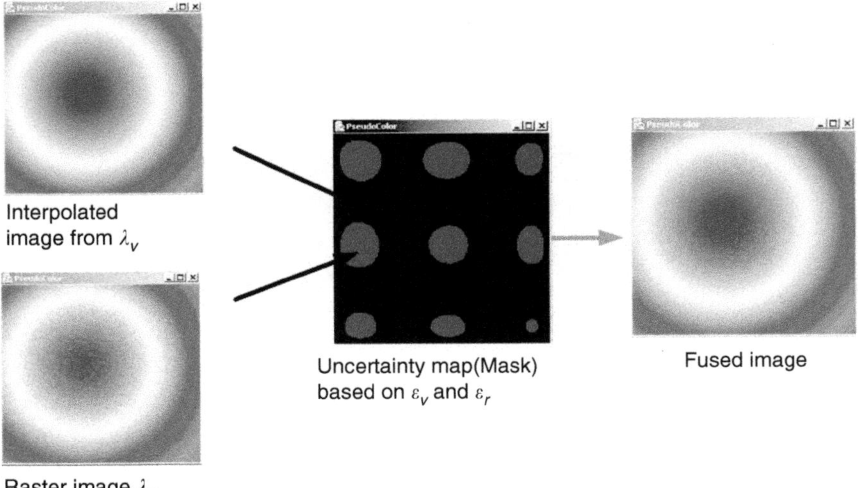

Figure 19.9
Vector and raster data fusion process using an uncertainty mask.

Figure 19.10 shows simulated fusion experiments with varying amount of raster noise. Other simulations with varying sample spacing, types, and amounts of sensor noise, measurement variable transformations for any underlying physical model, lead to design of more optimal data fusion systems.

19.5 Summary

The underlying motivation behind this chapter was the fact that using multiple data sources together leads to more information than using either of them separately. We have therefore overviewed (1) spatial interpolation with kriging, (2) raster and raster data integration (shallow integration of maps), and (3) raster and vector data fusion (fusion of images and point sets). We outlined spatial kriging as one of many methods used for vector data interpolation in a GIS domain. As a part of raster and raster data integration, we emphasized the importance of an appropriate quality metric that is usually driven by application requirements. As a part of raster and vector data integration, we showed how uncertainty analysis could be used for theoretical modeling of raster and vector data fusion.

Based on the reading of the material presented here, one can immediately understand why the development of automated or semiautomated data fusion systems is an open problem in general. Our ultimate goal is to develop a "fully automated" data fusion system, where a user specifies N input files with georeferencing information and the software would output a dataset with maximum spatial and temporal information content, and maximum accuracy and minimum uncertainty per output value. To achieve this goal, we have introduced theoretical frameworks for addressing the raster–raster and raster–vector data integration problems and illustrated the necessary links to such system implementations. Mathematical frameworks are needed to rigorously model each data fusion step since a final model of data fusion processes is an ordered sequence of preprocessing and decision-making operations on input datasets. By incorporating application requirements and any

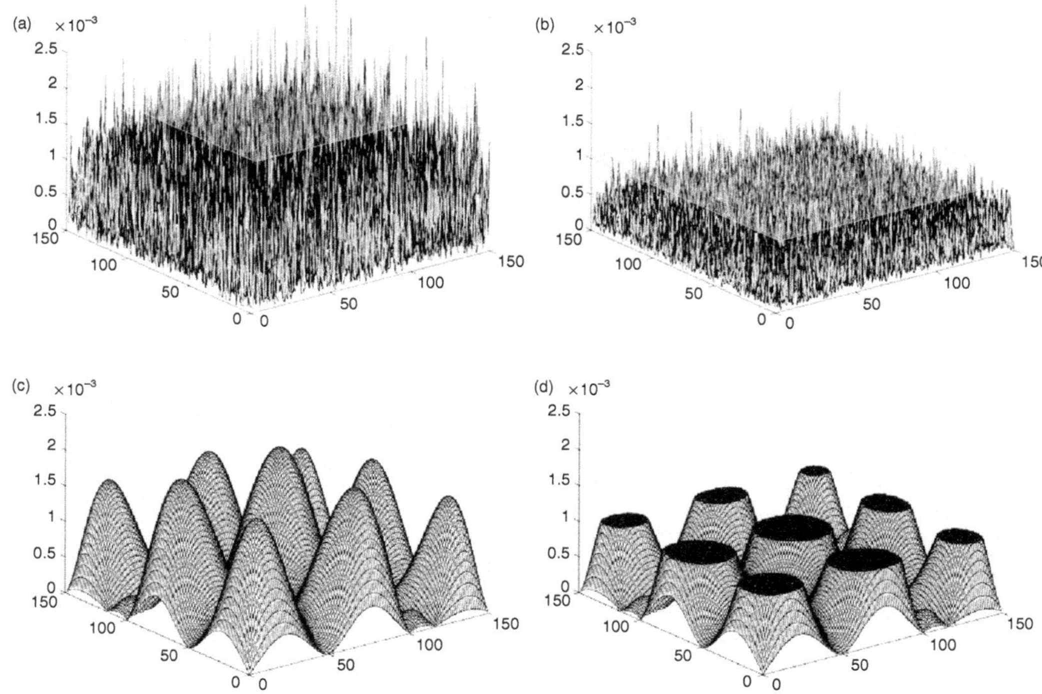

FIGURE 19.10
Simulations of the uncertainty of raster data with Gaussian sensor noise — (a) $3\sigma_r = 0.002123$ and (b) $3\sigma_r = 0.0010$. The uncertainty of B-spline interpolated vector data (c) with spacing 50 times larger than the raster data and zero sensor noise. The fusion mask (d) is obtained by thresholding the uncertainty of vector data at 0.001 since 99.73% of noisy raster values fall within $0 \pm 3\sigma_r$ [36].

computational constraints (e.g., real-time processing, limited CPU resources) into the data fusion model, one can aim at designing optimal data fusion systems.

References

1. Space Imaging, Extensive collection of remote sensing imagery for a fee: http://www.spaceimaging.com.
2. National Aeronautics and Space Institute (NASA), 2004. Freely available LandSAT data, URL: https://zulu.ssc.nasa.gov/mrsid/ (last date accessed: March 24, 2004).
3. Federal Geographic Data Committee (FGDC), 1998. Content standard for digital geospatial metadata, URL: http://www.fgdc.gov/standards/documents/standards/metadata/v2_0698.pdf (last date accessed: March 24, 2004).
4. United States Geological Survey (USGS), 2004. Imagery for U.S. states and principalities, URL: http://www.usgs.gov (last date accessed: March 24, 2004).
5. Markowitz, V.M., J. Cambell, I.A. Chen, A. Kosky, K. Palaniappan, and T. Topaloglou, 2003. Integration Challenges in Gene Expression Data Management – chapter 10. In Lacroix Z. and T. Critchlow, editors, Bioinformatics: Managing Scientific Data, Managing Scientific Data, Morgan Kaufmann (Elsevier Science), San Francisco, CA, pp. 277–301, September 2003, p. 300.

6. Jagadish, H.V. and F. Olken, Database managament for life science research: summary report of the workshop on data managament for molecular and cell biology at the National Library of Medicine, Bethesda, MD, February 2–3, 2003. *OMICS Journal of Integrative Biology*, 7, 2003, pp. 131–137.
7. Oxley, M.E. and S.N. Thorsen, Fusion or integration: what's the difference?, in *Proceedings of the 7th International Conference on Information Fusion*, Stockholm, Sweden, June 28–July 1, 2004, pp. 429–434.
8. Klein, L. A., Sensor and Data Fusion, A tool for information assessment and decision making. SPIE Press, Belligham, Washington, 2004, p. 318.
9. Hall, D. and J. Llinas, An introduction to multisensor data fusion, *IEEE Proceedings: Special Issue on Data Fusion*, 85, 1997, 6–23.
10. Shao, G., H. Zhu, and W.L. Mills, Jr., An algorithm for automated map mosaicing prior to georegistration, *Geographic Information Sciences*, 6, 2000, 97–101.
11. Lee, S. and P. Bajcsy, Multisensor raster and vector data fusion using uncertainty modeling, in *IEEE International Conference on Image Processing*, Singapore, 2004.
12. Lee, S. and P. Bajcsy, Multi-instrument analysis from point and raster data, Technical report NCSA-ALG-04-0002, February 2004.
13. Bajcsy, P., J. Han, L. Liu, and J. Young, Survey of bio-data analysis from data mining perspective, in Jason T.L. Wang, Mohammed J. Zaki, Hannu T.T. Toivonen, and Dennis Shasha, Eds., *Data Mining in Bioinformatics*, Springer-Verlag, Heidelberg, 2004, pp. 9–39, chap 2.
14. Urban, M., P. Bajcsy, R. Kooper, and J.-C. Lementec, Recognition of arm gestures using multiple orientation sensors: repeatability assessment, in *Proceedings of the 7th International IEEE Conference on Intelligent Transportation Systems*, Washington, DC, October 3–6, 2004, pp. 553–558.
15. Lementec, J.-C. and P. Bajcsy, Recognition of arm gestures using multiple orientation sensors: gesture classification, in *Proceedings of the 7th International IEEE Conference on Intelligent Transportation Systems*, Washington, DC, October 3–6, 2004, pp. 965–970.
16. Kak, A. and A. Kosaka, Multisensor fusion for sensory intelligence in robotics, in *Proceedings of Workshop on Foundation of Information/Decision Fusion: Applications to Engineering Problems*, Washington, DC, August 7–9, 1996.
17. ESRI Web site: http://www.esri.com/.
18. ERDAS IMAGINE software package by Leica Geosystems, URL: http://gis.leica-geosystems.com/.
19. McIDAS (Man computer Interactive Data Access System) by the University of Wisconsin-Madison's Space Science and Engineering Center, URL: http://www.ssec.wisc.edu/software/mcidas.html.
20. Research Systems, Inc., The Environment of Visualizing Images (ENVI), URL: http://www.rsinc.com/envi/index.cfm.
21. Bajcsy, P., et al., Image to knowledge, documentation at URL: http://alg.ncsa.uiuc.edu/tools/docs/i2k/manual/index.html.
22. Ahmed, S. and de Marsily, G., Comparison of geostatistical methods for estimating transmissivity using data on transmissivity and specific capacity. *Water Resources Research*. Vol. 23, pp. 1717–1737, 1987.
23. Krige, D.G., A statistical approach to some basic mine valuation problems on the Witwatersrand. *Journal of the Chemical, Metallurgical and Mining Society of South Africa*, Vol. 52, pp. 119–139, 1951.
24. Cressie, Noel, A.C. *Statistics for Spatial Data*, John Wiley & Sons Inc., New York, 1993.
25. American Society of Civil Engineers (ASCE), American Congress on Surveying and Mapping (ACSM), and the American Society for Photogrammetry and Remote Sensing (ASPR), *Glossary of Mapping Sciences*, 1994.
26. Kokar, M.M. and J. Tomasik, Towards a formal theory of sensor/data fusion, Technical report: COE-ECE-MMK-1/94, May 1994.
27. Hall, D.L., *Mathematical Techniques in Multisensor Data Fusion*, Artech House, Boston, London, 1992.

28. Ren, C.L. and G.K. Michael, Multisensor integration and fusion in intelligent systems, *IEEE Transactions on Systems, Man and Cybernetics*, 19, 1989, 901–931.
29. Chang, C.C. and Keisler, H.J. *Model Theory*, North Holland, Amsterdam, New York, Oxford, Tokyo, 1992.
30. Alumbaugh, T.J. and P. Bajcsy, Map mosaicking with dissimilar projections, spatial resolutions, data types and number of bands, in *ASPRS 2004 Annual Conference*, Denver, CO, May 23–28, 2004.
31. Flugel, W.A. and B. Muschen, Applied remote sensing and GIS integration for model parameterization (ARSGISIP), in *Proceedings of the 27th International Symposium on Remote Sensing of Environment*, Tromso, Norway, June 8–12, 1998, pp. 354–357.
32. Bouman, C. and M. Shapiro, Multispectral image segmentation using a multiscale image model, in *Proceeding of IEEE International Conference of Acoustics, Speech and Signal*, San Francisco, CA, March 23–26, 1992, pp. III-565–III-568.
33. Bajcsy, P. and S. Saha, A new thermal infrared camera calibration approach using wireless MEMS sensors, in *Communication Networks and Distributed Systems Modeling and Simulation Conference (CNDS 2004)*, San Diego, CA, January 19–22, 2004.
34. Krypton System, K610 CMM, by Krypton Inc., URL: http://www.krypton.be.
35. Grey-field polariscope photoelastic strain measurement system, by Stress Photonics Inc., URL: http://www.stressphotonics.com.
36. Kenney, J.F. and E.S. Keeping, The standard deviation and calculation of the standard deviation. 6.5-6.6, in *Mathematics of Statistics, Part 1*, 3rd ed., Princeton, Van Nostrand, NJ, 1962, pp. 77–80.
37. Kerr, J.R., D.P. Pond, and S. Inman, Infrared-optical multisensor for autonomous landing guidance, *Synthetic Vision for Vehicle Guidance and Control*, Vol. 2463, in *Proceedings of SPIE*, 1995, pp. 38–45.

20

Feature Extraction

Peter Bajcsy

CONTENTS

20.1 Introduction

At this point, all datasets have been acquired, spatially registered, and temporally synchronized, and possibly fused into an integrated dataset. We are looking for features that are critical for our application but are not directly measured, and therefore cannot be retrieved from raw data without processing the data. Our goal in this chapter is to overview the problems of feature extraction from a variety of data sets.

The meaning of the word "feature" (or attribute) is usually understood as a physical measurement or category associated with a spatial location at a certain time instance or interval. Nonetheless, every application domain has a different definition for features, regions of interest, or objects in question. For example, a feature extraction definition could be adapted according to the ERDAS Glossary (URL: http://support.erdas.com/Glossary.htm) as "The process of studying and locating areas and objects on the ground and deriving useful information from images." Another definition could be found in the Photonics Dictionary (URL: http://www.photonics.com/dictionary/), where feature extraction is defined as "in image processing and machine vision, the process in which an initial measurement pattern or some subsequence of measurement patterns is transformed to a new pattern feature." In image pattern recognition, features often contain information relative to gray shade, texture, shape, or context." The feature definition for this chapter could be formulated as follows: A feature is a cluster of points or a boundary/region of pixels that satisfy a set of predefined criteria. The criteria can be based on any descriptive quantities, such as *n*-dimensional shapes, spectral characteristics, orientations, and spatial distributions of points, boundaries, or regions. We will primarily focus on extracting features from point and boundary/raster data types.

Features are classified based on (1) their physical meaning as temporal, structural, geometrical, spectral, frequency-domain, and statistical features (see Section 3.2, Reference 1), (2) their values as continuous (elevation) and categorical (forest labels), and (3) their dimensionality as points, lines, or regions. There are many examples of these features. For instance, wildlife biologists are interested in density distributions of nests of eagles and hawks derived from point measurements, as well as, plant biologists are interested in distributions of tree types (spatial/geometrical features). Researchers in socioeconomics need to relate information about geographical variables (e.g., elevation statistics over man-made boundaries) with economics variables to understand development of our society and predict future changes. Similarly, it is important for any government to learn the temporal rate of a spreading disease (temporal features) from temporal sequences of point occurrences in order to establish preventive measures, for example, spread of soybean rust in agriculture or spread of AIDS from large cities to rural areas. As an example of spectral features, seasonal changes of agricultural field appearance computed from hyperspectral image are of interest to farmers since they indicate the crop health and future crop yields.

Additional aspects of feature extraction include heterogeneity of input data and computational complexity. One could mention the problem of analyzing complex maps consisting of raster data, textual legends, map symbols, and hand annotations as an example of heterogeneity of input data. Computational complexity of feature extraction could be due to a large feature search space or due to a complex feature model. While we gave only Geographical Information System (GIS) related examples, there exist many more feature examples from other domains, such as shape descriptors to classify characters in Optical Character Recognition (OCR), circular features to represent SAR scattering centers in object recognition from SAR imagery, arcs to localize semiconductor wafers in computer chip manufacturing, and so on.

From an application viewpoint, the motivation for feature extraction could be (1) data compression, (2) data analysis, or (3) data synthesis. Clearly, storing a road network extracted from aerial ortho-photographs is more efficient than storing the raw photographs. Data analysis and synthesis could be viewed as data characterizations in order to discover, understand, and predict physical phenomena. For example, clustering satellite spectral imagery leads to discovering spatial regions with similar properties. Classifying clusters into known categories could provide land use and land cover maps, and hence land understanding that is so important for farmers and urban developers. In next phase of understanding various phenomena, feature characteristics learned from data analyses can be used for modeling and prediction, for example, predicting hazardous tornadoes or earthquakes. It is also very frequent that the motivation for feature extraction is to perform a preprocessing step for further analysis. This type of motivation is found in visual tracking during searches of large collections of datasets (efficient retrieval), or in visualization (reduction of visual clutter). In general, the motivation of feature extraction is to transform measurements from one into another in order to (1) compress data or (2) characterize data.

The list of applications using feature extraction could be very long. We selected only those that would allow us to illustrate the multifaceted spectrum of possibly encountered feature extraction issues. First, is the issue of data quality, for example, when extracting features from historical maps for restoration projects [2,3]. Second, the issue of real-time performance requirements, for example, defect detection in semiconductor wafer analysis and the challenges in terms of processing time. Third, feature extraction from large datasets as encountered in remote sensing. Fourth, the issue of large cost found in earthquake structural engineering simulations. Fifth, the issue of measurement and processing uncertainty, for example, 3D volume reconstruction from microscopy images of cross sections in plant biology or medicine. Finally, the issue of feature complexity and corresponding feature extraction robustness, as frequently addressed in robotics and machine vision applications. The last issue could be illustrated by considering the problem of straight-line extraction for a known line orientation from images acquired under constant illumination in comparison with the problem of curvilinear feature extraction for lines with arbitrary orientation, curvature, and images acquired under varying illumination.

In terms of existing solutions for feature extraction from multiple types of input data, there does not exist a software package that would meet requirements of all applications. One of the most commonly used software solutions for image feature extraction would be Adobe Photoshop. In the GIS domain, it would be Environmental Systems Research Institute (ESRI) ArcGIS [4]. Other considerations about feature extraction software would include automation and software maturity. For example, the feature extraction software for OCR could be viewed as well as automated and mature, while software for face recognition is still in a research stage due to the problem complexity.

This chapter is organized according to the feature dimensionality classification and with respect to input data type. This organization is preferred for algorithmic design since feature extraction algorithms have to overcome dimensionality and input data type specific requirements. For instance, how to derive water quality features differs depending on if it is from point measurements (a set of water gauges) or from boundary measurements (characteristics of soil penetration by water) or from raster measurements (images of vegetation greenness). Sections 20.2 and 20.3 present a few specific examples of feature extraction problems from point and raster data to outline their scientific challenges and describe several approaches.

20.2 Feature Extraction from Point Data

Given a set of point data, one has to know the formula for extracting features. The formula could be a mathematical expression or a descriptive feature that provides discrimination for clustering or classification. We give examples of such feature formulas.

Let us consider the problem of extracting temporally changing horizontal, vertical, and diagonal stress and shear (features). The features are defined by mathematical formulas and the input dataset consists of a temporal sequence of point locations (see Figure 20.1).

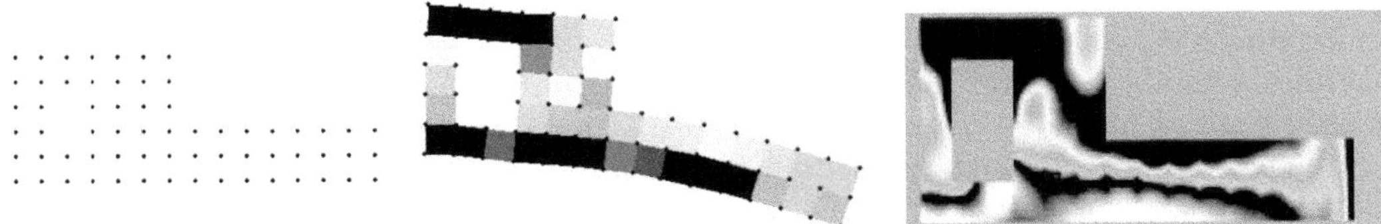

FIGURE 20.1
An example of a temporal sequence of point locations (left) that can be used for deriving stress and shear features according to mathematical formulas. The sensors are attached to an L-shaped structure that is being pulled down as illustrated by sensor locations in the middle. The features are computed from discrete measurements for four spatially neighboring points (middle) and are interpolated to form continuous features by a B-spline model (right).

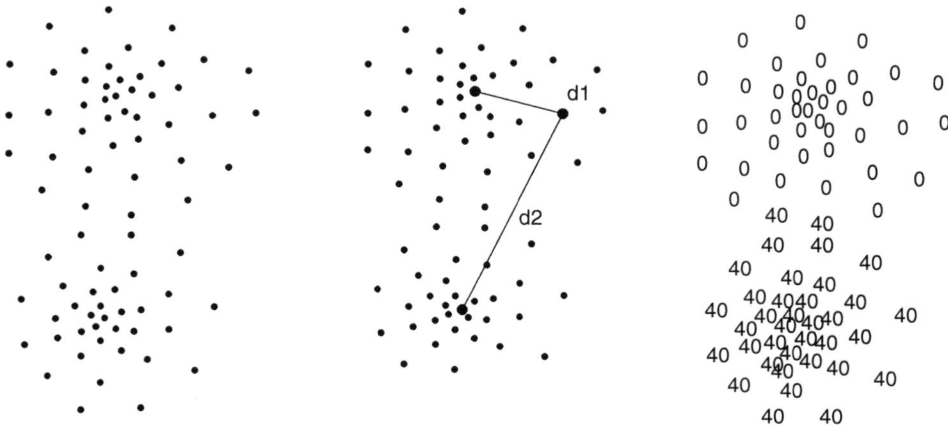

FIGURE 20.2
An example of two clusters (left) where a centroid of each cluster could be the representative point (center) and the distance of any point from the two centroids becomes the feature for label assignment (right). The label assignments are obtained by using a centroid based clustering method [5].

The location measurements are obtained from a set of point sensors attached to a 2D surface of a concrete structure under an increasing structural load. The computation of stress and shear features is straightforward assuming that the material loading is piecewise continuous. The challenges arise from the nature of these features that require (1) C^2 spatial continuity, (2) knowledge of neighboring points, (3) consistent definition of point coordinate systems, and (4) statistical noise models of point sensors. The definitions of stress and shear features are based on derivatives of point locations and therefore spatial continuity of variables is assumed. However, this assumption is violated at places where cracks and fissures occurred during an experiment. Thus, the feature extraction is undefined at those locations.

Next, let us assume that we are looking for descriptive features of a set of points. The most common descriptive feature used in pattern recognition would be (1) a distance of any point from a set of reference points (e.g., representative points of clusters), (2) mutual distances of all pairs of points, and (3) point density. Figure 20.2 and Figure 20.3 show examples of 2D point patterns, where clusters of points could be separated based on the descriptive feature type (1) and (2). There also exist more complex 2D point patterns that require finding and extracting descriptive features for (1) overlapping clusters of points, (2) clusters with orientation dependent point density, or (3) clusters of different shapes [6,7,11]. A few examples of more complex 2D point patterns are illustrated in Figure 20.4 and Figure 20.5.

While extracting descriptive features, a distance would be defined by one of the standard norms, such as Minkowski (see equation [20.1]) and Minkowski's special cases known as Euclidean ($p = 2$), and Manhattan ($p = 1$, also called City block), Mahalanobis (weighted Euclidean with weight equal to inverse covariance matrix R, see equation [20.2]), Bhattacharyya (generalized Mahalanobis distance with unequal covariance matrices, see equation [20.3]), and Chernoff (generalized Bhattacharyya distance with weighting influence of unequal covariance matrices, see equation [20.4]) [1, Section 3.2]. In these equations, X and Y are vectors defined as $X = (x_1, x_2, \ldots, x_n)$; $Y = (y_1, y_2, \ldots, y_n)$; R is the covariance matrix; R_X, R_Y are unequal covariance matrices for vectors X and Y, s is the parameter that allows weighted influence of unequal covariance matrices, and p is the order of Minkowski's

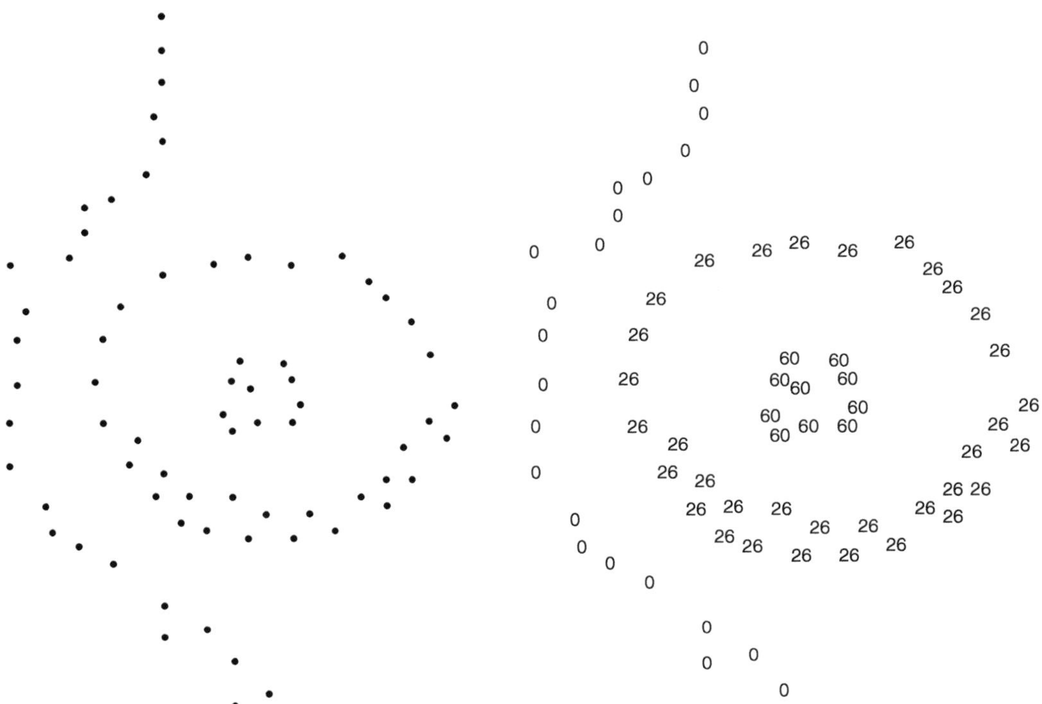

FIGURE 20.3
An example of three clusters (left), where mutual distances of all pairs of points become the features for label assignment (right). The label assignments are obtained with a clustering method using similarity analysis [6].

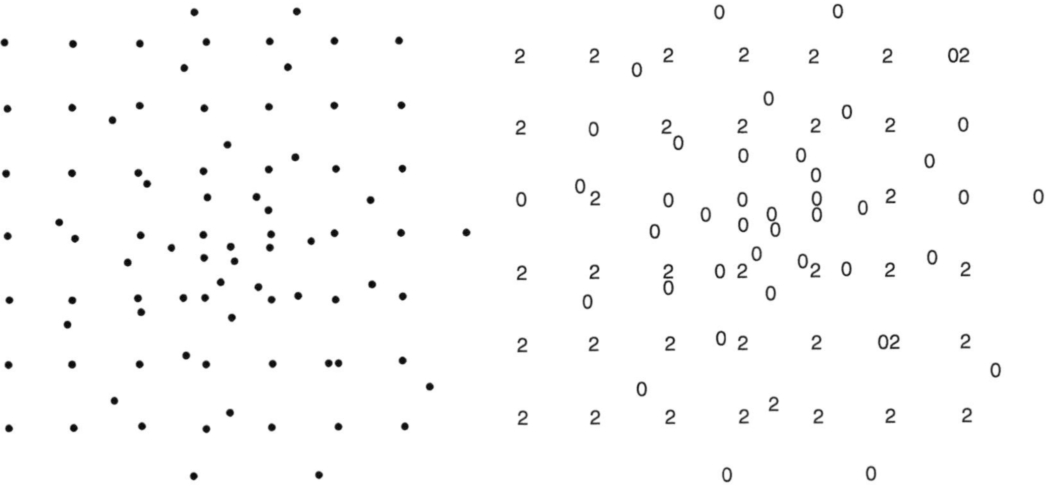

FIGURE 20.4
A spatially overlapping regular grid with a set of points distributed according to a Gaussian PDF (left). The label assignments (right) are obtained with a clustering method using similarity analysis [6].

FIGURE 20.5
Other challenging examples of 2D point patterns where features have to be defined based on our visual perception. Features that discriminate two clusters with orientation dependent density changes (left) or shape features (right).

metric. Choosing an appropriate distance norm for a given problem would be one of the user's decisions.

$$\text{Minkowski:}\quad D(X,Y) = \{(X-Y)^p\}^{1/p} = \left[\sum_{i=1}^{n} |x_i - y_i|^p\right]^{1/p} ;\quad p \in [1,\infty) \tag{20.1}$$

$$\text{Mahalanobis:}\quad D(X,Y) = (X-Y)^T R^{-1}(X-Y) \tag{20.2}$$

$$\begin{aligned}\text{Bhattacharyya:}\quad D(X,Y) &= \frac{1}{8}(X-Y)^T\{[R_X+R_Y]/2\}^{-1}(X-Y)\\ &+ \frac{1}{2}\ln\{[R_X+R_Y]/2\}/\{|R_X|^{1/2}+|R_Y|^{1/2}\}\end{aligned} \tag{20.3}$$

$$\begin{aligned}\text{Chernoff:}\quad D(X,Y) &= \frac{1}{2}s(1-s)(X-Y)^T\{sR_X+(1-s)R_Y\}^{-1}(X-Y)\\ &+ \frac{1}{2}\ln\{|sR_X+(1-s)R_Y|\}/\{|R_X|^s|R_Y|^{1-s}\};\quad s \in (0,1)\end{aligned} \tag{20.4}$$

Another consideration for users might be the computational cost of feature extraction and clustering. For example, for N points, there are $N(N-1)/2$ distances to compute. Thus, this number increases proportional to N^2 and might require significant computational resources for large N. Among other challenges, one should be aware of the fact that visually perceived features might be hard to define and compute. It is also hard to visualize high-dimensional point patterns and to apply feature extraction algorithms in higher dimensional spaces. Finally, a parameter optimization has to be considered as a part of the feature computation. For instance, optimization of an area size for point density estimation or selection of similarity values when multiple cluster solutions are plausible.

20.3 Feature Extraction from Raster Data

Let us consider 2D images as input raster data that contain information about objects of interest. The goal of feature extraction is to characterize these objects by (1) direct raster values (raster location or attribute), (2) derived raster measurements (statistical attribute values, shape, or distance measurements), or (3) a combination of both. Although there is

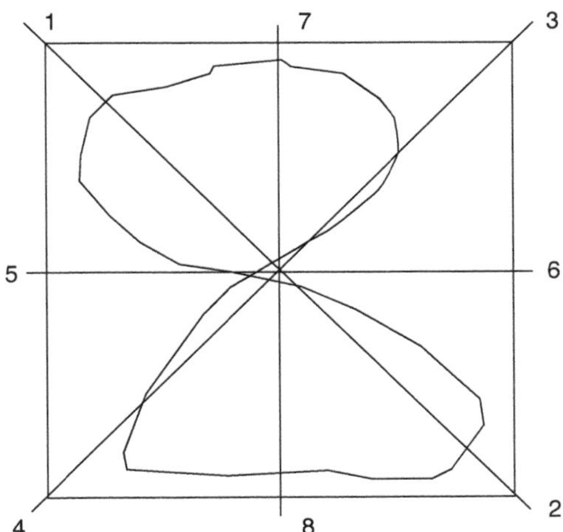

FIGURE 20.6
Features extracted from a segmented number 8.

a gamut of possible features to extract from raster data, we selected four distinct types of features that could be used for characterizing objects of interest in raster data and would illustrate the spectrum of problems. These four feature types are (1) distance measurements, (2) point measurements, (3) curvilinear characteristics, and (4) region characteristics. To explain the four feature types, we provide four applications including (1) OCR, (2) cell density for plant biology, (3) isocontour extraction from historical maps, and (4) region statistics for decision support. We will present these applications together with feature extraction problems from Section 20.3.1 to 20.3.4.

20.3.1 Distance Measurements

Given an image of a text document, our goal is to recognize individual characters. A typical OCR program would segment out characters and extract features describing each character for classification. For example, one of the test datasets for clustering algorithms is the 80X dataset. This dataset contains eight distance measurements computed according to Figure 20.6 and extracted from each of the three characters 8, 0, and X.

20.3.2 Point Measurements

In plant biology, it is important to extract information about plant cell density from plant cross sections. One of the frequent questions would be about the number of cells in a given cross section. Assuming that cells are color homogeneous, this information can be extracted by segmenting images and counting the number of segments while excluding segments originating from the noisy background. If it is desirable to obtain information about spatially varying cell density then segments can be replaced by their centroids (point features) and clustered based on their density [7]. This process is illustrated in Figure 20.7.

20.3.3 Curvilinear Characteristics

Let us assume that an environmental restoration project requires learning about the past terrain relief, and historical paper maps are available for this purpose [2]. In this case, the goal

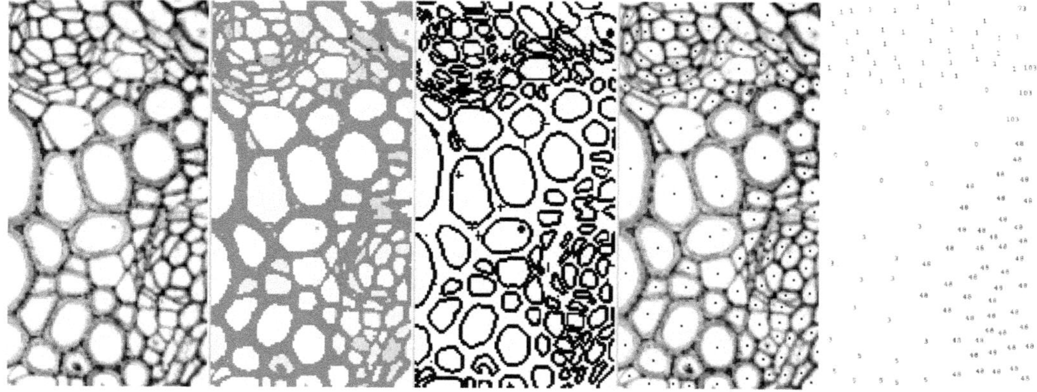

FIGURE 20.7
A photograph of a plant cross section (left) that is segmented into color homogeneous regions (second from left). For each segment illustrated with its contour (third from left), the centroid feature is computed and overlaid with the original image (second from right). Finally, the points formed from centroid features are clustered and labels are assigned to each point.

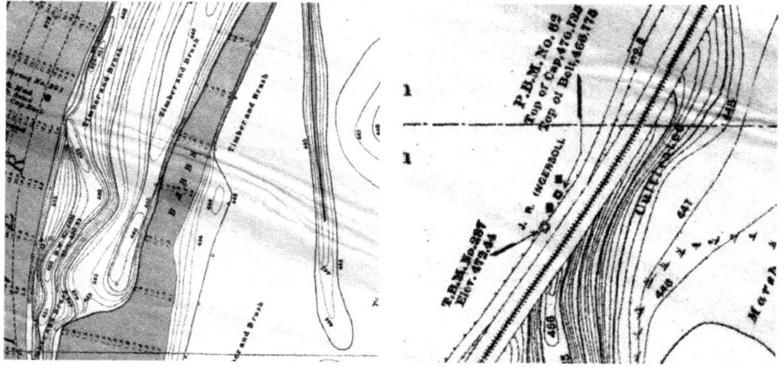

FIGURE 20.8
Examples of map subareas illustrating the algorithm robustness challenges.

is to automate an extraction of isocontours from the historical maps that are scanned into a digital format. The isocontours represent horizontal cross sections of 3D terrain at equal elevations [8] and can be classified as curvilinear image characteristics [5]. This problem becomes challenging since an isocontour extraction algorithm has to be robust to (1) within contour class variations (curvature variations, contour gaps, intensity noise, or spatially dense isocontours), (2) across class similarities of isocontours with other curvilinear features (dashed lines in a river bed, railroad lines, latitude, and longitude straight lines or edges of a georegistered map), and (3) other background clutter conditions (touching text or other map symbols, intersecting contours with other lines or map artifacts, such as shading due to creases and folds). Examples of challenges for robust isocontour extraction are shown Figure 20.8.

One possible approach to the problem is to decompose it into (1) finding an initial contour point and contour direction, (2) tracing a contour, and (3) extracting all isocontours in large historical maps automatically as proposed in Reference 3. The proposed method is based on a piecewise linear approximation of isocontours and can operate in semiautomated (user selects initial points) or fully automated mode (Figure 20.9). However, the trade-offs

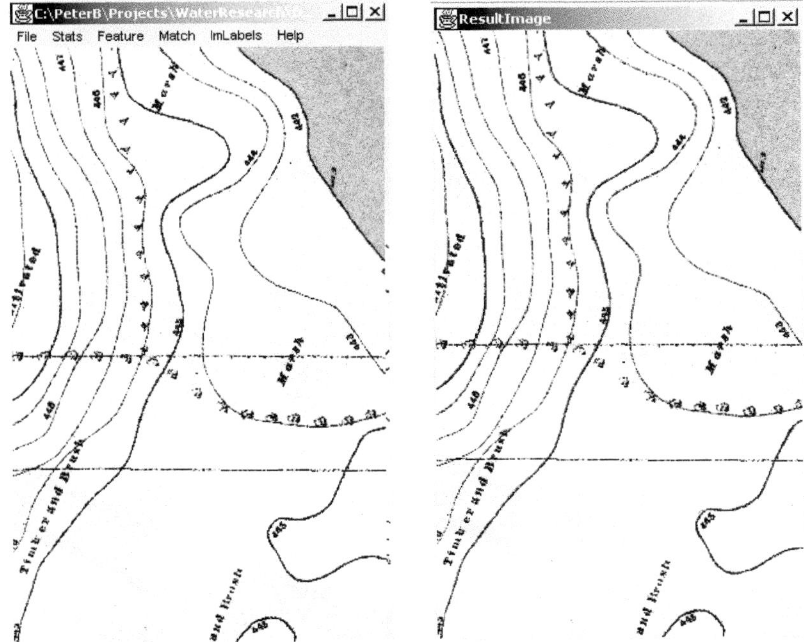

FIGURE 20.9
Example results obtained by using a semiautomated isocontour extraction method (a user selected an initial contour point). A subarea of scanned historical map (left) and the color coded isocontours as extracted by the algorithm (right) in Reference 3.

between maximum curvature and robustness to background clutter remain one part of the algorithmic optimization.

20.3.4 Region Characteristics

Our task is to support decision makers as described in the motivation example at the beginning of Chapter 14. For every region defined as a polygon, we have to extract statistics about elevation and forest cover, which can be obtained by georegistering region boundaries with digital elevation maps and with forest label maps, and then computing statistics from all map values that belong to each region.

This seemingly simple statistical feature extraction task needs careful consideration of all processing steps. First, region boundary points might be defined using a different geographic projection than the elevation and forest label maps. Additional challenge might arise due to different file formats and conflicting (or missing) georeferencing information in those files. Second, given coregistered map and boundary datasets, map point membership with respect to each region boundary has to be established (see [13], Section 3.6). Although there exist multiple valid mathematical approaches to this problem, according to our knowledge there is only one implementation solution that can robustly assign map pixel membership for convex and nonconvex region boundaries. This approach is based on incremental painting of all parts of a region (exterior region boundary and boundaries of holes) and computing the membership mask based on the number of painted layers. The membership mask is an image of the same size as the input map with each pixel labeled with a unique, region-specific color (see Figure 20.10 and Figure 20.11). Third, depending

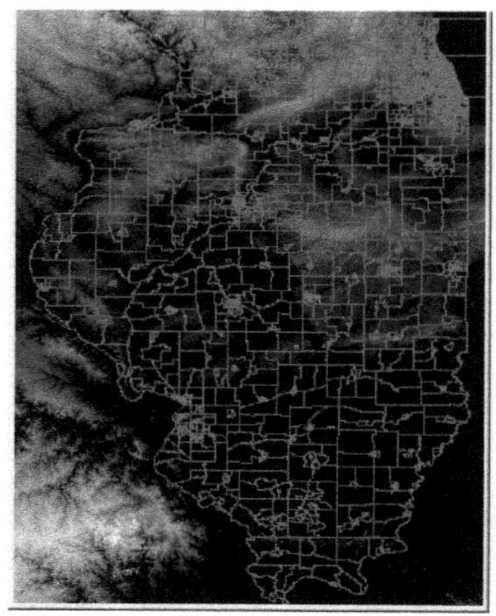

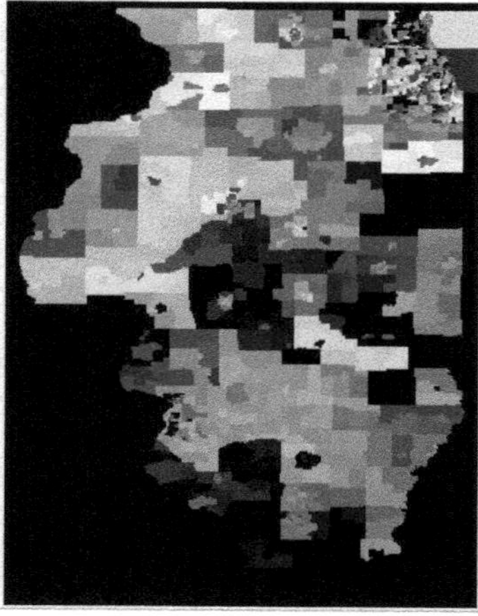

FIGURE 20.10
Georeferenced raster file of elevations with boundaries of Illinois U.S. Census Bureau Tracks (left) and the corresponding label image (right) used for computing statistics.

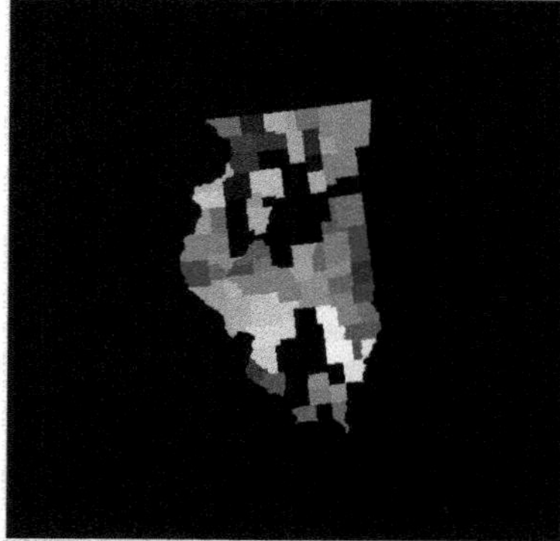

FIGURE 20.11
Georeferenced raster file of forest categories with boundaries of Illinois counties (left) and the corresponding label image (right) used for computing statistics.

on whether the input map represents a continuous variable (e.g., elevation) or a categorical variable (e.g., forest category label), the output statistics are computed as central moments for the continuous variable case (see Figure 20.12) or frequencies of label occurrences for the categorical case (see Figure 20.13).

20.4 Summary

In this chapter, we could not cover the whole spectrum of feature extraction problems because most of the time feature extraction problems are very much application domain specific. It is often the case that feature extraction is tightly coupled with the next processing step, such as clustering, classification, or other type of modeling or decision support. We provided classification of features based on the dimensionality and type of input data and presented a few specific examples of feature extraction problems from point and raster data.

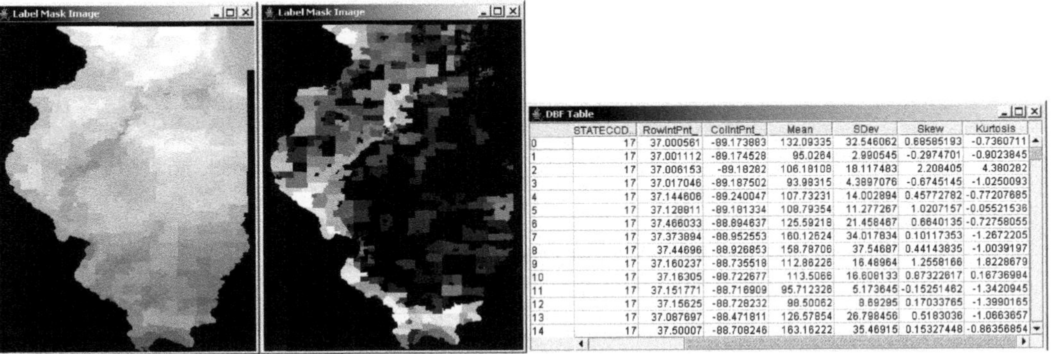

	STATECOD..	RowIntPnt_	ColIntPnt_	Mean	SDev	Skew	Kurtosis
0	17	37.000561	-89.173883	132.09335	32.546062	0.68585193	-0.7360711
1	17	37.001112	-89.174528	95.0264	2.990545	-0.2974701	-0.9023845
2	17	37.006153	-89.18282	106.18108	18.117483	2.208405	4.380282
3	17	37.017046	-89.187502	93.98315	4.3897076	-0.6745145	-1.0250093
4	17	37.144606	-89.240047	107.73231	14.002894	0.45772782	-0.77207685
5	17	37.128811	-89.181334	108.79354	11.277267	1.0207157	-0.05521536
6	17	37.466033	-88.894637	125.59218	21.458467	0.6640135	-0.72758055
7	17	37.373894	-88.952553	160.12624	34.017834	0.10117353	-1.2672205
8	17	37.44696	-88.926853	158.78706	37.54687	0.44143835	-1.0039197
9	17	37.160237	-88.735518	112.86226	16.48964	1.2558166	1.8228679
10	17	37.16305	-88.722677	113.5066	16.608133	0.87322617	0.16736984
11	17	37.151771	-88.716909	95.712326	5.173645	-0.15251462	-1.3420945
12	17	37.15625	-88.728232	98.50062	8.69295	0.17033765	-1.3990165
13	17	37.087697	-88.471811	126.57854	26.798456	0.5183036	-1.0663657
14	17	37.50007	-88.708246	163.16222	35.46915	0.15327448	-0.86356854

FIGURE 20.12
Average elevation (left) and standard deviation (middle) statistics computed from the elevation map shown in Figure 20.10 and saved in a tabular form (right).

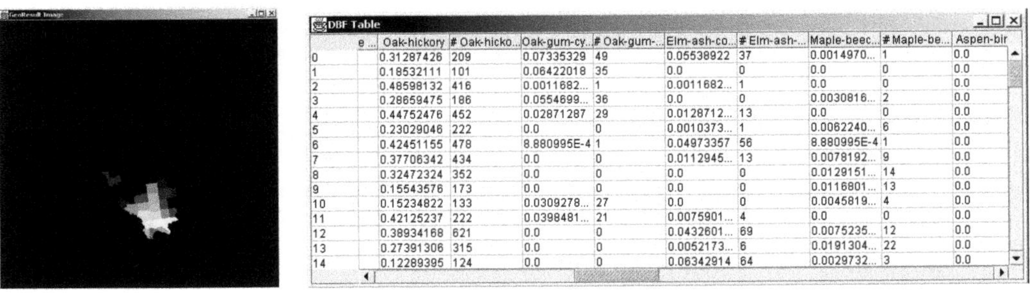

	e ...	Oak-hickory	# Oak-hicko...	Oak-gum-cy...	# Oak-gum-...	Elm-ash-co...	# Elm-ash-...	Maple-beec...	# Maple-be...	Aspen-bir
0		0.31287426	209	0.07335329	49	0.05538922	37	0.0014970...	1	0.0
1		0.18532111	101	0.06422018	35	0.0	0	0.0	0	0.0
2		0.48598132	416	0.0011682...	1	0.0011682...	1	0.0	0	0.0
3		0.28659475	186	0.0554699...	36	0.0	0	0.0030816...	2	0.0
4		0.44752476	452	0.02871287	29	0.0128712...	13	0.0	0	0.0
5		0.23029046	222	0.0	0	0.0010373...	1	0.0062240...	6	0.0
6		0.42451155	478	8.880995E-4	1	0.04973357	56	8.880995E-4	1	0.0
7		0.37706342	434	0.0	0	0.0112945...	13	0.0078192...	9	0.0
8		0.32472324	352	0.0	0	0.0	0	0.0129151...	14	0.0
9		0.15543576	173	0.0	0	0.0	0	0.0116801...	13	0.0
10		0.15234822	133	0.0309278...	27	0.0	0	0.0045819...	4	0.0
11		0.42125237	222	0.0398481...	21	0.0075901...	4	0.0	0	0.0
12		0.38934168	621	0.0	0	0.0432601...	69	0.0075235...	12	0.0
13		0.27391306	315	0.0	0	0.0052173...	6	0.0191304...	22	0.0
14		0.12289395	124	0.0	0	0.06342914	64	0.0029732...	3	0.0

FIGURE 20.13
Occurrence statistics computed from the forest image shown in Figure 20.11, presented graphically for oak hickory (left), and saved in a tabular form (right).

There are also feature extraction problems from vector data, for example, extracting Fourier descriptors, curvature, inflection points, and so on, that were not mentioned here. Furthermore, many image-based object recognition systems rely on image segmentation and clustering techniques to extract image regions as important features [7,9]. Extracting "meaningful" regions is a very essential preprocessing step of object recognition, but it poses significant automation challenges. The automation challenge is caused by a semantic gap between an aggregation of image pixels and the meaning as interpreted by a human. The "interpretation" by a computer algorithm has to be performed (1) with lower dimensional data (e.g., recognition of 3D objects from a 2D image projection), (2) without well-defined criteria of a region (e.g., texture definition consistent with human perception is still unclear [10, 12]), and (3) with a large set of rules about region formation based on human knowledge and experience. Thus, the expectations on feature extraction results have to be adjusted according to the problem complexity or even feasibility (e.g., ill-posed problems).

Summarizing, the difficulty of feature extraction varies a lot and we described some of the scientific challenges in this chapter. It has been known that data preparation including feature extraction usually takes the largest amount of time during data analysis. Thus, automation and systematic processing should be considered carefully beforehand.

References

1. Klein, L. A., Sensor and Data Fusion. *A Tool for Information Assessment and Decision Making*. SPIE Press, Belligham, Washington, 2004, p. 318.
2. Marlin, J.C., Backwater restoration opportunities: Illinois River, The Governor's Conference on the Management of the Illinois River, October 4, 2001, http://www.heartlandwater-resources.org/page7.html.
3. Bajcsy, P., Automatic extraction of isocontours from historical maps, *Proceedings of the 7th World Multiconference on Systemics, Cybernetics and Informatics (SCI 2003)*, Orlando, FL, July 27–30, 2003.
4. ESRI Web site: http://www.esri.com/.
5. Duda, R., P. Hart, and D. Stork, 2001. *Pattern Classification*, 2nd ed., Wiley-Interscience, New York, p. 654.
6. Bajcsy, P. and N. Ahuja, Hierarchical clustering of points using similarity analysis *IEEE Transactions on Pattern Analysis and Machine Intelligence*, 20, 1011–1015, September 1998.
7. Bajcsy, P. Hierarchical segmentation and clustering using similarity analysis. Ph.D. Thesis, University of Illinois at Urbana-Champaign, IL, 1997, p. 182.
8. Smith, B. and D.M. Mark, Ontology and geographic kinds, *Proceedings of International Symposium on Spatial Data Handling (SDH'98)*, Vancouver, Canada, July 12–15, 1998.
9. Bajcsy, P. and N. Ahuja, A new framework for hierarchical segmentation using homogeneity analysis, *Proceedings of the 1st International Conference on Scale-Space Theory in Computer Vision*, The Netherlands, pp. 319–322, July 1997.
10. Bajcsy, P. and N. Ahuja, Hierarchical texture segmentation using dictionaries, *Proceeding of the 3rd ACCV'98*, Hong Kong, pp. 291–298, January 8–11, 1998.
11. Ahuja, N. and Tuceryan, M., Extraction of early perceptual structure in dot patterns: Integrating region, boundary, and component gestalt. *Computer Vision, Graphics, and Image Processing*, 48: 304–356, December 1989.
12. Tuceryan, M. and Jain, A.K., Texture Segmentation Using Voronoi Polygons, *IEEE Trans on Pattern Analysis and Machine Intelligence*, Vol. *PAMI-12*, pp. 211–216, February 1990.
13. Foley J.D., A. van Dam, S.K. Feiner, and J.F. Hughes, *Computer Graphics, Principles and Practice*. Second edition, Addison-Wesley Publishing Company, Reading, MA, 1990, p. 1175.

21

Feature Selection and Analysis

Peter Bajcsy

Contents

21.1 Introduction

In many real cases, application engineers would like to understand and model phenomena. However, they do not know exactly what sets of input and output features should be part of the model and how they are related to each other. Although some of the features might represent variables that are governed by, for instance, physics or chemistry laws, there are many complex phenomena that have not been described by mathematical formulas. For example, water quality measurements can depend on a variety of variables, such as, temperature, season, soil type, land use, velocity and turbidity of water streams, elevation, etc. [1]. In order to establish a model, one has to extract features from multiple datasets, explore subsets of potential input and output features, and evaluate multiple relationships between input and output features in terms of their modeling accuracy. Thus, our current task is to select those features that could be used for accurate modeling. Our goal in the first part of this chapter is to overview the issues related to feature selection and outline a few approaches to solve the hyperspectral band selection problem as a specific case of the feature selection problem.

In the second part, we describe feature analysis and decision support for the problem introduced as a motivation example for data processing section in Chapter 14. We focus on the problems of (1) boundary aggregation given a set of boundary features, and (2) evaluation of boundary aggregations (also called geographic territorial partitions). Solutions to these two problems support a decision-making process with qualitative measures and visualization tools.

21.2 General Feature Selection Problem

While we motivate feature selection by our efforts to understand and model complex phenomena, there are also other reasons for feature selection. It has been known in knowledge discovery and data mining domain that irrelevant features and redundant features introduce noise into data-driven models. Frequently, feature selection improves data quality by removing features that might contain no information due to random noise. Furthermore, data mining methods are much slower with large number of features and large sized datasets. Finally, a good selection of features enhances the comprehensibility of any modeling results, which is crucial for application engineers without data mining background.

In general, we formulate the problem of feature selection as follows. Given a set of candidate features, select the best subset of input features that models output features. The modeling is represented by either prediction of continuous output features or classification into classes of categorical output features. The goal of feature selection is to pick the essential features and discard the redundant and adverse features from the total set of features. It is usually the case that the output features are given and we are primarily focusing on input feature selection from a set of candidate features. The input features are also sometimes denoted as explanatory variables and the output features as response variables [2].

A simple approach to the problem of feature selection is elimination of redundant and irrelevant features. Feature elimination can be achieved by applying common feature measures to assess (1) individual features, for example, information content with entropy measure, (2) pairs of features based on their (a) similarity, for example, distance similarity measure, (b) statistical independence, for example, correlation measure, or (c) consistency, for example, regression measure, and (3) multiple features, for example, model based dependency including linear combinations of features. The feature dependency is also frequently assessed by feature transformations, for instance, by using the Principal Component Analysis (PCA). The results of these assessments can be used by designing elimination rules, such as, eliminate a feature if (1) the feature is a linear combination of two other features, or (2) it has a negligible similarity distance to another feature, or (3) its entropy is close to zero. The assessments of input feature quality without the use of output features are denoted as unsupervised modeling techniques for feature selection.

A more general approach to feature selection would probably consist of three steps. First, take repeatedly a subset of all input features for testing. Second, compute an accuracy score for the subset of input features based on the modeling results using output features as ground truth labeling. In this step, one has to make a choice of a modeling technique and optimize its algorithmic parameters. Third, select the input feature subset that is producing the highest modeling accuracy score for a given set of output features. The modeling techniques that use input and output features are denoted as supervised modeling techniques for feature selection.

The feature selection problem can be encountered in several application domains, such as, remote sensing, medicine, business, or semiconductor manufacturing. In remote sensing, one would relate remotely sensed data with ground measurements. Agricultural and environmental engineers would relate soil and water quality measurements with aerial and satellite data. Decision makers in urban and rural development would model land use and cover for housing and farming development. Similarly, forestry scientists, wildlife biologists, atmospheric scientists, and earthquake engineers would explore very complex systems in the nature by sifting through input and output variables. In medicine, complexity of human body poses feature selection challenges for disease preventions and disease diagnoses. Well-known applications of feature selection in business are stock market modeling and production forecasting. Finally, in semiconductor manufacturing, we could mention analyses of chip, scribe, and packaging defects. These examples cover only a small fraction of all applications for feature selection and are provided just to illustrate a few uses of the presented material. There are large communities of researchers in pattern recognition, statistics, data mining, and remote sensing that are working on the problem of optimal feature selection [3–5].

There exist commercial solutions that could be applied to Geographical Information System (GIS) data. For example, the ENVI package [6] offers PCA and correlation techniques, or the ArcGIS software provides many feature manipulation operations for feature selection purposes. One can also find research solutions, for example, MultiSpec software from LARS at Purdue University [7], or I2K software from NCSA at University of Illinois in Urbana–Champaign [8].

21.3 Spectral Band Selection Problem

In Sections 21.3.1 to 21.3.4, we will focus on a specific case of hyperspectral and multispectral band selection, where the input features are spectral bands and the output features are continuous or categorical variables describing an object of interest. The underlying assumption in this type of modeling is that advanced hyperspectral sensors can better discriminate classes of objects than the standard Electro-Optical (EO) and Infrared (IR) sensors due to their higher spectral resolution. In addition, hyperspectral sensors provide noninvasive and nonintrusive reflectance measurements that will generate imagery at a fine spectral resolution, for example, 10 nm using NASA AVIRIS sensor [5], at much lower cost in a long term than the current periodic manual data collection.

Analyses of hyperspectral imagery have been applied in several GIS application areas [5,4] including environmental monitoring [9–12], sensor design [13,14], geological exploration [15–17], agriculture [18], forestry [19], security [20], cartography, and military [21,22]. Common problems in the area of hyperspectral analysis involving data relevancy include optimal selections of wavelength, number of bands, and spatial and spectral resolution [13,14,23]. Additional problems include the modeling issues of scene, sensor, and processor contributions to the measured hyperspectral values [11], finding appropriate classification methods [16], and identifying underlying mathematical models [15]. Every problem formulation is usually also associated with multiple application constraints.

For example, constraints related to communication bandwidth, data storage, discrimination or classification accuracy, minimum signal-to-noise ratio, sensor, and data acquisition cost must be addressed.

In almost all application areas, the basic goal of hyperspectral image analysis is to classify or discriminate objects. Driven by classification or discrimination accuracy, one would expect that, as the number of hyperspectral bands increases, the accuracy of classification should also increase and there would be no need for any band selection. Nonetheless, this is not the case in a model-based analysis [15,16]. Redundancy in data can cause convergence instability of models. Furthermore, variations due to noise in redundant data propagate through a classification or discrimination model. The same is true of spectral information that is not related to the feature being classified in the underlying mathematical model. Such information is the same as noise to any statistical model, even if it is unique and accurate. Thus, processing a large number of hyperspectral bands can result in higher classification inaccuracy than processing a subset of relevant bands without redundancy. In addition, computational requirements for processing large hyperspectral datasets might be prohibitive and a method for selecting a data subset is therefore sought. Although a method for band selection leads to data compression, the performance objective of data compression is based on data size (communication bandwidth) while the objective of band selection is classification or discrimination accuracy.

21.3.1 Band Selection Problem Formulation

In the general introduction of feature selection problem, we described two approaches to feature selection using unsupervised and supervised modeling techniques. Next, we will address the issue of hyperspectral band and modeling method selection using unsupervised and supervised techniques driven by modeling accuracy and computational cost. The problem is formulated as follows [24,25]: Given N unsupervised band selection methods and M supervised modeling methods, how would one obtain the optimal number of bands and the best performing pair of methods that maximize modeling accuracy and minimize computational requirements? This formulation is a variant of the problem definition in Reference 26.

In the past, this problem has been researched for a single band selection method by multiple researchers [11,17,21,22,27,28]. Many individual methods have been evaluated, for example, stepwise discriminant analysis in Reference 27, maximum likelihood classification in Reference 21 spatial autocorrelation analysis in Reference 11, or PCA jointly with Artificial Neural Network (ANN) analysis in Reference 19. However, the formulation includes also the problem of modeling method selection in addition to the problem of feature selection. It is apparent that these two selections are coupled and one cannot obtain an optimal solution without considering and optimizing both selections.

The reason for including modeling methods into the feature selection problem is based on the No-Free-Lunch (NFL) theorem [3], which states that no single supervised method is superior over all problem domains; methods can only be superior for particular datasets. There is usually no prior knowledge of the underlying structure of the data, and there is no universally accepted "best" supervised method by the NFL theorem. Thus, it is desirable to experiment over a range of methods and implementations to find which one is superior for hyperspectral band selection. To limit the computational complexity of trying all $\sum_{i=1}^{nb}\binom{nb}{i}$ combinations of bands with each supervised method, where nb is the number of bands, unsupervised methods can be used to trim the search space and include only the bands that the various unsupervised methods, respectively deem most informative and least redundant. It is also possible to form new features, such as, create eigenvectors (PCA), average adjacent bands, or search for basis functions in a subspace of data intrinsic dimensionality [29]. Nonetheless, new feature formation adds another degree of freedom in the feature selection analyses and the search space might become computationally prohibitive.

21.3.2 Band Selection Methodology

The trade-off between accuracy and computational requirements is related to the choice of bands and modeling methods. We are looking for a methodology for choosing hyperspectral bands that would provide sufficient, but not redundant, information to classification or prediction algorithms using a practical amount of computational resources. According to Reference 24, one can use unsupervised methods to compute rank ordered lists of bands in a computationally efficient way thereby prefilter bands based on their redundancy and their information content. Although direct comparisons of scores obtained by unsupervised methods are not valid due to different score scales, the rank ordered list of bands serves as a search pruning mechanism for supervised methods. The predictive accuracy of supervised methods with the top ranked bands from the unsupervised methods indicates both the quality of the top ranked bands and, indirectly, the quality of the unsupervised methods. Furthermore, supervised methods can be applied to a variable number of top ranked bands obtained from unsupervised methods. The trend of model errors as a function of the processed number of ranked bands demonstrates local (or global) minima that will identify the optimal number of bands S maximizing model accuracy. Finally, the problem of selecting a supervised method is resolved by choosing the method that forms the most accurate model with respect to the training examples of input bands and output ground measurements.

21.3.3 Cross-Validation

Cross-validation is frequently used to evaluate a supervised method given a set of bands and algorithmic control parameters [3]. The process of n-fold cross-validation involves splitting the dataset into n nonoverlapping, exhaustive subsets, and then building n models, one for each subset being withheld from training. Averaging the error calculated for each withheld set then scores each method. The error is defined as the difference between the actual values and the predicted values of the set's respective model. This assigned error is a function of not only the method and dataset but also any control parameters of the method. To account for this, one has to perform an optimization of the control parameters for each combination of top ranked band sets and supervised methods.

Cross-validation contributes to the list of computational requirements and one has to consider the choice of cross validation parameters carefully. For instance, the following selection of cross-validation parameters requires a day of computing on a cluster of eight processors: 12-fold cross-validation to assign the error to a subset of 120 bands based on the mean absolute error with 300 different randomly selected control parameter sets at each band set evaluation, and 8-fold cross-validation of the control parameter set to assign the final error after a subset of bands was selected. Once the model error is computed for band count varying from 1 to N, where N is the maximum number of bands, one would evaluate a Discrimination Measure (DM) as defined in Equation 21.1 with Error(S) indicating the error measure of the top S bands and Δ representing a small increment/decrement of the number of bands. The DM therefore establishes the confidence that S is the true optimal value for the number of top ranked bands by comparing the error using S bands to the error using $(S + \Delta)$ and $(S - \Delta)$ bands.

$$\text{DM} = \|\text{Error}(S) - \text{Error}(S + \Delta)\| + \|\text{Error}(S) - \text{Error}(S - \Delta)\|. \tag{21.1}$$

21.3.4 Band Selection Outcomes

Briefly, the entire process can be described as running unsupervised methods to rank the best bands followed by testing those band choices with the supervised methods to see which combinations are best for a particular application. The same process should also reveal the optimal number of bands for the application in question. The outcome of this band selection problem is expected to answer basic questions about which wavelength ranges should be used given a hyperspectral image and a specific application, as well as, what modeling methods (a combination of unsupervised and supervised methods) would lead to the most accurate model for input bands and output ground measurements.

21.4 Overview of Band Selection Methods

The band selection methodology described in Section 21.3.2 involves two types of band selection methods, unsupervised and supervised. Unsupervised methods order hyperspectral bands without any training and the methods are based on generic information evaluation approaches. Unsupervised methods are usually very fast and computationally efficient. These methods require very little or no hyperspectral image preprocessing. For instance, there is no need for image georeferencing or registration using geographic referencing information, which might be labor-intensive operations.

In contrast to unsupervised methods, supervised methods require training data in order to build an internal predictive model. A training dataset is obtained via registration of calibrated hyperspectral imagery with ground measurements. Supervised methods are usually more computationally intensive than unsupervised methods due to an arbitrarily high model complexity and an iterative nature of model formation. Another requirement of supervised methods is that the number of examples in a training set should be sufficiently larger than the number of attributes (bands, in this case). This requirement might be hard to meet as the number of hyperspectral bands grows and the collection of each ground measurement has an associated real-world cost. If taken alone, the unsupervised methods can, at best, be used to create classes by clustering of spectral values followed by assigning an average ground measurement for each cluster as the cluster label. Supervised methods therefore provide more accurate results than unsupervised methods.

In general, the input and output features can be either continuous or discrete (also commonly referred to as numeric and categorical, or scalar and nominal). For example, spectral reflectance values of each band are continuous input features. Ground measurements, such as, soil conductivity or grass category, would be either continuous or categorical output features. We describe next seven unsupervised methods including entropy, contrast, 1st and 2nd spectral derivative, ratio, correlation, and principal component analysis ranking (PCAr)-based algorithms that apply to continuous band values. Since one can encounter both types of output features (ground measurements in this case), we outline three supervised methods for continuous output features and three methods for modeling categorical features. The list of these six methods includes regression; instance based (k nearest neighbor) and regression tree algorithms for the continuous case, and naive bayes, instance based, and C4.5 decision tree algorithms for the categorical case. All methods form models for prediction of continuous input/continuous output variables or classification of continuous input/categorical output variables with global, local, and hybrid modeling approaches. These methods are described in many machine learning and pattern recognition texts, such as Reference 3.

21.4.1 Unsupervised Band Selection Methods

Information entropy: This method is based on evaluating each band separately using the information entropy measure [30, chapter 3] defined below.

$$H(\lambda) = -\sum_{i=1}^{m} p_i \ln(p_i) \tag{21.2}$$

where H is the entropy measure λ is the central wavelength, p is the probability density function of reflectance values in a hyperspectral band, and m is the number of distinct reflectance values. The probabilities are estimated by computing a histogram of reflectance values. Generally, if the entropy value H is high then the amount of information in the data is large. Thus, the bands are ranked in the ascending order from the band with the highest entropy value (large amount of information) to the band with the smallest entropy value (small amount of information).

First spectral derivative: The bandwidth, or wavelength range, of each band is a variable in a hyperspectral sensor design [13,14]. This method explores the bandwidth variable as a function of added information. It is apparent that if two adjacent bands do not differ greatly then the underlying geospatial property can be characterized with only one band. The mathematical description is shown later, where I represents the hyperspectral value, x is a spatial location, and λ is the central wavelength. Therefore, if D_1 is equal to zero then one of the bands is redundant. In general, the adjacent bands that differ significantly should be retained, while similar adjacent bands can be reduced.

$$D_1(\lambda_i) = \sum_{x} \|I(x, \lambda_i) - I(x, \lambda_{i+1})\|. \tag{21.3}$$

Second spectral derivative: Similar to the first spectral derivative, this method explores the bandwidth variable in hyperspectral imagery as a function of added information. If three bands are adjacent, and the two outside bands can be used to predict the middle band through linear interpolation, then the band is redundant. The larger the deviation from a linear model, the higher the information value of the band. The mathematical description of this method is shown in Equation 21.4, where D_2 represents the measure of linear deviation, I is a hyperspectral value, x is a spatial location, and λ is the central wavelength.

$$D_2(\lambda_i) = \sum_{x} \|I(x, \lambda_{i-1}) - 2I(x, \lambda_i) + I(x, \lambda_{i+1})\|. \tag{21.4}$$

Contrast measure: This method is based on the assumption that each band could be used for classification purposes by itself. The usefulness of a band would be measured by a classification error achieved by using only one particular band and minimizing the error. In order to minimize a classification error, it is desirable to select bands that provide the highest amplitude discrimination (image contrast) among classes. If the class boundaries were known *a priori* then the measure would be computed as a sum of all contrast values along the boundaries. However, the class boundaries are unknown *a priori* in the unsupervised case. One can evaluate contrast at all spatial locations instead assuming that each class is defined as a homogeneous region (no texture variation within a class). The mathematical description of the contrast measure computation is shown here for a discrete case.

$$\text{Contrast}M(\lambda) = \sum_{i=1}^{m} \|f_i - E(f)\|^* f_i, \tag{21.5}$$

where f is the histogram (estimated probability density function) of all contrast values computed across one band by using Sobel edge detector [30], $E(f)$ is the sample mean of the histogram f, λ is the central wavelength, and m is the number of distinct contrast values in a discrete case. The equation includes the contrast magnitude term and the term with the likelihood of contrast occurrence. In general, bands characterized by a large value of ContrastM are ranked higher (good class discrimination) than the bands with a small value of ContrastM.

Spectral ratio measure: In many practical cases, band ratios are effective in revealing information about inverse relationship between spectral responses to the same phenomenon (e.g., living vegetation using the normalized difference vegetation index [5]. This method explores the band ratio quotients for ranking bands and identifies bands that differ just by a scaling factor. The larger the deviation from the average of ratios E(ratio) over the entire image, the higher the Ratio$M$ value of the band. The mathematical description of this method is shown, where Ratio$M$ represents the measure, I is a hyperspectral value, x is a spatial location, and λ is the central wavelength.

$$\text{Ratio}M(\lambda_i) = \sum_x \left\| \frac{I(x,\lambda_i)}{I(x,\lambda_{i+1})} - E\left(\frac{I(x,\lambda_i)}{I(x,\lambda_{i+1})}\right) \right\| \tag{21.6}$$

Correlation Measure: One of the standard measures of band similarity is normalized correlation [3]. The normalized correlation metric is a statistical measure that performs well if a signal-to-noise ratio is large enough. This measure is also less sensitive to local mismatches since it is based on a global statistical match. The correlation based band ordering computes the normalized correlation measure for all adjacent pairs of bands similar to the spatial autocorrelation method [11] applied to all ratios of pairs of image bands. The mathematical description of the normalized correlation measure is shown below, where Cor$M$ represents the measure, I is a hyperspectral value, x is a spatial location, and λ is the central wavelength. E denotes an expected value and σ is a standard deviation.

$$\text{Cor}M(\lambda_i) = \frac{E(I(\lambda_i)^* I(\lambda_{i+1})) - E(I(\lambda_i))^* E(I(\lambda_{i+1}))}{\sigma(I(\lambda_i))^* \sigma(I(\lambda_{i+1}))}. \tag{21.7}$$

After selecting the first least correlated band based on all adjacent bands, the subsequent bands are chosen as the least correlated bands with the previously selected bands. This type of ranking is based on mathematical analysis of Jia and Richards [21], where spectrally adjacent blocks of correlated bands are represented in a selected subset.

Principal component analysis ranking (PCAr): PCA has been used very frequently for band selection in the past [5]. The method transforms a multidimensional space to one of an equivalent number of dimensions where the first dimension contains the most variability in the data, the second the second most, and so on. The process of creating this space gives two sets of outputs. The first is a set of values that indicate the amount of variability each of the new dimensions in the new space represents, which are also known as eigenvalues (ε). The second is a set of vectors of coefficients, one vector for each new dimension, that define the mapping function from the original coordinates to the coordinate value of a particular new dimension. The mapping function is the sum of the original coordinate values of a data point weighted by these coefficients. As a result, the eigenvalue indicates the amount of information in a new dimension and the coefficients indicate the influence of the original dimensions on the new dimension. According to [10], PCA based ranking system (PCAr) could make use of these two facts by scoring the bands (the "original" dimensions in the above discussion) as follows.

$$PCAr(\lambda_i) = \sum_j |\varepsilon_j c_{ij}| \tag{21.8}$$

λ_i is the central wavelength, ε_j is the eigenvalue for the jth principal component, and c_{ij} is the mapping coefficient of the ith central wavelength in the jth principal component. As the procedure for computing the eigenvalues and coefficients is both complex and available in most data analysis texts [3], it is omitted.

21.4.2 Supervised Prediction Methods for Continuous Output

Regression: The regression method is based on a multivariate regression [4,31] that is used for predicting a single continuous variable Y given multiple continuous input variables $\{X_1, \ldots, X_n\}$. The model building process can be described as follows. Given a set of training examples T, find the set of coefficients $\beta = \{\beta_0, \ldots, \beta_n\}$ that gives the minimum value of $g(T)$, where

$$g(T) = \min \sum_{e \in T} (Y_e - Y'_e)^2, \tag{21.9}$$

where Y_e is the observed output variable of a training example e and

$$Y'_e = \beta_0 + \beta_1 X_1^e + \beta_2 X_2^e + \cdots + \beta_n X_n^e, \tag{21.10}$$

therefore, Y'_e is the predicted value for Y_e given values for $\{X_1^e, \ldots, X_n^e\}$ which, in this case, are reflectance values at varying wavelengths for the training example e. The problem as stated can be solved numerically using well-known matrix algebra techniques. Further details for finding $\beta = \{\beta_0, \ldots, \beta_n\}$ are therefore omitted for the sake of brevity.

Instance-based method: The instance based method uses inverse Euclidean distance weighting of the k-nearest neighbors to predict any number of continuous variables [4,32]. To predict a value Y' of the example being evaluated e, the k points in the training dataset with the minimum distance (see Equation 21.11) to the point e over the spectral dimensions $\{X_1, \ldots, X_n\}$ are found.

$$d = \sqrt{\sum_{i=1}^{n} (X'_i - X_i)^2} \tag{21.11}$$

The weighted average of the observed Y values of these k closest training points is then computed where the weighting factor is based on the inverse of the distance from each of the k points to the point e according to Equation 21.12. Furthermore, the weighting factor is raised to the power w. Altering the value of w therefore influences the relationship between the impact of a training point on the final prediction and that training point's distance to the point being evaluated. The user must set the values of the control parameters k and w.

$$Y' = \frac{\sum_{i=0}^{k} (1/d_i^w) Y_i}{\sum_{i=0}^{k} (1/d_i^w)}. \tag{21.12}$$

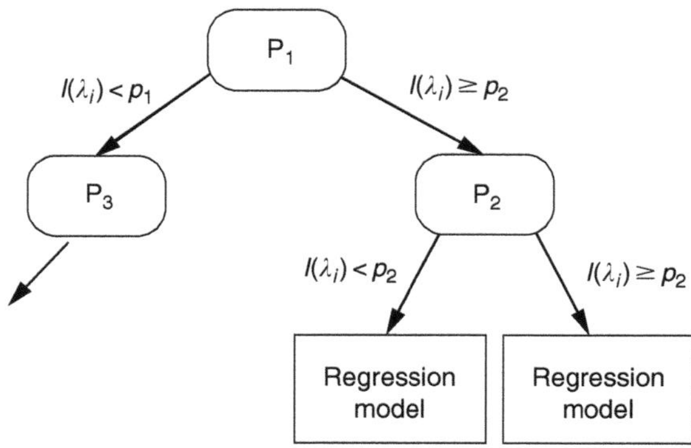

FIGURE 21.1
A simple regression tree where spectral values determine the path and the leaves contain regression models.

Regression tree: A regression tree is a decision tree that is modified to make continuous valued predictions [33]. They are akin to binary search trees where the attribute used in the path-determining comparison changes from node to node. The leaves then contain a distinct regression model used to make the final prediction.

To evaluate (or test) an example using a regression tree, the tree is traversed, starting at the root, by first comparing the reflectance value at a single wavelength requested by the node and compared to the split-point (in Figure 21.1, the P values). Particular wavelengths may be used by several nodes or none at all. If the reflectance value of the example at the appropriate wavelength is less than the split point, the left branch is taken, if greater than or equal to the split point, the right. This splitting procedure based on reflectance values continues until a tree leaf is encountered, at which time the prediction can be made based on data in the leaf.

To build a model, one must select the bands and the reflectance values for those bands are necessary to split the examples into sets that have similar target variables. To do this, a greedy approach could be employed based on minimization of the target variable's variance (defined in Equation 21.13). More precisely, at every node, find central wavelength λ and corresponding split point p such that the average variance of the targets of the two portions of the dataset s after being split is minimized. This average variance is weighted based on how many training examples take the left or right branch, respectively (see Equation 21.13).

$$\text{BestSplit}(s) = \min_{\lambda,p}\left(\frac{\text{var}(t,n_l)/|n_l| + \text{var}(t,n_r)/|n_r|}{|n_l| + |n_r|}\right), \tag{21.13}$$

where $n_l = \{\text{examples } e : l_{e(\lambda)} \leq p\}$, $n_r = \{\text{examples } e : l_{e(\lambda)} > p\}$ and the variance of the variable Y, in the set of examples s is given by

$$\text{var}(Y,s) = \sum_{i=0}^{|s|}(Y_i - \bar{Y})^2. \tag{21.14}$$

The algorithm halts when one of two criteria is met. The first is that the number of examples that evaluate to a node falls below m, the minimum allowed examples per node. The other is that the improvement (reduction) in variance that would be obtained by doing the best possible split is below some improvement threshold, t. In either case, the node at which the halting criteria are met is marked as a leaf and a regression model is built on the training examples that evaluate to that node. Both t and m are control parameters and have to be optimized too.

21.4.3 Supervised Prediction Methods for Categorical Output

Naive Bayes: Bayes law Equation 21.15 provides the posterior probability of an event C_i occurring given that event Λ has occurred based on the prior probabilities of C_i and Λ, as well as the posterior probability of event Λ given C_i. Here, this provides a means of calculating the probability of each possible class C_i given a spectral signature Λ and then selecting the class with the highest probability $P(C_i/\Lambda)$ as the prediction. $P(C_i)$ can easily be estimated from the set of training examples and $P(\Lambda)$, which is constant between classes, can be ignored as the classifier scheme is simply comparing the probabilities of different classes. To calculate the value of $P(\Lambda/C_i)$, conditional independence amongst attributes (here, spectral bands) is assumed (hence the name "Naïve" Bayes), which allows the use of Equation 21.16.

$$P(C_i/\Lambda) = \frac{P(\Lambda/C_i)P(C_i)}{P(\Lambda)} \tag{21.15}$$

$$P(\Lambda/C_i) = \prod_k P(I(\lambda_k)/C_i). \tag{21.16}$$

Continuous variables $I(\lambda_k)$ are usually binned, and estimated probabilities based on training data are stored in a histogram for every $(C_i, I(\lambda_k))$ pair for use in Equation 21.16. This introduces the need for control parameters for the binning method. The first parameter is a switch to select either binning by width or binning by depth. Binning by width takes a single interval size that all bins are given, with the lower bound of the first bin being the minimum value of the training set. In binning by depth, all bins are required to have an equal number of training examples, and the interval size is therefore variable between bins. The second parameter is therefore either the interval size or number of examples per bin, depending on which method is indicated by the first parameter. These parameters have to be optimized during model building process.

Instance-based method: Instance-based classifiers, sometimes called k-nearest neighbors classifiers [4,32], make a prediction for a test case based on the classes of the k training examples that have the smallest Euclidean distance to that test case. The training stage of model building is therefore nothing more than storing the training examples. During prediction the distances to all n training examples must be calculated for each test case, and the k smallest (where k is a user defined control parameter) are selected. Often, the prediction is made by a simple majority-rules vote of these k-nearest neighbors. Here, however, one can bias the votes by the inverse of the distance to the test case, raised to the power w (another control parameter). This gives training examples with a smaller distance a higher weight in the voting. The weighted "vote" for each possible class C_i is therefore given by:

$$V(C_i) = \sum_{e \in \{e : C(e) = C_j\}} \frac{1}{d_e^w}, \tag{21.17}$$

where $C(e)$ is the class of training example e, and d_e is the Euclidean distance $d_e = \sqrt{\sum_k (I(\lambda_k) - I(\lambda_k^e))^2}$ from the training example to the test case in the spectral space. The number of neighbors k and the exponent weight w have to be optimized during the model building process.

C4.5 Decision tree: A decision tree is a recursive search structure that can take on one of two forms: (1) a leaf, which has an associated class, or (2) a node that contains a test on a single attribute of the examples, and a branch and sub-tree for each possible outcome of that test [34].

C4.5 is widely considered the standard implementation of a classification decision tree. The learning process of a C4.5 decision tree involves finding the optimal test at each node to base the split on (or decide that the node should be a leaf). It exhaustively tries every reasonable test criterion at each node and selects the test based on some information gain criteria (see later). In the case of discrete attributes, this simply means creating a branch and subtree for every possible value of the attribute. For continuous attributes (the category spectral data falls into), C4.5 tries all $(m-1)$ possible values to perform a binary split for each attribute (less than evaluates to the left, greater than, or equal to evaluates to the right), where m is the number of training examples that have evaluated to the node in question. Because all attributes are tested at each node, the algorithm can become quite expensive for large numbers of attributes.

The information gain indicates the decrease in variability of the classes in each of the subtrees. That is, it measures the uniformity of the class labels of the examples in the child nodes as compared with the parent. The information of a node, given in terms of the set T of training examples it contains, is given by:

$$H(T) = -\sum_j p(C_j/T)\ln[p(C_j/T)], \tag{21.18}$$

where the probability $p(C_j/T)$ is simply

$$p(C_j/T) = \frac{|e : e \in T, C(e) = C_j|}{|T|}. \tag{21.19}$$

Finally, the information *gain* of a potential split S is given as the information of the parent minus the summation of the information content of its k children:

$$\text{Gain}(S) = H(T) - \sum_k \frac{|T_k|}{|T|} H(T_k), \tag{21.20}$$

where T_k is the subset of T and a set of examples that evaluate to the same child node. The potential split with the highest gain is selected and the algorithm is repeated on the children. A node is declared to be a leaf if either a minimum information gain threshold τ_i is not satisfied by the best potential split or similarly if the number of training examples in the node is less than the minimum examples per leaf τ_e. Both τ_i and τ_e are user defined parameters and have to optimized during model building process.

21.5 Conducting Band Selection Studies

In this section, we would like to discuss challenges of band selection studies and elaborate on expected band selection results from theoretical and experimental viewpoints.

In order to perform spectral band selection as described in Section 21.4, data preparation usually takes a significant amount of effort. For example, one has to collect ground measurements and aerial imagery of the same geographic location and ideally at the same

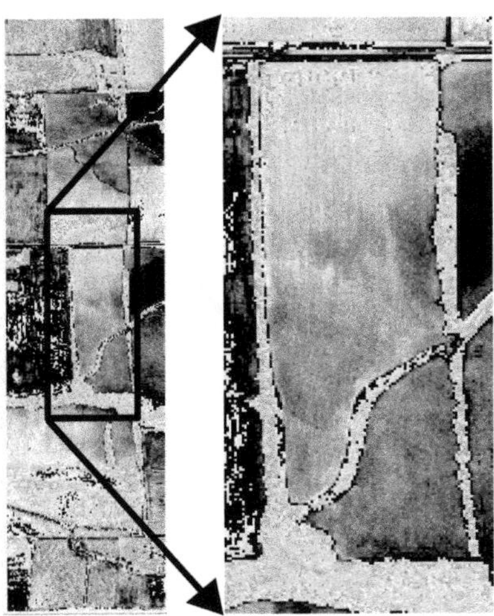

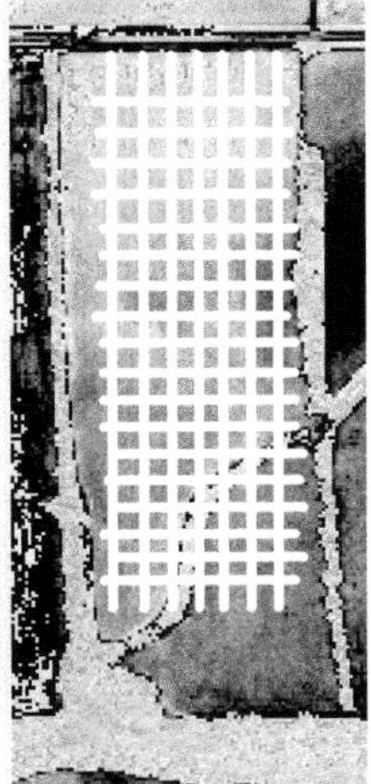

FIGURE 21.2
A hyperspectral image (left) obtained in April 26, 2000 at 4000 m altitude with a Regional Data Assembly Centers Sensor (RDACS), model hyperspectral (H-3), which is a 120-channel prism grading, push-broom sensor developed by NASA. The 120 bands correspond to the visible and infrared range of 471 to 828 nm, recorded at a spectral resolution of 3 nm. The area of interest is the Gvillo field located near the city of Columbia in the central part of Missouri (middle) and the associated ground measurements of soil electrical conductivity collected at the grid-based locations (right). The display shows combined bands with central wavelengths 471, 660, and 828 nm. HS images were provided by Spectral Visions, IL, a nonprofit research organization funded by the NASA Commercial Remote Sensing Program, and the ground measurements were collected, postprocessed, and provided for research purposes by the Illinois Laboratory Agricultural Remote Sensing (ILARS), UIUC.

time. Then, georeference both datasets and extract corresponding image spectral values to match ground measurements so that a set of examples with input and output features can be constructed. Thus, it is wise to plan these efforts by discussing any data acquisition, file format, data storage, and analysis requirement issues with all members of a band selection study. For instance, data acquisition at the wavelength range 400–900 nm would be driven by agricultural applications, where the 400–900 nm wavelength range corresponds to plant characteristics very well [26] and has been used for vegetation sensing in the past [18]. By selecting this wavelength range, the data analysis avoids issues related to water absorption bands (1400 and 1900 nm). Similarly, ground variables should be significant to agricultural applications, such as, soil electrical conductivity is an important characteristic considered for crop yield prediction and indirectly characterizes several important soil characteristics including soil texture (the relative amount of sand-silt-clay) and salinity (affects the crops ability to acquire water). An example of such experimental dataset is shown in Figure 21.2.

Next, let us assume that each unsupervised method sorts all bands based on band redundancy in ascending order. From a theoretical viewpoint, the dependency between modeling error obtained by evaluating supervised methods and the number of top ranked bands is specific for each supervised method. For example, we could analyze the three supervised methods for continuous output feature prediction. First, the regression-based supervised method is using a global modeling approach where very few bands (insufficient information) or too many bands (redundant information) will have a negative impact on the model accuracy. Thus, one could expect the trend of a parabola with one global minimum. Second, the instance-based method exploits local information and adding more bands will either decrease an error or preserve it constant. The expected trend is a down-sloped staircase curve with several plateau intervals. The beginning of each plateau interval can be considered as a local minimum for selecting the optimal number of bands (see crosses

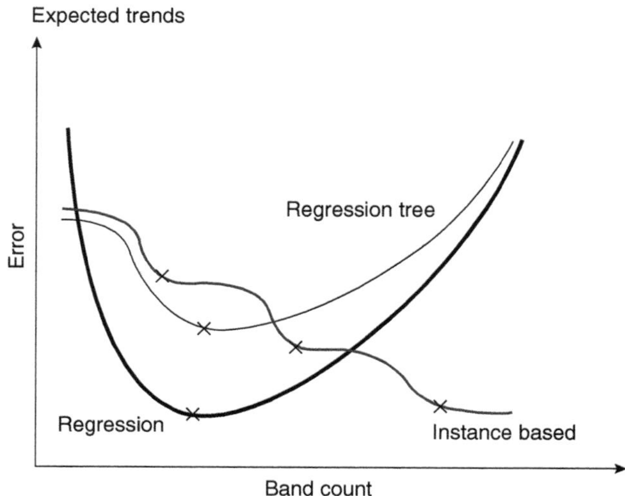

Figure 21.3
Expected trends based on the models of three supervised methods for continuous output feature prediction.

in Figure 21.3). Finally, the regression tree based method uses a hybrid approach from a standpoint of local vs. global information. It is expected to demonstrate a trend of the instance-based method for a small number of processed bands (band count) and a trend of the regression-based method for a large number of processed bands.

From an experimental viewpoint, it is recommended to compare band selection results obtained with unsupervised methods against the results obtained with a randomly ordered sequence of bands or an ordered sequence by pattern-based sampling (e.g., $1, N, N/2, N/4, 3N/4$, etc.). The qualitative comparison can reveal how much gain is achieved by ranking bands based on "smart" assessment with the unsupervised methods as opposed randomly ordering bands.

Finally, it is beneficial to obtain an insight about the band selection from a theoretical viewpoint based on the underlying physics. For example, as it was shown in Reference 24, the seven most significant hyperspectral bands that lead to the most accurate prediction of soil electrical conductivity are bands from the red visible wavelength range. One can expect such a result not only based on our empirical observations (soils with ferrites would appear reddish), but also based on studying electromagnetic theory (phenomenological and atomic models according to Balanis, [35]) and the dependency of electric conductivity on wavelength derived from Maxwell equations. By anticipating the relationships between hyperspectral values (reflected part of the Electro-Magnetic (EM) waves in the wavelength range [471 nm, 828 nm]) and surface/field characteristics that change electric and magnetic properties according to the EM theory of wave propagation [35], one can better plan experiments.

21.6 Feature Analysis and Decision Support Example

In general, feature analysis and decision support are very much application driven. The objective of feature analysis and decision support is to make the most optimal planning or management decisions. These decisions frequently include geographical information and discovering relationships in the presence of multiple heterogeneous datasets [36].

We now present analysis and decision support for the problem introduced as a motivation example for data processing section in Chapter 14. Our goal is to perform boundary aggregation based on a given set of boundary features, and evaluate the goodness of obtained geographic regions. The goodness evaluation is one component of the top level decision that is concerned with partitioning a geographic area into minimum number of geographic territories, sharing similar features and being spatially formed from one of the man-made boundaries (e.g., U.S. Census Bureau boundaries or U.S. postal zip code boundaries) into contiguous regions. The objective of the analysis of boundary aggregations (also called geographic territorial partitions) is to support a decision-making process with qualitative measures and visualization tools. We refer the reader to Chapter 14 for definitions of terms, such as, boundary, region, territory, and geographic territorial partition.

21.6.1 Feature Analysis for Boundary Aggregation Purposes

In general, boundary aggregation can be formulated as a clustering problem with no spatial constraints [4] or as a segmentation problem with spatial constraints [3] on the final boundary aggregations. Merging spatial boundaries into boundary aggregations (or geographical territorial partitions) is driven by maximizing intra-partition similarity of features and inter-partition dissimilarity of features. Spatial constraints are defined such that the final aggregations must form spatially contiguous geographical partitions.

An aggregation process can stop if (1) a certain number of final partitions have been found or (2) a maximum user-specified intra-partition dissimilarity value has been exceeded. The aggregation of two boundaries (any boundaries in a case of clustering or spatially adjacent boundaries in a case of segmentation) takes place when the Euclidean distance between their features is less than a discretely incremented threshold value. The aggregation continues until one of the exit criteria is met. The output can be the last partition of the incremental aggregation process or several partitions obtained at multiple feature similarity values. This bottom-up aggregation process and the multipartition output explains why the aggregation method is also called hierarchical.

We illustrate analysis using the two hierarchical aggregation methods, such as clustering (no spatial constraint) and segmentation (with spatial constraint). These methods are variants of the bottom-up *k*-means clustering and described in References 37 and 38. The result of aggregation is stored in a tabular form, for example, the column labeled as Clust_Label0 in Figure 21.4 (right), and presented to an end-user in a visual form with respect to the geographical location of each boundary (Figure 21.4, left). Results from clustering and segmentation can be easily visually compared as it is shown in Figure 21.5. Hierarchical results obtained as multiple intermediate outputs of the aggregation process can be presented as a movie (a stack of image frames) by showing a sequence of hierarchical results. Typical results from a hierarchical stack of results are shown in Figure 21.6.

21.7 Evaluation of Geographic Territorial Partitions and Decision Support

Once multiple geographic boundary aggregations have been obtained, there is a need to evaluate each geographic partition and make a decision about selecting the best partition according to some application-defined criteria. One can address this problem by error

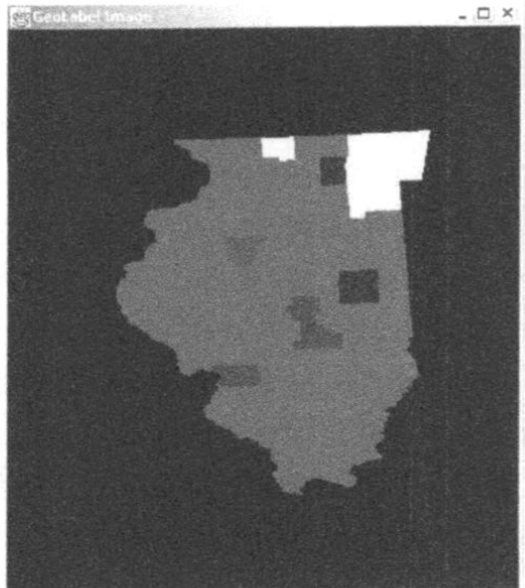

DBF Table

	RAPE94	ROBB94	ASSA94	BURG94	LARC94	MV_TH94	ARSO94	Clust_Label0
84	0	0	0	0	0	0	0	1
85	0	0	0	0	0	0	0	1
86	0	0	0	0	0	0	0	1
87	0	12	23	103	446	38	4	1
88	0	0	0	0	0	0	0	1
89	0	0	0	0	0	0	0	1
90	0	30	190	807	3215	290	38	5
91	0	10	96	546	2232	105	16	1
92	0	262	436	1194	3805	451	46	5
93	0	35066	41933	51673	150501	44093	1836	6
94	0	38	111	666	4096	134	32	5
95	0	715	1243	4834	9878	1306	59	5
96	0	0	0	0	0	0	0	1
97	0	0	0	0	0	0	0	1
98	0	0	0	0	0	0	0	1
99	0	0	0	0	0	0	0	1
100	0	394	863	2003	6325	863	78	2
101	0	0	0	0	0	0	0	1

FIGURE 21.4
Hierarchical clustering of FBI crime data with the exit criterion being the number of clusters and the clustered feature being auto theft in 2000. Clustering leads to six aggregations that are geographically depicted with six colors (left) and the labels are stored in a tabular form (right).

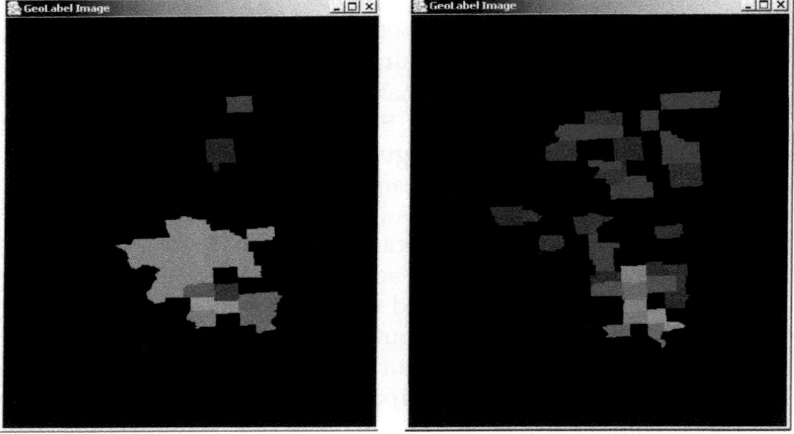

FIGURE 21.5
The results of hierarchical segmentation (left) and hierarchical clustering (right) of oak hickory occurrence feature with the exit criterion of 18 aggregations formed from Illinois county boundaries.

evaluation of each partition followed by comparative analysis of multiple partitions to support a decision making process.

21.7.1 Error Evaluation of New Geographic Territorial Partitions

In order to evaluate an error of a given territorial partition, it is assumed that the partition without any boundary aggregation had zero error. A partition formed by boundary aggregations will inherently introduce feature errors or deviations from the partition with no aggregations. The aggregated boundaries are represented by a new feature value that might differ from the original feature values of unaggregated boundaries. However, if the cost of using and maintaining a large number of partitions is too high then the trade-off between the number of partitions and the associated feature prediction error must be considered.

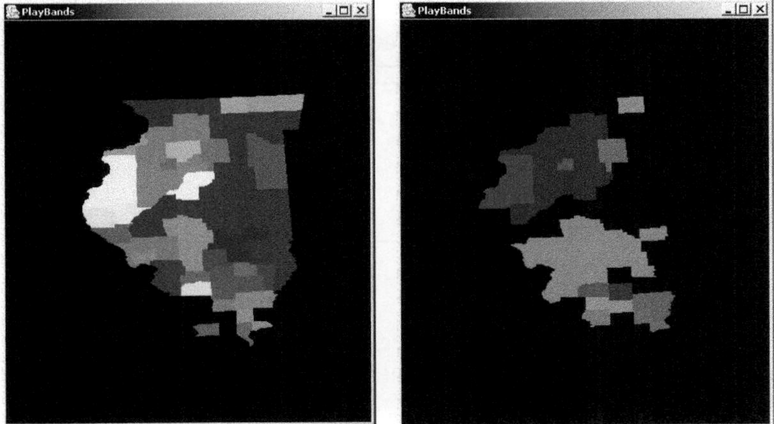

FIGURE 21.6
Hierarchical segmentation of extracted forest statistics (oak hickory occurrence) with two output partitions containing 43 (left) and 21 (right) aggregations formed from Illinois county boundaries.

An error evaluation can be conducted by using multiple error metrics depending on a specific application domain. We provide several metrics defined in Equations 21.21 to 21.24 in order to list a possible choice of metrics for interested end users. The list of metrics includes variance Equation 21.21, city block Equation 21.22, normalized variance Equation 21.23, and normalized city block Equation 21.24.

Variance metric:

$$Var^k = \sum_{r:label(r)=k} \sum_j (w(r)^* f_j(r) - \bar{f}_j^k)^2 = \sum_j Var(\{f_j^k\}). \tag{21.21}$$

City block metric:

$$CB^k = \sum_{r:label(r)=k} \sum_j \|w(r)^* f_j(r) - \bar{f}_j^k\| = \sum_j CB(\{f_j^k\}). \tag{21.22}$$

Normalized variance metric:

$$NormVar^k = \sum_j \frac{Var(\{f_j^k\})}{\bar{f}_j}. \tag{21.23}$$

Normalized city block metric:

$$NormCB^k = \sum_j \frac{CB(\{f_j^k\})}{\bar{f}_j}, \tag{21.24}$$

where values are defined as

$$\bar{f}^k = \left(\frac{\sum_{r:label(r)=k} f_1(r)^* w(r)}{\sum_{r:label(r)=k} w(r)}, \frac{\sum_{r:label(r)=k} f_2(r)^* w(r)}{\sum_{r:label(r)=k} w(r)}, \ldots \right) = (\bar{f}_1^k, \bar{f}_2^k,) \tag{21.25}$$

Table 21.1

Examples of data for error evaluation.

r	Name	$f_1(r)$	$label(r) = k$	$w(r)$
1	Zip code 1	10	1	1
2	Zip code 2	20	1	1
3	Zip code 3	40	2	1
4	Zip code 4	42	2	1

and the notation corresponds to: k is aggregation label, j is feature index, r is an index representing a region/boundary/territory, $w(r)$ is a weighting coefficient, and $\bar{f}_j$ is a sample mean of a feature j over all regions.

21.7.1.1 *Example of Error Evaluation*

Let us illustrate the error evaluation of new geographical territorial partitions using the data in Table 21.1. Given just one feature $f_1(r)$, we compute the sample means for the two clusters $k = 1$ and $k = 2$ ($\bar{f}_1^{k=1} = (10 + 20)/2 = 15$ and $\bar{f}_1^{k=2} = (40 + 42)/2 = 41$). Then, the four metrics can be computed by directly plugging all values into the corresponding equations above, for example, for $k = 1$ we obtain $Var^{k=1} = (10 - 15)^2 + (20 - 15)^2 = 50$, $CB^{k=1} = |10 - 15| + |20 - 15| = 10$, $NormVar^{k=1} = Var^{k=1}/\bar{f}_1^{k=1} = 50/15 = 3.\bar{3}$, and $NormCB^{k=1} = CB^{k=1}/\bar{f}_1^{k=1} = 10/15 = 0.\bar{6}$.

21.7.2 Decision Support by Comparative Analysis

In order to compare multiple partitions and support decision makers, one can consider calculating global or local predictive errors and visualizing the errors in terms of their geographic location. These two steps can be described as follows. First, given a feature or multiple features of interest, a global prediction error is computed per each partition and the global result error values are sorted for comparison. For example, the results of clustering and segmentation with the average terrain elevation feature of Illinois counties can be evaluated and compared as it is shown in Figure 21.7.

Second, any prediction error can be viewed in terms of its geographical error distribution. Given any set of boundary aggregations, the error varies with respect to a geographic location determined by the granularity of the underlying boundaries and their corresponding aggregation memberships. Thus, the geographic error distribution can be viewed as a sequence of georeferenced images where bright corresponds to a large error while dark means a small error. An example of geographic error distribution is shown in Figure 21.8.

In order to compare prediction errors computed for a specific region from several partitions, one could inspect computed errors from all partitions as a function of boundary indices. This type of comparative analysis is facilitated by the appropriate visualization and color-coding each partition differently. An example is illustrated in Figure 21.9. However, the tacit assumption for this type comparison is that the aggregations are defined over the same set of underlying boundaries. This assumption can be violated if aggregations are built with different underlying boundaries, for example, the U.S. Census Bureau boundaries and U.S. postal zip code boundaries. Then, prediction errors at a defined location by latitude and longitude cannot be directly compared by using boundary indices as

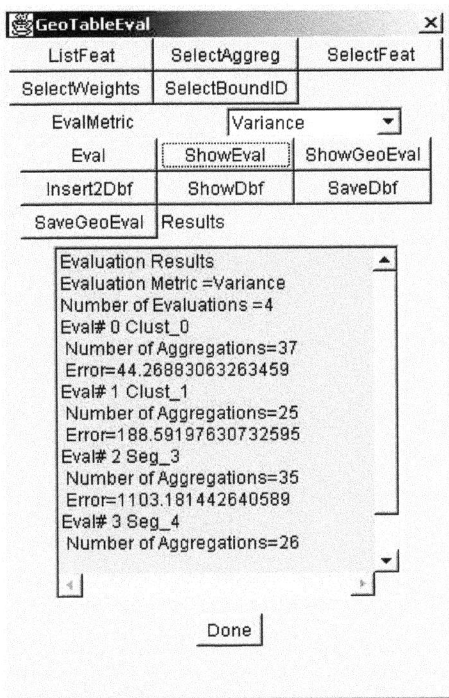

FIGURE 21.7
Global errors for multiple partitions by using the variance error metric. The partitions were obtained by clustering and segmentation of average terrain elevation feature and formed from boundaries of Illinois counties.

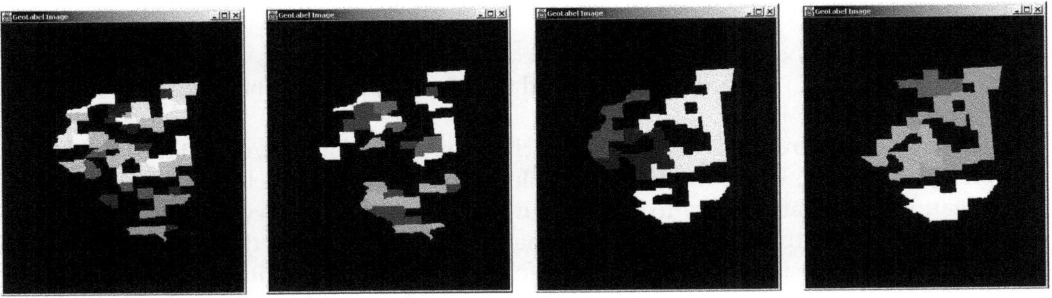

FIGURE 21.8
Geographic error distribution for the results presented in Figure 21.7. From left to right, geographical error distribution for the partitions with 37 (Eval# 0 Clust_0), 25 (Eval# 1 Clust_1), 35 (Eval# 2 Seg_3), and 26 (Eval# 3 Seg_4) aggregations. The error magnitude is illustrated from the darkest (low error value) to the brightest (large error value) intensity.

the common key. In this case, the prediction errors could be remapped from the spatial resolution of irregular boundary shapes to another resolution with common underlying boundary shapes. Given the fact that there exist many software tools converting boundary shapes into raster images, one could choose a raster image with a square pixel as the underlying common boundary shape. Then, the remapping would be accomplished by a simple image subtraction of two images that were obtained by converting boundary shapes with associated errors into raster images (see Figure 21.8 with geographical error distributions).

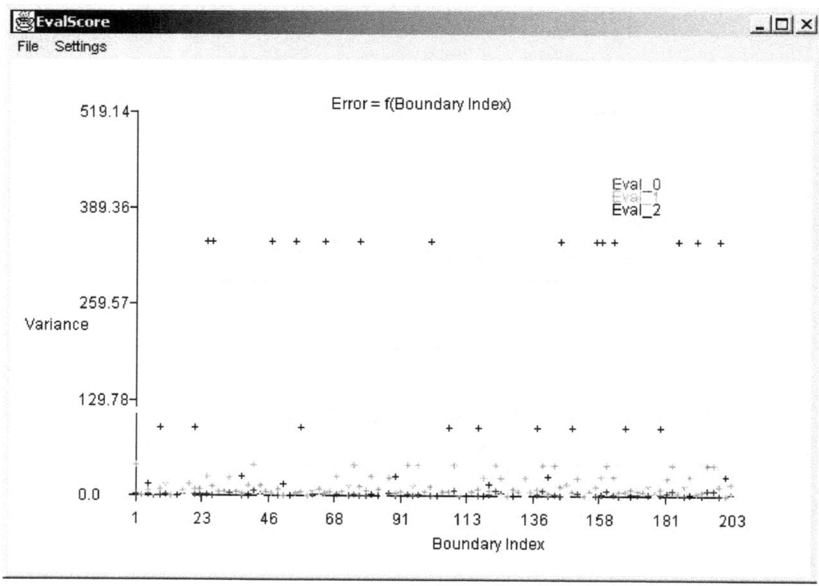

FIGURE 21.9
Variance error visualization per boundary for four different partitions (Eval 0, Eval 1, Eval 2, and Eval 3). If desired then the boundary index could be replaced by the name of a county. This type of visualization allows (a) comparison of multiple partitions based on color-coded values and (b) comparison of multiple boundaries within one partition based on the magnitude of variance error.

21.8 Summary

In this chapter, we focused on (1) hyperspectral band selection as a specific example of feature selection, and (2) feature analysis and decision support for the motivation example presented in Chapter 14.

We described feature selection problem driven by our effort to understand and model complex phenomena. While modeling complex phenomena, we are looking for a subset of input feature that relate closely to a set of output feature. In contrary to feature extraction that focuses on constructing new features as descriptors of input data, feature selection concentrates on choosing input and output features that form a relationship. Our primary application example was spectral band selection problem that could be encountered in a remote sensing domain, and was specifically applied to agriculture applications. In addition to describing all supervised and supervised band selection methods, we addressed some of the challenges in band selection studies.

The problem of feature selection is encountered more and more often because data producers might not know about all application domains that would receive and analyze their data. Therefore, researchers and engineers gather datasets that were not collected for their chosen analysis. These datasets might contain relevant and partially relevant features, or might have missing values (sensor failure) or widely ranging uncertainties of features. As it was explained in this chapter, a model that includes highly uncertain relevant features could lead to a less accurate prediction than a model without that feature. The methods presented here can help in feature quality assessment as well.

In the second part of this chapter, we presented an example of geographic territorial aggregation, which is a general problem that can be applied in many application domains.

There are many solutions to the problem of territorial partition/aggregation and there has to be a clearly defined optimality metric in order to make the most optimal decision. The goal of the second part was to illustrate how the material presented in all previous chapters of Section IV could be used for a particular analysis and decision support according to the motivation example in Chapter 14. In order to perform boundary aggregation based on average terrain elevation, oak hickory occurrence or auto theft features as shown in Figure 21.4, Figure 21.5, and Figure 21.7 of this chapter, one had to process raw data prior to the final analysis. First, digital elevation maps, forest label maps, FBI tabular reports and U.S. Census Bureau files with boundary definitions had to be gathered and well understood (Chapter 15). Second, all datasets had to be loaded to digital data representations that allowed efficient data storage and information retrieval (Chapter 16). Third, multiple datasets were spatially registered (Chapters 17 and 18), and integrated (Chapter 19), for example, FBI tabular reports with U.S. Census Bureau files, or multiple DEM and forest label files. Fourth, statistical features were extracted for each boundary from multiple maps (Chapter 20) and all features were assessed for inclusion into the final analysis (Chapter 21). Finally, in Chapter 21, an example of decision support based on aggregation, evaluation and comparison of multiple geographic territorial partitions demonstrated the utility of all previous data processing steps for specific class of applications and their end users.

References

1. Helsel, D.R. and R.M. Hirsch, 1992. *Statistical Methods in Water Resources*, Elsevier, New York, p. 522.
2. Ott, R.L. 1993. *An Introduction to Statistical Methods and Data Analysis*, Duxbury Press, Belmont, CA, p. 1051.
3. Duda, R., P. Hart, and D. Stork, 2001. *Pattern Classification*, 2nd ed., Wiley-Interscience, New York, p. 454.
4. Han, J. and M. Kamber, 2001. *Data Mining: Concepts and Techniques*, Morgan Kaufmann Publishers, San Francisco, CA, p. 550.
5. Campbell, James B., 1996. *Introduction to Remote Sensing*, 2nd ed. The Guilford Press, New York, chaps. 16.6 and 17.7.
6. Research Systems, Inc., The Environment of Visualizing Images (ENVI), URL: http://www.rsinc.com/envi/index.cfm.
7. MultiSpec software for hyperspectral and multispectral analyses, Developed at LARS Purdue University, West Lafayette, Indiana, documentation at URL: http://dynamo.ecn.purdue.edu/~biehl/MultiSpec/documentation.html.
8. Bajcsy, P. et al., Image to knowledge, documentation at URL: http://alg.ncsa.uiuc.edu/tools/docs/i2k/manual/index.html.
9. Csillag, F., L. Pasztor, and L. Biehl, 1993. Spectral band selection for the characterization of salinity status of soils, *Remote Sensing of Environment*, 43, 231–242.
10. Yamagata, Y., 1997. Advanced remote sensing techniques for monitoring complex ecosystems: spectral indices, unmixing, and classification of wetlands, Ph.D. thesis, University of Tokyo, Tokyo, Japan, also available as Report R141 (1999), National Institute for Environmental Studies, Ibaraki, Japan, p. 148.
11. Warner, T., K. Steinmaus, and H. Foote, 1999. An evaluation of spatial autocorrelation-based feature selection, *International Journal of Remote Sensing*, 20, 1601–1616.
12. Merényi, E., W.H. Farrand, L.E. Stevens, T.S. Melis, and K. Chhibber, 2000. Mapping Colorado River ecosystem resources in Glen Canyon: Analysis of hyperspectral low-altitude AVIRIS imagery, in *Proceedings of ERIM, 14th International Conference and Workshops on Applied Geologic Remote Sensing*, November 04–06, Las Vegas, Nevada, pp. 44–51.

13. Wiersma, D.J. and D.A. Landgrebe, 1980. Analytical design of multispectral sensors, *IEEE Transactions on Geoscience and Remote Sensing*, GE-18, 180–189.
14. Price, J.C., 1994. Band selection procedure for multispectral scanners, *Applied Optics*, 33, 3281–3288.
15. Hughes, G.F., 1968. On the mean accuracy of statistical pattern recognizers, *IEEE Transactions on Information Theory*, IT-14(1), pp. 55–63.
16. Benediktsson, J.A., J.R. Sveinsson, and K. Arnason, 1995. Classification and feature extraction of AVIRIS data, *IEEE Transactions on Geoscience and Remote Sensing*, 33, 1194–1205.
17. Merényi, E., R.B. Singer, and J.S. Miller, 1996. Mapping of spectral variations on the surface of Mars from high spectral resolution telescopic images, *ICARUS 1996*, 124, 280–295.
18. Gopalapillai, S. and L. Tian, 1999. In-field variability detection and yield prediction in corn using digital aerial imaging, *Transactions of the ASAE*, 42, 1911–1920.
19. Pu, R. and P. Gong, 2000. Band selection from hyperspectral data for conifer species identification, in *Proceedings of Geoinformatics'00 Conference*, June 21–23, Monterey Bay, CA, pp. 139–146.
20. Healey, G. and D.A. Slater, 1999. Invariant recognition in hyperspectral images, in *IEEE Proceedings of CVPR99*, June 23–25, Fort Collins, CA, pp. 438–443.
21. Jia, X. and J.A. Richards, 1994. Efficient maximum likelihood classification for imaging spectrometer datasets, *IEEE Transactions on Geoscience and Remote Sensing*, 32, 274–281.
22. Withagen, P.J., E. den Breejen, E.M. Franken, A.N. de Jong, and H. Winkel, 2001. Band selection from a hyperspectral data-cube for a real-time multispectral 3CCD camera, in *Proceedings of SPIE AeroSense, Algorithms for Multi-, Hyper, and Ultraspectral Imagery VII*, April 16–20, Orlando, FL (SPIE, Bellingham, Washington), 4381, 84–93.
23. Jasani, B. and G. Stein, 2002. *Commercial Satellite Imagery: A Tactic in Nuclear Weapon Deterrence*, Springer Praxis Publishing Ltd., Chichester, UK, p. 229.
24. Bajcsy, P. and P. Groves, 2004. Methodology for hyperspectral band selection, *Photogrammetric Engineering and Remote Sensing Journal*, 70, 793–802.
25. Groves, P. and P. Bajcsy, 2003. Methodology for hyperspectral band and classification model selection, IEEE Workshop on Advances in Techniques for Analysis of Remotely Sensed Data, Washington DC, October 27.
26. Swain, P.H. and S.M. Davis, 1978. *Remote Sensing: The Quantitative Approach*, McGraw-Hill, New York, chaps. 3–8.
27. Fung, T., F. Ma, and W.L. Siu, 1999. Band selection using hyperspectral data of subtropical tree species, in *Proceedings of Asian Conference on Remote Sensing, Poster Session 3*, 22–25 November, 1999, Hong-Kong, China (Asian Association on Remote Sensing), URL: http://www.gisdevelopment.net/aars/acrs/1999/ps3/ps3055pf.htm, last accessed February 12, 2004.
28. Shettigara, V.K., D.O'Mara, T. Bubner, and S.G. Kempinger, 2000. Hyperspectral band selection using entropy and target to clutter ratio measures, *Proceedings 10th Australasian Remote Sensing and Photogrammetry Conference*, 21-25 August, Adelaide, Australia (Proceedings are on a CD).
29. Bruske, J. and E. Merényi, 1999. Estimating the intrinsic dimensionality of hyperspectral images, in *Proceedings of European Symposium on Artificial Neural Networks*, 21–23 April, Bruges, Belgium (IEEE Region 8), pp. 105–110.
30. Russ, J.C., 1999. *The Image Processing Handbook*, 3rd ed., CRC Press LLC, Boca Raton, FL, p. 771.
31. Gill, P.E., W. Murray, and M.H. Wright, 1991. *Numerical Linear Algebra and Optimization*, Vol. 1, Addison-Wesley Publishing Company, Redwood City, CA, p. 410.
32. Witten, L. and E. Frank, 1999. *Data Mining: Practical Machine Learning Tools and Techniques with Java Implementations*, Morgan Kaufmann Publishers, San Francisco, CA, p. 416.
33. Breiman, L., J.H. Friedman, R.A. Olshen, and C.J. Stone, 1984. *Classification and Regression Trees*, Wadsworth International Group, Monterey, CA, p. 358.
34. Quinlan, J. R., 1993. *C4.5: Programs for Machine Learning*, Morgan Kaufmann Publishers, San Francisco, CA.
35. Balanis, C.A., 1989. *Advanced Engineering Electromagnetics*, John Wiley & Sons, New York, p. 981.

36. Miller, H.J. and J. Han, 1999. Discovering geographic knowledge in data-rich environments, Specialist Meeting report, National Center for Geographic Information and Analysis Project Varenius, March 18–20, Redmond, WA.
37. Bajcsy, P., P. Groves, S. Saha, T.J. Alumbaugh, and D. Tcheng, 2003. A system for territorial partitioning based on GIS raster and vector data, Technical report NCSA-ALG-03-0002, February.
38. Bajcsy P., 1997. Hierarchical segmentation and clustering using similarity analysis. Ph.D. thesis, University of Illinois at Urbana-Champaign, IL, p. 182.

Section V

Soft Computing

22

Statistical Data Mining

Amanda B. White and Praveen Kumar

Contents

Data mining encompasses various techniques that are used to statistically analyze data in order to extract relationships, patterns, and general information present, yet hidden, in very large datasets with many variables (Han and Kamber, 2001; Hastie et al., 2001). These techniques are different from traditional statistical methods, which have been designed for small datasets with few variables. Data mining methods are used for classification, prediction, and analysis and can be divided into two basic categories: supervised and unsupervised learning. Learning is used to describe data mining techniques because the relationships extracted from the data are considered to be translatable by an expert to knowledge, and thus "learned." Generally speaking, supervised learning methods use a guide to distinguish patterns in a data and unsupervised learning techniques do not. The following sections provide a broad introduction and overview of commonly used techniques in both categories such that the reader can confidently implement each of the methods. Examples in the realm of hydrology are also provided to allow the reader to become familiar with the type of problems for which these methods can be applied.

22.1 Supervised Learning

Supervised learning techniques are designed to predict, classify, and analyze problems using a response variable as a guide, or supervisor. In general, supervised techniques use $\mathbf{X}$ to predict $\mathbf{Y}$ where $\mathbf{X}$ is an $n \times r$ matrix, n being the number of input attributes and r being the number of points, or records, in the dataset, and $\mathbf{Y}$ is an $o \times r$ matrix, o being the number of output attributes or response variables. A model is trained on a portion of the data and tested on the remainder of the data. In this way, a new response matrix $\hat{\mathbf{Y}}$ is predicted for the test cases and an error associated with the prediction can be computed as a function of $\mathbf{Y} - \hat{\mathbf{Y}}$. The goal is then to minimize this error (Hastie et al., 2001). An overview of several supervised learning techniques is presented, including regression, tree-based methods, and nearest neighbor methods.

22.1.1 Regression Methods

Regression methods are powerful tools for analysis, classification, and prediction that can provide a wealth of information about the problem at hand. Linear models have been well-developed over the past 30 years and remain one of the most essential tools for scientists, engineers, economists, stock market analysts, and virtually any profession in which data is analyzed. Linear regression is essential to understand regression techniques, as multivariate and nonlinear regression are direct generalizations of linear models. Linear, multivariate, and nonlinear regression models will be discussed herein.

22.1.1.1 Linear Regression

Employing a linear regression model (Weisberg, 1980; Seber, 1984) requires making a large assumption that the variables in the dataset are linearly related. A simple optimization technique is used to minimize the distance from each point in the dataset to a line drawn through the data in a multidimensional space. The most popular optimization technique is the method of least squares. Assuming one input and one output variable, the linear model is as follows:

$$\mathbf{y} = \beta_0 + \beta_1 \mathbf{x}, \tag{22.1}$$

where $\mathbf{x}$ is a vector of inputs $\mathbf{x} = (x_1, x_2, \ldots, x_r)$, $\mathbf{y}$ is a vector of outputs $\mathbf{y} = (y_1, y_2, \ldots, y_r)$, r is the number of records in the dataset, β_0 is the intercept of the line, and β_1 is the slope of the line. The goal is to determine coefficients β_0 and β_1 such that sum of the square of the distance between the line and the output $\mathbf{y}$ (referred to as the residual sum of squares, RSS) is minimized:

$$\text{RSS} = \sum_{i=1}^{r} (y_i - \hat{y}_i)^2, \tag{22.2}$$

where $\hat{y}_i = \beta_0 + \beta_1 x_i$ is the predicted value on the line for each input value. Let us denote the previous equation in matrix notation:

$$\text{RSS}(\boldsymbol{\beta}) = (\mathbf{y} - \mathbf{X}\boldsymbol{\beta})^{\text{T}} (\mathbf{y} - \mathbf{X}\boldsymbol{\beta}), \tag{22.3}$$

where $\boldsymbol{\beta}$ is a 2×1 matrix containing β_0 and β_1, T refers to the transpose, $\mathbf{X}$ is an $r \times 2$ matrix containing ones in the first column and the transpose of vector $\mathbf{x}$ as the second column, and $\mathbf{y}$ is transposed to be an $r \times 1$ matrix ($\mathbf{y}^{\mathrm{T}}$). To determine the minimum of a function, the smallest slope is generally sought. Differentiation is used to determine the slope, thus differentiating RSS with respect to $\boldsymbol{\beta}$ and setting it equal to zero (to obtain the minimum slope) gives the normal equation:

$$\mathbf{X}^{\mathrm{T}}(\mathbf{y} - \mathbf{X}\boldsymbol{\beta}) = 0. \tag{22.4}$$

Solving for $\boldsymbol{\beta}$ gives:

$$\boldsymbol{\beta} = (\mathbf{X}^{\mathrm{T}}\mathbf{X})^{-1}\mathbf{X}^{\mathrm{T}}\mathbf{y}. \tag{22.5}$$

Solving this equation provides the optimized coefficients $\boldsymbol{\beta}$. As an aside, Equation 22.5 reduces to the matrix form of a line:

$$\mathbf{y} = \mathbf{X}\boldsymbol{\beta}. \tag{22.6}$$

An example of applying a linear regression model would be to use the discharge rate to determine the surface runoff rate. Typically, the flow rate of a stream is measured using a gage; however, the runoff rate is more complex due to its dependence on precipitation intensity, soil type, antecedent soil moisture, land use, etc. Hence, we want to determine a general model that will help in prediction of the surface runoff based on the measured flow rate. The discharge data used here is from a gaging station just upstream of Lake Decatur, Illinois, on the Sangamon River and the surface runoff at this location is computed using the PACE model (Durgunoglu et al., 1987; Markus et al., 2003). Linear regression can be performed in many math and spreadsheet software, including MATLAB, S-Plus, Mathematica, Excel, etc. In MATLAB, linear regression can be performed using simply:

```
beta = polyfit(x, y, 1);
y_hat = polyval(beta, x),
```

where the function *polyfit* takes the vectors $\mathbf{x}$ and $\mathbf{y}$ and determines $\boldsymbol{\beta}$ and *polyval* evaluates the linear fit. The input vector $\mathbf{x}$ is used to evaluate the linear fit in order to compare the predicted values $\hat{\mathbf{y}}$ to the original values. A plot showing the linear regression along with the equation of the line is in Figure 22.1.

The error associated with linear regression can be determined in various ways; however, the most popular method, coefficient of determination (R^2), is computed as:

$$R^2 = \frac{\sum_{i=1}^{r}(\hat{y} - \bar{y}_i)^2}{\sum_{i=1}^{r}(y_i - \bar{y})^2}, \tag{22.7}$$

where $\bar{y}$ is the mean of the output vector $\mathbf{y}$. The value of R^2 ranges from 0 to 1, 0 signifying no linear relationship and 1 signifying a perfect linear relationship. As can be seen from the R^2 value (0.815) associated with the discharge/runoff regression, the linear model is a good predictor.

22.1.1.2 Multivariate Linear Regression

Multivariate linear regression (Seber, 1984) is an extension of linear regression where multiple inputs are regressed to predict a single output. Multiple outputs should be considered

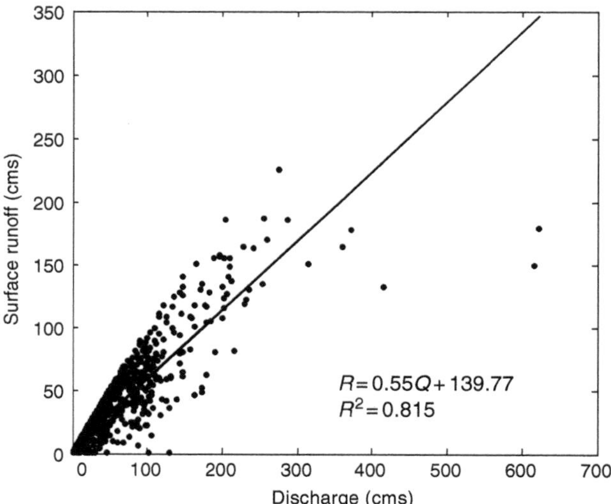

FIGURE 22.1
Linear regression where discharge Q (cubic meters per second, cms) is used to predict runoff R (cms). The equation shows the coefficients computed by MATLAB and the R^2 value illustrates the goodness of fit of the regression.

separately. The multivariate linear regression model is:

$$\mathbf{y} = \beta_0 + \beta_1\mathbf{x}_1 + \beta_2\mathbf{x}_2 + \cdots + \beta_n\mathbf{x}_n, \tag{22.8}$$

where $\mathbf{x}_i$ are vectors of inputs $\mathbf{x}_i = (x_{1i}, x_{2i}, \ldots, x_{ri})$, $\mathbf{y}$ is a vector of outputs $\mathbf{y} = (y_1, y_2, \ldots, y_r)$, r is the number of records in the dataset, n is the number of inputs, and β_n are coefficients. The method of least squares is also used to determine the best fit to the data, or the coefficients β_n, as in linear regression. Thus, following Equation 22.3:

$$\text{RSS}(\boldsymbol{\beta}) = (\mathbf{y} - \mathbf{X}\boldsymbol{\beta})^{\mathrm{T}}(\mathbf{y} - \mathbf{X}\boldsymbol{\beta}), \tag{22.9}$$

where $\boldsymbol{\beta}$ is an $(n+1) \times 1$ matrix containing $\beta_0, \beta_1, \beta_2, \ldots, \beta_n$, T refers to the transpose, and $\mathbf{X}$ is an $r \times (n+1)$ matrix containing ones in the first column and the transpose of vectors $\mathbf{x}_1, \mathbf{x}_2, \ldots, \mathbf{x}_n$ as the second, third, ..., $n+1$ columns, respectively. The problem is resolved in the same manner as for linear regression and the coefficient matrix $\boldsymbol{\beta}$ is determined. That is (per the normal equation — Equation 22.5):

$$\boldsymbol{\beta} = (\mathbf{X}^{\mathrm{T}}\mathbf{X})^{-1}\mathbf{X}^{\mathrm{T}}\mathbf{y}. \tag{22.10}$$

Continuing with the example from the previous section, the precipitation amount, in addition to the discharge rate, is used to determine the surface runoff rate. Thus, two inputs (precipitation and discharge) will be used to predict one output (runoff). Multivariate linear regression can be performed in several math software, including MATLAB, S-Plus,

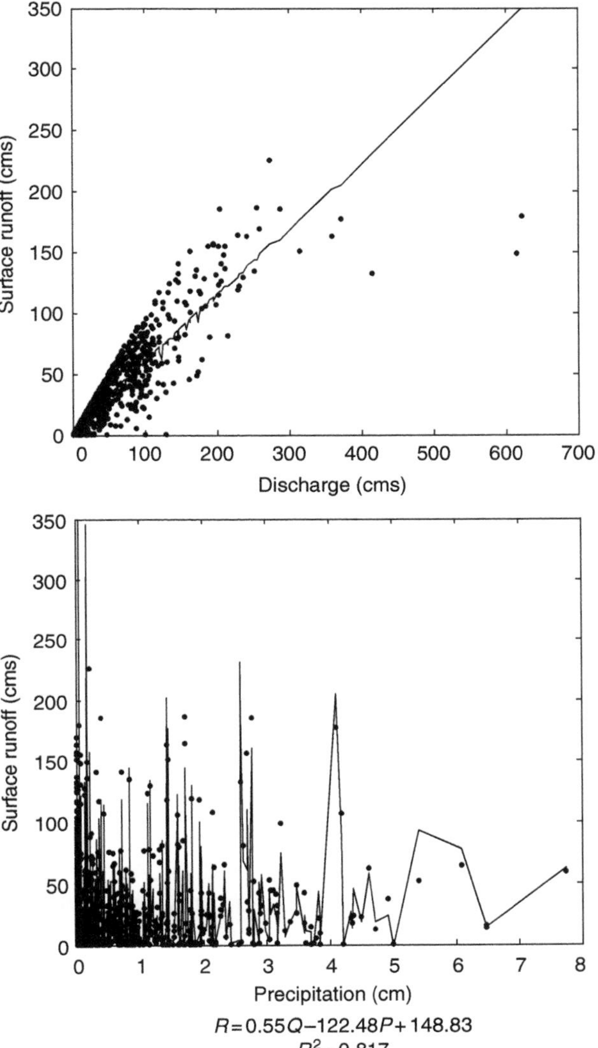

FIGURE 22.2
Multivariate linear regression where discharge Q (cubic meters per second, cms) and precipitation P (centimeters, cm) are used to predict runoff R (cms). The equation shows the coefficients computed by MATLAB and the R^2 value illustrates the goodness of fit of the regression.

and Mathematica. In MATLAB, multivariate linear regression can be performed using:

```
beta = x\y
y_hat = beta*x
```

where the first line solves Equation 22.6 for $\boldsymbol{\beta}$ using matrix $\mathbf{X}$ and vector $\mathbf{y}$ and the second line evaluates the fit by computing predicted values $\hat{\mathbf{y}}$ using the input data ($\mathbf{X}$) and the determined coefficient $\boldsymbol{\beta}$. Figure 22.2 shows plots of both inputs (precipitation and discharge) vs. the output (runoff) and the multivariate linear regression model. As can be seen by the R^2 value, the addition of precipitation to the regression produces a slightly better fit than using only discharge to predict the surface runoff.

22.1.1.3 Nonlinear Regression

Nonlinear regression (Bates and Watts, 1988) is similar to linear regression, in that a model is fit to the data and various coefficients are determined; however, the model for nonlinear regression must be specified, as it does not represent the equation of a line. Hence, the model takes the general form:

$$\mathbf{y} = f(\mathbf{X}, \boldsymbol{\beta}), \tag{22.11}$$

where $\mathbf{X}$ is an $r \times n$ matrix containing the transposed input vectors $\mathbf{x}_1, \mathbf{x}_2, \ldots, \mathbf{x}_n$ as the first, second, ..., n columns, respectively, $\mathbf{x}_i = (x_{1i}, x_{2i}, \ldots, x_{ri})$, $\mathbf{y}$ is a vector of outputs $\mathbf{y} = (y_1, y_2, \ldots, y_r)$, r is the number of records in the dataset, n is the number of inputs, and $\boldsymbol{\beta}$ is a matrix containing the coefficients of the function $f(\mathbf{X},\boldsymbol{\beta})$. The function can be polynomial, exponential, logistic, periodic, or a general nonlinear function, yet the goal remains the same as in linear regression — to determine the coefficient matrix $\boldsymbol{\beta}$ that best models the data. Iterative numerical methods are employed to solve the set of nonlinear equations simultaneously, including: interval bisection, fixed-point iteration, Newton's method, secant method, inverse interpolation, linear fractional interpolation, etc. (Heath, 2002).

Continuing again with the example from the previous two sections, the discharge rate is used to determine the surface runoff rate. Various forms of nonlinear equations are tested, yet all are nonlinear functions where one input (discharge Q) is used to predict one output (runoff R). In general, determining the form of the nonlinear model is either a trial and error process or the model is known previously, such as Manning's equation to determine the velocity in a pipe or open channel $v = (1/n)R^{2/3}S^{1/2}$ where n is Manning's roughness coefficient, R is the radius of the pipe or hydraulic radius of the channel, and S is the slope of the pipe or channel. Nonlinear regression can be performed in several math software, including MATLAB, S-Plus, and Mathematica. In MATLAB, nonlinear regression can be performed using (this form of the model was determined as the best fit to the data through trial and error):

```
beta = nlinfit(x, y, @nonlinear, beta)
function y = nonlinear(beta, x)
y = (exp(x.^ beta(2)).*beta(1)+beta(3)).* x
y_hat = (exp(x.^ beta(2)).*beta(1)+beta(3)).*x
```

where the first line performs the nonlinear regression (function *nlinfit*) and takes the matrix $\mathbf{X}$, vector $\mathbf{y}$, the name of the function containing the model (*nonlinear*), and an initial guess of the coefficient matrix $\boldsymbol{\beta}$. The second and third lines represent the model:

$$R = Q(\beta_1 e^{\beta_2 Q} - \beta_3) \tag{22.12}$$

and the fourth line evaluates the fit by computing predicted values $\hat{\mathbf{y}}$ using the input data ($\mathbf{X}$) and the determined coefficients $\boldsymbol{\beta}$. Figure 22.3 shows plots of the nonlinear regression model and, as can be seen by the R^2 value, a nonlinear model produces a better predictor of surface runoff than linear or multivariate linear models.

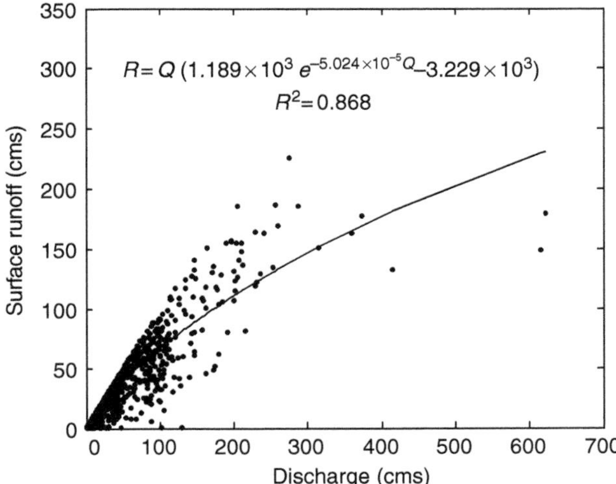

FIGURE 22.3
Nonlinear regression where discharge Q (cubic meters per second, cms) is used to predict runoff R (cms). The equation shows the coefficients computed by MATLAB and the R^2 value illustrates the goodness of fit of the regression.

22.1.2 Tree-Based Methods

Tree-based methods are supervised data mining techniques in which a tree-structured model is built on both input and output data and then used to classify, or predict, new, "unseen" data where the outcome is unknown. Tree-based techniques are useful when there are numerous variables in the database and when nonlinear relationships exist between the variables. This is due to absence of a prespecified model, such as in regression techniques. Classification and regression trees will be discussed, along with implementation issues and improvement techniques. The basic difference between classification and regression trees is the form of the output, which is discrete for classification trees and continuous for regression trees. For the reader to become acquainted with the terminology used in tree-based algorithms, an example is presented in Figures 22.4 and 22.5. Figure 22.4 is a classification tree and Figure 22.5 is the database from which this tree could have been formed.

22.1.2.1 Classification Trees

Classification trees (often referred to as decision trees) are named as such because they were initially created to classify data into discrete classes, or output (e.g., high/medium/low temperature, gaged/ungaged basin, total/partial/no cloud cover, etc.), using a set of discrete input attributes (e.g., high/medium/low precipitation, specific humidity, incoming radiation, etc.), as in the tree in Figure 22.4. ID3 (Quinlan, 1986) was the first classification tree induction algorithm of this kind. Its successor, C4.5 (Quinlan, 1993), is designed to handle continuous, or numerical, data as *input* (not output) and is currently the most widely used classification tree algorithm. In general, classification tree induction is performed by:

1. Placing the data (input and output) into a single node.
2. Computing the *information gain ratio* (discussed below) for each input attribute A based on a predefined *splitting criterion*.
3. Selecting the input attribute with the highest information gain ratio, A_{high}.

Input attributes				Class or Output
Discharge	Fish habitats?	Inhabitants upstream?	Endangered species?	Should A dam be removed?
High	True	X	True	NO
Medium	X	True	X	NO
High	True	X	False	YES
High	False	X	X	YES
Medium	X	False	X	YES
Low	X	X	X	YES

Example of the data that could be used to form the classification tree in Figure 22.5. Note that there are several input attributes, but only one output, or class; thus, the data will be *classified* as either YES or NO with respect to the question, "Should a dam be removed?" There are two values in the output, therefore there are two classes in this database. Each row is referred to as a record. The "X" signifies missing values (in a real database, this issue must always be dealt with).

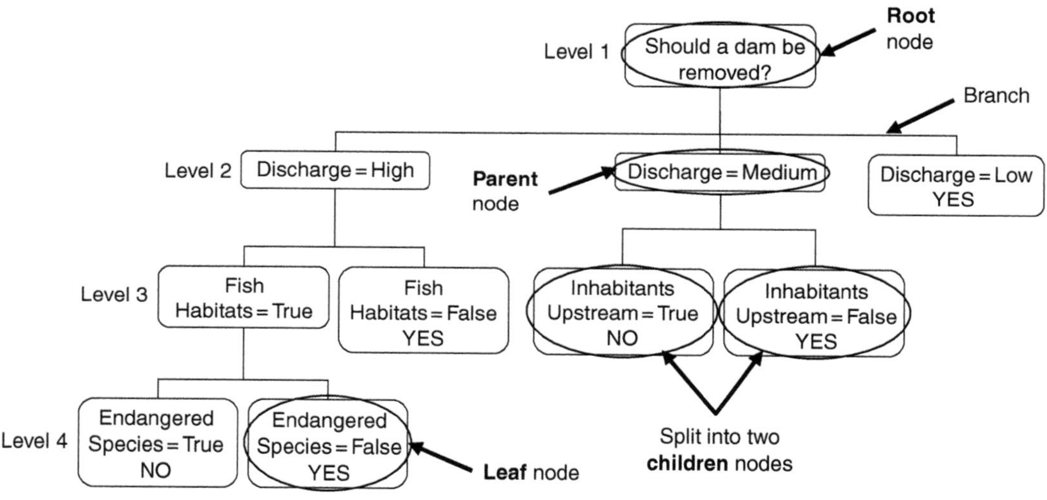

FIGURE 22.4
A classification tree has levels, branches, and nodes, and a split refers to the division between nodes. Each node has a population, which refers to the number of records in the node. The root node is the first node in a tree and the leaf nodes are the last nodes in a tree after following a branch to its end. When a parent node is split, the result is two or more children nodes. The depth of the tree refers to the number of levels and the breadth of the tree generally refers to the abundance of leaf nodes.

4. Splitting the data on input attribute A_{high} into f new, children nodes.
5. Repeating steps (2) through (4) for each of the f children nodes until a predefined *stopping criterion* is met.

Splitting criteria are different for discrete and continuous input attributes. If A_{high} is discrete, f is the number of instances of A_{high}. For example, if A_{high} contains entries of high, medium, and low temperatures, then $f = 3$ (i.e., node 1 contains records with temp = high, node 2 contains records with temp = medium, and node 3 contains records with temp = low). If A_{high} is continuous, f is typically two, where a splitting criterion such as the mean, median, or mode of the population in the node is used, although percentiles can also be used to give, say, four nodes if using quartiles. For example, if A_{high} is temperature and the database contains three entries: {20, 25, 30}, then $f = 2$ and node 1 contains records with temp > 25 and node 2 contains records with temp ≤ 25.

The *information gain* is a measure of the "goodness of split," or the amount of information realized by splitting on a particular input attribute, A. Note that, to perform the first split in the tree, the information gain is computed for the *entire* population, yet after the first split, the information gain is computed for the population *within the node*. Information is computed numerically using the concept of entropy, as in the field of information theory. The information gain Gain(A) is computed as the information attained by splitting on the output class $I(s_1, s_2, \ldots, s_m)$ minus the information attained by splitting on an input attribute $E(A)$:

$$\text{Gain}(A) = I(s_1, s_2, \ldots, s_m) - E(A), \tag{22.13}$$

where $I(s_1, s_2, \ldots, s_m)$ is:

$$I(s_1, s_2, \ldots, s_m) = -\sum_{i=1}^{m} p_i \log_2(p_i) \tag{22.14}$$

m represents the number of distinct values in the output (or classes C_i for $i = 1, \ldots, m$) in the population, p_i is the probability that an arbitrary record belongs to class C_i and is estimated as s_i/s, s_i being the number of records belonging to class C_i and s being the number of records in the population. $E(A)$ is defined as:

$$E(A) = \sum_{j=1}^{J} \frac{s_{1j} + \cdots + s_{mj}}{s} I(s_{1j}, \ldots, s_{mj}), \tag{22.15}$$

where J is the number of distinct values of input attribute A, and s_{ij} is the number of records belonging to class C_i in the population set j.

The *information gain ratio* is typically used to split the data because the *information gain* tends to favor attributes with a large number of instances, or values. The information gain ratio is computed as the information gain Gain(A) divided by the information contained by the split itself Split(A):

$$\text{Split}(A) = -\sum_{j=1}^{J} p_j \log_2(p_j), \tag{22.16}$$

where p_j is the probability that an arbitrary record within the population has an attribute value of j and is estimated as s_j/s, s_j being the number of records that have the attribute value j and s being the number of records in the population.

The data is typically partitioned into training and testing (unseen) sets to gauge the performance of the classification tree. The tree is built on the training data and tested on the testing data. The traditional method of dividing data into training and testing sets is a random 2/3 to 1/3 split, respectively (Han and Kamber, 2001). Other methods have been enlisted to perform the division, most of which require similar statistical attributes of the two datasets, such as ensuring that the training and testing sets have a similar mean and standard deviation. This type of statistical analysis should be performed on the two sets regardless of the division method in order to generate unbiased results. The testing error, which is considered the overall error of the tree, is computed by dividing the number of misclassified test records by the total number of test records.

There are a myriad of stopping criteria for a classification tree, the selection of which is somewhat arbitrary. These include: setting a minimum population of the leaf nodes, a maximum number of levels in the tree, a minimum error, or allowing the tree to develop until there is one record in each leaf node and then pruning the tree (removing branches from the tree) backward from the leaf nodes until a pruning criterion is met. The most popular and logical pruning criterion is to remove sets of leaf nodes, or subtrees, until a minimum overall error is reached (based on the test records). The process of pruning raises the issue of *over-fitting* a model to the data. One of the goals when building a classification tree is to find a model that best fits the data, or produces the least error, when used to classify unseen, test records. As the tree grows, the error will generally decrease until it reaches a minimum, at which point it begins to increase, or over-fit the model to the data. Over-fitting is undesirable because it not only increases the error, but also produces a specialized model that embodies the nuances of the data, whereas a more general model that can be used for accurate prediction of new data is usually the objective. Thus, to avoid over-fitting, stopping and pruning criteria are used.

The example presented here uses elevation, soil properties (percent sand, silt, and clay), and vegetation density to predict the land use. The study area is located in the Appalachian Mountains of Virginia in the United States. Elevation is obtained from the 1:250,000-scale U.S. Geological Survey (USGS) digital elevation model (DEM) and soil properties are computed using the 1:250,000-scale STATSGO database. The vegetation density is represented by the enhanced vegetation index (EVI) obtained from the MODIS-Terra satellite at 250-meter resolution for the growing season months (April to September). For each pixel, the biweekly EVI values are averaged over the month. The USGS 1:250,000-scale land use maps are acquired and all data are aligned on the same 250-meter grid where each pixel represents a record in the database. The D2K (Data to Knowledge) software (Welge et al., 1999) produced by the Automated Learning Group (ALG) at the National Center for Supercomputing Applications (NCSA) is used to build the classification tree, although C4.5 (Quinlan, 1993) is the underlying algorithm. The data was divided into training and testing sets according to the traditional 2/3 to 1/3 split, respectively.

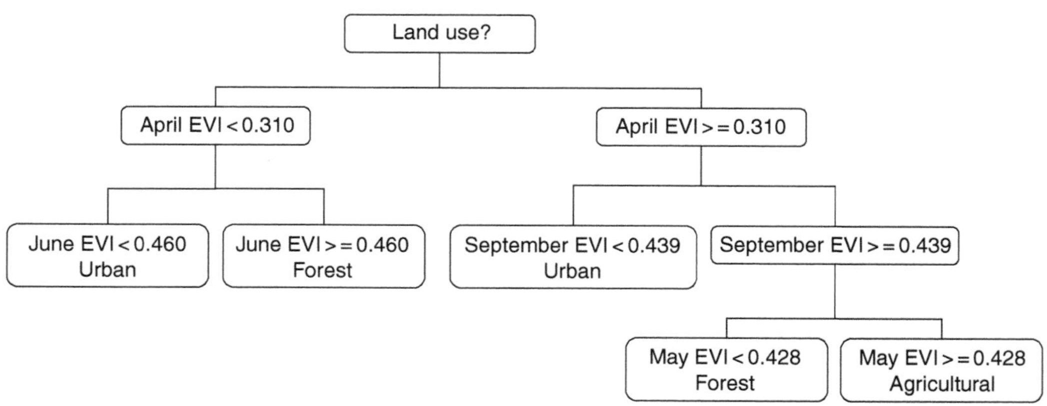

FIGURE 22.5
The classification tree built to determine the land use based on elevation, percent sand, silt, and clay, and vegetation density throughout the growing season. See Figure 22.7 for an example of the database used to create the tree and Figure 22.8 for the confusion matrix detailing the performance of the tree.

		Input attributes					Class or Output
Elevation	Percent sand	Percent silt	Percent clay	April EVI		September EVI	Land use
685	35	48	17	0.298	...	0.563	Forest
701	20	33	48	0.460		0.552	Agricultural
609	18	29	53	0.295		0.423	Urban

FIGURE 22.6
An example of the database used to build the classification tree in Figure 22.6. The classes are: forest, agricultural, and urban land use.

	Urban	Forest	Agricultural	Recall	Type II error
Urban	24	25	49	24.5%	75.5%
Forest	8	1067	130	88.5%	11.5%
Agricultural	4	108	607	84.4%	15.6%
Precision	66.7%	88.9%	77.2%		
Type I Error	33.3%	11.1%	22.8%		

FIGURE 22.7
Confusion matrix for the classification tree in Figure 22.6.

In Figure 22.6, the classification tree that was created is displayed, and in Figure 22.7, an example of the database is portrayed. A confusion matrix (Figure 22.8) is typically used to examine the accuracy of the model, in addition to the overall prediction/misclassification error, which is approximately 16.0% in this example. The identifiers at the top of the confusion matrix signify the class in which a record *should* be placed and the identifiers on the left signify the class in which a record is placed by the classification tree model. Thus, the values along the diagonal represent the records that were correctly classified. For each class, the precision is the number of correctly classified records (the diagonal value) divided by the total number of records that *actually* belong to the class (the summation of the column). The recall for each class is the number of correctly classified records (the diagonal value) divided by the total number of records *classified by the model* as belonging to the class (the summation of the row). Errors of type I and II are used to refer to the percent error associated with the precision and recall, respectively. The classification tree performed well with respect to the forested and agricultural land uses; however, the model performed poorly for the urban land use. This could be due to the absence of input attributes that identify urban settings such as population density, impervious area, temperature, etc. Also of interest are the attributes that were used to compose the tree. Here, the vegetation density was computed as being the most significant classifier, as no other attributes were used. In addition, the closer the attribute is to the root node, the more significant the attribute is, and hence the April EVI is the most significant attribute in this example because it is used to perform the first split of the root node.

22.1.2.2 Regression Trees

Regression trees (Breiman et al., 1984) are built in a similar manner to classification trees, yet regression trees allow for continuous output. In other words, the output of a classification tree is a class (i.e., temperature > 25 or temperature ≤ 25), whereas the output of a regression tree is represented by the mean and standard deviation of the population within the leaf

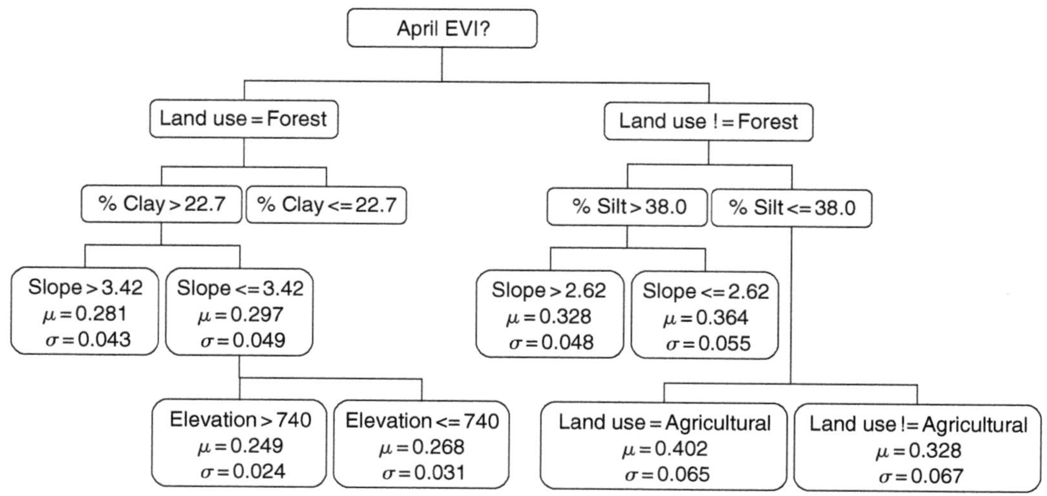

FIGURE 22.8
Regression tree using elevation, slope, soil properties (percent sand, silt, and clay), and land use to predict the vegetation density (EVI) in April. The mean (μ) and standard deviation (σ) of the April EVI over the population in each leaf node is shown. The symbol "!=" is used to state that the land use is not equal to a particular value.

node (i.e., temperature mean = 25 and temperature standard deviation = 5). Regression trees are advantageous because the output does not have to be discretized, or binned, which can lead to problems with the results, including: biases, large errors, and obscured information. The difference in the algorithm is simply that the best split is no longer chosen based on the maximum information gain ratio — it is chosen based on the minimum sum of the squared deviation from the mean of the population within the node, which is a measure of the *similarity* within the node:

$$\text{Similarity} = \sum_{i=1}^{r}(x_i - \bar{x})^2, \tag{22.17}$$

where x_i is the input attribute value, $\bar{x}$ is the mean of the input attribute values of the population within the node, and r is the number of records within the node. Again, the mean, median, mode, percentiles, or some other statistic of the input attribute is used to split the data into g nodes, although some regression trees search for the best splitting criterion. The error associated with each leaf node is the root mean square error (RMSE) between the testing set output values that fall within the leaf node after being filtered through the tree and the mean of the training set output values that fall within the leaf node. The overall error of the tree is the average RMSE of the leaf nodes.

A regression tree that was built using the same data from the example in Section 22.1.2.1 is presented in Figure 22.9. The only (yet crucial) difference from the classification tree is that the land use is an *input* attribute and the vegetation density in April (as represented by the continuous index, EVI) is the *output*. This is possible because a continuous attribute can be the output of a regression tree algorithm. Slope was also included in this analysis. The D2K software produced by the ALG at the NCSA is used (Tcheng et al., 1991). The error, or spread of the data, associated with each leaf node is shown in Figure 22.9 as the standard deviation, and the overall error of the regression tree is 0.00228 (variance) or 0.0477 (standard deviation).

22.1.2.3 Improvement Techniques

Several techniques have been designed to improve the accuracy of classification/prediction models, including windowing, bagging, and boosting. These methods will be discussed in the context of classification trees, although each can easily be modified to accommodate regression trees. The simplest technique to increase the accuracy of a tree-based model is to build several trees using different, randomly selected training/testing sets and choose the one with the least error. Windowing (Quinlan, 1993) is slightly more complex, such that:

1. The training set is randomly selected and a window, or subset, of the training set is uniformly selected (to create a balanced distribution of classes).
2. A model is built based on the data in the window and is used to classify the remainder of the training set (outside the window).
3. A portion of the misclassified records are added to the window.
4. Steps (2) and (3) are repeated until a reasonable number (noise and indeterminacy must be accounted for) of the records in the training set outside the window are correctly classified.

In bagging (Quinlan, 1996), several training sets of the same size as the original dataset are sampled *with replacement* from the original dataset (referred to as bootstrapping). Each of these training sets is used to build a tree-based model, and then *all* of the records in the original dataset are classified by each model. A count of the number of times a record is classified as belonging to a particular class is kept track of, and the final classification of a record is decided by the majority (referred to as majority voting) and ties are resolved arbitrarily. The underlying assumption of bagging is that the system represented by the data is unstable, where small perturbations (manifested as the sampling of the training sets) can produce entirely different classification models. This instability is "captured" by bagging and has been shown to improve the accuracy of the model.

Although conceptually similar to bagging, boosting (Quinlan, 1996) is more elaborate because it uses a weighing scheme such that, the higher the weight assigned to the record, the more influence the record has on the model. The process of boosting is as follows:

1. All of the records in the dataset are assigned weights equal to $1/N$, where N is the number of records in the dataset and the trial number t is set to 1.
2. A model is built based on a randomly selected training set, the error ε^t is computed based on the remaining records (i.e., the testing set) as the sum of the weights of the misclassified records, and:
   (a) If $\varepsilon^t \geq 0.5$ (a majority of the testing set is being misclassified), the trials are terminated and T is set to $t - 1$.
   (b) If $\varepsilon^t = 0$ (none of the testing set is being misclassified), the trials are terminated and T is set to t.
   (c) Else, the weights of the testing set are reassigned by multiplying the weights of the correctly classified records by the factor $\theta^t = \varepsilon^t/(1-\varepsilon^t)$. The weights are then renormalized for the whole dataset such that the sum of the weights is equal to 1.
3. The trial number is set to $t + 1$ and steps (2) and (3) are repeated until $t = T$ or the process is terminated (as in steps 2(a) and 2(b)).
4. All of the records are classified based on majority voting from all T models, where each vote is worth $\log(1/\theta^t)$.

Boosting also requires instability of the system represented by the dataset (manifested by the two termination requirements on ε^t: if $\varepsilon^t \geq 0.5$, the model is under-fitting the data, and if $\varepsilon^t = 0$, the model is over-fitting the data). As far as increasing the accuracy of the model, boosting generally outperforms bagging, although not consistently, as severe degradation has been observed in the performance of the model with boosting.

22.1.3 Nearest Neighbor Methods

One of the conceptually simplest methods of classifying, predicting, and analyzing data is the nearest neighbor method. The goal is to classify a record based on the "nearest" records, as defined by some distance measure such as the Euclidean distance:

$$d = \sqrt{\sum_i (x_i - y_i)^2}, \tag{22.18}$$

the city-block distance:

$$d = \sum_i |x_i - y_i|, \tag{22.19}$$

or the Minkowski distance:

$$d = \left[\sum_i |x_i - y_i|^p\right]^{1/p}, \tag{22.20}$$

where p is some arbitrary number. First, the data is divided into training and testing sets (this can be performed in various ways — see Section 22.1.2.1 on classification trees for more information). Then, the distance from each record in the testing set to each record in the training set is computed. Finally, for each testing record, the nearest training records are determined, and the testing record is assigned to the class to which the majority of the nearest training records belong. Ties are randomly broken.

Nearest neighbor techniques generally perform well on databases with continuous input attributes, for the distance between two numerical, or continuous, values is simple to compute. However, the concept of distance for discrete attributes is relative because one must specify the distance between the instances of the attribute, which can be challenging and can lead to biases in the results. For example, what is the distance between sandy clay and silty clay soil? In general, to perform nearest neighbor classification, continuous input attributes are most appropriate, thus using porosity, hydraulic conductivity, or some other numerical measure of the soil's properties may be more suitable.

There are several types of nearest neighbor techniques. The k-nearest neighbor approach (Fix and Hodges, 1951; Dasarthy, 1991; Ripley, 1996) is by far the most commonly used and is discussed in detail here; however, variations of this technique have been developed, such as adaptive methods (Short and Fukunaga, 1981; Myles and Hand, 1990; Friedman, 1994; Hastie and Tibshirani, 1996), which are better suited to handle high-dimensional classification problems.

22.1.3.1 k-Nearest Neighbor

The most popular nearest neighbor approach is the k-nearest neighbor. The k value is the number of training records that are considered to be "near" a testing record and is specified *a priori*. The algorithm is described in detail as follows. The distance between two points x and $x_{i'}$ is simply:

$$d(x_i, x_{i'}) = x_i - x_{i'}. \tag{22.21}$$

Now, consider a point x_i to be composed of a vector of inputs $\mathbf{x}_i = (x_{i1}, \ldots, x_{in})$, which corresponds to an output y_i where i represents the point, or record, and n is the number of inputs. The most common method to measure the distance between two vectors is the squared Euclidean distance. Thus, the distance between two vectors $\mathbf{x}_i$ and $\mathbf{x}_{i'}$ is:

$$d(\mathbf{x}_i, \mathbf{x}_{i'}) = \sum_{j=1}^{n} (x_{ij} - x_{i'j})^2, \tag{22.22}$$

where x_{ij} refers to the elements of $\mathbf{x}_i$. In the k-nearest neighbors algorithm, a "distance matrix" is computed containing the distances between each training and testing record, which creates a symmetric matrix (the distance between $\mathbf{x}_i$ and $\mathbf{x}_{i'}$ is typically equal to the distance between $\mathbf{x}_{i'}$ and $\mathbf{x}_i$) with a diagonal of zeros (the distance between $\mathbf{x}_i$ and $\mathbf{x}_i$ is zero). Once the distance matrix is computed, the k-nearest training records to each of the testing records is determined through sorting. If the output is discrete (which is traditionally the case), majority voting is used to establish the class to which a testing record belongs. Meaning, a testing record is assigned to the class to which the majority of the k-nearest training records belong and ties are broken at random. The percent of the testing records that are misclassified is used to determine the accuracy of the model. If the output is continuous, the mean, median, mode, etc., of the k-nearest training records can be used to predict the output of the testing records. In this case, the difference between the predicted value and the actual value will provide the accuracy of the model. The term model is used here, yet technically no model is built — there is no regression function or classification tree developed with the k-nearest neighbor technique, which is considered to be a disadvantage of this method.

When using the k-nearest neighbor technique, there are several important issues to consider. One is that, if a particular input attribute's values are much larger than the other input attributes' values, the distance between the two records may be dominated by the larger values. Therefore, the data should be normalized before performing the algorithm. Another issue is the large amount of memory required to store the distance matrix, particularly if the dataset is very large. In such a case, the distance matrix cannot be computed *a priori*, which will cause slow processing times and difficulties in determining which training records are nearest to the testing records. One way to avoid this problem is to store only the upper or lower triangular matrix, since the distance matrix is symmetric. Another aspect of the k-nearest neighbor approach that can slow processing time is the memory-intensive process of sorting, although this difficulty has become less significant due to advances in the field of database management (Flores, 1969; Knuth, 1981).

An important aspect of the k-nearest neighbor method that has been overlooked is the determination of k. An optimal k value can be found using trial-and-error over a range or by using an optimization technique such as the golden section search, Newton's method, or genetic algorithms (Goldberg, 1999; Heath, 2002). Typically, the error will increase with the k value, although this is not always the case (Figure 22.10).

Using the database described in Section 22.1.2.1, the elevation, slope, percent sand, silt, and clay, and the vegetation density (monthly-averaged EVI from April to September) are used to predict the land use. A traditional 2/3 to 1/3 split is performed to create the training and testing datasets, respectively, and MATLAB is used to execute the algorithm

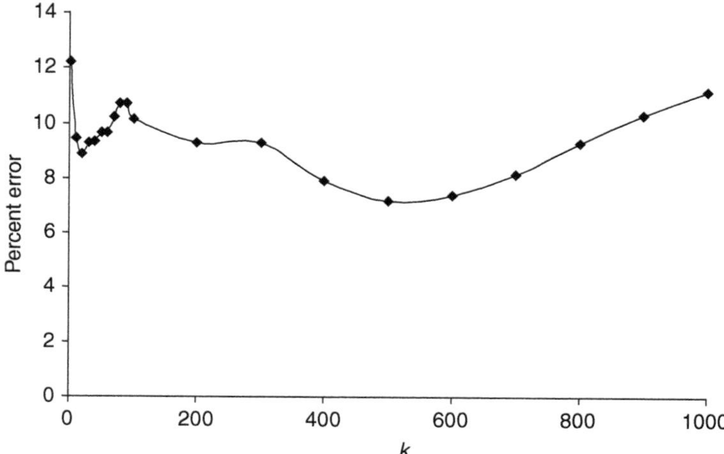

FIGURE 22.9
Illustration of the variance of accuracy with respect to k. This figure also portrays the results of the k-nearest neighbor example presented in this section.

obtained from The MathWorks, Inc., which is freely available and can be found on their Web site. The percent of misclassified records with respect to k is shown in Figure 22.10. It is interesting that this plot is bimodal, having a local minimum at $k = 20$ (8.9% error) and a global minimum at $k = 520$ (7.1% error). Thus, if one were to use an optimization technique that can be trapped by local minima or maxima (e.g., the golden section search), the optimal value for k might not have been discovered.

22.2 Unsupervised Learning

Like supervised learning techniques, unsupervised methods are designed to predict, classify, and analyze problems; however, unlike supervised techniques, a response variable is not used as a guide, or supervisor. In unsupervised methods, one has a set of observations **X** and the goal is to directly infer properties of the dataset without the assistance of a supervisor to provide the "correct" answers (where **X** is an $r \times n$ matrix, r being the number of points, or records, in the dataset, and n being the number of input attributes). An advantage of unsupervised techniques is that the attribute dimension of **X** is generally higher than for supervised techniques, as the algorithms no longer determine how the properties of **X** change with respect to another set of variables (the output **Y**). The disadvantage of eliminating the supervisor is that there is no straightforward way to measure success, or accuracy, thus making it difficult to draw reliable conclusions from the output of most unsupervised learning techniques (Hastie et al., 2001). In this section, an overview of two different unsupervised learning techniques is presented: association rules and cluster analysis.

22.2.1 Association Rules

Association rules first appeared as a method of mining commercial databases for marketing purposes such as stocking shelves, sales promotion, catalog design, etc. (Agrawal et al.,

1993). It has since been used for engineering management and design, scientific research, and many other purposes. The most attractive aspect of association rules is their simple and comprehensible nature, for the output is a set of rules such as:

IF (soil type = sand) AND (particle size = coarse) THEN (porosity = low).

This example rule is intuitive, yet association rule algorithms search the dataset to extract rules that are not immediately intuitive, thus providing information that the user can translate into knowledge. The specific goal of association rule algorithms is to find joint values of the attributes in **X** that appear most frequently in the dataset. For discrete data this task is tractable, but for continuous data it would be nearly impossible computationally; consequently, it is necessary to discretize the data into contiguous numerical ranges based on statistical properties of the attribute values. As discussed in the supervised learning section, there are disadvantages to binning data, yet without binning, the problem would be too difficult to solve. The most common association rule algorithm is the "*A priori*" algorithm (Agrawal and Srikant, 1994; Mannila et al., 1994; Agrawal et al., 1996). This algorithm will be discussed in detail, along with extensions to the algorithm and related issues.

22.2.1.1 A Priori *Algorithm*

In order to determine the most frequent joint values of attributes in a dataset, one must first define what is meant by "frequent." Many measures of frequency have been put forth, such as *support*, *confidence*, and *interest*:

$$support(A \Rightarrow B) = P(A \cup B) = \frac{\text{number of records with both } A \text{ and } B}{\text{total number of records in dataset}}, \tag{22.23}$$

$$confidence(A \Rightarrow B) = P(B|A) = \frac{\text{number of records with both } A \text{ and } B}{\text{number of records with } A}, \tag{22.24}$$

$$interest(A \Rightarrow B) = \frac{P(A \cup B)}{P(A)P(B)} = \frac{support(A \Rightarrow B)}{support(A)support(B)}, \tag{22.25}$$

where A and B are attribute values (i.e., soil type = sand) and $A \Rightarrow B$ refers to the if–then structure of a rule (if A, then B). The *interest* is related to the correlation between A and B such that: if *interest* = 1, A and B are independent and $P(A \cup B) \approx P(A)P(B)$; if *interest* <1, A and B are negatively correlated (the occurrence of one discourages the occurrence of the other); and if *interest* > 1, A and B are positively correlated (the occurrence of one encourages the occurrence of the other). The greater the magnitude of the *interest*, the more dependence exists between A and B. In setting a minimum frequency threshold t, only the rules that are considered "frequent" will be returned. Formally, the problem solved by the *A priori* algorithm is:

$$\{H|\text{support}(H) > t\}, \tag{22.26}$$

where h is a set of attributes with particular values such as {*soil type* = sand, *particle type* = coarse, *porosity* = low} and H consists of all sets h ($h \in H$) that can be formed from the dataset **X** that are above the threshold t. Due to memory restrictions associated with large datasets (i.e., not being able to hold the data in the computer's random access memory, RAM), the solution to this problem requires exploitation of the "*A priori* property." This property states that if a set h has *support* greater than the threshold t, any subset of h also has *support* greater than t. In this way, the number of times the data is read from memory is reduced and the algorithm can be performed on a large dataset in a reasonable amount of time.The *A priori* algorithm is performed as follows:

1. The number of attributes n in a set h is set equal to 1 and the threshold t is fixed.
2. The data is read and sets with n attributes are formed (if $n > 1$, the sets are formed from the existing sets that were not discarded).
3. The *support* of each set is computed and sets with *support* less than threshold t are discarded.
4. $n = n + 1$ and Steps (2) through (4) are repeated until no sets can be formed with *support* above the threshold t.

The sets formed in the last loop with *support* above the threshold t are returned and these frequent sets are translated into association rules. First, all possible rules from each set are formed. Consider the previously discussed set {*soil type* = sand, *particle type* = coarse, *porosity* = low}. There are three items in this set ($n = 3$), thus three unique, single-outcome rules can be formed:

IF (soil type = sand) AND (particle type = coarse) THEN (porosity = low)
IF (soil type = sand) AND (porosity = low) THEN (particle type = coarse)
IF (porosity = low) AND (particle type = coarse) THEN (soil type = sand).

The number of unique, multiple-outcome rules that can be formed is:

$$\text{Number of rules} = \sum_{r=1}^{n} \frac{n!}{r!(n-r)!}, \tag{22.27}$$

where $n!/r!(n-r)!$ is the combination of n items taken r at a time, or ${}_nC_r = \binom{n}{r}$. Once the rules for each set are formed, the rules with *confidence* and *interest* above a particular threshold are retained. Traditionally, the *confidence* is used to determine the most frequently occurring rules; however, a rule with high *support* and *confidence* does not necessarily mean that it is a strong rule. For example, say there are 10,000 records in a database and *soil type* = sand occurs in 6,000 records, *porosity* = low occurs in 7,500 records, and both occur in 4,000 records. If the minimum *support* is 30% and the minimum *confidence* is 60%, the following rule would be retained since the *support* and *confidence* thresholds are satisfied and it would be considered a strong association rule:

$$\text{IF (soil type = sand) THEN (porosity = low)}$$
$$[\text{support} = 4,000/10,000 = 40\%,\ \text{confidence} = 4,000/6,000 = 66\%].$$

Nevertheless, this rule is misleading since the probability of *porosity* = low occurring in the database is 7,500/10,000 = 75%, which is even larger than 66%. Consequently, this rule is considered neither strong nor interesting. Use of the *interest* measure helps to avoid these situations, since it determines the dependence of A on B in the rule $A \Rightarrow B$.

The number of rules that are extracted from a dataset depends on the threshold limits set for *support*, *confidence*, and *interest*, and typically a large number of rules are formed. To remedy this overabundance of rules and to find rules that are of interest, metarule data mining techniques have been developed (Klemettinen et al., 1994; Fu and Han, 1995).

Metarules are formed from the output of an association rule algorithm. For instance, if the following two rules are obtained in the output:

IF (soil type = sand) AND (particle size = coarse) THEN (porosity = low)
IF (soil type = sand) AND (particle size = very coarse) THEN (porosity = low)

this metarule could be formed:

IF (soil type = sand) AND (particle size = coarse to very coarse) THEN (porosity = low).

The metarule conveys the information in a more compact manner, thus saving the user from the often daunting task of sorting through each of the numerous rules. Metarule extraction methods are usually performed in database software such as Microsoft Access so that the user can query both the rules and the metarules to find information that is of interest to them.

The example presented here implements the *Apriori* algorithm to predict the vegetation density (EVI) during April using the elevation, slope, soil properties (percent sand, silt, and clay), and land use (analogous to the regression tree example in Section 22.1.3.2). The database used to develop the regression tree is also used here, although each of the attributes is binned in order to perform the algorithm (a drawback of association rule approaches). Each continuous attribute (elevation, slope, percent sand, silt, and clay, and the April EVI) is divided into quartiles and assigned as low (0–25th percentile), low-mid (25–50th percentile), mid–high (50–75th percentile), or high (75–100th percentile). A minimum *support* threshold of 10.0% and a minimum *confidence* threshold of 70% are used, resulting in the rules shown in Figure 22.11. The D2K software produced by the Automated Learing Group at the National Center for Supercomputing Applications is used. Upon examination of the rules, consistencies and trends can be seen, such as: in each rule one of the requirements is that the land use is agricultural, in the rules with the greatest *support* (first 5 rules) the soil properties appear, and in each rule the outcome is that the vegetation is very dense (*April EVI* = high). This set of rules implies that soils with low to mid sand content, low silt content, and mid to high clay content are more adept for agricultural purposes in this area, and that farming tends to occur in areas with low slopes and low to mid elevations.

IF	THEN	*S* (%)	*C* (%)
(*land use* = agricultural) AND (*percent sand* = low-mid)	EVI = high	13.13	75.09
(*land use* = agricultural) AND (*percent sand* = low-mid) AND (*percent silt* = low) AND (*percent clay* = mid-high)	EVI = high	12.19	77.24
(*land use* = agricultural) AND (*percent_sand* = low-mid) AND (*percent silt* = low)	EVI = high	12.19	77.24
(*land use* = agricultural) AND (*percent silt* = low) AND (*percent clay* = mid-high)	EVI = high	12.19	77.24
(*land use* = agricultural) AND (*percent sand* = low-mid) AND (*percent clay* = mid-high)	EVI = high	12.19	77.24
(*land use* = agricultural) AND (*slope* = low)	EVI = high	11.16	72.61
(*land use* = agricultural) AND (*elevation* = low-mid)	EVI = high	10.61	71.42

FIGURE 22.10
Set of association rules generated by the *A priori* algorithm. Elevation, slope, soil properties (percent sand, silt, and clay), and land use are used to predict the vegetation density (EVI) during April. S = *support* and C = *confidence*.

22.2.2 Cluster Analysis

Clustering generally refers to the grouping of a set of objects into subsets, or clusters, such that the objects within a cluster are more similar to one another than the objects in other clusters. In the context of data mining, the objects are records, or vectors $\mathbf{x} = (x_1, \ldots, x_n)$, where n represents the number of attributes, and the measure of similarity/dissimilarity is the distance between the vectors. Clustering is related to nearest neighbor methods, in that records are grouped based on the distance between them; however, clustering algorithms operate differently than nearest neighbor algorithms because they are unsupervised, whereas nearest neighbor algorithms are supervised. Several clustering techniques are presented here due to their popularity among the data mining methods. Note that, when implementing most clustering algorithms, outlier detection and normalization prior to implementation is of utmost importance, else the results may be biased.

22.2.2.1 *k-Means Clustering*

The k-means clustering algorithm was first developed by MacQueen (1967) and continues to be used today due to its speed, simplicity, and excellent performance. The goal of k-means is to partition the dataset in such a way that minimizes the within-cluster distances, or maximizes the between-cluster distances. A cluster center is represented by the mean $\bar{\mathbf{x}}$ of the vectors in the cluster and the within-cluster distance w is the sum of the distances to the mean:

$$w = \sum_{i=1}^{m} (\mathbf{x}_i - \bar{\mathbf{x}})^2, \tag{22.28}$$

where m is the number of vectors in the cluster and $\mathbf{x}_i$ represents the ith vector. In detail, the procedure is:

1. The number of clusters k is set and the center of each cluster is randomly generated.
2. The distance (Euclidean, city-block, Minkowski, etc.) from each cluster center to each vector is computed and the closest cluster center to each vector is identified and assigned to that vector.
3. Each cluster center is replaced by the mean of the closest vectors $\bar{\mathbf{x}}$.
4. Steps 2 and 3 are repeated until the distance from each vector to the closest cluster center is minimized (this occurs when the vectors are assigned to the same cluster through two or more iterations).

Generally, the algorithm is repeated many times for different k values and different randomly-generated cluster centers in order to minimize the overall within-cluster distance W:

$$W = \frac{1}{2} \sum_{j=1}^{k} \sum_{C(i)=j} (\mathbf{x}_i - \bar{\mathbf{x}}_j)^2, \tag{22.29}$$

where k is the number of clusters, $C(i)$ represents the cluster that the ith vector is assigned to, and $\bar{\mathbf{x}}_j$ represents the mean of the jth cluster and is a vector $\bar{\mathbf{x}}_j = (\bar{x}_{a1}, \bar{x}_{a2}, \ldots, \bar{x}_{an})$ where $\bar{x}_{an}$ represents the mean of the nth attribute within the jth cluster. Repetition of the algorithm is performed since a local-optimum, as opposed to the global-optimum, tends to be discovered.

Although *k*-means clustering has been shown to perform well on many datasets, there are several disadvantages to using *k*-means. One drawback is that *k*-means is not suitable for discovering clusters that are nonconvex in shape (i.e., not sphere-like, such as a crescent-moon shape that has a concave side) or clusters of very different size, although other clustering algorithms exist that are density-based (as opposed to distance-based) that work well on these types of clusters (DBSCAN by Ester et al., 1996; DENCLUE by Hinneburg and Keim, 1998; OPTICS by Ankerst et al., 1999). In the *k*-means algorithm, the number of clusters *k* must be specified *a priori*, which can be a disadvantage due to the trial-and-error process to determine the "best" number of clusters. The shared nearest neighbor (Ertz et al., 2001) clustering technique avoids this problem by determining the number of clusters automatically and also has the advantage of producing spatially-contiguous clusters if an attribute is spatial (i.e., varies in space). The *k*-means technique is sensitive to noise and outliers because the statistical mean can be significantly affected by noise and outliers. This can be prevented to some degree through outlier detection methods (Barnett and Lewis, 1984; Arning et al., 1996; Knorr and Ng, 1997, 1998) and normalization of the data, yet noise remains to be an issue. The *k*-means algorithm has also been shown to perform poorly on discrete data, although there are variations of *k*-means that can avoid this problem such as the *k*-modes (Huang, 1998) and the *k*-medoids (Kaufman and Rousseeuw, 1990) approaches. In *k*-modes clustering, the mode of the closest vectors is used to assign the cluster center, as opposed to the mean, and in *k*-medoids clustering, the medoid, or most centrally-located vector, is used. Both *k*-modes and *k*-medoids are more appropriate when using discrete data, and the *k*-medoids algorithm tends to be more robust than *k*-means due to its insensitivity to noise and outliers. Finally, *k*-means has been shown to be sensitive to the initial, randomly-selected cluster centers, thus the algorithm should be performed several times to determine the "best," most appropriate clustering.

Figure 22.12 illustrates the clustering of data using the *k*-means algorithm. Elevation, soil properties (percent sand, silt, and clay), land use, and the EVI for April are used as inputs (similar to the database used in Section 22.1.2.1), and the number of clusters *k* is

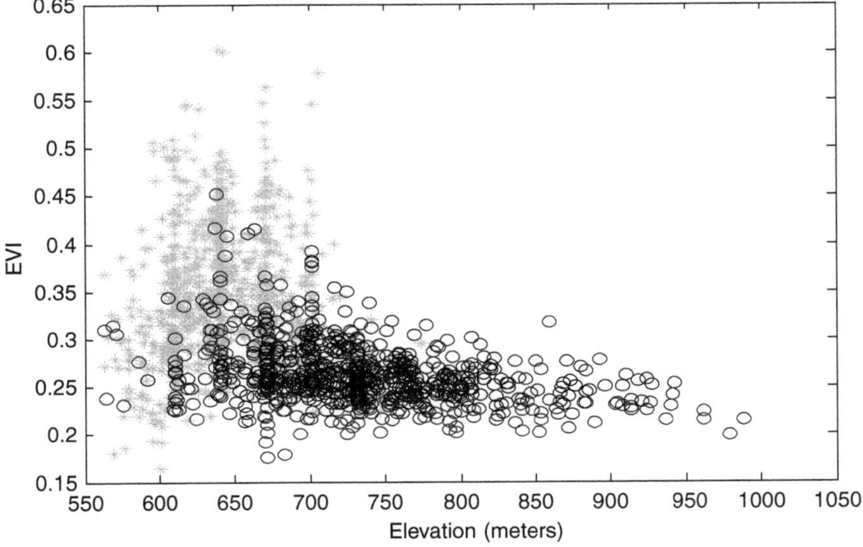

FIGURE 22.11
Two clusters produced using the *k*-means algorithm. Although many variables are used as input, elevation was plotted vs. the EVI for April due to space limitations.

set to two. Typically, a database will have many dimensions, thus one of the drawbacks of clustering techniques is the visualization of the clusters in all dimensions. For instance, the example presented here contains only 6 inputs, but can produce 15 two-dimensional graphs (${}_6C_2$), 20 three-dimensional graphs (${}_6C_3$), and most software cannot manage beyond three dimensions. Figure 22.12 is one of the many graphs that can be used to visualize the clusters and, as can be seen, the k-means algorithm performs fairly well in dividing the data into two coherent clusters.

22.2.2.2 Hierarchical Clustering

Hierarchical clustering forms a binary tree-structure (each split consists of only two nodes) in which each level represents a particular grouping of the data into disjoint clusters and the entire tree represents an ordered sequence of groupings. Once the tree is formed such that the root node contains the whole dataset and the leaf nodes contain only one record, the user must decide which level(s) represents the "best," most natural clustering of the data. There are two types of hierarchical clustering algorithms: agglomerative (bottom-up) and divisive (top-down). Agglomerative algorithms (such as AGNES, Kaufman and Rousseeuw, 1990) begin from the bottom, or the leaf nodes which contain one record, and recursively merge pairs of clusters with the minimum between-cluster distance into a single cluster. Divisive algorithms (such as DIANA, Kaufman and Rousseeuw, 1990) begin from the top, or the root node that contains all of the records, and recursively split the clusters with the maximum between-cluster distance into two clusters. In the following sections, the advantages and disadvantages of both types of hierarchical clustering techniques will be discussed; however, a disadvantage of both methods is that the set of clusters produced is not unique, thus the user must be aware of this and perform their analysis accordingly.

Agglomerative clustering

Agglomerative algorithms iteratively merge pairs of clusters that are the most similar, or closest distance-wise, beginning with the leaf nodes and working up to the root node. The three most commonly used techniques to compute the distance between clusters are single linkage, complete linkage, and average linkage. In single linkage, the distance between two clusters is determined by the distance between the two closest records (nearest neighbors) in the different clusters:

$$\begin{aligned} d_{\mathrm{SL}}[C(1), C(2)] &= \mathit{min}\, d(\mathbf{x}_{C(1),i}, \mathbf{x}_{C(2),j}) \text{ with} \\ i \in (1, \ldots, n_{C(1)}) &\quad \text{and} \quad j \in (1, \ldots, n_{C(2)}), \end{aligned} \tag{22.30}$$

where $C(i)$ represents the ith cluster, $\mathbf{x}_{C(i),j}$ is the jth record, or vector, in the ith cluster, and $n_{C(i)}$ is the number of records in the ith cluster. This method tends to string records together to form clusters, and the resulting clusters tend to represent long "chains." In complete linkage, the distance between two clusters is determined by the distance between the two farthest records (farthest neighbors) in the different clusters:

$$\begin{aligned} d_{\mathrm{CL}}[C(1), C(2)] &= \mathit{max}\, d(\mathbf{x}_{C(1),i}, \mathbf{x}_{C(2),j}) \text{ with} \\ i \in (1, \ldots, n_{C(1)}) &\quad \text{and} \quad j \in (1, \ldots, n_{C(2)}). \end{aligned} \tag{22.31}$$

This method performs quite well when the records naturally form distinct "clumps," yet if the clusters tend to be elongated, or "chain-like" in nature, this method is inappropriate.

In average linkage, the distance between two clusters is determined by the average of all pairs of records in the different clusters:

$$d_{\mathrm{AL}}[C(1), C(2)] = \frac{1}{n_{C(1)} n_{C(2)}} \sum_{i=1}^{n_{C(1)}} \sum_{j=1}^{n_{C(2)}} d(\mathbf{x}_{C(1),i}, \mathbf{x}_{C(2),j}) \text{ with}$$
$$i \in (1, \ldots, n_{C(1)}) \quad \text{and} \quad j \in (1, \ldots, n_{C(2)}). \tag{22.32}$$

Again, if the clusters are naturally elongated, this method is inappropriate. There are other techniques to compute the distance between two clusters, such as using a weighed average, the distance between the centroids of the clusters, the incremental sum of squares, or the increase in the within-cluster distance w (Equation 22.28) as a result of merging the two clusters.

The database used in the k-means clustering example is also used here to illustrate agglomerative clustering (see Section 22.1.2.1). Again, the elevation, soil properties (percent sand, silt, and clay), land use, and April EVI are clustered. Figure 22.13 shows the hierarchical trees, or dendrograms, created for each type of cluster distance measure (single, complete, and average linkage). Dendrograms are used in hierarchical clustering to determine the "best" clustering of the data. Each vertical line represents a cluster and the length of the line is a measure of the distance between the clusters linked by the horizontal lines. As expected, the single linkage distances are smaller than the complete linkage, and the average linkage distances are in between (refer to the y-axes of Figure 22.13). In order to determine a set of clusters from a tree diagram, one can either: (1) specify the number of clusters; (2) use trial and error by randomly choosing a cluster distance threshold; or (3) determine the cluster distance threshold by using the inconsistency coefficient υ, which characterizes a link by comparing its length with the mean length of other links at the same level in the tree. The lower the value of υ, the more similar the clusters connected by the link. The inconsistency coefficient at a particular level of a tree is:

$$\upsilon = \frac{\ell - \mu_\ell}{\sigma_\ell}, \tag{22.33}$$

where ℓ is the number of links, μ_ℓ is the mean length of the links, and σ_ℓ is the standard deviation of the length of the links. Figure 22.14 shows the clusters of elevation vs. April EVI created for each linkage type using a prespecified number of clusters ($k = 2$ in order to compare the agglomerative and k-means clustering algorithms). The average linkage measure produced a more natural clustering of the data than the single or complete linkage measures. Comparing the clusters produced by the k-means and agglomerative (average linkage) algorithms, the performance is roughly equivalent. However, on a more difficult problem agglomerative clustering will tend to outperform k-means, particularly with respect to the control the user has over the formation of the clusters in hierarchical algorithms.

Divisive clustering

Divisive algorithms iteratively split clusters into two clusters, which are the least similar, or farthest distance-wise, beginning with the root node and working down to the leaf nodes. One approach is to recursively implement the k-means, k-modes, k-medoids, or some other type of clustering algorithm where the number of clusters $k = 2$. This approach is subject to the shortcomings of the chosen clustering approach, which can be dependent upon the initial, randomly-selected cluster centers. Algorithms have been specifically developed to

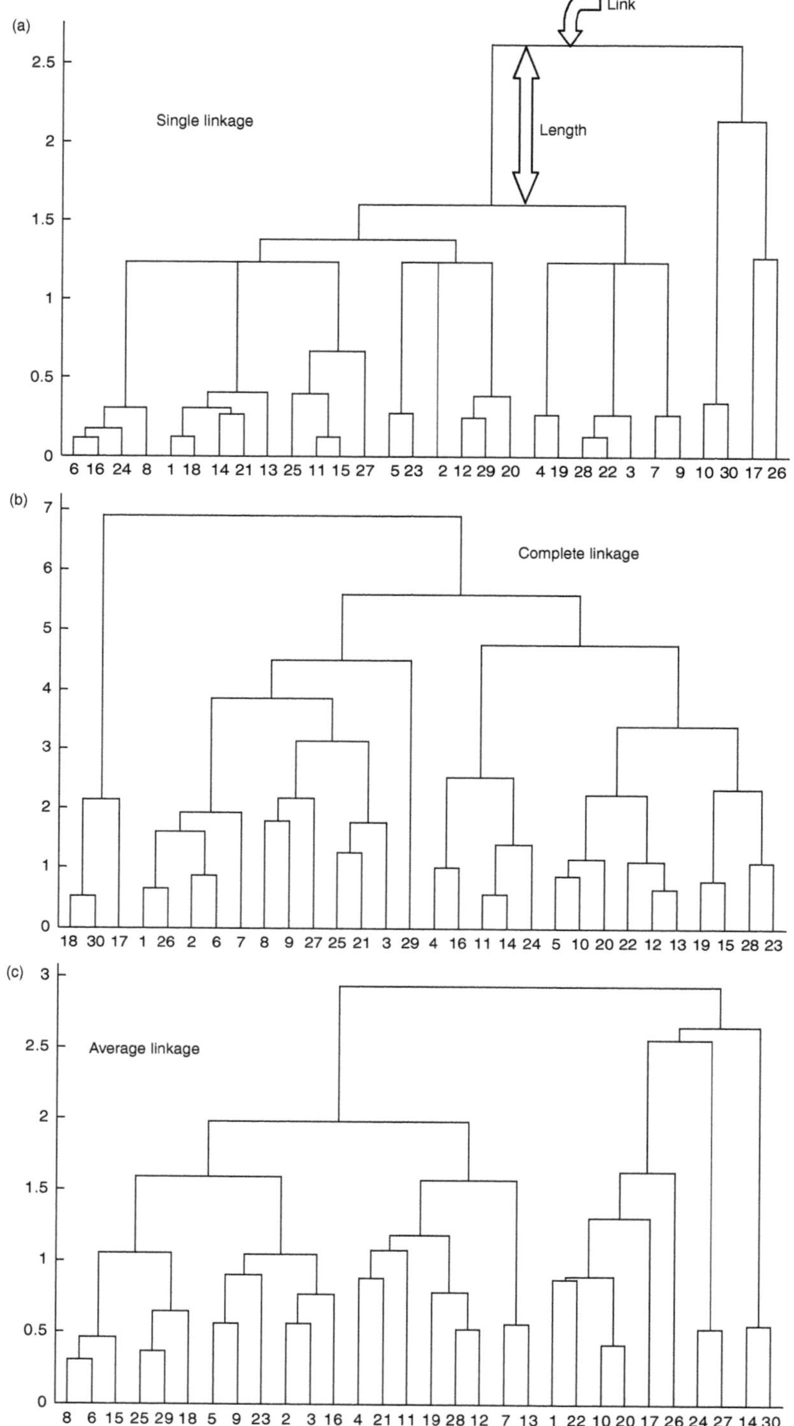

FIGURE 22.12
Hierarchical trees (dendrograms) from agglomerative clustering using (a) single, (b) complete, and (c) average linkage to compute the distance between clusters. The x-axis designates the leaf nodes and the y-axis is the distance between clusters.

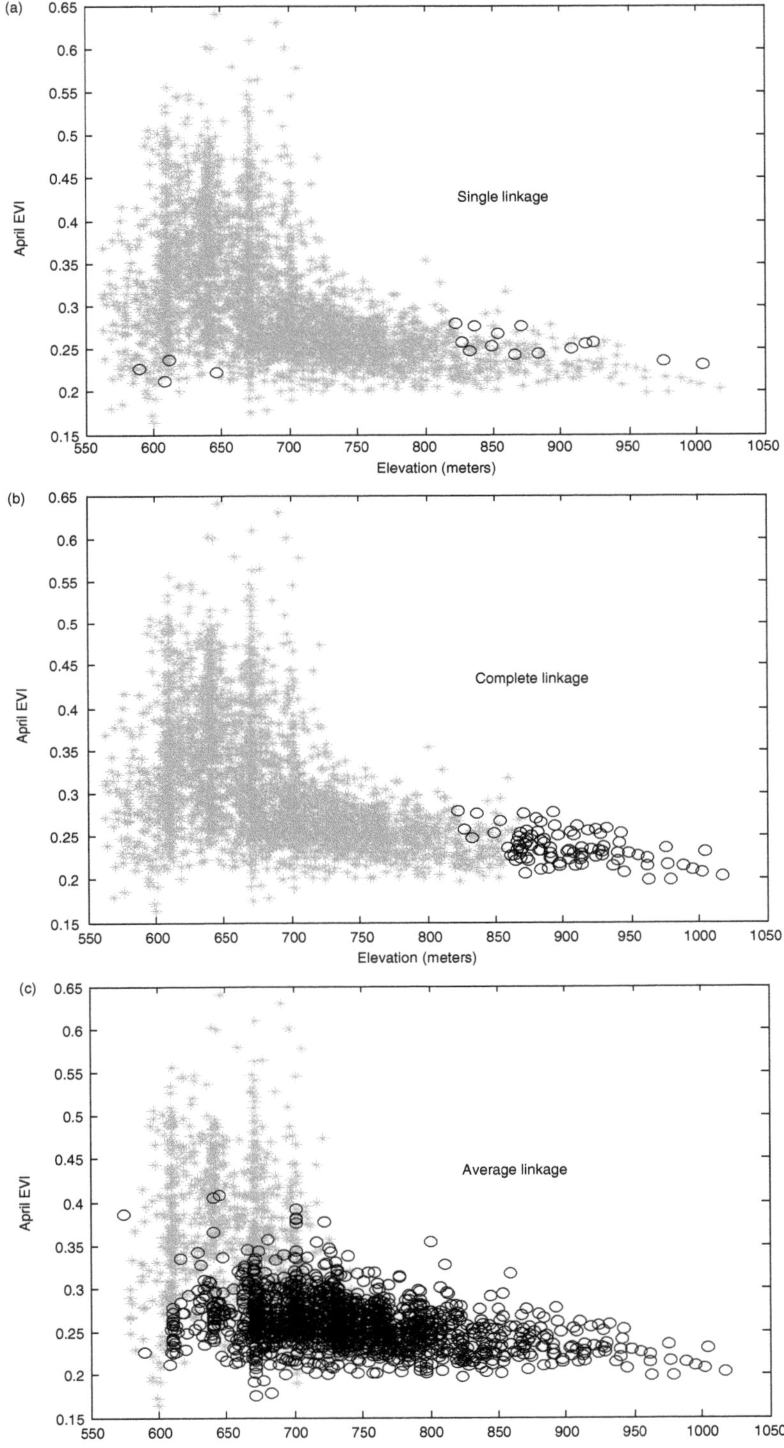

Figure 22.13
Clusters formed from agglomerative clustering using (a) single, (b) complete, and (c) average linkage to compute the distance between clusters. The elevation is plotted vs. the EVI for April (compare to Figure 22.12).

avoid this inheritance problem, such as Macnaughton-Smith et al. (1965), who proposed the following algorithm:

1. All records are placed in cluster $C(1)$.
2. The record with the largest average distance from the other records is placed in cluster $C(2)$.
3. The record in $C(1)$ with the largest [average distance from the records in $C(2)$ minus the average distance from the records remaining in $C(1)$] is placed in $C(2)$.
4. Step 3 is repeated until the difference in averages becomes negative, that is, there are no longer any records in $C(1)$ that are closer, on average, to those in $C(2)$.

Thus, the original cluster is split into two clusters. This process is applied iteratively and a threshold is used to determine whether or not a cluster should be split. Kaufmann and Rousseeuw (1990) proposed splitting the cluster with the greatest *diameter* (the maximum distance from one record to another within the cluster), but one could also choose to split the cluster with the largest *average* within-cluster distance. The procedure continues until the clusters contain a single record or the distance among the records within the clusters is zero.

References

1. Agrawal, R., T. Imielinski, and A.N. Swami, Mining association rules between sets of items in large databases, in P. Buneman and S. Jajodia Eds., *Proceedings of the 1993 ACM SIGMOD International Conference on Management of Data*, Washington, DC, pp. 207–216, May 26–28, 1993.
2. Agrawal, A., H. Mannila, R. Srikant, H. Toivonen, and A.I. Verkamo, Fast discovery of association rules, in U.M. Fayyad, G. Piatetsky-Shapiro, P. Smyth, and R. Uthurusamy Eds., *Advances in Knowledge Discovery and Data Mining*, AAAI Press/MIT Press, Cambridge, MA, pp. 307–328, 1996.
3. Agrawal, R. and R. Skirant, Fast algorithms for mining association rules, IBM Research report RJ9839, IBM Almaden Research Center, San Jose, CA, June 1994.
4. Ankerst, M., M. M. Breunig, H.-P. Kriegel, and J. Sander, OPTICS: ordering points to identify the clustering structure, in *Proceedings of the International Conference on Management of Data (SIGMOD)*, Philadelphia, PA, pp. 49–60, 1999.
5. Arning, A., R. Agrawal, and P. Raghavan, A linear method for deviation detection in large databases, in *Proceedings of the 2nd International Conference on Knowledge Discovery and Data Mining*, pp. 164–169, 1996.
6. Barnett, V. and T. Lewis, *Outliers in Statistical Data*, John Wiley & Sons, New York, 1984.
7. Bates, D. and D. Watts, *Nonlinear Regression Analysis and its Applications*, John Wiley & Sons, New York, 1988.
8. Breiman, L., J. Friedman, R. Olshen, and C. Stone, *Classification and Regression Trees*, Wadsworth International Group, Monterey, CA, 1984.
9. Dasarthy, B., *Nearest Neighbor (NN)Norms*, IEEE Computer Society Press, Washington, DC, 1991.
10. Durgunoglu, A., H.V. Knapp, and C.A. Changnon, PACE Watershed Model (PWM): Vol. 1, model development, Illinois State Water Survey Contract report No. 437, Champaign, IL, 1987.
11. Ertz, L., M. Steinbach, and V. Kumar, Finding topics in collections of documents: a shared nearest neighbor approach, in *Proceedings of Workshop on Text Mining, First SIAM International Conference on Data Mining*, Chicago, IL, 2001.

12. Ester, M., H. Kriegel, J. Sander, and X. Xu, A density-based algorithm for discovering clusters in large spatial databases with noise, in *Proceedings of the 2nd International Conference on Knowledge Discovery and Data Mining*, AAAI Press, Portland, OR, 1996.
13. Fix, E. and J.L. Hodges, Discriminatory analysis, nonparametric regression: consistency properties, Technical report, U.S. Air Force School of Aviation Medicine, 1951.
14. Flores, I., *Computer Sorting*, Prentice Hall, Englewood Cliffs, NJ, 1969.
15. Friedman, J.H., Flexible metric nearest neighbor classification, Technical report, Stanford University Statistics Department, No. 113, 1994.
16. Fu, Y. and J. Han, Metarule-guided mining of association rules in relational databases, in *Proceedings of the 1st International Workshop on Integration of Knowledge Discovery with Deductive and Object-Oriented Databases*, Singapore, pp. 39–46, December 1995.
17. Goldberg, D.E., *Genetic Algorithms in Search, Optimization, and Machine Learning*, Addison-Wesley, Reading, MA, 1999.
18. Han, J. and M. Kamber, *Data Mining: Concepts and Techniques*, Morgan Kaufmann, San Francisco, CA, 2001.
19. Hastie, T. and R. Tibshirani, Discriminant adaptive nearest neighbor classification, *IEEE Transactions on Pattern Analysis and Machine Learning*, 18, 607–616, 1996.
20. Hastie, T. and R. Tibshirani, and J. Friedman, *The Elements of Statistical Learning — Data Mining, Inference, and Prediction*, Springer-Verlag, New York, 2001.
21. Heath, M.T., *Scientific Computing, an Introductory Survey*, 2nd ed., McGraw Hill, New York, 2002.
22. Hinneburg, A. and D.A. Keim, An efficient approach to clustering in large multimedia databases with noise, in *Proceedings of the 4th International Conference on Knowledge Discovery and Data Mining*, AAAI Press, New York, pp. 224–228, 1998.
23. Huang, Z., Extensions to the k-means algorithm for clustering large datasets with categorical values, *Data Mining and Knowledge Discovery*, 2, 283–304, 1998.
24. Kaufman, L. and P.J. Rousseeuw, *Finding Groups in Data*, John Wiley & Sons, New York, 1990.
25. Klemettinen, M., H. Mannila, P. Ronkainen, H. Toivonen, and A.I. Verkamo, Finding interesting rules from large sets of discovered association rules, in *Conference on Information and Knowledge Management*, Gaithersburg, MD, pp. 401–407, November 1994.
26. Knorr, E.M. and R.T. Ng, A unified notion of outliers: properties and computation, in *Proceedings of the 3rd International Conference on Knowledge Discovery and Data Mining*, pp. 219–222, 1997.
27. Knorr, E.M. and R.T. Ng, Algorithms for mining distance-based outliers in large datasets, in *Proceedings of the 24th International Conference on Very Large Data Bases*, pp. 392–403, 1998.
28. Knuth, D.E., *The Art of Computer Programming*, 2nd ed., Addison-Wesley, Boston, MA, 1981.
29. Macnaughton-Smith, P., W. Williams, M. Dale, and L. Mockett, Dissimilarity analysis: a new technique of hierarchical subdivision, *Nature*, 202, 1034–1035, 1965.
30. MacQueen, J., Some methods for classification and analysis of multivariate observations, in L.M. Le Cam and J. Neyman Eds., *Proceedings of the 5th Berkeley Symposium on Mathematical Statistics and Probability*, Vol. 1, University of California Press, Berkeley, CA, pp. 281–297, 1967.
31. Mannila, H., H. Toivonen, and A.I. Verkamo, Efficient algorithms for discovering association rules, in *AAAI Workshop: Knowledge Discovery in Databases*, July 1994.
32. Markus, M., C. Tsai, and M. Demissie, Uncertainty of weekly nitrate-nitrogen forecasts using artificial neural networks, *Journal of Environmental Engineering*, 129, 267–274, 2003.
33. Myles, J.P. and D.J. Hand, The multi-class metric problem in nearest neighbor discrimination rules, *Pattern Recognition*, 23, 1291–1297, 1990.
34. Quinlan, J.R., Induction of decision trees, *Machine Learning*, 1, 181–106, 1986.
35. Quinlan, J.R., *C4.5: Programs for Machine Learning*, Morgan Kaufman, Los Altos, CA, 1993.
36. Quinlan, J.R., Bagging, boosting, and C4.5, in *Proceedings of the 13th National Conference on Artificial Intelligence*, Menlo Park, CA, 1996.
37. Ripley, B.D., *Pattern Recognition and Neural Networks*, Cambridge University Press, Cambridge, MA, 1996.
38. Seber, G., *Multivariate Observations*, John Wiley & Sons, New York, 1984.

39. Short, R.D. and K. Fukunaga, The optimal distance measure for nearest neighbour classification, *IEEE Transactions on Information Theory*, 27, 622–627, 1981.
40. Tcheng, D.K., B. Lambert, S.C.-Y. Lu, and L.A. Rendell, AIMS: an adaptive interactive modelling system for supporting engineering decision making, in *Proceedings of the 8th International Workshop on Machine Learning*, San Francisco, CA, pp. 645–649, 1991.
41. Weisberg, S., *Applied Linear Regression*, John Wiley & Sons, New York, 1980.
42. Welge, M., W.H. Hsu, L.S. Auvil, C. Bushell, J. Martirano, T.M. Redman, and D. Tcheng, Data to Knowledge (D2K): a rapid application development environment for knowledge discovery in databases, Technical report, National Center for Supercomputing Applications, University of Illinois at Urbana-Champaign, 1999, http://alg.ncsa.uiuc.edu/do/index.

23

Artificial Neural Networks

Momcilo Markus

CONTENTS

23.1 Introduction

Artificial Neural Networks (ANN) emerged with the advances in computers and flourished since the late 1980s as a powerful computational technique for computing input–output mappings. The ANN method has been applied for pattern recognition, classification, time-series analysis, and system control, in medical sciences, business, engineering, and numerous other fields. Although the impetus comes from modeling of cerebral activity, the focus of ANN is mathematical rather than biological. The paper by McCulloch and Pitts [1] has been widely considered as the first paper published on ANN. In this paper the authors proposed that a computational unit (node) should be triggered if it receives an input over some threshold from all the other nodes to which it is connected. This simple network facilitated further work on ANN, as the framework of ANN as consisting of nodes, weights, biases, and transfer functions has been established. Several decades later, various authors published more comprehensive introductions to ANN (e.g., [2,3]). Along with the abundant applications, researchers have been working on various theoretical aspects of ANN. Cybenko [4] and Ito [5] demonstrated that within a given interval, any continuous function can be approximated with a desired accuracy by a combination of sufficiently many sigmoids. Sandberg [6] provided more general approximation theorems for discrete-time systems, including the radial basis function.

The ANN can be defined as a system of "massively parallel interconnected networks of simple elements and their hierarchical organization which are intended to interact with the objects of the real world in the same way as biological nervous systems do" [7]. ANNs are capable of determining relationships between system inputs and system outputs, by a network of interconnecting nodes that adjust their connecting parameters based on training samples and discover the rules governing the association between the inputs and the outputs. ANN consists of one input layer, one or more hidden layers, and one output layer. Typical network architecture with n input nodes, h hidden nodes in one hidden layer, and m output nodes is presented in Figure 23.1. Input nodes receive data from sources external to the network and are the nodes without paths directed toward them. Hidden nodes receive data from input nodes and send data to output nodes. Output nodes produce data generated by the network.

Hydrologic applications of ANN also have a long record. There have been various applications of ANNs in evaluating a range of water resources problems. In numerous rainfall-runoff studies [8–10], the ANNs were compared favorably with traditional models. Back-Propagation (BP) ANNs also had promising results in water quality prediction [11,12]. Amenu et al. [13] used an online radial-basis/Kalman filtering neural network and achieved a better nitrate-N forecast accuracy. ANNs have also been used in contamination assessment [14], spatial interpolation [15], and drought analysis [16]. One of the most comprehensive analyses of the role of ANNs in hydrology was undertaken by the ASCE Task Committee

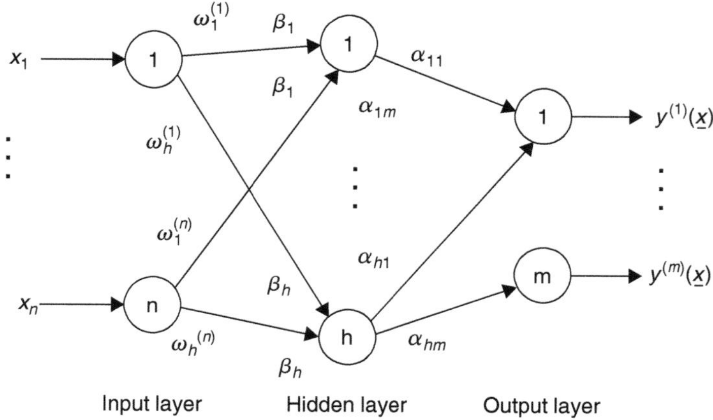

FIGURE 23.1
Typical ANN architecture with one input, one hidden, and one output layer. The input, the hidden, and the output layers have n, h, and m nodes, respectively.

on Artificial Neural Networks in Hydrology (ASCE Task Committee) [17,18] to serve as an introduction to ANNs for hydrologists, and also to summarize the merits and limitations of neural networks. Among rare attempts to quantify the uncertainties in the ANN models, Guan et al. [19] used a neural network example in a framework for uncertainty assessment of an ecological model. In addition, Salas et al. [20] ran a Monte Carlo experiment to quantify the parameter uncertainty associated with an ANN model. Aires [21] and Aires et al. [22,23] assessed the uncertainty of ANN weights and outputs, and the sensitivity of an output with respect to the inputs of the model, using Bayesian statistics with application to remote sensing.

Back Propagation [24] is considered the best-known training algorithm for neural networks. It has lower memory requirements than most algorithms, and usually reaches an acceptable error level quite quickly, although it can then be very slow to converge properly on an error minimum. Radial basis function network [25] units respond nonlinearly to the distance of points from the center represented by the radial unit. The response surface of a single radial unit is therefore a Gaussian (bell-shaped) function, peaked at the center, and descending outwards. Probabilistic Neural Networks (PNN) [26,27] learn to estimate a probability density function and are typically used in classification problems. Generalized Regression Neural Networks (GRNNs) work in a similar fashion to PNNs, but perform regression rather than classification tasks [28]. As with the PNN, Gaussian kernel functions are located at each training case. Self Organizing Feature Map (SOFM, or Kohonen) networks are used quite differently to the other networks. Whereas all the other networks are designed for supervised learning tasks, SOFM networks are designed primarily for unsupervised learning [29]. Recurrent neural networks [30] is a model with bidirectional data flow. While BP network propagates data linearly from input to output, recurrent networks also propagate data from later processing stages to earlier stages. Minimum Resource Allocation Network (MRAN) [31] is an online adaptive neural network that automatically configures the number of hidden nodes based on the input–output patterns presented to the network. MRAN uses the radial basis function to approximate the input–output patterns, and Kalman filtering for sequential estimation of the model parameters.

Section 23.2 describes offline (batch) BP networks. Section 23.3 compares autoregressive models and BP ANNs in synthetic data generation. Section 23.4 describes the online MRAN method.

23.2 Back-Propagation Neural Networks

The most frequently used supervised neural network method, Back Propagation (BP), approximates target (output) data with a nonlinear transformation of input data. The nonlinear transformation function is called activation function. Commonly used activation functions are the logistic function (sigmoid) (Equations 23.1 and 23.2) and the hyperbolic tangent (Equation 23.3).

Typical sigmoid is written as

$$\sigma(z) = \frac{1}{1+e^{-z}}. \tag{23.1}$$

A modified bipolar form of sigmoid can be defined as [20]

$$\sigma(z) = \frac{1-e^{-z}}{1+e^{-z}}. \tag{23.2}$$

The typical hyperbolic tangent is given by:

$$\tanh(z) = \frac{e^z - e^{-z}}{e^z + e^{-z}}. \tag{23.3}$$

Any function [4] can be approximated within desired accuracy using a combination of sufficiently many sigmoid functions. Consider a Multiple Input–Multiple Output (MIMO) system with n inputs, h hidden nodes in one hidden layer, and k outputs, as shown in Figure 23.1. The true relationship between the inputs and the outputs is not known, but in each time step t ($t = 1, \ldots, T$) the values of simultaneous inputs, $x_t^{(i)}$, $i = 1, n$, and outputs $y_t^{(k)}(x)$, $k = 1, m$ observations are known. Vector of inputs at time t, is denoted as $\underline{x}$ and is equal to $\underline{x} = [x_t^{(1)}, x_t^{(2)}, \ldots, x_t^{(n)}]^T$. Then, the observed outputs can be approximated using the sigmoidal transformation (Equation 23.2) of the observed inputs:

$$\hat{y}_t^{(k)}(\underline{x}) = \gamma_k + \sum_{j=1}^{h} \alpha_{jk}\sigma\left[-\beta_j + \sum_{i=1}^{n} \omega_j^{(i)} x_t^{(i)}\right]. \tag{23.4}$$

The symbols α, β, γ, and ω denote the ANN model parameters. The coefficient γ_k is associated with the output k; α_{jk} is associated with the hidden node j and output k; β_j with the hidden node j; and $\omega_j^{(i)}$ with the input i and hidden node j. The outputs can be approximated by the tangent hyperbolic functions or any other activation function. In case of the tangent hyperbolic function, Equation 23.4 would have to include $\tanh(\cdot)$ instead of $\sigma(\cdot)$.

The overall goal of the parameter estimation (training) methods is to find a set of model parameters that gives the minimum approximation error. The BP method is a supervised learning method that iteratively adjusts the model parameters (often called biases and weights, or sometimes, shifts and rotators) based on the approximation error. The parameters could be modified based on the gradient descent method. This method searches for the minimum of the MIMO error function method. For simplicity of illustration, let us consider a system with one input and one output in Equation 23.4:

$$\hat{y}_t(x) = \gamma + \sum_{j=1}^{h} \alpha_j \sigma[-\beta_j + \omega_j x_t]. \tag{23.5}$$

The approximation error can be written as a difference between the approximated and the observed outputs at time t:

$$e_t = \hat{y}_t(x) - y_t(x). \tag{23.6}$$

The total approximation square error for the entire dataset $t = 1, 2, \ldots, T$, can be expressed as:

$$E_t = \frac{1}{2}\sum_{t=1}^{T} e_t^2 = \frac{1}{2}\sum_{t=1}^{T} [\hat{y}_t(x) - y_t(x)]^2. \tag{23.7}$$

This error surface is a multidimensional surface that depends on the model parameters $\underline{\theta}$, where $\underline{\theta} = [\gamma, \alpha_j, \beta_j, \omega_j, j = 1, \ldots, h]$. For a system defined by Equation 23.5, the number of model parameters is equal to $p = 3h + 1$. The minimum value of this surface is determined by solving a system of p partial differential equations:

$$\frac{\partial E[\underline{\theta}|x_t, y_t, t = 1, \ldots, T]}{\partial \theta_i} = 0, \tag{23.8}$$

where $i = 1, \ldots, p$. The parameters $\theta_i, i = 1, \ldots, p$, correspond to $\gamma, \alpha_j, \beta_j, \omega_j, j = 1, \ldots, h$.

These equations are solved numerically. The initial iterations of the model parameters are assumed and those values are iteratively adjusted based on the error function using the steepest descent method. For example, parameter γ in Equation 23.5 is solved by assuming the initial value, $\gamma^{(0)}$, and by calculating the adjusted $\gamma^{(1)}$. The next iteration $\gamma^{(2)}$ is obtained using $\gamma^{(1)}$, etc., according to:

$$\begin{aligned}
\gamma^{(k+1)} &= \gamma^{(k)} - \eta \frac{1}{T} \sum_{t=1}^{T} \left. \frac{\partial E}{\partial \gamma} \right|_{\gamma=\gamma^{(k)}}, \qquad k = 1, \ldots, K \\
\alpha_j^{(k+1)} &= \alpha_j^{(k)} - \eta \frac{1}{T} \sum_{t=1}^{T} \left. \frac{\partial E}{\partial \alpha_j} \right|_{\alpha_j=\alpha_j^{(k)}}, \qquad k = 1, \ldots, K, \quad j = 1, \ldots, h \\
\beta_j^{(k+1)} &= \beta_j^{(k)} - \eta \frac{1}{T} \sum_{t=1}^{T} \left. \frac{\partial E}{\partial \beta_j} \right|_{\beta_j=\beta_j^{(k)}}, \qquad k = 1, \ldots, K, \quad j = 1, \ldots, h \\
\omega_j^{(k+1)} &= \omega_j^{(k)} - \eta \frac{1}{T} \sum_{t=1}^{T} \left. \frac{\partial E}{\partial \omega_j} \right|_{\omega_j=\omega_j^{(k)}}, \qquad k = 1, \ldots, K, \quad j = 1, \ldots, h,
\end{aligned} \tag{23.9}$$

where η is the learning rate, K denotes the total number of iterations. The partial derivative of the error function with respect to γ is given by:

$$\frac{\partial E_t}{\partial \gamma} = -[\hat{y}_t(x) - y_t(x)]. \tag{23.10}$$

Other parameters (α_j, β_j, ω_j, $j = 1, \ldots, h$) are calculated based on equations similar to Equation 23.9 and the following partial derivatives:

$$\frac{\partial E_t}{\partial \alpha_j} = [\hat{y}_t(x) - y_t(x)]\sigma(-\beta_j + \omega_j x_t), \tag{23.11}$$

$$\frac{\partial E_t}{\partial \beta_j} = [\hat{y}_t(x) - y_t(x)] \frac{2\alpha_j \exp[-(-\beta_j + \omega_j x_t)]}{\{1 + \exp[-(-\beta_j + \omega_j x_t)]\}^2}, \tag{23.12}$$

$$\frac{\partial E_t}{\partial \omega_j} = [\hat{y}_t(x) - y_t(x)] \frac{2\alpha_j x_t \exp[-(-\beta_j + \omega_j x_t)]}{\{1 + \exp[-(-\beta_j + \omega_j x_t)]\}^2}. \tag{23.13}$$

A summarized BP batch method is presented in the flowchart, Figure 23.2. The training procedure starts with observed input and output data, and randomly assumed model parameters. For each time step t, the parameters produce an approximation of the output, based

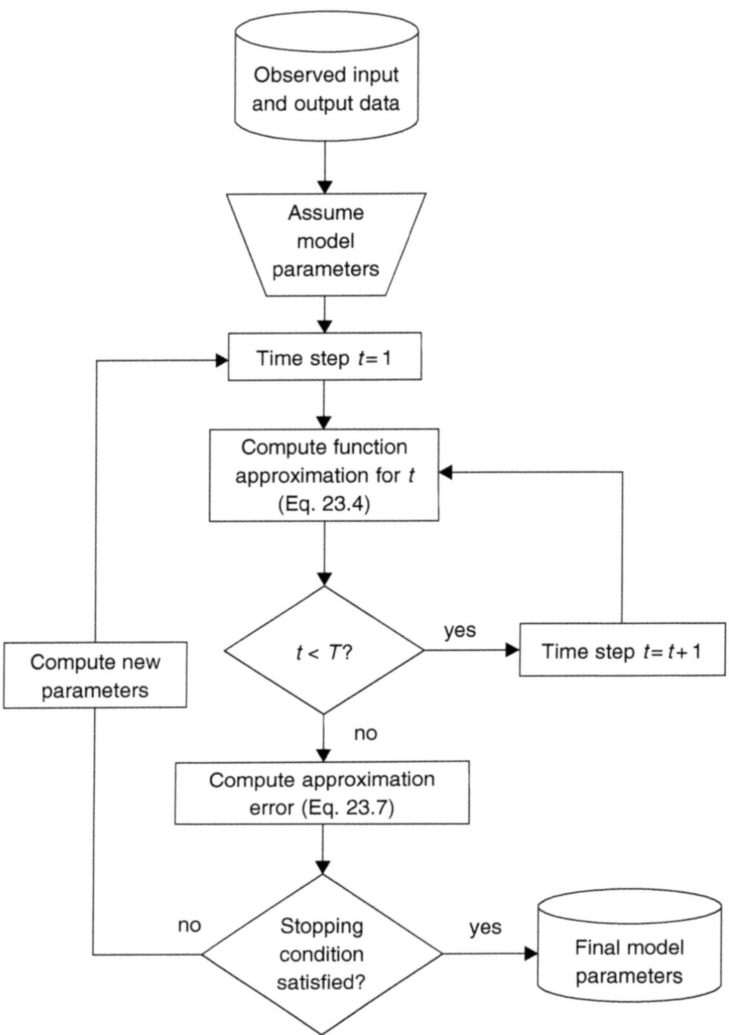

FIGURE 23.2
Flowchart of a batch BP neural network applied to a time series, $t = 1, \ldots, T$.

on Equation 23.4, or Equation 23.5 for simplified system. In general, the first approximation of the output considerably differs from the observed output, due to the random nature of the initial parameters. This difference is expressed by the error given by Equation 23.7. This error is then used for calculating the new updated parameters based on Equation 23.9. When the parameters produce an approximation with error satisfying the stopping condition, the calculation stops and the model parameters are considered final. The stopping condition for the approximation error for early strict-fitting neural networks was typically chosen as an arbitrary small number. In strict fitting, any desired accuracy can be achieved by increasing the number of hidden nodes [4]. However, for noisy input–output relationships, such as those in hydrology, the strict fitting is generally not possible. Our goal is to detect those relationships, and avoid fitting on the noise. An appropriate method for validating this kind of approximation could be the cross-validation method [32,33]. Cross-validation has been successfully applied to hydrologic relationships [12].

23.3 Synthetic Data Generation Based on Neural Networks

23.3.1 Introduction

Synthetic data generation is a major application of modeling hydrologic time series. Synthetic time series generation has a long history of hydrologic applications, ranging from hydrologic design to operation of water resources systems. Using the observed statistical properties of the observed series, future random sequences of the time series can be generated based on the Monte Carlo technique. This technique is applied to generate many realizations of the process statistically indistinguishable from the observed time series, where the observed time series is just one realization of a random process.

A random process is generated by solving the following discrete stochastic equation [34]:

$$p = F(x(t_n)|x(t_{n-1}), x(t_{n-2}) \ldots), \tag{23.14}$$

where $F(x(t_n)|x(t_{n-1}), x(t_{n-2}), \ldots)$ is the conditional Cumulative Distribution Function (CDF) of a random variable, $x(t_n)$ at time t_n conditional on the past values of the random variable. Given the CDF, a random sequence can be obtained by solving equation (23.14).

Salas et al. [35] presented a systematic and comprehensive overview of data generation methods and case studies. Along with other numerous methods they describe AutoRegressive Moving Average (ARMA) models [36]. The ARMA models are frequently used in numerous applications in synthetic time series generation. On the other hand, ANNs have not been used much for synthetic data generation. Among rare applications are study by Zou et al. [37] who embedded a neural network into a traditional Monte Carlo simulation, which resulted in a more efficient computation and uncertainty analysis of water quality modeling. In another application, Ochoa-Rivera et al. [38] developed a multivariate streamflow generation model based on a multiplayer feed-forward neural network with very promising results. Unlike neural networks, traditional models typically require normally distributed data and noise, and homoscedastic variance of the noise. In addition, discharge data are typically skewed and need to be normalized by applying an appropriate transformation. The final results need to be retransformed back to the original domain, which generally produces a bias in the results. ANNs can use nonnormally distributed data, and do not have the retransformation bias.

The following applications illustrate some of application of BP ANNs to synthetic data generation. First example compares an AR(1) model with a simple neural network in generation of annual discharge. Second example compares generation based on ANNs with the periodic autoregressive order-1 (PAR(1)) model. In the last example, PAR(1) and ANN were compared in data generation and one-step forecasting. The results show the accuracy of both methods.

23.3.2 Generation of Annual Discharge Time Series Based on Neural Networks

The model is first estimated based on the observed data. A neural network equivalent of the AR(1) model could be a network with one input x_t and one output $y_t(x)$:

$$y_t(x) = \gamma + \sum_{j=1}^{h} \alpha_j \sigma[-\beta_j + \omega_j x_t] + \varepsilon_t, \tag{23.15}$$

where symbols γ, α_j, β_j, and ω_j denote the model parameters, h is the number of hidden nodes and ε_t is the error term. Furthermore, Equation 23.15 was modified to represent a single time series $y_t, t = 1, \ldots, n$, where t is time in years, and n is the total number of years. Instead of using pairs $[x_t, y_t(x)]$, for training and testing, the single series uses pairs $[y_{t-1}, y_t]$. Accordingly, this above equation becomes:

$$y_t = \gamma + \sum_{j=1}^{h} \alpha_j \sigma[-\beta_j + \omega_j y_{t-1}] + \varepsilon_t. \quad (23.16)$$

The model parameters are estimated using the procedure described in Section 23.2, and the residual statistics $\varepsilon_t \sim (\mu_\varepsilon, \sigma_\varepsilon)$ and their distribution are determined based on the available training data. The synthetic data generation (Equation 23.16) arbitrarily assumes y_0 (often equal to the mean value) and randomly generates error term ε_1. The random error term is generated using the appropriate distribution and has to preserve the mean and the variance of the training residuals. For simplicity, it was assumed that the residuals are not serially correlated. The first generated term y_1 is a function of y_0 and ε_1, second term y_2 is a function of y_1 and ε_2, and so on. A synthetic time series of any desired length can be generated.

$$\begin{aligned}
y_1 &= \gamma + \sum_{j=1}^{h} \alpha_j \sigma[-\beta_j + \omega_j y_0] + \varepsilon_1 \\
y_2 &= \gamma + \sum_{j=1}^{h} \alpha_j \sigma[-\beta_j + \omega_j y_1] + \varepsilon_2 \\
&\vdots \\
y_t &= \gamma + \sum_{j=1}^{h} \alpha_j \sigma[-\beta_j + \omega_j y_{t-1}] + \varepsilon_t.
\end{aligned} \quad (23.17)$$

The nonrandom choice for the initial value and the nonrandom choice for the first generated random number create a bias in the initial values of the generated series. A way to eliminate this bias is to generate $m + n$ data points, and omit the first m data points. The generated time sequence beyond the first m values can be considered appropriate for practical applications.

23.3.2.1 Example of Annual Streamflow Time Series Generation Based on Neural Networks

The previously described method for data generation based on neural networks is applied to the modular annual discharges of the St. Lawrence River at Ogdensburg, NY. The St. Lawrence watershed covers 670,292 square kilometers (km^2), and the average discharge is 6,941 cubic meters per second (m^3/sec). Available data included 90 annual modular discharges from 1861 through 1950 (Figure 23.3). The original series had mean equal to 1.000, standard deviation 0.088, skewness coefficient of −0.210, lag-1, and lag-2 autocorrelation coefficients $R(1) = 0.690$ and $R(2) = 0.498$. The neural network model parameters were estimated based on the BP method, and the final values of the model parameters were $\gamma = -0.163$, $\alpha = 3.718$, $\beta = -0.067$, and $\omega = 0.392$. The noise variance was equal to 0.004. Throughout this computation, it was assumed that the number of hidden nodes h was equal

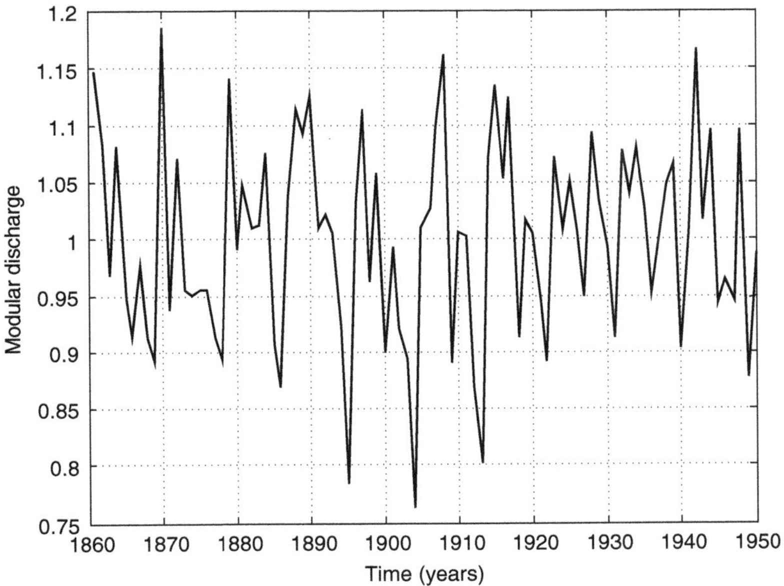

FIGURE 23.3
Modular annual discharges at Ogdensburg on the St. Lawrence River, NY, 1861–1950.

TABLE 23.1
Summary of generation of annual streamflows based on AR(1) and ANN.

	Samples generated by the AR(1) model					Samples generated by the ANN model				
Sample	**Mean**	**St.Dev.**	**Skew.**	**R(1)**	**R(2)**	**Mean**	**St.Dev.**	**Skew.**	**R(1)**	**R(2)**
1	1.008	0.091	−0.453	0.714	0.555	1.012	0.093	−0.472	0.723	0.566
2	0.962	0.099	−0.165	0.738	0.569	0.965	0.101	−0.163	0.747	0.581
—	—	—	—	—	—	—	—	—	—	—
30	1.070	0.102	0.305	0.757	0.492	1.076	0.102	0.291	0.759	0.495
μ	1.009	0.082	−0.073	0.616	0.352	1.014	0.083	−0.079	0.623	0.361
σ	0.021	0.013	0.326	0.112	0.159	0.022	0.013	0.326	0.111	0.158

to 1, and that the parameter λ was equal to 1.000, which gives a final set of equations for data generation based on the ANN model:

$$y_t = -0.163 + 3.718\sigma[-0.067 + 0.392y_{t-1}] + \varepsilon_t. \tag{23.18}$$

Data generation was performed 30 times in 90 years based on neural networks. For comparison, an AR(1) model was used for the same purpose. The AR(1) model had the autoregression parameter equal to 0.683, and noise variance equal to 0.004. Both ANN and AR(1) noise time series were normally distributed and serially independent. The same noise time series is used in both models in order to compare the two-generation techniques. The results presented in Table 23.1 demonstrated that results of the two methods are very similar. The table shows the statistics of the first, the second, and the last generated sample. The last two rows in the table present the mean and the standard deviation

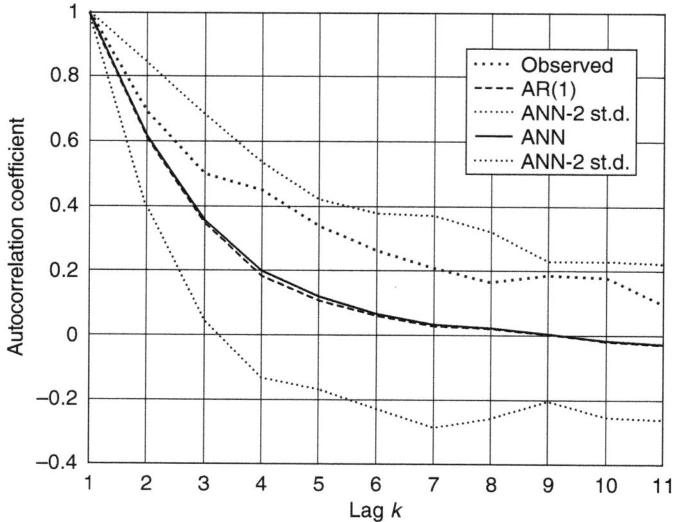

FIGURE 23.4
Lag-k autocorrelation for observed and generated annual discharges at Ogdensburg on the St. Lawrence River.

for the generated samples. The original series has mean equal to 1.000, standard deviation 0.088, skewness −0.210, lag-1, and lag-2 autocorrelation coefficients $R(1) = 0.690$ and $R(2) = 0.498$. Table 23.1 indicates that the earlier statistics of the synthetic samples based on PAR(1) and ANN are very similar. Small differences could be a result of a nonlinear nature of neural networks. Figure 23.4 shows the autocorrelation function (ACF) for ANN (including +/− two standard deviations), AR(1), and the ACF of the observed sample. The ACFs for ANN and AR(1) are also very similar.

23.3.3 Generation of Monthly Discharge Time Series Based on Neural Networks

Neural networks can be used for monthly streamflow data generation. Prior to data generation, the ANN model parameters and the statistical properties of the residuals must be estimated. As in the annual model, the model is represented as a sum of a deterministic and a stochastic part, where the deterministic part is a simple sigmoidal transformation of a previous realization of a process, and the stochastic part consists of a series of random numbers with same statistical properties as the residuals of the original model. However, in the monthly model, the statistical properties of the inputs, the outputs, and the residuals differ for each month.

The process of data generation is illustrated on a normally distributed monthly time series $y_{\nu,\tau}$, where ν denotes year and τ represents month. The original series $Q_{\nu,\tau}$ generally has nonnormal distribution. After a logarithmic or another appropriate transformation, series $Q_{\nu,\tau}$ is transformed to normally distributed series $y_{\nu,\tau}$. The condition of normality is not required for ANNs [39]. However, in this section we normalized data to facilitate comparison with the PAR(1) model. Generally, $y_{\nu,\tau}$ is expressed as a function of previous observation of the same series $y_{\nu,\tau-1}$:

$$y_{\nu,\tau} = \gamma_\tau + \sum_{j=1}^{h} \alpha_{j,\tau}\sigma[-\beta_{j,\tau} + \omega_{j,\tau} y_{\nu,\tau-1}] + \varepsilon_{\nu,\tau}. \tag{23.19}$$

Symbols $\alpha_{j,\tau}$, $\beta_{j,\tau}$, γ_τ, and $\omega_{j,\tau}$ are the model parameters, h is the number of hidden nodes, j is the hidden node and $\varepsilon_{\nu,\tau}$ is the error term. The data generation process starts with the assumed initial value for previous months discharge $y_{1,0}$ and the first generated random term for month 1, $\varepsilon_{1,1}$, as given in equation (23.20). When the term $\varepsilon_{\nu,\tau}$ is normally distributed, it can be expressed as a product of the standard deviation of the residuals for month τ, $\sigma_{\varepsilon\tau}$, and a standard normal noise term $\xi_{\nu,\tau}$. Each month's discharge is generated based on the previous month. The discharge for January in year ν is generated based on the discharge for December of the year $\nu - 1$.

$$
\begin{aligned}
y_{1,1} &= \gamma_1 + \sum_{j=1}^{h} \alpha_{j,1}\sigma[-\beta_{j,1} + \omega_{j,1} y_{1,0}] + \varepsilon_{1,1} \\
y_{1,2} &= \gamma_2 + \sum_{j=1}^{h} \alpha_{j,2}\sigma[-\beta_{j,2} + \omega_{j,2} y_{1,1}] + \varepsilon_{1,2} \\
&\vdots \\
y_{1,12} &= \gamma_{12} + \sum_{j=1}^{h} \alpha_{j,12}\sigma[-\beta_{j,12} + \omega_{j,12} y_{1,11}] + \varepsilon_{1,12} \\
y_{2,1} &= \gamma_1 + \sum_{j=1}^{h} \alpha_{j,1}\sigma[-\beta_{j,1} + \omega_{j,1} y_{1,12}] + \varepsilon_{2,1} \\
&\vdots
\end{aligned}
\tag{23.20}
$$

Following this procedure any desired number of years $M \times N$ can be generated. Symbol M denotes the number of generated samples and N represents the sample size.

23.3.3.1 Example of Monthly Streamflow Time Series Generation Based on Neural Networks

The method for data generation based on neural networks is applied to monthly mean discharge data observed at the Rio Grande near Del Norte, Colorado. The Rio Grande watershed covers 3418.8 km^2, and the average discharge is 25.7 m^3/sec. Data used in this example included the period between 1949 and 1988. For simplicity, it was assumed that the number of hidden nodes h equals 1. The statistics of the original data, and data generated by ANN were compared with statistics of data generated by the PAR(1) model. Equations (23.20) were used for data generation. The model parameters were estimated based on the historical observed data. Table 23.2 shows the ANN parameters and the residual standard deviation for each month.

These parameters were used for generation of 1×100 years. For illustration, a ten-year period of monthly observed discharges was presented along with arbitrary 10-year sequences of monthly discharges generated by PAR(1) and ANN models (Figure 23.5). The segments depicted in Figure 23.3 illustrate the observed monthly and annual variability as well as the variability of the data generated by the PAR(1) and ANN models.

The comparison of the data generation results with the observed monthly data is presented in Figure 23.6. All four statistical parameters shown in the figure show that artificial neural networks reproduced the observed data more accurately than the PAR(1) model.

TABLE 23.2

Parameters of ANN model for each month.

Month τ	$\alpha_{1,\tau}$	$\beta_{1,\tau}$	γ_{τ}	$\omega_{1,\tau}$	$\sigma_{\varepsilon\tau}$
1	1.658	−0.402	−0.261	1.375	0.458
2	7.242	0.248	0.883	0.269	0.384
3	6.561	0.008	0.026	0.211	0.713
4	8.374	0.889	3.455	0.127	0.887
5	4.984	0.359	0.858	0.227	0.835
6	7.265	0.182	0.655	0.178	0.761
7	1.494	−0.129	−0.108	1.805	0.400
8	6.512	0.429	1.346	0.283	0.511
9	1.137	−1.062	−0.300	2.278	0.640
10	2.506	0.155	0.193	0.659	0.660
11	7.505	0.538	1.977	0.257	0.505
12	4.431	−0.512	−1.059	0.468	0.451

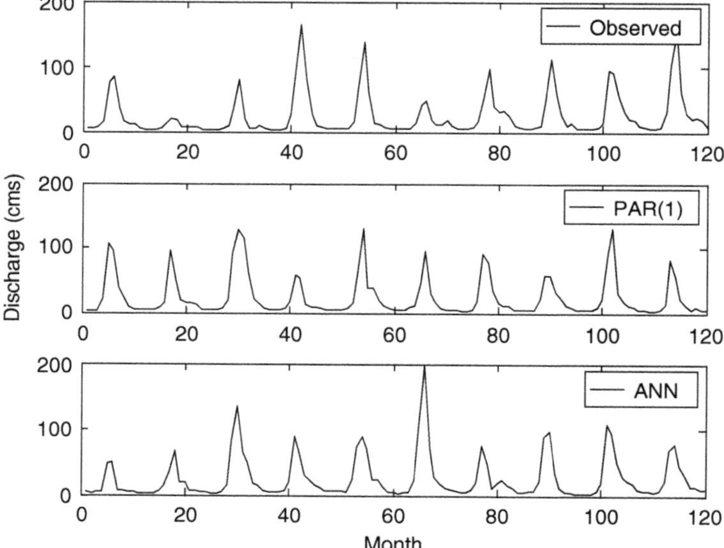

FIGURE 23.5
Ten-year sequence of observed monthly discharges, generated by the PAR(1) model and generated by the ANN model.

23.3.4 Comparison between ANN and PAR(1) Models in Data Generation and Forecasting

Statistical distributions of various hydrologic data types, such as discharge, sediment, or nutrients, typically have positive skewness, caused by large but rare events that occasionally take place. Most traditional methods, applied to hydrologic data simulation, generation or forecasting require normally distributed data with skewness equal to zero. To satisfy this requirement the original nonnormal data has to be transformed to normally distributed data. After the modeling with the normally distributed data is performed, the final results are retransformed back to the observed domain. The transformation–retransformation

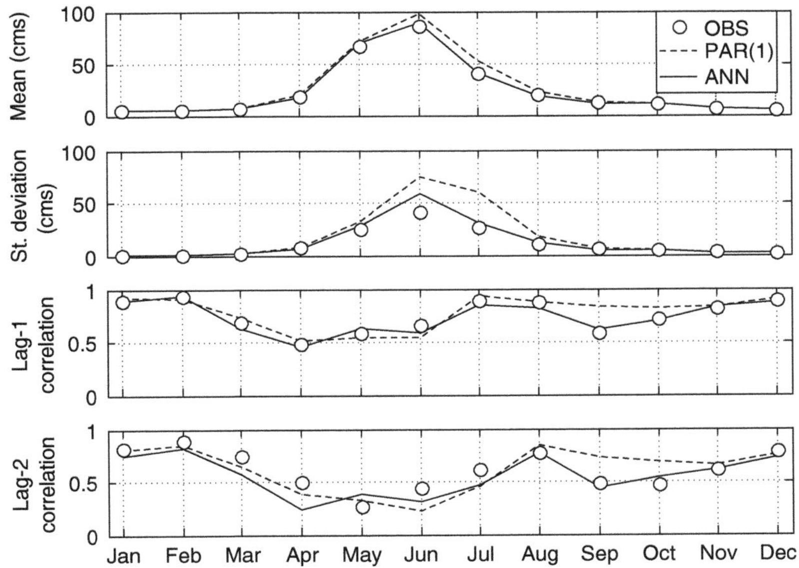

FIGURE 23.6
Monthly mean, standard deviation, lag-1 correlation coefficient and lag-2 correlation coefficient for observed data and data generated by PAR(1) and ANN.

method typically generates a bias, in the literature referred to as the retransformation bias [40].

Artificial neural networks, on the other hand, offer an alternative approach, not bound by the restrictions of the traditional methods. Not having to deal with transformation and retransformation, ANNs potentially offer a more accurate modeling tool. To test this hypothesis, we designed an experiment that generated two samples and then applied various one-step forecasting models to each of the generated samples. The first generated monthly time series was based on a PAR(1) monthly model. The second sample was generated based on an ANN monthly model. Next, we tested various ANN-based and PAR(1)-based approaches applied to both generated data samples, with the following goals: (1) to compare the accuracy of the ANN and PAR(1) models, (2) to examine the effect of the retransformation bias, and (3) to compare the robustness of the ANN and PAR(1) models.

For better comparison with PAR(1), the ANN model assumed the architecture with minimal number of parameters. The PAR model had order equal to 1, while the ANN model had one input, one output, and the number of hidden nodes equal to 1. The monthly parameters assumed for the PAR(1): the mean, the standard deviation, the autoregressive lag-1 coefficient, and the residual standard deviation are presented in Table 23.3. The parameters assumed for the ANN model, $\gamma, \alpha_j, \beta_j, \omega_j$ and the residual standard deviation are presented in Table 23.4. Once the parameters are known, the synthetic monthly data generation could start. The details of data generation were described by equations (23.20).

23.3.4.1 Models Tested on PAR(1) Generated Data

The first test evaluated the performance of neural networks and periodic autoregressive model on data generated by a PAR(1) model (Table 23.3). To evaluate the accuracy of neural networks in predicting a true PAR(1) process was of particular interest. Both PAR(1) and

TABLE 23.3

Parameters of the PAR(1) model used for synthetic data generation.

Month τ	μ_τ (m^3/sec)	σ_τ (m^3/sec)	Φ_τ	$\sigma_{\varepsilon\tau}$
January	1.26	0.22	0.86	0.51
February	0.92	0.29	0.90	0.44
March	0.66	0.30	0.90	0.44
April	0.57	0.31	0.82	0.57
May	0.64	0.29	0.87	0.48
June	0.92	0.21	0.74	0.67
July	1.88	0.23	0.39	0.92
August	2.85	0.17	0.68	0.74
September	2.99	0.18	0.76	0.65
October	2.57	0.23	0.74	0.68
November	1.97	0.22	0.88	0.47
December	1.50	0.17	0.88	0.48

TABLE 23.4

Parameters of the ANN model used for synthetic data generation.

Month τ	α_τ	β_τ	γ_τ	ω_τ	$\sigma_{\varepsilon\tau}$
January	1.16	−1.53	−0.79	2.11	0.51
February	5.86	0.55	1.46	0.36	0.44
March	6.98	0.34	1.13	0.26	0.44
April	−2.42	−2.39	1.56	−1.29	0.57
May	−3.06	−1.33	1.58	−0.86	0.48
June	2.41	−0.04	0.77	2.09	0.67
July	1.61	−0.90	−0.34	1.11	0.92
August	6.29	0.36	1.10	0.23	0.74
September	2.10	−0.18	−0.11	0.95	0.65
October	6.67	0.30	1.13	0.23	0.68
November	4.88	0.19	0.43	0.41	0.47
December	5.30	0.20	0.33	0.36	0.48

ANN models were applied to the PAR(1)-generated data, and their forecast accuracy was compared. The procedure started with synthetic generation of 10 sequences of 50 years of monthly data based on the assumed PAR(1) model. In each sequence, the first 30 years were chosen for parameter estimation, last 20 for testing, regardless of the model. The model performance for both models was tested using RMSE.

Four models tested in this case (Figure 23.7) (A1, A2, A3, and A4) were applied to each generated sample, as follows:

A1: In the transformed domain, the neural networks is fitted on the first 30 years of each sample, and tested on the last 20 years. Resulting forecasts are retransformed back to the original domain.

A2: The generated data is transformed back to original domain and then the ANN model is fitted on the first 30 years of each sample, and tested on the last 20 years. Resulting forecasts are in original domain.

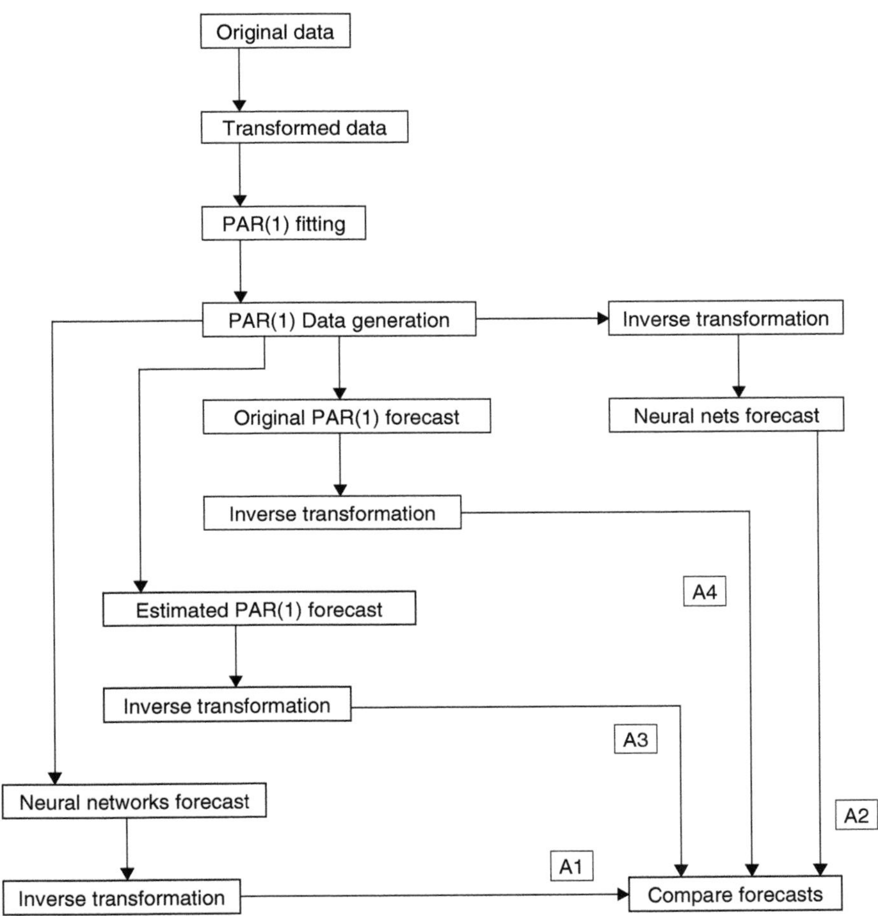

FIGURE 23.7
Schematic presentation of the models A1, A2, A3, and A4 tested on the PAR(1) generated monthly time series.

A3: In the transformed domain, the PAR(1) is fitted to first 30 years of each sample, and tested on the last 20 years. Resulting forecasts are retransformed back to the original domain.

A4: In the transformed domain, the parameters of the original PAR(1) model are used for forecast of the last 20 years of the record. Resulting forecasts are retransformed back to the original domain.

23.3.4.2 *Models Tested on ANN*

A "true" ANN process with one input one output, and one hidden node was obtained by generating ten samples of 50 years each, using the parameters given in Table 23.4. Four PAR-based and ANN-based forecasting models were tested on one-step forecasting of this ANN process. The models tested here are named N1, N2, N3, and N4 (Figure 23.8), and their description is as follows:

N1: The generated data (original domain) is first normalized by a transformation and then the PAR(1) model is estimated on the first 30 years of each sample, and

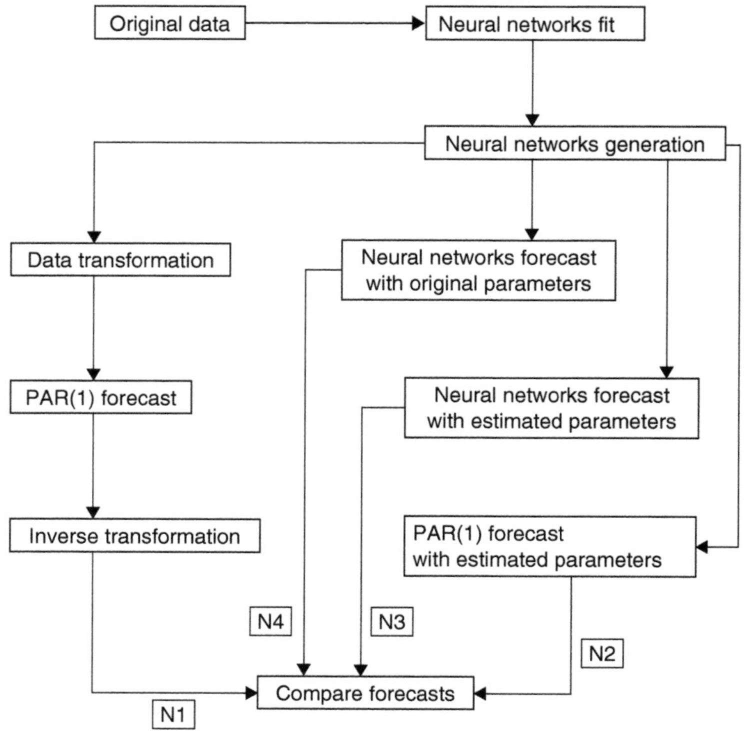

FIGURE 23.8
Schematic presentation of the models N1, N2, N3, and N4 tested on the ANN generated monthly time series.

tested on the last 20 years. Resulting forecasts are transformed back to the original domain.

N2: The PAR(1) model was estimated on the first 30 years of each sample, and tested on the last 20 years. The entire computation was done in original domain under an assumption that the data in original domain is normally distributed.

N3: The neural networks is estimated on the first 30 years of each sample, and tested on the last 20 years. Whole computation is in the original domain, and neither transformation, nor retransformation was needed.

N4: The parameters of the original ANN model are used for forecasting of the last 20 years of the record in the original domain. Resulting forecasts are in original domain. Neither transformation nor retransformation was necessary.

23.3.4.3 Discussion

The errors (RMSE) for the four models tested on PAR(1) generated data (A_1, A_2, A_3, and A_4) and the four models tested on ANN generated data (N_1, N_2, N_3, and N_4) are shown in Figure 23.9. The results demonstrated the following.

When the data are generated by PAR(1), the most accurate forecasting model, in terms RMSE, is the original PAR(1) model (A4). In model A4 parameters are equal to original parameters that were used to generated data, and no parameter estimation was performed during model comparison, and such a result was not surprising. Second most accurate model was model another PAR(1)-based model (A3), followed by the models based on

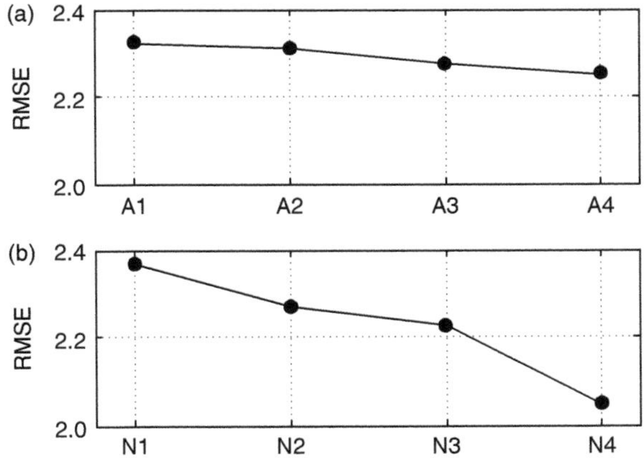

FIGURE 23.9
RMSE in m^3/sec for models (a) A1, A2, A3, and A4 and (b) N1, N2, N3, and N4.

neural networks (A1 and A2). Between those two models, model A2 was more accurate than model A1. The difference between A1 and A2 represents how much accuracy is lost when the retransformation is applied after the calculations. The neural networks models are reasonably close to PAR models in terms of forecasting accuracy.

On the data generated by ANN the most accurate forecasting model was, as expected, obtained by the original ANN (model N4). Second best model for forecasting the data generated by neural networks is model N3. The models PAR(1) with transformation (N1) and without transformation (N2) were less accurate than neural networks. The difference in accuracy between N1 and N2 may be viewed as a loss of accuracy due to the transformations.

The ANN model was able to mimic the PAR(1) process and predict it with insignificant loss of accuracy, (2.2–3.1%). Conversely, the PAR(1) model applied to the ANN process was (6.0–13.4%) less accurate than the ANN model itself (Figure 23.9). This example illustrates the robustness of ANNs, supporting their perception as "the second best model."

23.4 Radial Basis Neural Networks — Minimal Resource Allocation Networks

23.4.1 Introduction

Most of the earlier applications of ANNs were based on the BP feed-forward method. Such networks start training with assumed weights and biases, the number of hidden nodes, learning rate, momentum, and various other parameters. Finding the final architecture is often lengthy, and can be subjective, as well. The process is typically based on a trial-and-error method, and numerous runs need to be implemented. In addition, BP computation is often too time consuming, particularly in the case of large size problems. Many attempts to overcome the shortcomings of the traditional applications have been presented in recent years. The Minimum Resource Allocation Network (MRAN) is one example of such effort. This adaptive method automatically increases or decreases the number of hidden nodes based on the data presented to the network. MRAN is an online method with a moving window and uses the Radial Basis Function (RBF). On the contrary, the BP method uses a sigmoid as activation function, and is usually in a batch mode.

The MRAN has been applied to function approximation, pattern classification, chaotic time series forecasting, nonlinear dynamic system identification, etc., and produced a more compact network with the same or smaller errors [31,41,42]. Because of the more compact network with the same or smaller errors, the network is called Minimal Resource Allocation Network (MRAN). The adjective "minimal" is used in a loose sense, without any rigorous mathematical proof. Proving the minimality of MRAN is an open problem in approximation theory.

The online nature of the MRAN method permits it to deal with nonstationary problems, that is, with variable complexity of input–output relationships in a single run. Such significant variations, seasonal or diurnal, are characteristic for hydrologic relationships. The purpose of this chapter is to present the main attributes of the MRAN technique and illustrate it through a hydrologic application. More detailed hydrologic application can be found in Amenu et al. [13].

23.4.2 MRAN Algorithm

The MRAN algorithm is a sequential learning algorithm for Gaussian RBF neural networks [41]. After the BP networks, the RBF networks comprise one of the most used network models. They have recently drawn much attention due to their good generalization ability and simple network structure that avoids lengthy calculations as compared with the BP networks. The RBF networks are three-layer, feed-forward networks, whose middle layer uses an RBF. Figure 23.10 shows a typical Gaussian RBF network structure with a MIMO system.

The activation function of the hidden units is radially symmetrical in the input space, and the output from each hidden unit depends only on the radial distance between the input vector $\underline{x}$ and the center parameter μ of that hidden unit. The response of each hidden unit is scaled by its connection weight α to the output units. The overall network response for

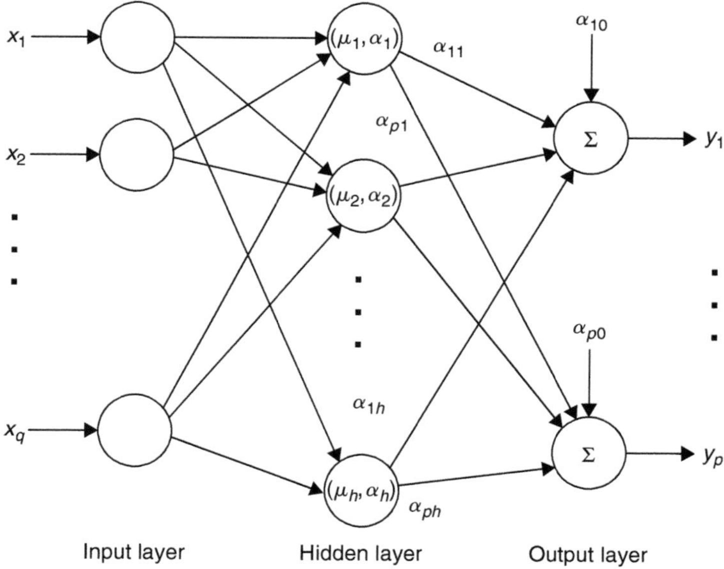

FIGURE 23.10
Structure of Gaussian RBF neural networks.

mapping $R^q \rightarrow R^p$ (MIMO case) is:

$$\underline{y}_i = \underline{f}(\underline{x}_i) = \underline{\alpha}_0^i + \sum_{k=1}^{h} \underline{\alpha}_k^i \Phi_k(\underline{x}_i), \tag{23.21}$$

where $\underline{x}_i \in R^q$, and $\underline{y}_i \in R^p$, are input and output vectors. Symbol q represents the number of inputs, and p is the number of outputs. The bias vector $\underline{\alpha}_k^i$ can be expressed as

$$\alpha_0^i = [\alpha_{10}^i, \ldots, \alpha_{l0}^i, \ldots, \alpha_{p0}^i]^T. \tag{23.22}$$

The coefficient vector $\underline{\alpha}_k^i$ is the connecting weight vector of the kth hidden unit to output layer, which is in the vector form of

$$\underline{\alpha}_k^i = [\alpha_{1k}^i, \ldots, \alpha_{lk}^i, \ldots, \alpha_{pk}^i]^T. \tag{23.23}$$

Thus, the coefficient matrix of the network can be expressed as:

$$A_{p\times h}^i = \begin{bmatrix} \alpha_{11}^i & \cdots & \alpha_{1k}^i & \cdots & \alpha_{1h}^i \\ \cdots & & \cdots & & \cdots \\ \alpha_{l1}^i & \cdots & \alpha_{lk}^i & \cdots & \alpha_{lh}^i \\ \cdots & & \cdots & & \cdots \\ \alpha_{p1}^i & \cdots & \alpha_{pk}^i & \cdots & \alpha_{ph}^i \end{bmatrix}. \tag{23.24}$$

For MISO case, the overall output $f(\underline{x})$ from the network is a linear combination of the outputs from its hidden units, and is expressed as

$$f(\underline{x}_i) = \alpha_0^i + \sum_{k=1}^{h} \alpha_k^i \phi_k(\underline{x}_i), \tag{23.25}$$

where $\phi_k(\underline{x})$ is the response (activation function) of the kth hidden unit, α_k is the connection weight between the kth hidden unit and the output unit, and α_o is the bias term.

Typical selections for RBF are thin-plate-spline $\Phi(r) = r^2 \log(r)$, multiquadratic $\Phi(r) = (r^2 + \beta^2)^{1/2}$, inverse multiquadratic $\Phi(r) = [(r^2 + \beta^2)^{1/2}]^{-1}$, and Gaussian $\Phi(r) = \exp(-r^2/\beta^2)$. Symbol r represents the Euclidean distance between a center $\underline{\mu}$ and the data points $\underline{x}$, and β is a real variable. The MRAN activation function (RBF) is Gaussian and is given by:

$$\phi_k(\underline{x}) = \exp\left(-\frac{\|\underline{x} - \underline{\mu}_k\|^2}{\sigma_k^2}\right), \tag{23.26}$$

where $\underline{\mu}_k$ is the center vector for the kth hidden unit, and σ_k is the width of the Gaussian function of the kth hidden unit. The symbol $\|\cdot\|$ is the Euclidian norm, equal to the length of vector $x = (x_1, x_2, \ldots, x_n)$ and expressed as:

$$\|x\| = \sqrt{|x_1|^2 + |x_2|^2 + \cdots + |x_n|^2}, \tag{23.27}$$

that gives a distance from the origin to the point x on R^n.

23.4.2.1 Allocation of New Hidden Nodes

The MRAN learning process involves allocation of new hidden neurons and adjusting the network parameters [43]. The network initially has no hidden units, and, as new observations are received, the network grows based on selected growth criteria, as described later. The following three criteria are used to determine whether a new hidden unit should be added after a new input–output observation $(\underline{x}_i, y_i)$ at the ith instant is presented to the network:

$$\|\underline{x}_i - \underline{\mu}_{nr}^i\| > \varepsilon_1, \tag{23.28}$$

$$\|\underline{e}_i\| = \|\underline{y}_i - f(\underline{x}_i)\| > \varepsilon_2, \tag{23.29}$$

and

$$e_{\text{rms}}^i = \sqrt{\frac{\sum_{j=i-(n_w-1)}^{i} \left(\underline{e}_j^T \underline{e}_j\right)}{n_w}} > \varepsilon_3, \tag{23.30}$$

where ε_1, ε_2, and ε_3 are thresholds to be selected appropriately, $\underline{\mu}_{nr}$ is the center vector of the hidden unit closest to the input vector $\underline{x}_i$, and n_w is a user-selected sliding data window. The first criterion compares the distance between the new observation and all the existing nodes. If the distance is larger than ε_1, a new node is added. The second criterion is an error check, in which ε_2 represents the desired approximation accuracy of the instantaneous network output. The third criterion checks whether the network meets the required sum-squared error specification for the past n_w network outputs. This criterion is used mainly to control the effect of noise, which can lead to overfitting of hidden neurons.

The pruning algorithm begins with $\varepsilon_1 = \varepsilon_{1\max}$, the largest scale of interest in the input space, which decays until it reaches $\varepsilon_{1\min}$:

$$\varepsilon_1 = \max\{\varepsilon_{1\max}\gamma^i, \varepsilon_{1\min}\}, \tag{23.31}$$

where γ is a decay constant that ranges from 0 to 1. The decaying distance criterion allows fewer basis functions of larger widths initially and more basis functions of smaller widths as the number of observations increases to finetune the approximation.

When all three criteria are satisfied, a new hidden unit is added to the network. Parameters associated with the new hidden unit in the MIMO case at instant i are as follows:

$$\underline{\alpha}_{h+1}^i = \underline{e}_i, \tag{23.32}$$

$$\underline{\mu}_{h+1}^i = \underline{x}_i, \tag{23.33}$$

and

$$\sigma_{h+1}^i = \kappa\|\underline{x}_i - \underline{\mu}_{nr}^i\|, \tag{23.34}$$

where $\underline{e}_i$ is the vector of p approximation errors, $\underline{x}_i$ is the vector of q input observations presented to the network, $\underline{\mu}_{nr}^i$ is the center of the hidden unit closest to $\underline{x}_i$, and κ is a factor that determines the overlap of the responses of the hidden units in the input space.

In case of MISO, $p = 1$ and Equation 23.32 becomes:

$$\alpha_{h+1} = e_i. \tag{23.35}$$

23.4.2.2 Extended Kalman Filter

When the observation $(\underline{x}_i, y_i)$ does not meet the three criteria for adding a new hidden unit, the network parameters are adjusted using the Extended Kalman Filter (EKF) to fit that observation. The total network parameters at the ith instant are represented by the vector $\underline{w}_i = [\underline{\alpha}_0^{i^T}, \underline{\alpha}_1^{i^T}, \underline{\mu}_1^{i^T}, \sigma_1^i, \ldots, \underline{\alpha}_h^{i^T}, \underline{\mu}_h^{i^T}, \sigma_h^i]^T$. The dimension of this vector is equal to $z \times 1$, where z represents the total number of parameters, equal to $z = p + h(p + q + 1)$. For example, if the number of hidden nodes, $h = 2$, number of inputs $q = 3$, and number of outputs, $p = 1$, the number of parameters will be $z = 1 + 2(1 + 3 + 1) = 11$. For MISO, $\underline{w}_i = [\alpha_o, \alpha_1, \underline{\mu}_1^T, \sigma_1, \ldots, \alpha_h, \underline{\mu}_h^T, \sigma_h]^T$, and $z = 1 + h(q + 2)$.

The EKF approach obtains the posterior estimate $\underline{w}_i$ from its prior estimate $\underline{w}_{i-1}$ and its prior error covariance estimate P_{i-1} as follows:

$$\underline{w}_i = \underline{w}_{i-1} + K_i \underline{e}_i, \tag{23.36}$$

where K_i is the Kalman gain ($z \times p$) and $\underline{e}_i$ ($p \times 1$) is the approximation error vector. The Kalman gain is vector given by:

$$K_{i(z\times 1)} = P_{i-1}B_i[R_i + B_i^T P_{i-1} B_i]^{-1}. \tag{23.37}$$

In this equation the variable R_i is the variance of the measurement noise, and $B_i = \nabla_w f(\underline{x}_i)$ is the gradient of the function $f(\underline{x}_i)$ with respect to the parameter vector $\underline{w}$ evaluated at $\underline{w}_{i-1}$. For the structure in Figure 23.10, B_i becomes:

$$\begin{aligned} B_{i(z\times p)} = \Bigg[& I_{(p\times p)}, \phi_1(\underline{x}_i) I_{(p\times p)}, \phi_1(\underline{x}_i) \frac{2\underline{\alpha}_1^i}{(\sigma_1^i)^2}(\underline{x}_i - \underline{\mu}_1^i)^T, \phi_1(\underline{x}_i)\frac{2\underline{\alpha}_1^i}{(\sigma_1^i)^3}\|\underline{x}_i - \underline{\mu}_1^i\|^2, \\ & \ldots, \phi_h(\underline{x}_i) I_{(p\times p)}, \phi_h(\underline{x}_i)\frac{2\underline{\alpha}_h^i}{(\sigma_h^i)^2}(\underline{x}_i - \underline{\mu}_h^i)^T, \phi_h(\underline{x}_i)\frac{2\underline{\alpha}_h^i}{(\sigma_h^i)^3}\|\underline{x}_i - \underline{\mu}_h^i\|^2 \Bigg]^T \end{aligned} \tag{23.38}$$

The error covariance matrix P_i is:

$$P_{i(z\times z)} = (I_{(z\times z)} - K_i B_i^T) P_{i-1} + q_o I_{(z\times z)}, \tag{23.39}$$

where $I_{(z\times z)}$ is an identity matrix and q_o is a scalar coefficient. When a new hidden unit is allocated, the dimensions of P_i increase with new rows and columns. Under this condition, P_i becomes:

$$P_i = \begin{bmatrix} P_{i-1} & 0 \\ 0 & p_o I_{(z_1\times z_1)} \end{bmatrix}, \tag{23.40}$$

where p_o is a parameter that reflects the uncertainty of the initial values assigned to the parameters, and z_1 is the number of new parameters introduced by the addition of the new hidden unit. For the network in Figure 23.10, $z_1 = p + q + 1$. More detailed discussion on the EKF approach can be found in Sum et al. [43].

23.4.2.3 Pruning Strategy

If the network is allowed to grow only according to the growth criteria described earlier, some hidden neurons, although active initially, subsequently may end up contributing little to the network output. A more compact network topology can be realized by removing inactive hidden units as learning progresses. This is achieved in the MRAN algorithm by incorporating a pruning strategy, which checks the weight for each hidden unit for every iteration, and those units with weights below a certain threshold value are removed.

The pruning strategy has two main steps. First, at each iteration (observation), outputs from each hidden unit are normalized with respect to the maximum value of the outputs of all the hidden units according to:

$$r_{lk}^{i} = \frac{\|o_{lk}^{i}\|}{\|o_{l,\max}^{i}\|} \tag{23.41}$$

$$\underline{o}_{k}^{i} = \underline{\alpha}_{k}^{i}\Phi_{k}(\underline{x}_{i}), \tag{23.42}$$

and

$$o_{l,\max}^{i} = \max(\|o_{l1}^{i}\|, \ldots, \|o_{lk}^{i}\|, \ldots, \|o_{lh}^{i}\|), \tag{23.43}$$

where $\underline{o}_k$ is the output of the kth hidden unit, and r_k is its normalized output value. The normalized value of each hidden neuron is compared with a user-defined threshold δ. If the normalized value falls below this threshold for n_w consecutive observations, it indicates that the kth hidden neuron makes an insignificant contribution to the network output and will be removed from the network. Consequently, EKF dimensionality is updated to suit the reduced network.

The MRAN learning algorithm can be summarized by the flowchart given in Figure 23.11. The main steps are as follows:

- The network begins with no hidden units.
- As observations are received, the network grows by using some of them as the new hidden units. A new hidden unit can be added based on new data (sliding window).
- Parameters are updated based on EKF.
- A hidden node can be removed based on new data using the pruning algorithm.

In general, the algorithm automatically computes the five main types of parameters in the network: the number of hidden neurons (h), the center positions (μ) for all hidden neurons, the widths (σ) of the Gaussian functions, the connection weights (α) between hidden layer and output layers, and the bias (α_o). However, the user needs to specify certain parameters prior to the model run. These include $\varepsilon_{1\max}$, $\varepsilon_{1\min}$, ε_2, ε_3, γ, κ, q_o, p_o, δ, n_w, and s_w. The MRAN algorithm uses network growth criteria to add neurons, and a pruning strategy to prune the network to obtain a minimal RBF neural network.

23.4.3 Application

The MRAN algorithm was applied to calculate weekly nitrate-N forecasts on the Sangamon River near Decatur in Illinois, USA, using observed hydrologic and meteorological data for

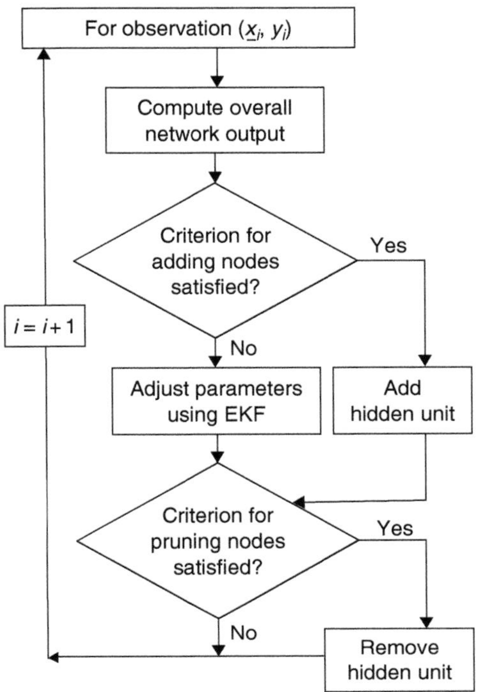

FIGURE 23.11
Flowchart of MRAN algorithm.

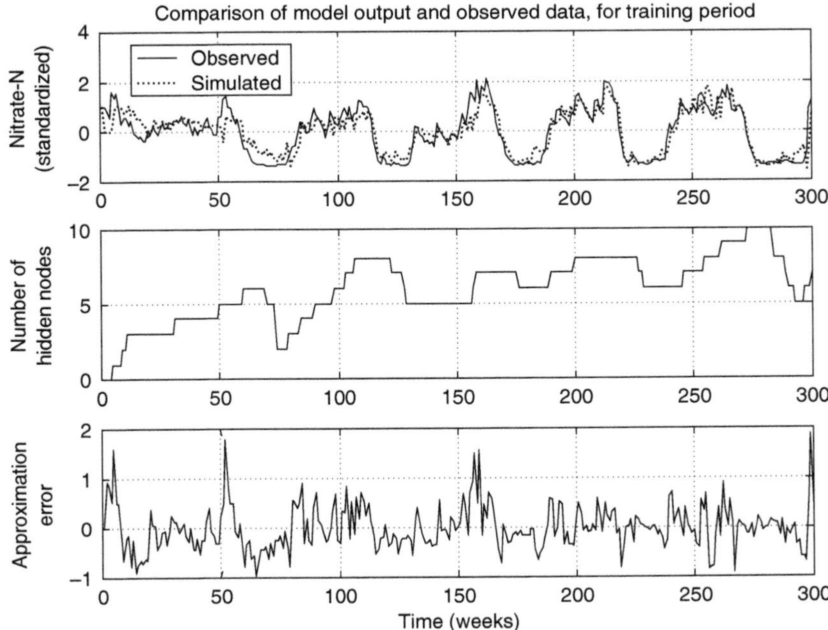

FIGURE 23.12
The MRAN predictions of weekly nitrate-N data on the Upper Sangamon River near Monticello, Illinois, based on previous: nitrate-N, temperature, precipitation, and discharge.

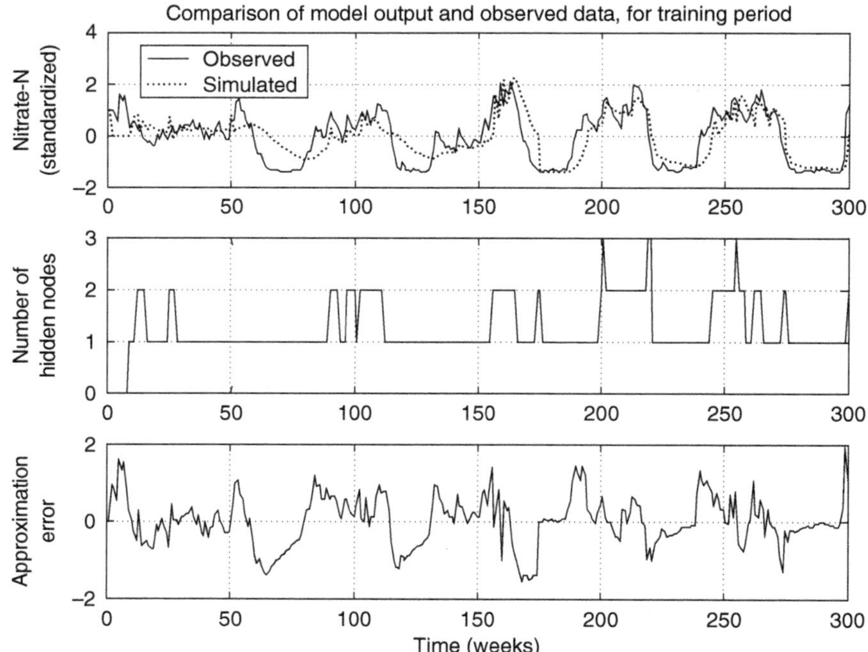

FIGURE 23.13
The MRAN predictions of weekly nitrate-N data on the Upper Sangamon River near Monticello, Illinois, based on previous discharge.

TABLE 23.5

User-defined MRAN parameters for the SISO and MISO networks, described in Section 23.4.

Symbol	Description of the variable	SISO	MISO
q	Number of network inputs	1	4
p	Number of network outputs	1	1
$\varepsilon_{1\,\max}$	Max threshold for criterion #1 for adding neuron	1.20	2.20
$\varepsilon_{1\,\min}$	Min threshold for criterion #1 for adding neuron	1.0	0.30
γ	Decay constant for criterion #1	0.5	0.999
δ	Threshold for pruning neuron	0.4	0.05
ε_2	Threshold for criterion #2 for adding neuron	0.39	0.10
ε_3	Threshold for criterion #3 for adding neuron	0.425	0.30
κ	Overlap factor	0.84	0.88
q_o	Coefficient for random walk in EKF	0.003	0.005
p_o	Estimate of parameter uncertainties in EKF	0.090	0.90
n_w	Size of sliding window for adding hidden neurons	8	15
s_w	Size of sliding window for removing hidden neurons	3	15

January 1994–April 1999 [44]. Such predictions are important in water supply planning and in reducing nitrate-N concentration during the critical periods [12].

Two models with different complexity were tested. The first model uses current week's average nitrate-N concentration, total precipitation, average river discharge, and average air temperature to predict next week's nitrate-N concentration. Herein, this model is referred to as the MISO model. Figure 23.12 shows weekly observed and predicted nitrate-N concentration in the standardized form, the change in the number of hidden nodes with time, and the approximation error, of the MISO model. The second model (Figure 23.13) is a SISO model and uses only current week's discharge to predict next week's nitrate-N concentration. The user-defined parameters for the models are shown in Table 23.5. Figure 23.12 and Figure 23.13 show how the difference in the complexity of the models affects the learning process, the number of hidden nodes, and the accuracy of approximation. The more complex MISO model had more accurate approximation, with RMSE = 0.58, than the simpler SISO model, with RMSE = 1.05.

References

1. McCulloch, W.S. and W. Pitts (1943). A logical calculus of the ideas immanent in nervous activity, *Bulletin of Mathematical Biophysics*, 5, 115–133.
2. Wasserman, P.D. (1989). *Neural Computing Theory and Practice*, Van Nostrand Reinhold, New York, p. 230.
3. Hertz, J., A. Krogh, and R. Palmer (1991). *Introduction to the Theory of Neural Computation*, Addison-Wesley Publishing Company, Redwood City, CA.
4. Cybenko, G. (1989). Approximation by superposition of a sigmoidal function, *Mathematical Control, Signals and Systems*, 2, 303.
5. Ito, Y. (1991). Approximation of functions on a compact set by finite sums of a sigmoid function without scaling, *Neural Networks*, 4, 817–826.
6. Sandberg, I.W. (1991). Approximation theorems for discrete-time systems, *IEEE Transactions on Circuits and Systems*, 38, 564–566.
7. Kohonen, T. (1988). An introduction to neural computing, *Neural Networks*, 1, 3–16.
8. French, M.N., W.F. Krajewski, and R.R. Cuykendal (1992). Rainfall forecasting in space and time using a neural network, *Journal of Hydrology*, 137, 1–37.
9. Hsu, K., H.V. Gupta, and S. Sorooshian (1995). Artificial neural network modeling of the rainfall-runoff process, *Water Resources Research*, 31, 2517–2530.
10. Tokar, A.S. and M. Markus (2000). Precipitation-runoff modeling using artificial neural networks and conceptual models, *Journal of Hydrologic Engineering*, 5, 156–161.
11. Maier, H. and G. Dandy (1996). The use of artificial neural networks for the prediction of water quality parameters, *Water Resources Research*, 32, 1013–1022.
12. Markus, M., C.W.-S. Tsai, and M. Demissie (2003). Uncertainty of weekly nitrate-nitrogen forecasts using artificial neural networks. *Journal of Environmental Engineering*, 129, 267–274.
13. Amenu, G., M. Markus, P. Kumar, and M. Demissie (2005). Hydrologic applications of Minimal Resource Allocation Network (MRAN), *Journal of Hydrological Engineering* (under revision).
14. Ray, C. and K.K. Klindworth (2000). Neural networks for agrichemical vulnerability assessment of rural private wells, *Journal of Hydrologic Engineering*, 5, 162–171.
15. Rizzo, D.M. and D.E. Dougherty (1994). Characterization of aquifer properties using artificial neural networks: neural kriging, *Water Resources Research*, 30.
16. Shin, H.-S. and J.D. Salas (2000). Regional drought analysis based on neural networks, *Journal of Hydrologic Engineering*, 5, 145–155.
17. American Society of Civil Engineers, Task Committee on Application of Artificial Neural Networks in Hydrology (2000a). Artificial neural networks in hydrology. I: Preliminary concepts, *Journal of Hydrological Engineering*, ASCE, 5, 115–123.
18. American Society of Civil Engineers, Task Committee on Application of Artificial Neural Networks in Hydrology (2000b). Artificial neural networks in hydrology. II: Hydrologic applications, *Journal of Hydrological Engineering*, ASCE, 5, 124–137.
19. Guan, B.T., G.Z. Gertner, and P. Parysov, (1997). A framework for uncertainty assessment of mechanistic forest growth models: a neural network example, *Ecological Modeling*, 98, 47–58.

20. Salas, J.D., M. Markus, and A.S. Tokar (2000). Streamflow forecasting based on artificial neural networks, in R.S. Govindaraju and A.R. Rao, Eds., *Artificial Neural Networks in Hydrology*, Kluwer Academic Publishers, The Netherlands, pp. 23–51.
21. Aires, F. (2004). Neural network uncertainty assessment using Bayesian statistics with application to remote sensing: 1. Network weights, *Journal of Geophysics Research*, 109, D10303.
22. Aires, F., C. Prigent, and W.B. Rossow (2004). Neural network uncertainty assessment using Bayesian statistics with application to remote sensing: 2. Output errors, *Journal of Geophysics Research*, 109, D10304.
23. Aires, F., C. Prigent, and W.B. Rossow (2004). Neural network uncertainty assessment using Bayesian statistics with application to remote sensing: 3. Network Jacobians, *Journal of Geophysics Research*, 109, D10305.
24. Rumelhart, D.E., G.E. Hinton, and R.J. Williams (1986). Learning internal representations by error propagation, in D.E. Rumelhart and J.L. McClelland, Eds., *Parallel Distributed Processing: Explorations in the Microstructure of Cognition. Vol. 1: Foundations*, MIT Press, Cambridge, MA.
25. Broomhead, D.S. and D. Lowe (1988). Multivariable functional interpolation and adaptive networks, *Complex Systems*, 2, 321–355.
26. Parzen, E. (1962). On estimation of a probability density function and mode. *Annals of Mathematical Statistics*, 33, 1065–1076.
27. Specht, D.F. (1990). Probabilistic neural networks, *Neural Networks*, 3, 109–118.
28. Specht, D.F. (1991). A general regression neural network, *IEEE Transactions on Neural Networks*, 2, 568–576.
29. Kohonen, T. (1982). Self-organized formation of topologically correct feature maps, *Biological Cybernetics*, 43, 59–69.
30. Williams, R. and D. Zipser, A learning algorithm for continually running fully recurrent neural network, *Neural Computation*, 1, 270–280.
31. Sundararajan, N., P. Saratchandran, and L. Ying Wei (1999). *Radial Basis Function Neural Networks with Sequential Learning, MRAN and Its Applications*, World Scientific, Singapore.
32. Stone, M. (1974). Cross-validatory choice and assessment of statistical predictions, *Journal of the Royal Statistical Society*, Series B, 36, 111–147.
33. Geisser, S. (1975). The predictive sample reuse method with applications, *Journal of the American Statistical Association*, 70, 320–328.
34. Brass, R.L. and I. Rodriguez-Iturbe, *Random Functions and Hydrology*, Addison-Wesley, Reading, MA, 1985.
35. Salas, J.D., J.R. Delleur, V. Yevjevich, and W.L. Lane (1980). *Applied Modeling of Hydrologic Time Series*, Water Resources Publications, Littleton, CO.
36. Box, G.E.P. and G.M. Jenkins, *Time Series Analysis Forecasting and Control*, Holden-Day, San Francisco, CA, 1976.
37. Zou, R., W.S. Lung, and H. Guo (2002). Neural network embedded monte carlo approach for water quality modeling under input information uncertainty, *Journal of Computing in Civil Engineering*, 16, 135–142.
38. Ochoa-Rivera, J.C., R. Garcia-Bartual, and J. Andreu (2002). Multivariate synthetic streamflow generation using a hybrid model based on artificial neural networks, *Hydrology and Earth System Sciences*, 6, 641–654.
39. Maier, H.R. and G.C. Dandy (2000). Application of artificial neural networks to forecasting of surface water quality variables: issues, applications and challenges, in R.S. Govindaraju and A.R. Rao Eds., *Artificial Neural Networks in Hydrology*, Kluwer Academic Publishers, The Netherlands, pp. 287–309.
40. Cohn, T.A., L.L. DeLong, E.J. Gilroy, R.M. Hirsch, and D.K. Wells (1989). Estimating constituent loads, *Water Resources Research*, 25, 937–942.
41. Yingwei, L., N. Sundararajan, and P. Saratchandran (1998). Performance evaluation of a sequential minimal radial basis function (RBF) neural network algorithm, *IEEE Transactions Neural Networks*, 9, 308–318.
42. Jianping, D., N. Sundararajan, and P. Saratchandran (2000). Complex-valued minimal resource allocation network for non-linear signal processing, *International Journal of Neural Systems*, 10, 95–106.

43. Sum, J., C.-S. Leung, G.H. Young, and W.-K. Kan (1999). On the Kalman filtering method in neural-network training and pruning, *IEEE Transactions on Neural Networks*, 10, 161–166.
44. Keefer, L. and M. Demissie (2000). *Watershed Monitoring for the Lake Decatur Watershed 1998–1999*, Illinois State Water Survey, Champaign, IL.

24

Genetic Algorithms

Barbara Minsker

CONTENTS

24.1 Introduction

John Holland proposed the first genetic algorithm (GA) in 1975 [1], but the approach became popular primarily after publication of David Goldberg's classic textbook in 1989 [2]. Since that time, GAs have been used for numerous hydroinformatics applications where optimization (i.e., searching for solutions, represented by a set of decision variable values, that meet a specified objective function and constraint) is needed. Just in the last few years, an explosion of hydroinformatics applications using genetic algorithms have been published; examples include: model calibration [3–6]; groundwater monitoring and remediation design [7–10]; watershed, reservoir, and river management [11–14]; and water distribution system design [15–18]. GAs use an analogy to the processes of natural selection or "survival of the fittest" to search for good solutions to a specified problem. This chapter describes the basics of how simple GAs work (Section 24.2) and illustrates how a hydroinformatics application can be formulated as a GA problem (Section 24.3). Key theoretical concepts and guidelines for effective use of GAs are then introduced (Sections 24.4 and 24.5), along with approaches for overcoming computational barriers in applying GAs to real-world hydroinformatics problems (Section 24.6). Finally, a number of advanced GAs are introduced that offer promise for improving the solution of complex hydroinformatics applications (Section 24.7).

24.2 GA Basics

Genetic algorithms are among several types of optimization methods that use a stochastic approach to randomly search for good solutions to a specified problem, including simulated annealing [19], tabu search [20, 21], and numerous variations. These stochastic approaches (also called heuristic algorithms) use various analogies to natural systems to build from promising solutions, ensuring greater efficiency than completely random search. The advantage of using these types of approaches over traditional optimization methods, such as nonlinear programming, is that they can solve any type of problem without explicit specification of problem characteristics (e.g., derivatives of the objective function). This property is particularly important for hydroinformatics applications, where the optimization problem often involves integer decision variables or potential solutions that must be evaluated with complex existing simulation models, where derivative calculations would be difficult or impossible. Stochastic approaches also perform broad global search, unlike traditional nonlinear optimization approaches such as nonlinear programming that can converge to local minima when the optimization problem is multimodal. These benefits have resulted in widespread use and acceptance of these approaches among the hydroinformatics community. The disadvantages of these approaches are that they are not guaranteed to find the globally optimal solution and they can be substantially slower than traditional optimization methods for problems that can be solved using traditional approaches. Therefore, the author recommends that these methods only be used for problems that cannot be effectively solved using traditional optimization approaches, such as those with numerous integer decision variables, nonconvexities (a convex problem would have a bowl-shaped decision space to search), or other irregularities. When applied to such problems, these algorithms have been demonstrated to find substantially better solutions than could be found using trial-and-error search [22].

Numerous types and variations of genetic algorithms exist, but this chapter will provide details only on the most widely used approach called the binary simple genetic algorithm (SGA). Figure 24.1 shows the basic operations of an SGA. First, the decision variables are encoded into binary form (more on this in Section 24.3), called a "chromosome" (or sometimes "string") because it gives the genetic encoding ("genes" or "bits") describing each potential solution. Next, an initial "population" of potential solutions is created, usually by filling a set of chromosomes (population "members") with random initial values. Each member of the population is then evaluated to see how well it performs (i.e., its "fitness") with respect to the user-specified objective function and constraints ("fitness function"). Then the population is transformed into a new population (the next "generation") using three primary operations: selection, crossover, and mutation. A fourth operator, elitism,

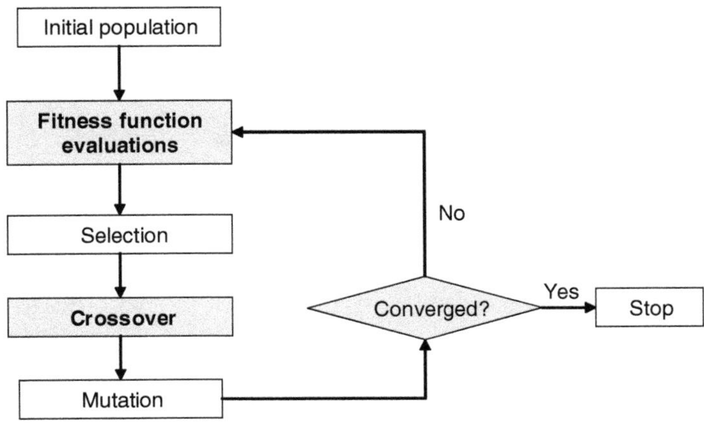

Figure 24.1
Simple genetic algorithm operations.

is also usually included to ensure that good solutions are not lost from one generation to the next. These operations are described in more detail in the following subsections. This transformation process from one generation to the next continues until the population converges to the optimal solution, which usually occurs when some percentage of the population (e.g., 90%) have the same optimal chromosome. For additional details on the SGA process, please consult Goldberg's textbook [2].

24.2.1 Selection

The selection operator involves randomly choosing members of the population to enter a mating pool. The operator is carefully formulated to ensure that better members of the population (with higher fitness) have a greater likelihood of being selected for mating, but that worse members of the population still have a small probability of being selected. Having some probability of choosing worse members is important to ensure that the search process is global and does not simply converge to the nearest local optimum.

Probably the most widely used and understood selection approach is called tournament selection. In tournament selection, several members of the population are chosen randomly. These members then participate in a tournament to enter the mating population, where their probability of winning the tournament is proportional to their fitness. For example, if two members compete in a tournament (called "binary" tournament selection) with fitness values of 50 and 100, the first member will have a 1/3 probability and the second member will have a 2/3 probability of winning the tournament.

If the population size is 100, the selection process is continued until the mating population contains 100 members. At this point, the crossover operation begins.

24.2.2 Crossover

During crossover, pairs of chromosomes (parents) are randomly selected from the mating population. With a user-specified crossover probability, P_c, genes from one parent chromosome are swapped with corresponding genes on the other parent chromosome to create two children. When the swap does not occur (probability $1 - P_c$), the two parents are transferred to the child population unchanged.

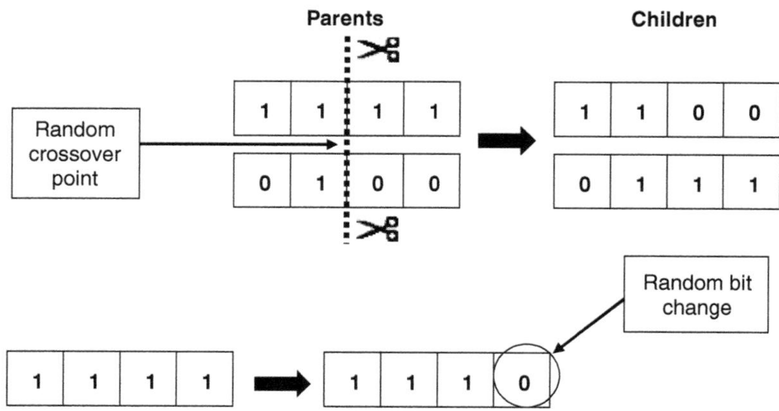

FIGURE 24.2
Crossover and mutation operations on four-gene chromosomes.

Figure 24.2 illustrates the crossover process for single-point crossover, where one point on the chromosome is randomly selected for gene exchange. In the case shown, the crossover point occurs after the second gene, resulting in exchange of the third and fourth genes in the children. In multipoint crossover, multiple locations on the chromosome are selected for gene exchange, each with probability P_c. The highest amount of exchange occurs during uniform crossover, where every gene has a probability P_c of being exchanged with its corresponding gene on the other parent chromosome.

24.2.3 Mutation

Once the children are created during crossover, the mutation operator is applied to each child. Figure 24.2 shows how mutation changes a chromosome with four genes. Each gene has a user-specified mutation probability, P_m, of being mutated from a value of 0 to a value of 1, or vice versa. In the case shown in Figure 24.2, only the 4th gene is changed from a 1 to a 0.

24.2.4 Elitism

The final operator, elitism, simply involves replacing one of the chromosomes in the children population with the best member of the parent population. This operator has been shown to increase the speed of convergence of the SGA because it ensures that the best solution found in each generation is retained. While this operator could be applied more broadly (e.g., retaining the two or three best solutions), overuse can lead to premature convergence to the incorrect solution.

24.3 Formulating Hydroinformatics Optimization Problems: A Case Study in Groundwater Monitoring Design

Before GAs can be applied to hydroinformatics optimization problems, each problem must first be formulated into a set of chromosomes and fitness function that the GA can process.

This section lays out the formulation process for a real-world optimization problem in groundwater monitoring design (Section 24.3.1) and then discusses two complications that arise in many hydroinformatics applications: continuous decision variables that must be converted to binary form (Section 24.3.2) and constraints (Section 24.3.3).

24.3.1 Formulation Process for the Hydroinformatics Case Study

Any optimization algorithm requires that the problem to be solved is expressed in terms of three components: (1) decision variables, (2) objective function, and (3) constraints. The decision variables are the solutions that are being sought, and are represented by the chromosome in the genetic algorithm. The objective function represents the goals of the optimization in terms of a mathematical equation to be minimized or maximized (called the fitness function in a GA). Its value can be computed once the values of the decision variables are determined. The constraints represent limitations on the decision variables in terms of mathematical equations. These equations are either incorporated into the fitness function or considered in the selection process within the GA (more on this later in Section 24.3.3). The process of creating these equations for a groundwater monitoring design case study is illustrated below.

The case study involves identifying an optimal set of monitoring wells to be sampled at a BP (formerly British Petroleum) site where a leaking gasoline pipeline contaminated the groundwater underneath the site with benzene, toluene, ethylbenzene, and xylene (BTEX). Active cleanup of the contamination has been completed and BP is now interested in identifying whether any of the 36 monitoring wells that are currently being sampled are redundant, meaning that the information they provide can be obtained from another well without significant increase in error, and can be eliminated from their sampling plan. The decision variables for the problem then become simple sampling flags (0/1) for all 36 wells. When a specific flag is 1 then the monitoring well at that location is used. Hence the chromosome for this problem consists of 36 binary genes, each of which indicates the sampling of the corresponding well.

This problem essentially has two objectives: minimize monitoring costs and minimize errors made with the collected data. To create a mathematical equation to represent each of these objectives, some assumptions and simplifications must be made. First, BP does not have detailed monitoring costs available that can be used to create a cost equation. Laboratory analytical costs are simple to estimate, but labor costs associated with sampling trips and data analysis are billed with other consultant costs and cannot be ascertained readily. Therefore, BP requested that the first objective be simply to minimize the number of samples collected, which assumes that the costs are roughly proportional to the number of samples collected.

Quantifying the second objective, minimizing errors made with the collected data, also requires some simplifications and assumptions. At this site (and many others), contaminant data are used to interpolate the chemical concentrations (or mass) throughout the contaminated area (called the contaminant plume) in each monitoring period. The interpolated concentrations are then used to assess how the contaminants are changing over time. Interpolating the concentrations requires an interpolation model to be fit to the data. In this case a detailed analysis comparing different analytical and geostatistical approaches was performed and quantile kriging was selected as the most accurate model (based on cross-validation tests). Given this approach for analyzing the chemical concentrations, the second objective can then be expressed as minimizing the differences between the interpolated BTEX and benzene plumes using all available data and the interpolated plumes using some subset of the data. Note that this approach ignores the inherent uncertainty in the data, which will be addressed subsequently in Section 24.7.2. It also assumes that a sampling plan created using the most recent data will be applicable in future monitoring periods, an assumption that is reasonable given that the contaminant plumes at this site are not moving significantly over time. For sites with migrating or fluctuating plumes, an alternative approach can be pursued in which the sampling plan is optimized to correctly interpolate multiple plumes from different historical time periods or, if a transport model is available, for predicted future plumes [23].

With these assumptions, the objective functions become:

1. Minimize the number of wells sampled,

$$\text{Minimize} \sum_{i=1}^{n} x_i, \tag{24.1}$$

where n is the total number of wells (36 in this case) and x_i has the value 0 for a well not sampled and 1 for a well sampled.

2. For the contaminants, minimize the interpolation error that occurs when a set of well locations K are removed from sampling. The interpolation error is expressed in terms of the maximum error between the actual concentrations and those estimated with a particular subset of K redundant well locations, scaled by the target error for each contaminant $E_{\text{tar,con}}$:

$$\underset{K}{\text{Minimize}} \ \text{error} = \left(\frac{\underset{i}{\text{Max}} \left| c_i^s - c_i^{\text{est}}(K) \right|}{E_{\text{tar,con}}} \right), \tag{24.2}$$

where c_i^s is the true sampled concentration at point i, and $c_i^{\text{est}}(K)$ is the concentration at point i estimated using only the monitoring wells included in design K. This error metric considers only the locations i for which there are measurements available that are not being used in the interpolation model, since those are the only locations where the concentrations are known with certainty. Note that the target error, $E_{\text{tar,con}}$, was set by BP to scale the errors to a level that would begin to be significant (5 ppb for benzene and 100 ppb for BTEX), but higher errors could still be acceptable.

The simple genetic algorithm assumes that there is only one objective, but this case study has three objectives (one for cost, one for benzene error, and one for BTEX error). Later we will see how this problem can be solved with all objectives simultaneously, but for now we can convert the error objectives into constraints that fix the maximum concentration errors to the target values $E_{\text{tar,con}}$. The constraint is formulated in the same way as the error objective in Equation 24.2, but now sets a fixed threshold on the allowable error:

$$\underset{i}{\text{Max}} \left| c_i^s - c_i^{\text{est}}(K) \right| \le E_{\text{tar,con}}. \tag{24.3}$$

24.3.2 Converting Decision Variables to Binary Form

In the groundwater monitoring design example, all of the decision variables were binary, which makes the process of creating a binary chromosome simple. In this case, the 36 binary digits are simply joined together into a string of 36 binary digits. Many decision variables in hydroinformatics problems are continuous, however. For example, if new wells were being considered in the monitoring design, continuous decision variables could be defined for the coordinates of the new wells. Such decision variables can either be handled using a real-coded genetic algorithm, which is specifically designed to handle real decision variables, or by converting the real variables to binary form. Real-coded genetic algorithms are more difficult to use and there is no theory to guide their use. If the particular application you are solving has hundreds of real variables then a real-coded genetic algorithm may be worth investigating, but otherwise a binary genetic algorithm is recommended.

To illustrate how the conversion is made from a real decision variable, consider a decision variable C_x^1 that represents the x-coordinates of the first new well. Suppose that the coordinates can vary between 0 and 1000 and we want to use five binary digits to represent the coordinates. In this case, 2^5, or 32, possible combinations of the five binary digits are possible and can represent 32 different well coordinates. The first coordinate would be 00000 and would represent a 0 coordinate. The second coordinate would be represented in binary by 00001 and would have a coordinate of $(1000/32) = 31.25$. Similarly, the third coordinate would be represented in binary by 00010 and would have a coordinate of $(2 * 1000/32) = 62.5$. This process continues to the last coordinate, which would be represented in binary by 11111 and would have a coordinate of 1000. In this way, the real-valued coordinate decision variable is discretized into discrete form. If the coordinates are not sufficiently accurate, then more digits can be used to represent each coordinate until the representation is sufficiently accurate. In general, to calculate the number of binary digits k needed to represent a real decision variable U with upper bounds U^{max} and U^{min}, respectively, and a particular level of precision dU, use the following:

$$2^k = \frac{U^{\text{max}} - U^{\text{min}}}{dU} + 1. \tag{24.4}$$

24.3.3 Constraint Handling

The SGAs, like most other nonlinear optimization methods, cannot handle constraints such as the maximum error constraint in the monitoring design problem (Equation 24.3) without special handling. Simple bounds on decision variables, such as the well coordinates in the monitoring design problem, can be handled through the encoding of the binary chromosomes. For more complex constraints, most SGAs used in hydroinformatics applications handle constraints using penalty functions. Many types of penalty functions exist; a linear penalty function for the monitoring problem would be as follows:

$$\underset{K}{\text{Penalty}} = W \sum_{i,\text{con}} \{E_{\text{tar,con}} - (c_i^s - c_i^{\text{est}}(K))\}. \tag{24.5}$$

This penalty, which represents the amount that the constraint shown in Equation 24.3 is violated (summed across all wells that are removed in a particular plan K), is only invoked when the maximum error target is exceeded. The parameter W is a penalty weight. For a quadratic penalty, another commonly used penalty function, the same equation would apply but the term in brackets would be squared before it is summed.

Once the penalty shown in Equation 24.5 is calculated, it is then added to the objective function given in Equation 24.1 to create the fitness function for the SGA:

$$\text{Minimize} \sum_{i=1}^{n} x_i + \text{Penalty}. \tag{24.6}$$

Note that some SGA solvers assume that the fitness function requires maximizing fitness. This can be solved by multiplying Equation 24.6 by (−1) and maximizing the result, which is equivalent to minimizing Equation 24.6.

Properly implementing an SGA with a penalty function requires that the weight W given in Equation 24.5 be carefully adjusted by trial-and-error. If the penalty is too low, then the constraint will not be satisfied. However, if the penalty is too high, then the SGA operations will be so heavily weighted toward satisfying the constraints that it will not be able to find the optimal solution to the original objective function. The best approach to finding the appropriate penalty weight is to start with a small weight and increase it until the constraints are satisfied in the final population. This trial-and-error process can be time-consuming, especially when there are multiple constraints with quite different orders of magnitudes. One solution is to scale all of the constraints and the objective function to be of similar order of magnitude. Chan Hilton and Culver [24] tested a number of different penalty-based approaches for groundwater remediation design. Another solution is to use a more sophisticated constraint handling approach, such as those that have recently begun emerging from the genetic and evolutionary computation literature. Zavislak [25] tested two new constraint-handling approaches on two groundwater remediation design case studies and found that both approaches were as effective as the penalty approach given by equation (24.6), but required no trial-and-error experimentation.

24.4 GA Theory

Using an SGA requires setting four parameters: population size, crossover probability, mutation rate, and criteria for stopping the optimization. Setting the parameters correctly can make the difference between success and failure of the optimization process. For example, if the population size is too low or the crossover probability is too high, SGAs can prematurely converge to the wrong solution. Insufficient crossover or incorrect population sizes can lead to drift, where the chromosomes fluctuate and eventually randomly converge to a nonoptimal solution. Figure 24.3 shows an example comparing drift with normal convergence. Normal convergence with appropriate parameter settings leads to average fitness that increases over time. Drift leads to fluctuating fitness values that ultimately converge to suboptimal solutions.

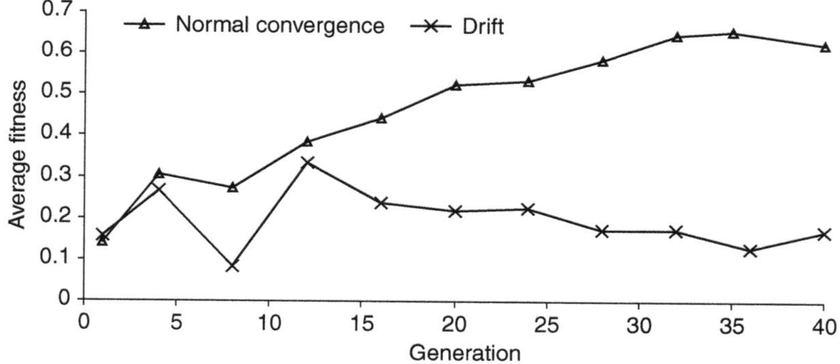

FIGURE 24.3
Comparison of normal SGA convergence and drift.

The standard approach to ensuring SGA performance is to perform numerous iterative runs with different parameter settings (e.g., one hydroinformatics application used 60 trial runs to determine parameter settings [26].) However, with some understanding of GA theory, the number of runs can be dramatically reduced, usually to 2 to 4 runs. This section provides a brief overview of the essentials of SGA theory needed to use SGAs effectively (more details can be found in several GA books [2,27]). The next section presents a design methodology that uses this theory to identify appropriate SGA parameters.

SGAs work by mixing highly fit pieces of solutions (chromosomes), called "building blocks," during crossover. Over time, the best building blocks are selected more and more often and are eventually assembled into the optimal chromosome. Crossover and selection together provide the force of innovation, creating completely new solutions, and provide most of the searching work during the initial generations. Mutation and selection together perform local search in the area near the current solution. As the search gets close to convergence, the population becomes more similar and crossover does not significantly change the population (since crossing two identical chromosomes will always produce the same chromosome), so mutation then does most of the searching work. Mutation is an inefficient local search algorithm, which is why SGAs tend to converge slowly (more on this later).

Given this understanding of how SGAs operate, two requirements are needed to ensure that an SGA does find the optimal solution: (1) an adequate supply of initial building blocks that can be recombined during crossover, which requires a sufficiently large population size, and (2) the ability to discern which are the best building blocks, which requires sufficient crossover for adequate mixing but not so much that good combinations of building blocks are "disrupted" (broken up) before they can take hold in the population. The next section describes an SGA design methodology that can be used for any application to ensure that these conditions are met.

24.5 Design Methodology for SGA Parameter Setting and Finding the Optimal Solution

A detailed three-step design methodology for selecting SGA parameter values and finding the optimal solution has been published that can be used for any application [28]. However, some of the population-sizing parameters given in that methodology can be difficult to estimate. Therefore, this chapter gives a simplified version of the three-step design methodology for easier implementation, described as follows and summarized in Figure 24.4.

24.5.1 Step 1: Preliminary Analysis

Before beginning to implement GAs to solve any hydroinformatics application, first analyze the problem to determine how feasible it will be to solve with an SGA. For most hydroinformatics applications, the most time-consuming part of the SGA will be evaluating fitness, especially when the fitness evaluation requires running a computationally-intensive model. The maximum number of fitness evaluations, FE, to solve an SGA is no more than:

$$\mathrm{FE} = tN, \tag{24.7}$$

Step 1: Preliminary analysis

- Represent all decision variables in binary
- Compute chromosome length l (sum of binary digits for each decision variable)
- Minimum population size $N=1.4l$
- Maximum number of generations $t=2l$
- Number of fitness evaluations $\mathrm{FE}=tN$
- Calculate time for each fitness evaluation and total time
- If total time is infeasible, reformulate

Step 2: Parameter setting

- Crossover probability

$$P_c=\frac{s-1}{s} \qquad (s=\text{tournament size})$$

- Mutation probability $P_m=\max\left\{\frac{1}{N},\frac{1}{l}\right\}$

Step 3: Sequential runs

- Run GA with minimum population size N
- Double population size
- Inject optimal solution from previous run into initial population
- Rerun GA
- If optimal solution is the same, stop
- Otherwise, repeat (2) through (5)

FIGURE 24.4
Summary of the SGA design methodology.

where t is the number of generations to convergence and N is the population size. An upper bound on the number of generations to convergence, t, is

$$t \sim 2l, \tag{24.8}$$

where l is the string length [29]. To avoid drift, convergence must occur before the time when drift occurs, t_{drift}. Thierens et al. [29] showed that drift occurs at

$$t_{\text{drift}} \sim 1.4N. \tag{24.9}$$

Combining Equations 24.8 and 24.9 leads to the following lower bound population size to avoid drift:

$$N > 1.4l. \tag{24.10}$$

Substituting Equations 24.8 and 24.10 into Equation 24.7 gives an upper bound on the minimum number of fitness function evaluations required to solve the SGA.

For the monitoring design problem discussed previously, the string length l is 36. The maximum number of generations to convergence t would then be 72 (from Equation 24.8) and the minimum population size N to avoid drift would be $1.4(36) = 50$ (from Equation 24.10). Thus the estimated number of fitness evaluations in Equation 24.7 would be $\mathrm{FE} = (72)(50) = 3600$. A test run of the fitness function evaluation can then be completed to estimate whether the total computational time will be reasonable. In the case of this application, the fitness function can be evaluated in seconds and 3600 evaluations can be completed in a few minutes. If the fitness function evaluations are predicted to require more computational time than would be reasonable (sometimes weeks or months when the fitness function is particularly intensive), see Section 24.6 for more information before proceeding to the next step of the SGA design methodology.

24.5.2 Step 2: Parameter Setting

The next step in the design methodology is to set the parameters needed for the SGA. The first parameter is crossover probability, P_c. Recall that crossover, when combined with selection, is an essential operation that recombines existing solutions to find new solutions and create the global search capabilities of the SGA. However, too much crossover can cause good solutions to be disrupted (broken up) before they have a chance to take hold in the population. Thierens [30] found that, to avoid disruption, the crossover probability for tournament selection should be set as follows:

$$P_c \leq \frac{s-1}{s}, \tag{24.11}$$

where s is the tournament size ($s = 2$ for binary tournament selection). To maximize the amount of mixing that occurs, which leads to faster convergence, P_c should be set as large as possible, which leads to:

$$P_c = \frac{s-1}{s}. \tag{24.12}$$

Next, the mutation probability, P_m, is set. DeJong [31] recommended:

$$P_m = \frac{1}{N}, \tag{24.13}$$

where N is the population size. More recently, others have recommended:

$$P_m = \frac{1}{l}. \tag{24.14}$$

In our hydroinformatics applications, we have found that the larger of the two values from Equations 24.13 and 24.14 generally works well. The important thing is that the mutation rate should be relatively low compared to the crossover rate to ensure that the focus is on global search. Excessive mutation can lead to premature convergence to a local optimum, since the mutation operation is a local process.

24.5.3 Step 3: Sequential Runs

The final step of the design methodology is to identify the optimal solution using sequential SGA runs with increasing population sizes, starting with the minimum population size identified in Step 1 (50 in the case of the monitoring design problem). Once the first population size has been run to convergence, the population size is doubled (see Reed et al. [28] for details on why a doubling of the population is recommended). For best results, the optimal solution from the first run should be "injected" into the initial population of the second run. This leads to faster convergence in later runs [32]. This doubling continues until the optimal solution remains unchanged from one run to the next, which typically requires no more than 3 to 4 runs total.

24.6 Overcoming Computational Limitations

For many hydroinformatics applications with complex fitness functions (e.g., those that require running computationally-intensive numerical models), the preliminary analysis described in the previous section may indicate that months or even years may be required for the SGA run. A recently completed groundwater transport optimization demonstration [33] encountered this difficulty, using transport simulation models that could require up to two hours for each fitness evaluation. In that project, a number of heuristic approaches were used to overcome computational limitations, including:

1. *Running fewer generations.* Recall that the relationships used in the preliminary analysis described in the previous section give the maximum number of generations to convergence, but the optimal solution is often identified much earlier.
2. *Limiting the number of decision variables or the accuracy with which they are represented.* For example, in the case of the monitoring design problem, this could involve representing the coordinates in the monitoring design problem with fewer digits, leading to shorter chromosomes. Recall that the chromosome length directly affects both the population size and the number of generations required to convergence.
3. *Decomposing the optimization problem into sequential subproblems that are easier to solve.* In the case of the monitoring design problem, this could involve initially sampling all well locations and simply identifying promising new sampling locations, then fixing the new locations and identifying which existing and new wells should be sampled. Decomposing the problem means that each run has a much shorter chromosome, which allows smaller population sizes and smaller number of generations.
4. *Creating surrogate or "response" functions that replace simulation models, or other computationally-expensive fitness functions, during the optimization.* In this approach, data from fitness function evaluations (usually simulation model runs) are used to train a surrogate function that is subsequently used to perform rapid fitness evaluations. Response-function approaches using neural networks have been widely applied and tested for groundwater applications [25,34,35].

Such approaches may result in some loss of accuracy in finding the globally-optimal solution, but have still been found to be more effective than trial-and-error search for the best solution [32].

Of the approaches for addressing computational limitations, two that have been most widely applied and tested on hydroinformatics problems have been parallel implementations and response-function approaches. Parallel implementations, in which the optimization is done on multiple processors running in parallel, are most easily applied with optimization approaches that can be easily decomposed into independent processes, such as genetic algorithms. No existing off-the-shelf optimization software currently exists that uses parallel computing, but this capability may develop in the next few years. Response-function approaches have been implemented in at least one off-the-shelf groundwater optimization software, called SOMOS [36].

Emerging research in this area focuses on efficient response-function creation during the optimization process (an "online" approach), avoiding numerous upfront simulation runs to train the response function. SOMOS includes such an approach that has not yet been published [37]. Yan and Minsker [38] demonstrated that such an approach can reduce the number of simulation model runs required for optimization by more than 85%, using a field scale groundwater remediation optimization case study at Umatilla Army depot.

24.7 Advanced GAs

The SGA is the traditional type of GA that has been in use for decades now, but it is not always the best choice. This section summarizes a few more advanced GAs that have been applied to hydroinformatics applications and provides references for more detailed information.

24.7.1 Multiobjective GAs

Hydroinformatics applications often involve multiple, conflicting objectives. For example, the groundwater monitoring design problem described previously involves simultaneously minimizing the number of samples and minimizing interpolation errors. There are two traditional approaches to solving optimization problems with multiple objectives. The first is to create a constraint from one of the objectives, as shown for the monitoring design in Equation 24.3. The right hand side of the constraint, in this case the target error for each constituent $E_{tar,con}$, can then be varied to try to capture the full set of tradeoffs between the objectives. The second traditional approach is to combine the conflicting objectives into a single objective. This can be done by weighing the objectives by their importance and then adding them, or by more sophisticated multiattribute utility approaches that seek to quantify the decision maker's preferences among the different objectives [39]. These types of approaches are difficult to apply to many hydroinformatics applications, where numerous stakeholders are involved in the decision-making that may have quite different preferences. Instead, the approach that is usually taken is to identify optimal tradeoffs between the objectives, called the nondominated or Pareto front, and then allow decision makers to negotiate after they see the tradeoffs.

Figure 24.5 gives an example of a Pareto front showing the tradeoffs between the number of samples and interpolation error. Each diamond symbol on the figure represents a monitoring design with the lowest interpolation error for a given level of sampling or, conversely, the least amount of sampling for a given interpolation error. With most optimization approaches, the Pareto front must be generated by repeatedly solving the optimization

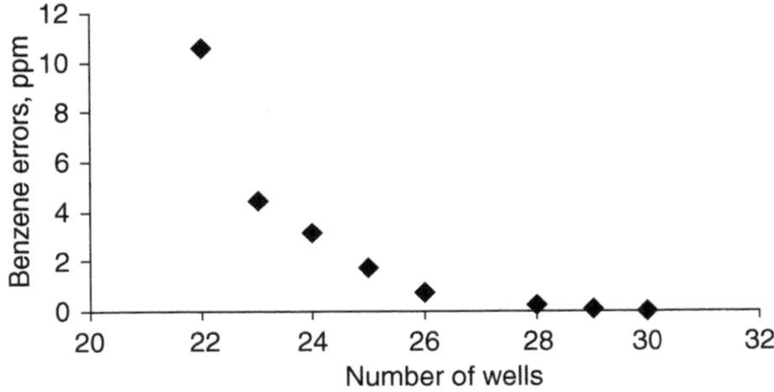

FIGURE 24.5
Tradeoffs between number of wells sampled and benzene interpolation errors in the multiobjective monitoring design application. Each diamond represents a particular monitoring design.

problem for different, fixed levels of one objective (e.g., minimizing number of samples for different levels of allowed interpolation error). When the optimization is computationally intensive, or more than two objectives and uncertainty must be considered, this process can become unmanageable. In this case, multiobjective genetic algorithms have been shown to be more efficient than other optimization algorithms at identifying optimal tradeoffs, which they can accomplish with only a single optimization run. Several books on multiobjective genetic algorithms can be consulted for more on the issues and approaches associated with these problems [40,41].

24.7.2 Considering Uncertainty with GAs

Hydroinformatics applications often have substantial inherent uncertainty. For example, for the groundwater monitoring design problem described previously, heterogeneities in the subsurface environment introduce uncertainties in the plume interpolation models and the optimization models that rely upon them for making decisions. In most real-world applications of optimization to date, uncertainty has not been considered during the initial optimization process. Instead, post-optimization sensitivity analyses are used to identify how robust the optimal solutions are to potential errors in the models. The results are then used to adjust the recommended optimal solution to provide an ad hoc safety factor (e.g., by increasing the number of wells sampled or by reducing the acceptable interpolation error and reoptimizing). Such approaches are simple to implement, but may not provide substantial reliability in performance of the optimized system, or may overdesign the system.

An alternative to this approach is to use stochastic optimization, which usually involves generating multiple "realizations," or potential scenarios of different outcomes (e.g., multiple interpolation models created using conditional simulation [42]) and searching for designs that meet the objectives for all, or a significant fraction, of the realizations. Several approaches have been developed for incorporating uncertainty into genetic algorithms. Chan Hilton and Culver [43] developed an approach that tests each member of the population on a different realization in each generation and then selects those solutions that have survived the selection process for a number of generations over newer solutions, since older solutions would have been exposed to more realizations and hence should be more robust. Beckford and Chan Hilton [44] extended their approach to multiple objectives. Smalley et al. [45] tested an approach called a noisy genetic algorithm, which was first proposed by Miller [46], for groundwater remediation design and found highly reliable solutions by testing each chromosome on only a few realizations per generation. Gopalakrishnan and Minsker [47] developed guidelines for using noisy genetic algorithms, combining guidelines on SGA parameter settings (such as those described in Section 24.5) with new guidelines on the amount of sampling that should be done for each chromosome, based on theory developed by Miller [46]. Singh and Minsker [48] extended their work to multiobjective genetic algorithms, but found that the standard noisy genetic algorithm was not sufficient to obtain reliable solutions with multiple objectives. Instead, they developed multiobjective probabilistic selection and extended sampling approaches that produced highly reliable designs with only a few realizations per generation. Maier et al. [49] have proposed a different approach to incorporating uncertainty into genetic algorithms using a first-order reliability method. This approach does not require sampling and is hence computationally faster, but is only applicable to certain types of uncertainty.

Prior to embarking on an effort to incorporate uncertainty in a hydroinformatics application, a few precautions should be kept in mind. First, stochastic optimization approaches can only incorporate the uncertainty that can be *quantified*. In some hydroinformatics applications, other uncertainties may be equally or more important and can significantly reduce the effective robustness of optimal solutions, even those found using rigorous stochastic approaches. For example, in groundwater optimization, numerous study has focused on uncertainty in hydraulic conductivity, but errors in site conceptual models can be equally or more problematic and are difficult to quantify. Second, sensitivity analyses using deterministic optimization models should first be undertaken to identify which aspects of any model embedded in the optimization most affect the performance of the optimal solutions found. Additional data collection efforts can then focus on those aspects of the conceptual model or specific parameters that are most critical to ensuring robust performance of the optimal design. It is likely that a broader effort using relatively simple approaches to quantify uncertainty in all of the key aspects will be more informative than major investments in rigorous stochastic modeling approaches for handling uncertainty of only one parameter (e.g., hydraulic conductivity). Simpler uncertainty approaches can include soliciting expert opinions to identify a set of plausible modeling scenarios, using available data to fit simple probability distributions. For example, Singh and Minsker [48] used triangular distributions, which require only minimum, maximum, and most likely values, to characterize hydraulic conductivity values at Umatilla Army Depot based on pumping test results. Guan and Aral [50] utilize fuzzy approaches with genetic algorithms to capture uncertainty in groundwater remediation design.

24.7.3 Linkage-Learning GAs

Finally, a new type of GA is emerging called the linkage-learning GA that offers promise for faster solution of hydroinformatics applications. The concept behind linkage-learning GAs is to reduce the disruptive nature of the crossover operation by learning important linkages among chromosomes with high fitness and ensuring that those linkages are not broken. For example, in the monitoring design problem, there may be a few wells that are particularly critical to obtaining plume maps with low interpolation error. A linkage-learning GA will learn which wells are most important and ensure that children generated during crossover retain those wells. Arst [51] tested two linkage-learning GAs, extended compact GA [52] and Bayesian optimization algorithm [53], on a groundwater remediation design test case and found that both methods converged substantially faster than the SGA with little or no loss in accuracy. Research is ongoing to test these algorithms more rigorously, including testing a newer approach called hierarchical Bayesian optimization algorithm [54].

References

1. Holland, J.H., *Adaptation in Natural and Artificial Systems*, University of Michigan Press, Ann Arbor, MI, 1975.
2. Goldberg, D.E., *Genetic Algorithms in Search, Optimization, and Machine Learning*, Addison-Wesley, New York, 1989.
3. Gwo, J.-P., In search of preferential flow paths in structured porous media using a simple genetic algorithm, *Water Resources Research*, 37(6), 1589–1602, 10.1029/2000WR900384, 2001.
4. Karpouzos, D.K. et al., A multipopulation genetic algorithm to solve the inverse problem in hydrogeology, *Water Resources Research*, 37(9), 2291–2302, 10.1029/2000WR900411, 2001.
5. Sciortino, A. et al., Experimental design and model parameter estimation for locating a dissolving dense nonaqueous phase liquid pool in groundwater, *Water Resources Research*, 38(5), 10.1029/2000WR000134, 2002.

6. Kapelan, Z.S. et al., Multiobjective sampling design for water distribution model calibration, *Journal of Water Resources Planning and Management*, 129, 466, 2003.
7. Hsiao, C.-T. and L.-C. Chang, Dynamic optimal groundwater management with inclusion of fixed costs, *Journal of Water Resources Planning and Management*, 128, 57, 2002.
8. Maskey, S. et al., Groundwater remediation strategy using global optimization algorithms, *Journal of Water Resources Planning and Management*, 128, 431, 2002.
9. Reed, P.M. and B.S. Minsker, Striking the balance: long-term groundwater monitoring design for conflicting objectives, *Journal of Water Resources Planning and Management*, 130, 140, 2004.
10. Aksoy, A. and T.B. Culver, Impacts of physical and chemical heterogeneities on aquifer remediation design, *Journal of Water Resources Planning and Management*, 130, 311, 2004.
11. Huang, W.-C. et al., Linking genetic algorithms with stochastic dynamic programming to the long-term operation of a multireservoir system, *Water Resources Research*, 38, 1304, 2002.
12. Vink, K. and P. Schot, Multiple-objective optimization of drinking water production strategies using a genetic algorithm, *Water Resources Research*, 38(3), 10.1029/2001WR000365, 2002.
13. Srivastava, P. et al., Watershed optimization of best management practices using AnnAGNPS and a genetic algorithm, *Water Resources Research*, 38, 2002.
14. Harrell, L.J. and S.R. Ranjithan, Detention pond design and land use planning for watershed management, *Journal of Water Resources Planning and Management*, 129, 98, 2003.
15. Cui, L.-J. and G. Kuczera, Optimizing urban water supply headworks using probabilistic search methods, *Journal of Water Resources Planning and Management*, 129, 380, 2003.
16. Tolson, B.A. et al., Genetic algorithms for reliability-based optimization of water distribution systems, *Journal of Water Resources Planning and Management*, 130, 63, 2004.
17. Prasad, T.D. et al., Booster disinfection of water supply networks: multiobjective approach, *Journal of Water Resources Planning and Management*, 130, 367, 2004.
18. Ostfeld, A. and E. Salomons, Optimal layout of early warning detection stations for water distribution systems security, *Journal of Water Resources Planning and Management*, 130, 377, 2004.
19. Metropolis, N.A. et al., Equations of state calculations by fast computing machines, *Journal of Chemistry and Physics*, 21, 1087, 1953.
20. Glover, F., Future paths for integer programming and links to artificial intelligence, *Computing and Operations Research*, 5, 533–549, 1986.
21. Glover, F., Tabu search — Part I, *ORSA. Journal of Computing*, 1, 190–206, 1989.
22. Minsker B. et al., *Final Technical Report for Applications of Flow and Transport Optimization Codes to Groundwater Pump and Treat Systems*, Environmental Security Technology Certification Program, 2003. Available at http://www.frtr.gov/estcp.
23. Reed, P.M., B.S. Minsker, and A.J. Valocchi, Cost effective long-term groundwater monitoring design using a genetic algorithm and global mass interpolation, *Water Resources Research*, 36, 3731–3741, 2000.
24. Chan Hilton, A.B. and T.B. Culver, Constraint handling for genetic algorithms in optimal remediation design, *Journal of Water Resources Planning and Management*, 126, 128, 2000.
25. Zavislak, M., *Constraint Handling in Groundwater Remediation Design with Genetic Algorithms*, M.S. thesis, University of Illinois, Urbana, IL, 2004.
26. Aly, A.H. and R.C. Peralta, Comparison of a genetic algorithm and mathematical programming to the design of groundwater cleanup systems, *Water Resources Research*, 35, 2415–2425, 1999.
27. Goldberg, D.E., *The Design of Innovation: Lessons from and for Competent Genetic Algorithms*, Kluwer Academic Publishers, Boston, MA, 2002.
28. Reed, P.M., B.S. Minsker, and D.E. Goldberg, Designing a competent simple genetic algorithm for search and optimization, *Water Resources Research*, 36, 3757–3761, 2000.
29. Thierens, D., D.E. Goldberg, and A.G. Pereira, Domino convergence, drift, and the temporal-salience structure of problems, in *The 1998 IEEE International Conference on Evolutionary Computation Proceedings*, IEEE Press, New York, 1998, pp. 535–540.
30. Thierens, D., Analysis and design of genetic algorithms, Doctoral dissertation, Katholieke Universiteit Leuven, Leuven, Belgium, 1995.

31. DeJong, K.A., *An Analysis of the Behavior of a Class of Genetic Adaptive Systems*, Doctoral dissertation, University of Michigan, Ann Arbor, MI, 1975.
32. Espinoza, F., B.S. Minsker, and D.E. Goldberg, An adaptive hybrid genetic algorithm for groundwater remediation design, *Journal of Water Resources Planning and Management*, 131, 14–24, 2005.
33. Minsker, B., Y. Zhang, R. Greenwald, R. Peralta, C. Zheng, K. Harre, D. Becker, L. Yeh, and Yager, K., *Final Technical Report for Application of Flow and Transport Optimization Codes to Groundwater Pump and Treat Systems, Environmental Security Technology Certification Program (ESTCP)*, 2003. Available at http://www.frtr.gov/estcp/.
34. Aly, Alaa H., and Richard C. Peralta, Optimal design of aquifer cleanup systems under uncertainty using a neural network and a genetic algorithm, *Water Resources Research*, 35, 2523–2532, 1999.
35. Rogers, Leah L., Farid U. Dowla, and Virginia M. Johnson, Optimal field-scale groundwater remediation using neural networks and the genetic algorithm, *Environment Science Technology*, 29, 1145–1155, 1995.
36. Systems Simulation/Optimization Laboratory and HydroGeoSystems Group. *Simulation/Optimization MOdeling System (SOMOS) Users Manual*, SS/OL, Deptartment of Biological and Irrigation Engineering, Utah State University, Logan, UT, p. 457, 2001.
37. Peralta, R., personal communication.
38. Yan, S. and B.S. Minsker, A dynamic meta-model approach to genetic algorithm solution of a risk-based groundwater remediation design model, *American Society of Civil Engineers (ASCE) Environmental and Water Resources Institute (EWRI) World Water and Environmental Resources Congress 2004 and Related Symposia*, Salt Lake City, UT, 2004. Available at http://cee.uiuc.edu/research/emsa.
39. de Neufville, R., *Applied Systems Analysis: Engineering Planning and Technology Management*, McGraw-Hill Publishing Company, New York, 1990.
40. Deb, K., *Multi-Objective Optimization Using Evolutionary Algorithms*, John Wiley & Sons Ltd, New York, 2001.
41. Coello, C.C. et al., *Evolutionary Algorithms for Solving Multi-Objective Problems*, Kluwer Academic, New York, 2002.
42. Goovaerts, Pierre, *Geostatistics for Natural Resources Evaluation*, Oxford University Press, New York, 1997.
43. Chan Hilton, A.B. and T.B. Culver, Groundwater remediation design under uncertainty using genetic algorithms, *Journal of Water Resources Planning and Mgnagement*, 131, 25–34, 2005.
44. Beckford, O. and A.B. Chan Hilton, Development of robust genetic algorithms to multi-objective groundwater remediation design under uncertainty, in *Proceedings of the First Annual Environmental and Water Resources Systems Analysis (EWRSA) Symposium*, May 19–22, 2002, Roanoke, VA, ASCE, 2002.
45. Smalley, J.B., B.S. Minsker, and D.E. Goldberg, Risk-based *in situ* bioremediation design using a noisy genetic algorithm, *Water Resources Research*, 36, 3043–3052, 2000.
46. Miller, B.L., *Noise, Sampling, and Efficient Genetic Algorithms*, Ph.D. thesis, University of Illinois, Urbana, IL, 1997 (Also IlliGAL Report No. 97001, http://www-illigal.ge.uiuc.edu/).
47. Gopalakrishnan, G. and B. Minsker, Optimal sampling in a noisy genetic algorithm for risk-based remediation design, *Journal of Hydroinformatics*, 5, 11–25, 2003.
48. Singh, A. and B.S. Minsker, Uncertainty based multi-objective optimization of groundwater remediation at the Umatilla Chemical Depot, *American Society of Civil Engineers (ASCE) Environmental and Water Resources Institute (EWRI) World Water and Environmental Resources Congress 2004 and Related Symposia*, Salt Lake City, UT, 2004. Available at http://cee.uiuc.edu/research/emsa.
49. Maier, H.R., B.J. Lence, B.A. Tolson and R.O. Foschi, First-order reliability method for estimating reliability, vulnerability, and resilience, *Water Resources Research*, 37, 779–790, 2001.
50. Guan, J. and M.M. Aral, Optimal design of groundwater remediation systems using fuzzy set theory, *Water Resources Research*, 40, W01518, 2004.

51. Arst, R. *Which are Better, Probabalistic Model-Building Genetic Algorithms (PMBGAs) or Simple Genetic Algorithms (SGAs)? A Comparison for an Optimal Groundwater Remediation Design Problem*, M.S. thesis, University of Illinois, Urbana, IL, 2002. Available at http://cee.uiuc.edu/research/emsa.
52. Harik, G., *Linkage Learning via Probabilistic Modeling in the ECGA*, Illinois Genetic Algorithms Laboratory Technical Report, No. 99010, Department of General Engineering, University of Illinois, http://www-illigal.ge.uiuc.edu, January 1999.
53. Pelikan, M., D.E. Goldberg, and E. Cantú-Paz, BOA: The Bayesian Optimization Algorithm Joint meeting; 8th July, Orlando, FL, *Proceedings of the International Conference on Genetic Algorithms*, GECCO-99, 525–532, 1999.
54. Pelikan, M. and D.E. Goldberg, *Hierarchical BOA Solves Ising Spin Glasses and MAXSAT*, Illinois Genetic Algorithms Laboratory Technical Report, No. 2003001, Department of General Engineering, University of Illinois, http://www-illigal.ge.uiuc.edu, 2003.

25

Fuzzy Logic

Lydia Vamvakeridou-Lyroudia and Dragan Savic

Contents

25.1 Introduction

Fuzzy logic (FL), or more generally, fuzzy reasoning (FR) is a machine intelligent mathematical way of dealing with imprecision, uncertainty, vagueness, ambiguity, and subjective interpretation. Almost 40 years have passed since L.A. Zadeh [1] first introduced fuzzy sets and FL. During these years, the number and range of Artificial Intelligence applications, using FR, has grown considerably, to the point that it now includes a wide variety of fields, such as industrial applications, engineering, hydrology, and environmental modeling, techno-economical studies, decision support systems (DSS), pattern recognition and data mining, medicine, chemistry, archaeology, social and behavior studies, politics, and finance.

25.1.1 What Is a Fuzzy Set?

In classical Boolean logic, sets there are only two possible states for any set, condition, variable, or constraint: "true–false," "yes–no," "1–0," "accepted–not accepted," as opposed to the real world, where the human mind can think, operate, classify, select, and decide using not only the black–white distinction, but also various shades of grey. Fuzzy logic, as an Artificial Intelligence (AI) technique, enables the implementation of "greyness" within the context of mathematical simulation, attributed to input data, process simulation, or process outcome.

In classic boolean logic, sets have clear and crisp boundaries. Items may belong (= 1) or not belong (= 0) to a crisp set (e.g. the crisp set of "apples" contains only apples; an orange does not belong (membership grade = 0) to this set. Fuzzy sets are sets with unclear boundaries, the degree or grade of belonging expressed by a membership function. The higher the membership function value, the greater the degree of belonging to the fuzzy set, and vice versa. One such example is the fuzzy set of "young people." Clearly a teenager belongs to this set (membership function = 1), while an octogenarian does not (membership function = 0). In between, however, the boundaries are not clear, even subjective. What membership grade should be assigned to a 35 year old? The answer is not only fuzzy in itself, it is affected by subjective interpretation — a teenager would think anyone over 20 as "old," while 35-year-old people would not answer the same for themselves, let alone parents who think of their children as "young," even if they are over 50. Yet, despite ambiguity and subjectivity, the human mind can deduce, operate, and make decisions in a fairly orderly and productive way; an asset for which computers are ill-fitted.

25.1.2 Difference between Fuzziness and Randomness

Suppose you are about to meet a person, you have never seen before. Before the meeting you learn from other sources (= information) that the person in question has a 95% chance (probability) of being under 35 years of age. Another independent source tells you that this same person has a membership grade of 0.95 to the fuzzy set of young people (without explicitly defining to you "young" people). Which information has a greater value, is more precise and more reliable to act upon? Interestingly enough, the fuzzy logic information is more accurate. Having a 95% probability of being under 35, means that there is a 5% probability of the person in question being 80 years old, whereas the 0.95 membership grade to the fuzzy set of young people, not only excludes overage persons but additionally excludes children and teenagers (which would have a membership grade equal to the young people set, whatever the subjectivity of the source that provided you with the information). So, according to the fuzzy set information, you definitely expect to see a young adult, whereas with the statistical information the person might be anything, from a baby to 100 years old. Another example from literature [2] is more illustrative. Suppose there are two glasses in front of you. The liquid in the first glass is described as having 95% chance of being healthy and good. The liquid in the second glass has a 0.95 membership grade to the fuzzy set of "healthy and good" liquids. Which glass would you choose, keeping in mind that the glasses could theoretically also contain poison?

There is also one additional important remark. Once you have met the person and asked about his/her age, or analyzed the contents of the glass, the issue of probability is solved. The prior probability of 95% becomes posterior probability of 1 or 0, that is, the person is the right age or not, the liquid is good or not. However, the fuzzy membership grade of 0.95 remains with the person or the liquid, no matter whether you know his/her age or analyze the contents of the glass. This is the clear distinction between fuzziness and probability: fuzziness describes the ambiguity of an event, while probability (randomness) describes the uncertainty in the occurrence of the event. In other words, fuzzy logic and fuzzy sets are linked with the *possibility* of an event, as opposed to the probability of an event.

25.1.3 Types of Uncertainty

Apart from ambiguity, fuzzy sets and fuzzy logic can also be used to describe vagueness, imprecision, and uncertainty, not related with randomness: vagueness is used to describe types of uncertainty associated (related) usually to linguistic information (e.g., this hotel is "good," the proposal is "acceptable"), whereas imprecision is usually associated with countable data, described by linguistic definition (the temperature is "high") or by measurement (the temperature is "about 40°C"). Even if the numeric value 40°C had been used alone, without the "about" description, a degree of vagueness is nevertheless attached to it, due to the imprecision of the instrument used for measuring it, for example, 1%. Mathematic formulae, used for simulating physical phenomena in computer models, tend to treat all data and information in a deterministic way, regardless of the uncertainty contained in them. Take for instance a widely used mathematical formula (Hazen–Williams) for calculating the head loss h in a pipe, as a function of the rate of flow: $h = f(C, D, L, Q)$, where Q is the rate of flow, D the pipe diameter, L the pipe length, and C a pipe roughness coefficient. Even if D and L can be measured with sufficient accuracy, and therefore the relevant information assumed to be deterministic, this is not the case for the roughness coefficient C. For C only wild guesses can be made, based on the material, age and general condition of the pipe and no probabilistic model can help in this case. Therefore any head loss calculation obtained by the Hazen–Williams formula is bound to be uncertain, to a greater or lesser degree, no matter how precise, complex, or accurate the mathematical formula actually is. Fuzzy sets and fuzzy logic can handle reasoning using uncertain mathematical formulae, successfully including them in decisions support schemes, involving for example, replacement of a pipe.

Additionally, there are other kinds of uncertainty not related to randomness: humans can read bad handwriting, or recognize objects and faces in blurred photographs, tasks which computers can only approach through FR.

25.1.4 Linguistic Variables and Inference Rules

Linguistic descriptions (e.g., "Speed cars are generally more expensive," "Very cold weather is dangerous for old people") are widely used in everyday speech. Humans can understand, process and act on them, while conventional computer mathematical models cannot. This kind of linguistic descriptions can be introduced in mathematical modeling with the use of FL. Within fuzzy models they can be defined as *linguistic parameters* or *linguistic variables* and used in a similar way, as any other numeric parameter or variable.

Thus, linguistic inference rules can also be introduced. Simple inference rules, also called implications in predicate logic (fuzzy or not), have the following structure: *IF "a,"* THEN *"b,"* where a is the hypothesis (premise) or antecedent and b relatively the conclusion (outcome, action) or consequent. For instance, the rule:

IF temperature is greater than 40°C THEN sound the alarm

is a Boolean (crisp) inference rule, comprising of crisp premise (*temperature is greater than 40°C*) and crisp outcome/action (*sound the alarm*). Fuzziness may be introduced in any (or both) sides of the inference rule, using linguistic variables:

- Fuzzy premise — Crisp outcome/action *(IF it is hot THEN sound the alarm)*
- Crisp premise — Fuzzy outcome *(IF the temperature is greater than 40°C THEN risk is high)*
- Fuzzy premise — Fuzzy outcome *(IF it is hot THEN risk is high)*

25.2 Fuzzy Sets Essentials

25.2.1 Definition of Fuzzy Sets and Membership Functions

Let X be the universal set or universe of discourse. A fuzzy set A in X is defined by its membership function $\mu_A(x)$ denoting the membership grade of each $x \in X$ to the fuzzy set A. For instance, let X be the universal set of real numbers. The fuzzy set A, described as "*real numbers close to 100*" is defined by the membership function $\mu_A(x)$, as follows:

$$\mu_A(x) = \frac{1}{1 + (x - 100)^2}. \tag{25.1}$$

The graphic representation of this membership function is shown in Figure 25.1. Obviously the maximum membership grade occurs for $x = 100$. The nearer x is to 100, the higher the membership grade of x to A, and vice versa. Unlike probability density functions (PDFs), fuzzy membership function values are not restricted to the [0,1] interval. Theoretically any value, larger or smaller than one, can be chosen as maximum, but, usually the range of values for membership functions is between 0 and 1, inclusive, in which case, the fuzzy set is called *normal*. Mathematically it can be said that membership functions are mappings of variables $x \in X$ to the range [0,1]. Each and every fuzzy set is uniquely defined by its own specific membership function. As a consequence, the terms "fuzzy set" and "membership function" end up meaning the same thing, often used indiscriminately. For instance the definition "linear fuzzy set" means "fuzzy set defined by a linear membership function," etc.

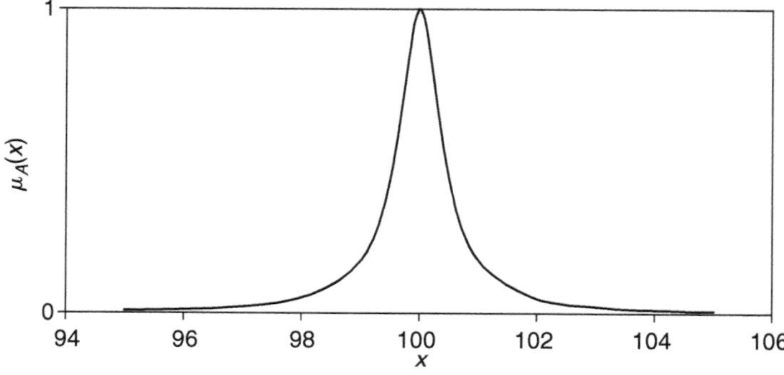

FIGURE 25.1
Membership function for the fuzzy set "numbers close to 100."

There are no restrictions to the shape of membership functions and fuzzy sets. Membership functions may be linear, nonlinear, trapezoidal, triangular, convex, or non-convex (Figure 25.2). The maximum value for the membership function is called the *height* of the fuzzy set (the height of normal fuzzy sets is 1); the range of values for which $\mu(x) \neq 0$ is called the *support* of the fuzzy set; the range of values, for which membership function values are equal to the height, is called the *core* of the fuzzy set; whereas the range of values, for which $0 < \mu(x) < height$ are the *boundaries* of the fuzzy set. For instance in Figure 25.2(a), the support of the fuzzy set is the range $[a, d]$, the core is the range $[b, c]$ and both ranges $[a, b)$ and $(c, d]$ are the boundaries. Triangular fuzzy sets, where the core consists of a single point, are often called *single point* fuzzy sets. Shouldered fuzzy sets present no upper (or lower) bound for their core. They are useful for defining fuzzy sets related to a meaning similar to "no greater than" (left shouldered — no lower bound) or "no less than" (right shouldered — no upper bound). For instance, the fuzzy set of "young people" (see Section 25.1.1) should be defined by a fuzzy set like in Figure 25.2(f), with variable x representing age. Even classic crisp sets, defined by binary Boolean logic (yes–no), may be represented as a special case of fuzzy sets, where the range of both boundaries is equal to zero, as in Figure 25.2(g). If the core of a crisp set consists of a single point, it is called a *singleton*. In fuzzy terms, this set consists of a single point with membership function equal to one, while for all other values of x, the membership grade is zero.

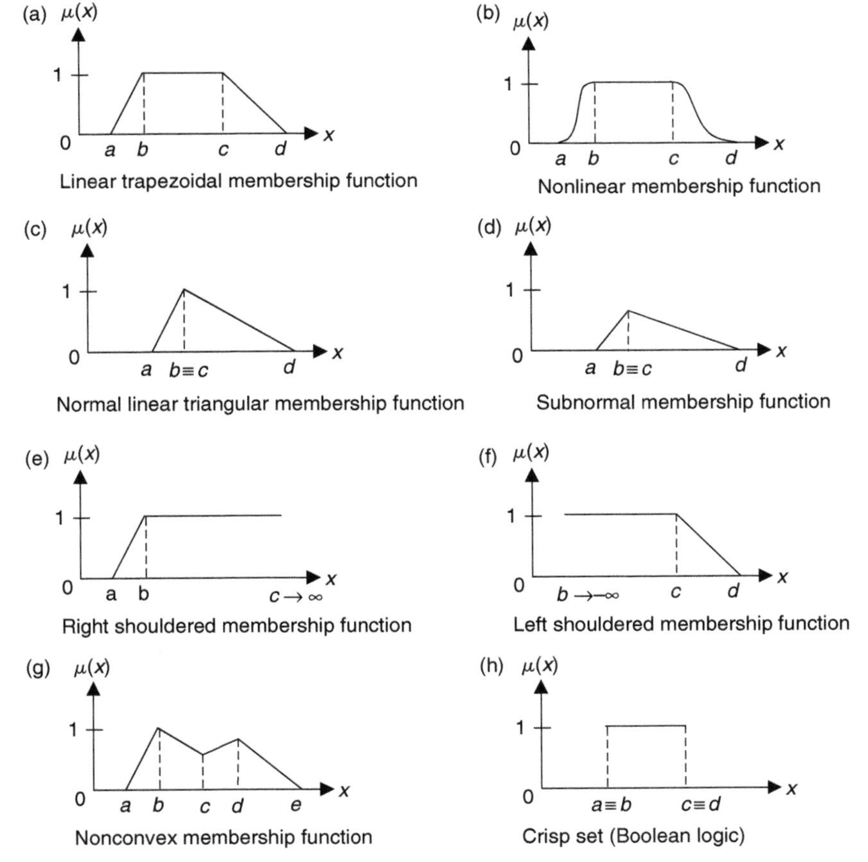

FIGURE 25.2
Various types of fuzzy sets and membership functions.

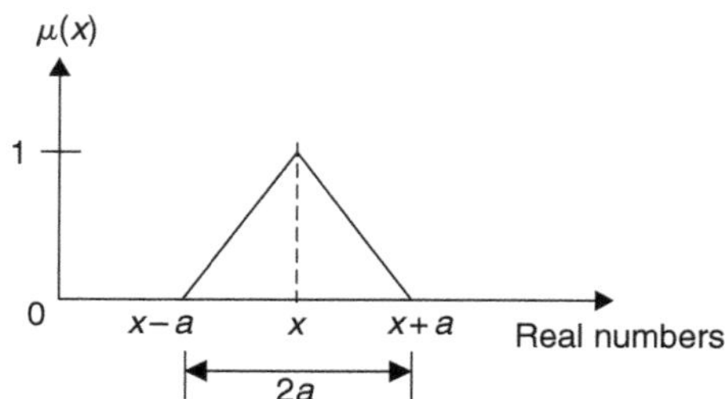

Figure 25.3
Fuzzy number.

A *fuzzy number* is a single point convex normal fuzzy set, defined on the "real numbers" universe of discourse. The fuzzy set shown in Figure 25.1 represents the "fuzzy number 100." Usually (but not necessarily) fuzzy numbers are symmetrical, namely the boundaries in both the uphill and downhill side are equal, while linear triangular membership functions are easier to apply than nonlinear ones (Figure 25.3). Fuzzy numbers can be used for the mathematical simulation of numerical data (meter readings and other similar input signals), where a degree of uncertainty exists due to inaccuracy or noise.

Fuzzy sets may also be defined by noncontinuous membership functions, even by discrete points. Let us consider a universe of discourse X consisting of five states (points/items) as follows: $X = \{a, b, c, d, e\}$. For instance this universe of discourse may comprise five available locations for installing a rain gauge. For each of these five points a membership value to the fuzzy set A of "acceptable locations" is assigned as follows: $\mu_A(x) = \{0.2, 0.7, 1, 0.8, 0.5\}$. There is no need to define a continuous membership function, because only the membership grades at the points are needed. The ensuing discrete fuzzy set is usually written in the following way:

$$A = \left\{ \frac{0.2}{a} + \frac{0.7}{b} + \frac{1}{c} + \frac{0.8}{d} + \frac{0.5}{e} \right\}. \tag{25.2}$$

Usually, only points with nonzero membership grades (the support of the fuzzy set) are written. This notation, widely used in literature, is called Zadeh's notation for discrete fuzzy sets [3] and should not be confused with conventional addition. Sometimes even brackets may be omitted [4], while the left-hand side of Equation 25.2 might be replaced by the fuzzy set's membership function. The general form for discrete, countable fuzzy sets is:

$$A = \frac{\sum_{i=1}^{n} \mu_i}{x_i}. \tag{25.3}$$

An *alpha cut* or *a-cut* A_a of a fuzzy set A is a crisp (not fuzzy) set, containing all x, for which the membership grade exceeds a (Figure 25.4):

$$A_a = \{x | \mu_A(x) \geq a\}. \tag{25.4}$$

For instance, the alpha cut for $a = 0.75$ for the fuzzy set in Equation 25.2 is (omitting zeros): $A_{0.75} = \{1/e + 0.8/d\}$. Depending on whether $\geq$ (greater or equal) or $>$ (greater) are used, the a-cut is respectively called *weak* or *strong* [5]. Theoretically any fuzzy set can be divided into an infinite number of a-cuts. Sometimes, in literature, alpha cuts are also called *lambda cuts* or *λ-cuts* [3]. They are extremely useful for fuzzy controllers and defuzzification (i.e., the last step of a fuzzy model returning crisp outcomes, see also Section 25.2.4).

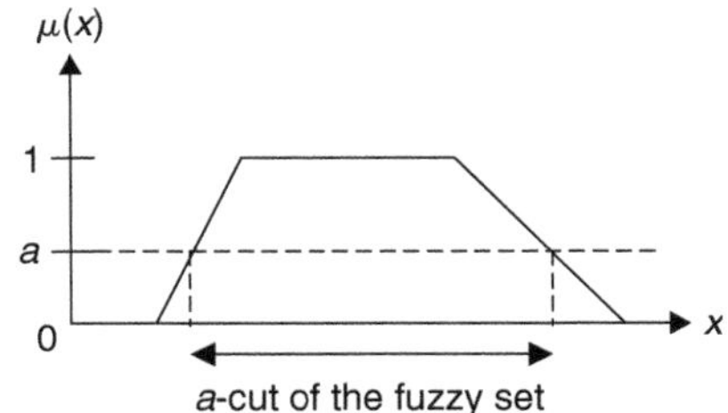

Figure 25.4
Alpha cut (a-cut) of a fuzzy set.

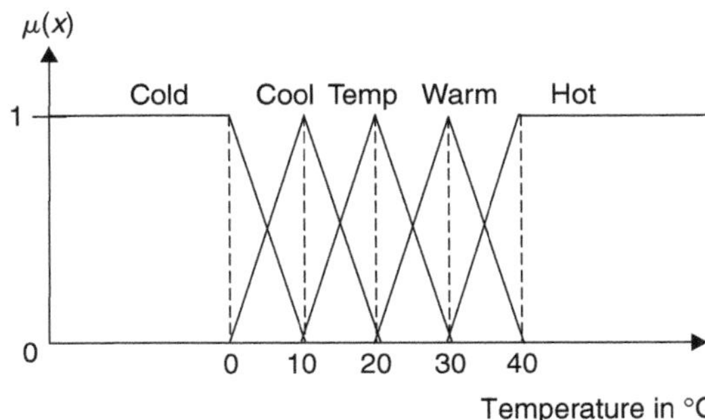

Figure 25.5
Fuzzy partition for temperature.

In order to fully describe the linguistic variables associated with many real world variables, a single fuzzy set is many times not enough. Take for instance the variable "air (or system) temperature": it can be cold, cool, temperate, warm, or hot. Therefore, five fuzzy sets have to be defined, their membership functions being related and overlapping. In order to achieve this, the universe of discourse is divided into segments, each defining the core of one fuzzy set, while boundaries are accordingly arranged, so that each possible temperature value may belong to at least one, but no more than two adjacent fuzzy sets. This process is called *fuzzy partition* or *frame of cognition* of a variable [6]. In Figure 25.5 such a partition is shown for temperature, based on five linguistic definitions. The degree of overlapping in this example is high (50%), but it could also be smaller. Nonlinear membership functions can also be used, instead of the triangular linear ones. It should be noted that fuzzy partitions are different from Boolean partitions, where each variable value belongs to only one crisp set. Fuzzy partitions are important tools for fuzzy systems, fuzzy controllers, and fuzzy cognition.

Determining suitable membership functions is one of the hardest tasks for fuzzy models. There are many ways to achieve this, and apart from some general rules and tips, in most cases it is problem specific.

25.2.2 Basic Operators on Fuzzy Sets

The *complement* of a fuzzy set A is a fuzzy set A', complying with the notion "not A" (e.g., the complement of the fuzzy set "young people," is a fuzzy set called "not young" people — but not "old" people, which is a different fuzzy set). In the case the fuzzy set is a normal

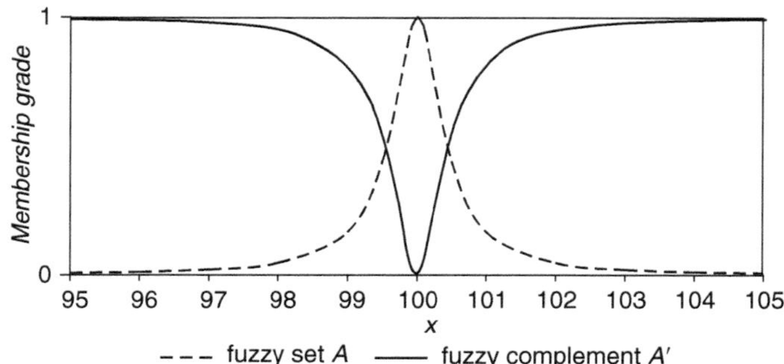

FIGURE 25.6
Fuzzy complement of the fuzzy set "numbers close to 100."

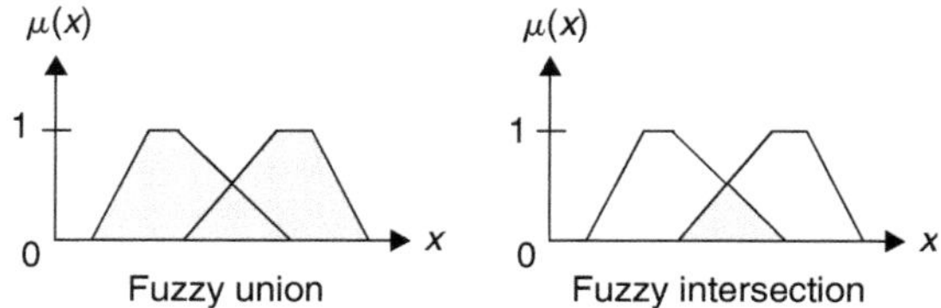

FIGURE 25.7
Fuzzy union and intersection of two fuzzy sets.

one, with membership function values in the range [0,1], the complement is defined as:

$$\mu_{A'}(x) = 1 - \mu_A(x), \qquad x \in X. \tag{25.5}$$

Any crisp set and its complement are mutually exclusive, that is, $A \cap A' \equiv \emptyset$. This is not the case for fuzzy sets, where any element may belong to both A and A', with different membership grades (Figure 25.6). The *fuzzy intersection* $A \cap B$ of two fuzzy sets A and B in the same universe of discourse X, also called the *min operator*, is defined as (Figure 25.7):

$$\mu_{A\cap B} = \min\{\mu_A(x), \mu_B(x)\}. \tag{25.6}$$

Accordingly, the *fuzzy union* $A \cup B$ of two fuzzy sets A and B, also called the *max operator* is defined as (Figure 25.7):

$$\mu_{A\cup B} = \max\{\mu_A(x), \mu_B(x)\}. \tag{25.7}$$

Equations 25.5, 25.6, and 25.7 show only the most widely used operators for fuzzy complement, intersection, and union respectively. There exist a large number of other operators, also used to represent complement, intersection, or union, but it is beyond the scope of this book to present them in detail [7].

Another important operator for fuzzy sets is the so-called *extension principle*, which allows the application of numerical operations on fuzzy sets, in similar ways as classical numerical operations, by "extending" fuzziness from input sets to output sets. Suppose the "close to 30°C temperature" fuzzy set A has been determined as follows: $A = \{0.2/28 + 1/30 + 0.2/32\}$. In case we need to use this fuzzy set for numerical operations within a function that contains temperature readings in Fahrenheit, there is no need to define a new fuzzy set. The transform function from °C to °F is $t_F = f(t_C) = 9(t_C + 32)/5$. To define the resulting fuzzy set B, of the same property in °F we need to compute algebraically the new points.

Therefore $B = \{0.2/f(28)+1/f(30)+0.2/f(32)\} = \{0.2/82.4+1/86+0.2/89.6\}$. Mathematically the new fuzzy set $B = f(A)$ is defined as:

$$B(y) = f[A(x)] = \left\{\frac{\mu_1}{f(x_1)} + \frac{\mu_2}{f(x_2)} + \cdots + \frac{\mu_n}{f(x_n)}\right\}. \tag{25.8}$$

In a similar way the arithmetic product or addition of two or more fuzzy sets can be obtained. Suppose the intensity I (mm/h) of a specific rainfall is defined as a fuzzy set $A = \{0.4/3 + 1/4 + 0.5/5\}$, while the duration $t(h)$ of the same rainfall is also a fuzzy set $B = \{0.3/2 + 1/3 + 0.6/4\}$. Fuzzy set C representing rainfall depth h (mm), can be obtained by multiplying intensity by duration, that is, $h = I \times t$, or in fuzzy set terms $C(h) = A(I) \times B(t)$. The support of fuzzy set A is $I = \{I_1, I_2, I_3\} = \{3, 4, 5\}$, while for the second fuzzy set B the support is $t = \{t_1, t_2, t_3\} = \{2, 3, 4\}$. The support of the new fuzzy set $A \times B$, representing rainfall depth, should consist of the following points: $d = I \times t = \{6(= I_1 \times t_1),\ 9(= I_1 \times t_2),\ 12(= I_1 \times t_3),\ 8(= I_2 \times t_1),\ 12(= I_2 \times t_2),\ 16(= I_2 \times t_3),\ 10(= I_3 \times t_1),\ 15(= I_3 \times t_2),\ 20(= I_3 \times t_3)\}$. For the first point $(6 = I_1 \times t_1)$, it stands $\mu_A(I_1) = 0.4$ and $\mu_B(t_1) = 0.3$. Usually (but not necessarily) for the relative membership function of the combined fuzzy set, the minimum value would be chosen: $\mu_{A\times B}(I_1 \times t_1) = \min\{\mu_A(I_1),\ \mu_B(t_1)\} = \min\{0.3, 0.4\} = 0.3$. In the case of points resulting more than once by the mapping configuration (i.e., point 12 results twice $12 = I_2 \times t_2$ and $12 = I_1 \times t_3$), the maximum of partial membership function values is taken: for example, $\mu_{A\times B}(12) = \max[\min\{\mu_A(I_2), \mu_B(t_2)\}, \min\{\mu_A(I_1), \mu_B(t_3)\}] = \max[\min\{1,\ 1\}, \min\{0.4,\ 0.6\}] = \max[1,\ 0.4] = 1$. Therefore fuzzy set C representing rainfall depth is defined as: $C = A \times B = \{0.3/6 + 0.3/8 + 0.4/9 + 0.3/10 + 1/12 + 0.5/15 + 0.6/16 + 0.5/20\}$.

Mathematically the extension principle for discrete fuzzy sets is expressed as follows: let $A_1, A_2, \ldots, A_n$ be fuzzy sets defined on the universe of discourses $X_1, X_2, \ldots, X_n$ and $B(y) = f(A_1, A_2, \ldots, A_n)$ a fuzzy set produced by a mapping (relation) among $A_1, A_2, \ldots, A_n$, defined on the universe of discourse Y, where $y \in Y$. The membership function of B is given as:

$$\mu_B(y) = \max_{y=f(x_1,x_2,\ldots,x_n)} \{\min[\mu_{A_1}(x_1), \mu_{A_2}(x_2), \ldots, \mu_{A_n}(x_n)]\}. \tag{25.9}$$

If function f is continuous, the max (maximum) operator is replaced by the sup (supremum, lower upper bound) operator. Similar operators are applied for general fuzzy relations or implications, as described in the next chapter.

25.2.3 Fuzzy Logic Principles

Logic is a mathematical system of *reasoning*, comprising the methods and principles of reasoning, in all its possible forms. Widely used over the centuries by philosophers and mathematicians, logic consists of propositions (antecedents, premises) combined by logical functions to yield logical results (consequents). For instance: Men are mortal/Socrates is a man/then Socrates is mortal. A simple proposition has the form: x is P, where x stands for any subject in the universe of discourse ($x \in X$) and the *predicate* P represents a function, defined on X, which for each value of x forms a proposition.

In classical Boolean predicate logic any proposition is either "true" (1) or "false" (0). However this binary black-and-white way is not always successful in real-life problems and reasoning. A system of logic is needed, where for any proposition there can be at least three answers: "True/Yes," symbolically represented as (1), "False/No," represented as (0) and "Uncertain/maybe/we do not know," represented as (0.5). This is formally called

TABLE 25.1

Primitives of Boolean and Lucasiewicz three-valued logic.

		Boolean logic					**Lucasiewicz three-valued logic**				
P	Q	$P \wedge Q$	$P \vee Q$	$-P$	$P \rightarrow Q$	$P \leftrightarrow Q$	$P \wedge Q$	$P \vee Q$	$-P$	$P \rightarrow Q$	$P \leftrightarrow Q$
0	0	0	0	1	1	1	0	0	1	1	1
0	$\frac{1}{2}$						0	$\frac{1}{2}$	1	1	$\frac{1}{2}$
0	1	0	1	1	1	0	0	1	1	1	0
$\frac{1}{2}$	0						0	$\frac{1}{2}$	$\frac{1}{2}$	$\frac{1}{2}$	$\frac{1}{2}$
$\frac{1}{2}$	$\frac{1}{2}$						$\frac{1}{2}$	$\frac{1}{2}$	$\frac{1}{2}$	1	1
$\frac{1}{2}$	1						$\frac{1}{2}$	1	$\frac{1}{2}$	1	$\frac{1}{2}$
1	0	0	1	0	0	0	0	1	0	0	0
1	$\frac{1}{2}$						$\frac{1}{2}$	1	0	$\frac{1}{2}$	$\frac{1}{2}$
1	1	1	1	0	1	1	1	1	0	1	1

three-valued logic. Three valued logic (and its extension to multiple *n-valued* logic) has been introduced before fuzzy logic, in the 1920s and 1930s. For a logic system to be fully determined, basic operations, called *primitives*, have to be defined. These primitives are:

1. Conjuction ($\wedge$), the logical AND ($P \wedge Q$)
2. Disjuction ($\vee$), the logical OR ($P \vee Q$)
3. Negation ($-$), the logical NOT ($-P$)
4. Implication ($\rightarrow$) (P implies Q: $P \rightarrow Q \equiv (-P) \vee Q$)
5. Equivalence ($\leftrightarrow$) (If $P \rightarrow Q$ and $Q \rightarrow P$, then $P \leftrightarrow Q$)

Various multiple valued logic systems have been developed, depending on the way those primitives have been defined [4]. Table 25.1 shows the primitives for classic Boolean logic and the most known three valued logic system: Lucasiewicz's logic.

Defining fuzzy logic primitives can be done in similar ways as Boolean and three-valued logic: fuzzy logic conjunction (AND), disjunction (OR), and negation (NOT) may be defined, using fuzzy intersection (Equation 25.6), fuzzy union (Equation 25.7), and fuzzy complement (Equation 25.5) respectively. However fuzzy implication ($A \rightarrow B$: IF x is A, THEN y is B), which is the main inference engine of fuzzy rules, expands from "true" and "false" consequents to "quasi true" and "quasi false" ones, which can be obtained using various operators on fuzzy sets. In reality fuzzy implication represents a fuzzy relation between two fuzzy sets A and B and their elements $x \in A$ and $y \in B$. The most usual one is the minimum implication (also called Mamdani's implication):

$$\mu_{A\rightarrow B}(x, y) \equiv \min\{\mu_A(x), \mu_B(y)\}, \tag{25.10}$$

and the product implication

$$\mu_{A\rightarrow B}(x, y) \equiv \{\mu_A(x) \bullet \mu_B(y)\}, \tag{25.11}$$

but there exist numerous others (see Reference 3, Chapter 7, for a detailed description). What makes fuzzy reasoning fascinating is the ability to invent, determine, and apply one's own fuzzy logic primitives for the mathematic simulation of any real life problem, depending on the problem, the point of view, the psychology (pessimism or optimism), and the purpose of the model. Indeed, provided your own original operator is consistent with some basic mathematic requirements [4], fuzzy logic provides the mathematical framework to simulate almost anything, from extremely crisp to extremely fuzzy, setting stable inference rules that work, and drawing conclusions using (dumb, binary) computers, in similar ways as the (clever, linguistic) human mind, thus belonging to the highly valuable group of *artificial intelligence* (AI) techniques.

25.2.4 Fuzzy Systems and Fuzzy Reasoning

Logic inference within a fuzzy system (*fuzzy reasoning*) may be seen as the extension of n-valued logic for $n \to \infty$. Indeed fuzzy logic and fuzzy reasoning deal with linguistic variables (HIGH, SHORT, SMALL, BIG) defined through fuzzy sets and membership functions, which may take whatever value within the [0,1] range. Another term for fuzzy logic is *approximate reasoning*, where both propositions (antecedents) and consequents are linguistic fuzzy variables, such as: *IF temperature is HIGH, THEN snowmelt is SIGNIFICANT*, or *IF rain is VERY INTENSE and hill slope is STEEP, THEN flood risk is HIGH.*

Any fuzzy reasoning system includes a number of fuzzy inference rules, called the "IF–THEN" rules, through which approximate reasoning is, for example, IF x is A AND y is B THEN z is C. The set of rules represents the core of any fuzzy system. This type of reasoning, defining antecedents and consequents in linguistic terms by using fuzzy sets, comes closer to human thinking and reasoning (e.g., IF it is cold THEN wear a coat). Moreover it resembles human thinking in another aspect: knowledge about the rules may be derived by sources other than numerical data, such as human empirical knowledge and intuition. It is however broader than strict expert systems that accept knowledge only from "experts." In this aspect, expert systems may be seen as a subset of fuzzy inference systems [8].

The general form for a fuzzy rule r is as follows: let us assume $x_1, x_2, \ldots, x_n$ as fuzzy antecedents and $A_{1r}, A_{2r}, \ldots, A_{nr}$ their respective fuzzy sets for rule r ($r = 1, \ldots, N_r$). The general form of a rule is:

$$\text{IF } x_1 \text{ is } A_{1r} \text{ AND } x_2 \text{ is } A_{2r} \text{ AND} \ldots x_n \text{ is } A_{nr} \text{ THEN } y_r \text{ is } B_r, \tag{25.12}$$

where y_r is the output of the rule (fuzzy variable) and B_r its respective fuzzy set. Fuzzy reasoning using inference techniques, usually involves two steps: in the first step a number of individual fuzzy sets, each referring to a specific fuzzy variable, have to be aggregated. The most frequent types of aggregation involve AND/OR operators. In the second step an implication operator (THEN) is needed, as described by Equations 25.10 and 25.11. The output for rule r is a fuzzy set B_r defined by its membership function $\mu_{Br}(y_r)$.

Due to the fact that any fuzzy system consists of Nr rules, the final output for variable y will be an aggregated fuzzy set of usually (but not necessarily) disjunctive (OR) type:

$$\mu_B(y) = \max[\mu_{B_1}(y_1), \mu_{B_2}(y_2), \ldots, \mu_{BNr}(y_{Nr})]. \tag{25.13}$$

Using aggregation and implication leads to complex shapes for the output fuzzy set. If a crisp numerical result y^* is needed, a *defuzzification* process should be applied, to assign expected numeric values to fuzzy regions. There exist various methods of defuzzification [3]. The most common and widely used method is the so-called *centroid method* (Figure 25.8[a]), where the centroid of the fuzzy region is calculated and used as the representative defuzzified value y^*. Other types of defuzzification are the maximum value method (Figure 25.8[b]), which may be applied if the combined fuzzy region presents

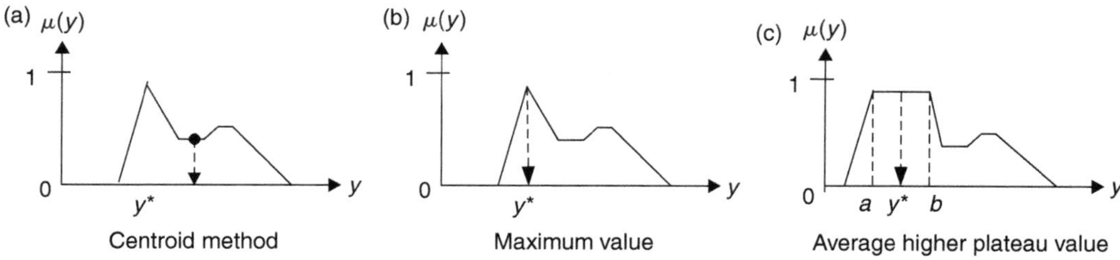

FIGURE 25.8
Defuzzification.

a distinct maximum value, and the average higher plateau value (Figure 25.8[c]), which is recommended as an alternative to the maximum value method for fuzzy regions with no distinct higher value. Selection of the most suitable defuzzification method is generally problem specific and results can vary considerably. It is however beyond the scope of this chapter to analyze further the merits and weak points of each method, a subject that is thoroughly discussed in literature [5].

25.3 Fuzzy Modeling

25.3.1 Types of Models

Any mathematical model consists of three parts: input data, computational process, and output. This structure is universally applied for all kinds of models, whatever their special application field or purpose. For instance a rainfall-runoff model would receive input data for rainfall and regional topography, processing them by a series of mathematical operations, in order to produce runoff estimations as output. A real-time flood-warning model receives input signals about water level and rainfall, which together with the river spatial topography data are processed by a controller to produce flood forecasts, warnings, and actions (e.g., opening of a sluice gate).

Within the general application area of hydrology, hydraulics, and hydroinformatics, models can be divided into two large categories: parametric and black box models. For parametric models, each equation and parameter involved should have a physical relevance (e.g., using Saint Venant equations for flood routing). Black box models are quite the opposite: only input and output are relevant, the model structured and calibrated so as to fit the data, with no concern about the physical meaning of the equations involved (e.g., artificial neural networks). In both types uncertainties concerning parameters are treated using stochastic techniques (e.g., Monte Carlo simulation).

Fuzzy models do not belong to either of those two large categories: they cannot be considered "black box," since fuzzy rules and fuzzy operators are not "blind"; there is always a natural physical meaning justifying their structure. On the other hand, algebraic equations, the core of all parametric models, are replaced by "relations" and "inference rules," while uncertainties are treated using fuzzy sets and linguistic parameters. They can be labeled "colored box" models, or "greyness shaded models," lying between the two large categories, combining elements from both, but belonging, strictly speaking, to none. Using Zadeh's [1] own words: fuzzy logic can be used for "precise computations in an imprecise world," using "relations instead of computations."

25.3.2 Engineering/Hydroinformatics Application Areas and Types of Applications

The FL and FR can be applied and included at numerous and various mathematical models, extending practically to almost any scientific and engineering field. Within the engineering/hydroinformatics field, there are many applications, involving FL or FR, which roughly can be categorized as follows:

- *Modeling*, where fuzzy inference rules are used to replace (partially or in full) numerical computations in parametric models, thus simulating uncertainties both for parameters and mechanisms. Such techniques have been applied, that is, for rainfall-runoff conceptual models [9], including snowmelt [10], for flood forecasting models, where fuzzy inference rules replace conventional time series analysis techniques [11], for groundwater flow [12], or for water network analysis [13].
- *Multiobjective and single objective optimization*, where FL, as a mathematical method aiming at compromising, is appropriate for combining multiple conflicting criteria, vague/linguistic constraints, or qualitative assessment of alternatives, for example, for multiple purpose reservoir planning [14,15], water resources development [16], environmental modeling [17], or water distribution network optimization [18].
- *Data management*. FL is very useful for all kinds of applications involving handling and processing of large amounts of data, sometimes in different formats or coming from various sources. In such cases, FL can greatly improve the use and processing of databases: *Searching* using fuzzy criteria, *data mining and knowledge discovery* applying fuzzy decision trees and classification rules, *pattern recognition*, and *ranking/ordering*. Classic stochastic-statistical analysis is not sufficient in case of erratic, very short-term records, or in case the input database is diversified, its structure complex and unknown. FR may help by making sense and linking diversified information, offering a different "soft approach" to information retrieval and data mining, by introducing fuzzy inference rules [19–21].
- *Geospatial information systems* and *GIS applications*. Digital maps are a special case of data management worth mentioning, because of numerous hydroinformatics applications using FL, such as watershed simulation [22], floodplain analysis [23], maps [24], and snow coverage monitoring [25].
- *Decision Support Systems (DSS)*, a currently growing field, where multiple criteria, multiple stakeholders, and uncertainties due to data, and past, present, and future assessments, can be simulated using FL, usually including multiobjective optimization or combined with GIS. DSS are distinguished from "pure" multiobjective models, because they need to manipulate a much broader input database, with hydroinformatics FL and FR applications ranging from flood [26] and groundwater management [27], or water quality models [28], to water resources planning [29], and the evaluation of water companies' assets [30].
- *Expert systems*, can be considered a special case of DSS, where decisions are automatically updated and modified by current input data. Linguistic variables and fuzzy inference rules may replace conventional stochastic methods, for example, for flood frequency analysis [31].
- *Real-time* models, a special case of expert systems with feedback, resulting in (crisp) actions automatically undertaken and executed by the model (e.g., closing a pump or opening a sluice gate). Real-time models using fuzzy controllers are abundant in industrial applications; it can be said that this is the engineering field where the usefulness of FL was first recognized and adopted. Engineering/hydroinformatics

applications (sometimes combined with remote sensing or SCADA systems) include control and management of water supply systems [32], sewer systems [33], and flood water level control [34].

- *Combined with Genetic Algorithms (GA)*. There are two types of combined GA–FL models. The first type uses GA for multiobjective optimization, applying FL for the assessment and ranking of solutions, for example, for hydropower management [35] or for water network optimization [36]. The second type is using GA for the calibration of parametric models, where FL is used to identify the optimal parameter set, for example, to determine the Muskingum parameters within a rainfall-runoff model [37], tuning fuzzy inference rules for flood frequency expert systems [31], or finding classification rules for data mining [38].
- *Combined with Artificial Neural Networks (ANN)* for example, for signal processing and modeling hydrological time series, for either data generation or forecasting [39]. It is also used for real-time models, for example, for web-based models for water level prediction [40] and flood control [41]. This type of applications is called neuro-fuzzy computing, namely a fuzzy system with rules trained by a learning algorithm, derived by ANN theory.
- *Combined with Cellular Automata (CA)*. The CA are discrete parallel dynamic models, dividing the *d*-dimensional space by a lattice grid to "cells," each labeled by a "state" from a limited range of possible values. Change of state in time is performed by rules, thus simulating highly complex dynamic systems (such as complex flows). Fuzzy rules may replace deterministic rules and the first fuzzy hydroengineering applications include percolation and porous media flow [41] and urban planning/environmental models [42]. It is the type of applications expected to extend and increase in the years to come, that is, for river sedimentation, underground flow, water quality, etc.

25.4 Fuzzy Reasoning Tutorial — An Example

25.4.1 Introduction

All fuzzy models, irrespective of their specific purpose or field of application, share some common features and characteristics. The purpose of this tutorial is to present in brief the parts, mechanism, and environment of simple fuzzy models. It has been designed so that no prior knowledge of fuzzy systems is needed. After going through this chapter, readers should be able to think in "fuzzy terms" about modeling their own problems, or understanding other papers presenting specific applications in detail.

Any fuzzy model, however complex, consists of the following phases (or steps/parts):

A. *Inputs*: The model receives a set of inputs (data), as its knowledge of the outside world. Inputs can be anything: crisp numerical values, fuzzy numbers, strings, linguistic parameters, simple fuzzy sets, or complex fuzzy regions (coming as output from other fuzzy models), arranged as scalars, vectors, or arrays, constant or varying in time and space.

B. *Fuzzification* (preprocessor): Where fuzzy variables are defined, inputs are assigned to proper membership functions and arithmetic operations on fuzzy sets are performed.

C. *Fuzzy inference process*: The model executes a series of fuzzy propositions or rules (i.e., if x is A, then y is B) to arrive at conclusions. The rules are included in the application code, forming the core of the model. All rules are executed in a parallel way, meaning that all rules contribute to the final outcome; solving a system of equations would be the equivalent for classic numerical methods. The rules are considered known (by whatever method: experts, observations, etc.) prior to the modeling process. Should the rules be unknown, they might be estimated automatically, for example, by applying a neuro-fuzzy model. The number of rules should be sufficient to provide a consistent output whatever the input values or combinations.

D. *Defuzzification* (metaprocessor): Fuzzy inference usually results in complex fuzzy regions or membership functions for the fuzzy solution variables involved. In this case a crisp numeric or logical outcome is needed (e.g., for executing a control action, like closing a pump or opening a sluice gate, giving a warning, or painting the digital map a color), a defuzzification process is needed, to assign expected numeric values to fuzzy regions. This phase may be omitted if the fuzzy outcome is to be used as input to another model.

25.4.2 A Simple Fuzzy Model

In order to make the phases clear, let us assume the following fuzzy problem: monitoring and simulating snow cover and snowmelt in mountainous regions involves uncertainties, both about the data (i.e., coming from remote sensing/satellites) and the process itself. In the case temperature rises snow melting will start. If in addition to relatively high temperature in winter, there is also (heavy) rain, snowmelt will increase, affecting flow levels at the watershed river, which is in risk of flooding. This physical process, although logical, commonplace, and frequent, is hard to simulate using classical numerical modeling, because of a number of obvious uncertainties and ambiguities involved: How "higher than usual" should temperature be? How "heavy" should the rain be? Will the process affect the whole of the watershed catchment's area, or part of it? What about previous conditions, that is, is snow coverage uniform? Is it "thick" or "thin"? Suppose now that there is a continuous monitoring of the rain and air temperature in the area, using land gauging stations, and remote sensing, using satellites, for snow coverage. All real time signals (data) are transmitted to the control centre's database, together with short term weather forecasts, coming from an external source (another agency), by that meaning that weather forecasts should be taken as granted or as input data. The objective of the control centre is to estimate snowmelt, thus deciding whether risk for flooding is high for the next day and give warnings (short-term real-time forecasting, using FL). In the numerical example that follows a (simplified) model simulation for one time step (one day) will be carried out.

According to the phases (parts) of any FL model, the steps to be taken are:

A. *Inputs*: This involves sorting out, converting, and categorizing data from various origins: rain and temperature from each gauging station, weather forecasts, current flow level, current snow coverage. Rain and temperature data series from each station may be considered crisp or fuzzy numbers, but snow coverage should be defined as a fuzzy set, as well as water flow at the river, because although water levels come from reading meters as numerical values, conversion to rate of flow can only be done with a degree of imprecision and uncertainty, hence a fuzzy set is needed. Weather forecasts usually are defined as linguistic parameters

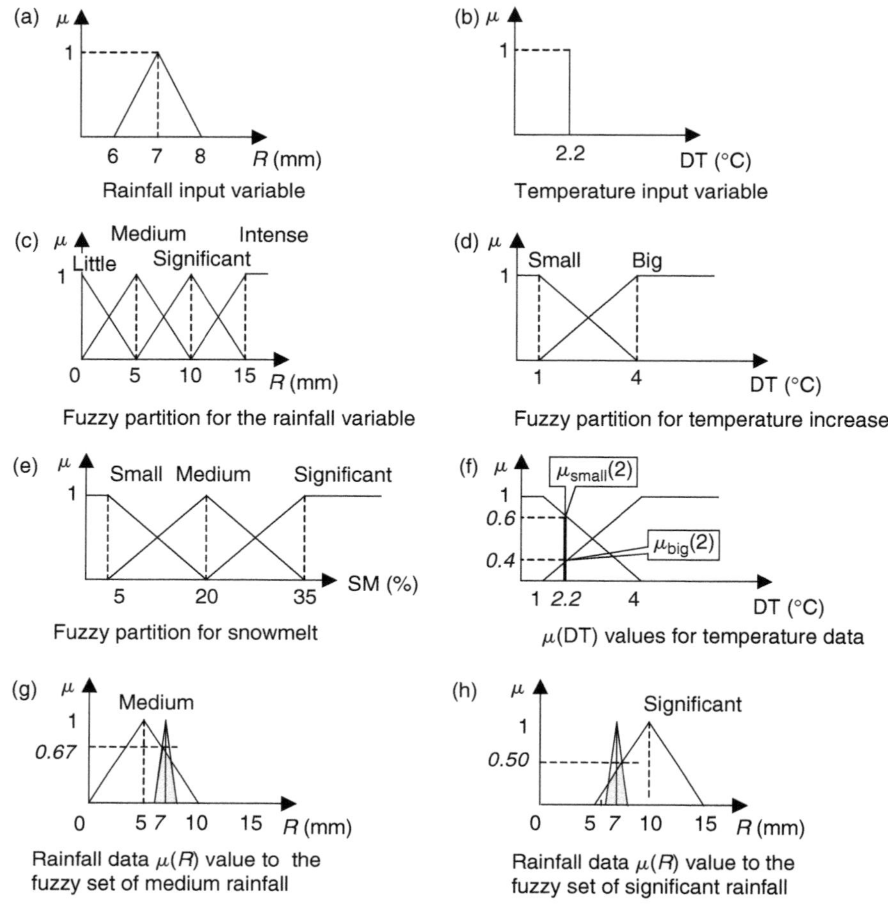

FIGURE 25.9
Numerical example. Fuzzy sets, partitions, and arithmetic operations.

(qualititative), that is, fair, rainy, heavy rain, etc. The objective of this phase is to determine which data are needed, what type (crisp or fuzzy) they should be, and what would be their format (e.g., array, time step) within the model. The general problem, as specified in the previous section, includes a large number of fuzzy and crisp variables. In order to make the numerical part of the example easier to understand and manipulate, let us assume that the fuzzy model consists of two input variables only: rainfall and temperature and the time step will be one day. This assumption simplifies the "real world" model considerably, but is considered necessary in order to provide an easy introduction to fuzzy models.

B. *Fuzzification*: Once data have been defined and classified, appropriate fuzzy sets must be defined, ready to be used as inputs to the model. The type and shape of the fuzzy sets should be selected, as well as the necessary fuzzy partitions, based on linguistic definitions for each variable (Figure 25.9). *Snowmelt* (SM) is easier to model as the "percentage of existing snow to melt" as shown in Figure 25.9(e). *Rainfall* (*R*) in mm, represents the mean daily rainfall in the area, estimated with the help of various preprocessors, such as radars, sensors, and GISs. In fact this variable is the output of another (fuzzy) model. The rainfall input variable will therefore be in the form of a fuzzy set, for example, a fuzzy number, which is the simplest type of membership function. Let us assume that the input value for the numerical example is the fuzzy number 17 mm, as shown in Figure 25.9(a). Instead of using the "temperature" variable, it might be better to use the "increase of temperature" variable (Figure 25.9[d]). Therefore DT in °C, stands for the *increase in temperature* measured in the area. In a real world problem, this variable should also be the output of another model, most probably assuming the form of a fuzzy set. However, both for simplicity reasons and for showing an example of handling crisp input variables, let us assume this variable is crisp (a singleton), that is, a numerical value equal to 2.2°C, as shown in Figure 25.9(b). The DT data input numerical value is assigned a membership function values to the partition fuzzy sets of "small" ($\mu_{\text{small}}(2.2) = 0.6$) and "big" ($\mu_{\text{big}}(2.2) = 0.4$) temperature increase (Figure 25.9[f]). These values are computed by linear interpolation. In a similar way, the rainfall input variable is assigned membership function values to the rainfall fuzzy linguistic partitions, by calculating the fuzzy intersection between "fuzzy number 7" on one hand, and "medium rainfall" (max $\mu_{\text{medium}}(\text{fuzzy\_7}) = 0.67$) or "significant rainfall" (max $\mu_{\text{significant}}(\text{fuzzy\_7}) = 0.5$) on the other, as shown in Figures 25.9(g) and (h), respectively. The membership function value of the rainfall data to the "little" and "intense" rainfall fuzzy sets is zero.

C. *Inference*: A set of inference rules must be established, relating input variables (rainfall R, temperature DT) and output (snowmelt SM). Using fuzzy logic symbols, the implication form is $\text{DT} \wedge R \to \text{SM}$. For simplicity, let us assume that the inference system consists of only three rules, obtained by experts:

   1. IF temperature increases DT is "small" AND (rainfall is "little" OR "medium") THEN snowmelt is "small"
   2. IF temperature increases DT is "big" AND rainfall is "significant" THEN snowmelt is "medium"
   3. IF rainfall is "intense" THEN snowmelt is "significant"

   The temperature variable is not excluded from the third rule, because the rule holds for any value of DT (i.e., $\mu(\text{DT}) = 1, \forall \text{DT} \in \Re$). Out of the three rules above, only the first two will "fire" (i.e., are significant) for the input data values of this numerical example, because the rainfall input value (R = "fuzzy 7") has zero membership to the fuzzy set "intense rainfall." The first rule has an internal part for the rainfall, which will be calculated first:

$$\begin{aligned} R \text{ is "little" OR "medium"} &\equiv \mu_{\text{little}}(R) \vee \mu_{\text{medium}}(R) \\ &= \max[\mu_{\text{little}}(R), \max(\mu_{\text{medium}}(R))] \\ &= \max(\mu_{\text{medium}}(R)) = 0.67. \end{aligned}$$

   According to the Mamdani min operator implication formula (equation [25.10]), the implications $\text{DT} \wedge R \to \text{SM}$ for the first and second rules are:

$$\text{Rule1: } \mu_{\text{small}}(\text{SM}) = \min[\mu_{\text{small}}(\text{DT}) \wedge \max(\mu_{\text{medium}}(R))] = \min(0.6, 0.67) = 0.6,$$

$$\text{Rule2: } \mu_{\text{medium}}(\text{SM}) = \min[\mu_{\text{big}}(\text{DT}) \wedge \max(\mu_{\text{significant}}(R))] = \min(0.4, 0.5) = 0.4.$$

   The graphical interpretation of the rules is shown in Figures 25.10(a) and (b), respectively. The final output for snowmelt, shown graphically in Figure 25.10(c), will be the fuzzy region given by the disjunctive fuzzy union of the two implications

(a)
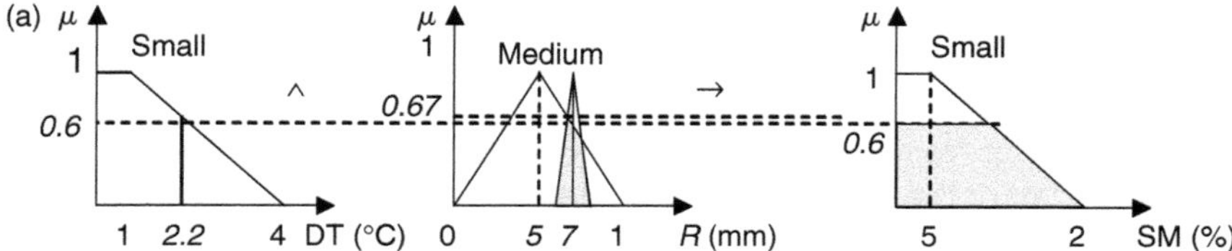

Rule 1: If DT is "small" and R is ("small" or "medium") THEN SM is "small"

(b)
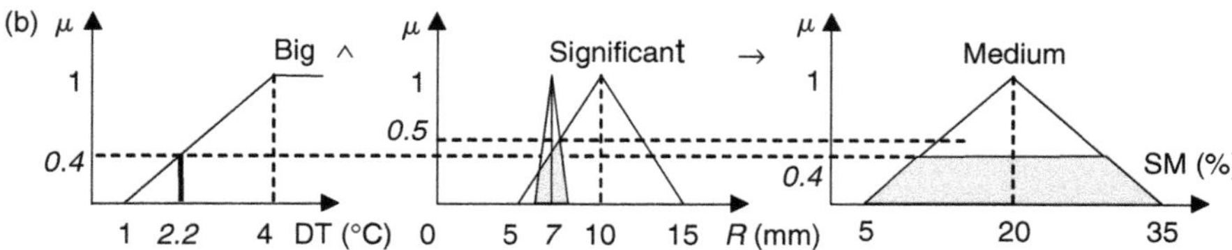

Rule 2: If DT is "big" and R is "significant" THEN SM is "medium"

(c)
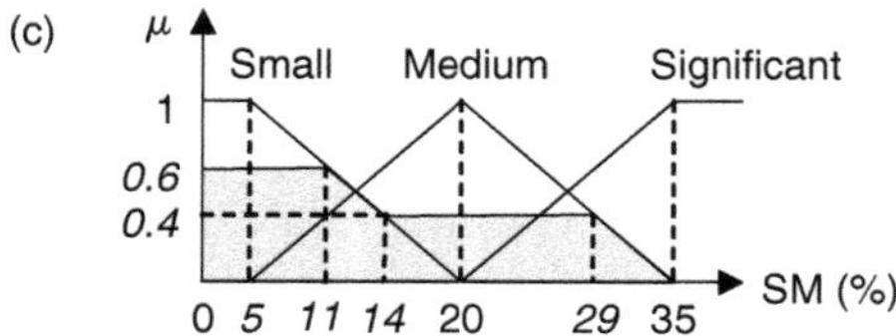

Combined disjunctive fuzzy region for snowmelt SM as output

(d)
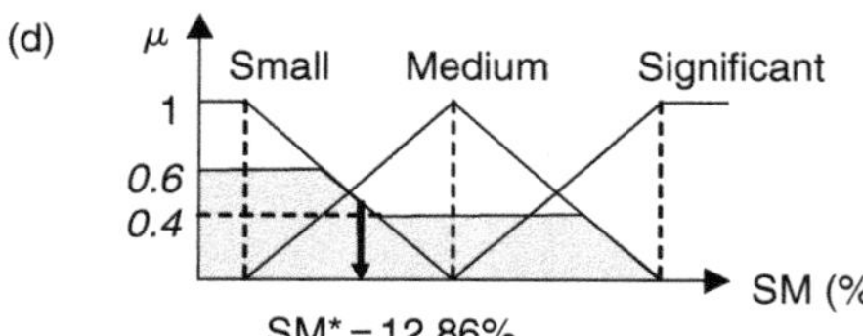

Defuzzification by the centroid method

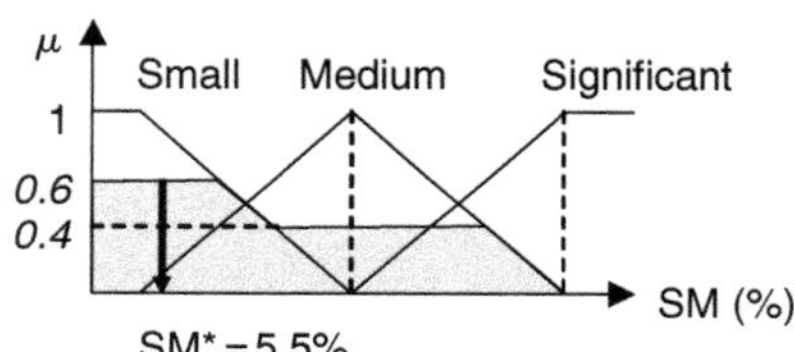

Defuzzification by the higher plateau average method

FIGURE 25.10
Numerical example. Graphical representation of fuzzy inference.

(equation [25.13]):

$$\mu(\text{SM}) = \max[\mu_{\text{small}}(\text{SM}), \mu_{\text{medium}}(\text{SM})].$$

D. *Defuzzification*: If the system is attached to an alarm device, alarm warnings should be issued, in case the snowmelt exceeds a pre-specified risk level. Accordingly a defuzzification method must be selected and applied to obtain a final numeric result, which will represent snowmelt. Defuzzification is carried out using two methods: the centroid method (Figure 25.10[d]) and the average higher plateau method (Figure 25.10[e]). It is obvious that results differ considerably. Selection of the most appropriate defuzzification method is problem specific, usually tuned by calibration.

This numerical example of a fuzzy system, although simplified in order to make introduction to fuzzy reasoning, calculations, and graphic representations easy, includes all main parts and techniques applied to more complicated systems: fuzzy and crisp inputs, fuzzy partitions, linguistic definitions of fuzzy sets, basic fuzzy operators for fuzzy union, intersection and implication, inference rules, and defuzzification. A "real world" problem would involve more variables, more partitions for fuzzy sets, possibly linguistic hedges, maybe other operators and certainly many more inference rules. However the overall structure of the fuzzy model and the computational steps (A to D) would be similar.

References

1. Zadeh, L.A., Fuzzy sets, *Information and Control*, 8, 338–353, 1965.
2. Bezdek, J., Fuzzy models — what are they, and why?, *IEEE Transactions on Fuzzy Systems*, 1, 1–5, 1993.
3. Ross, T.J., *Fuzzy Logic with Engineering Applications*, McGraw Hill International Editions, New York, 1995.
4. Klir, G.J. and Folger T.A., *Fuzzy Sets, Uncertainty and Information*, Prentice-Hall, New York, 1988.
5. Cox, E., *The Fuzzy Systems Handbook*, AP Professional, Academic Press Inc., 1994.
6. Pedrycz, W., *Fuzzy Control and Fuzzy Systems*, Research Studies Press Ltd., England, 1993.
7. Klir, G.J. and Yuan, B., On nonspecificity of fuzzy sets with continuous membership functions, *Proceedings of the IEEE International Conference on Systems, Man and Cybernetics*, 1, 25–29, 1995.
8. Ross, T.J., Booker, J.M., and Parkinson, W.J., *Fuzzy Logic and Probability Applications. Bridging the Gap*, ASA SIAM, 2002.
9. Ozelkan, E.C. and Duckstein, L., Fuzzy conceptual rainfall-runoff models, *Journal of Hydrology*, 253, 41–68, 2001.
10. Blaskova, S. and Beven, K., Flood frequency estimation by continuous simulation of subcatchment rainfalls and discharges with the aim of improving dam safety assessment in a large basin in the Czech Republic, *Journal of Hydrology*, 292, 153–172, 2004.
11. Xiong, L., Shamseldin, A.Y., and O'Connor, K.M., A non-linear combination of the forecasts of rainfall-runoff models by the first-order Takagi-Sugeno fuzzy system, *Journal of Hydrology*, 245, 196–217, 2001.
12. Schulz, K. and Huwe, B., Water flow modelling in the unsaturated zone with imprecise parameters using a fuzzy approach, *Journal of Hydrology*, 201, 211–229, 1997.
13. Revelli, R. and Ridolfi, L., Fuzzy approach for analysis of pipe networks, *Journal of Hydraulic Engineering ASCE*, 128, 93–101, 2002.
14. Chang, N.-B., Wen, C.G., Chen, Y.L., and Yong, Y.C., A grey fuzzy multiobjective programming approach for the optimal planning of a reservoir watershed. Part A: theoretical development, *Water Research*, 30, 2329–2334, 1996.
15. Chang, N.-B., Wen, C.G., Chen, Y.L., and Yong, Y.C., A grey fuzzy multiobjective programming approach for the optimal planning of a reservoir watershed. Part B: application, *Water Research*, 30, 2335–2340, 1996.
16. Bector, C.R. and Goulter, I., Multiobjective water resources investment planning under budgetary uncertainty and fuzzy environment, *European Journal of Operational Research*, 82, 556–591, 1995.
17. Yeh, S.-C. and Tung, C.-P., Optimal balance between land development and groundwater conservation in an uncertain coastal environment, *Civil Engineering and Environmental Systems*, 20, 61–81, 2003.
18. Xu, C. and Goulter, I., Optimal design of water distribution networks using fuzzy optimisation, *Civil Engineering and Environmental Systems*, 16, 243–266, 1999.
19. Mannila, H., Local and Global Methods in Data Mining: Basic Techniques and Open Problems, *Proceedings ICALP 2002*, Widmayer et al., Eds., Springer-Verlag, Berlin Heidelberg, pp. 57–68, 2002.

20. Savic, D.A., Davidson, J.W., and Davis, R.B., Data mining and knowledge discovery for the water industry, in D.A. Savic and G.A. Walters, Eds., *Water Industry Systems*, Vol. 2, Research Studies Press Ltd., UK, pp. 155–163, 1999.
21. Vamvakeridou-Lyroudia, L.S., Extracting information from urban water network malfunction records with fuzzy inference rules, in C. Maksimovic, D. Butler, and F.A. Memon, Eds., *Advances in Water Supply Management*, A.A. Balkema Publishers, Rotterdam, pp. 41–49, 2003.
22. Freer, J.E., McMillan, H., McDonnell, J.J., and Beven, K.J., Constraining dynamic TOPMODEL responses for imprecise water table information using fuzzy rule-based performance measures, *Journal of Hydrology*, 291, 254–277, 2004.
23. Tkach, R.J. and Simonovic, S.P., A new approach to multi-criteria decision making in water resources, *Journal of Geographic Information and Decision Analysis*, 1, 25–44, 1997.
24. Patil, G.P., Balbus, J., Biging, G., Jaja, J., Myers, W.L., and Taillie, C., Multiscale advanced raster map analysis system: definition, design and development, *Environmental and Ecological Statistics*, 11, 113–118, 2004.
25. Bordogna, G., Carrara, P., Rampini, A., and Spaccapietra, S., A time-travel tool for monitoring environmental phenomena by remote sensing techniques, *Proceedings of the AGILE Conference on GIS*, April 29–May 1, 2004, Heraklion, Greece, pp. 579–586, 2004.
26. Despic, O. and Simonovic, S.P., Aggregation operators for soft decision making in water resources, *Fuzzy Sets and Systems*, 115, 11–33, 2000.
27. Merino, G.G., Jones, D.D., Clements, D.L., and Miller, D., Fuzzy compromise programming with precedence order in criteria, *Applied Mathematics and Computation*, 134, 185–205, 2003.
28. Khadam, I.M. and Kaluarachchci, J.J., Multi-criteria decision analysis with probabilistic risk assessment for the management of contaminated ground water, *Environmental Impact Assessment Review*, 23, 683–721, 2003.
29. Bender, M.J. and Simonovic, S.P., A fuzzy compromise approach to water resource systems planning under uncertainty, *Fuzzy Sets and Systems*, 115, 35–44, 2000.
30. Langman, E.J. and De Rosa, S., Making defensible decisions in the face of uncertainty, in C. Maksimovic, D. Butler, and F.A. Memon, Eds., *Advances in Water Supply Management*, A.A. Balkema Publishers, Rotterdam, pp. 375–384, 2003.
31. Shu, C. and Burn, D.H., Homogeneous pooling group delineation for flood frequency analysis using a fuzzy expert system with genetic enhancement, *Journal of Hydrology*, 291, 132–149, 2004.
32. Vamvakeridou-Lyroudia, L.S. and Politaki, S., D.AN.A.I.S. — an original real time expert model controlling the water supply aqueduct of the Greater Athens area, in D. Savic and G. Walters, Eds., *Water Industry Systems*, Vol. 1, Research Studies Press Ltd., UK, pp. 487–499, 1999.
33. Klepiszewski, K. and Schmitt, T.G., Comparison of conventional rule based flow control with control processes based on fuzzy logic in a combined sewer system, *Water Science and Technology*, 46, 77–84, 2002.
34. Lobbrecht, A.H. and Solomatine, D.P., Control of water levels in polder areas using neural networks and fuzzy adaptive systems, in D. Savic and G. Walters, Eds., *Water Industry Systems*, Vol. 1, Research Studies Press Ltd., UK, pp. 509–518, 1999.
35. Basu, M., An interactive fuzzy satisfying method based on evolutionary programming technique for multiobjective short-term hydrothermal scheduling, *Electric Power Systems Research*, 69, pp. 277–285, 2004.
36. Vamvakeridou-Lyroudia, L.S., Walters, G.A., and Savic, D.A., Fuzzy multiobjective design optimisation of water distribution networks, *Journal Water Resources, Planning and Management ASCE*, 131, 2005 (In Press).
37. Cheng, C.T., Ou, C.P., and Chau, C.W., Combining a fuzzy optimal model with a genetic algorithm to solve multi-objective rainfall-runoff calibration, *Journal of Hydrology*, 268, 72–86, 2002.
38. Hu, Y.-C., Chen, R.-S., and Tzeng, G.-H., Finding fuzzy classification rules using data mining techniques, *Pattern Recognition Letters*, 24, 509–519, 2003.

39. Nayak, P.C., Sudheer, K.P., Rangan, D.M., and Ramasastri, K.S., A neuro-fuzzy computing technique for modelling hydrological time series, *Journal of Hydrology*, 291, 52–66, 2004.
40. Bazartseren, B. , Hilderbrandt, G., and Holz, K.-P., Short term water level prediction using neural networks and neuro-fuzzy approach, *Neurocomputing*, 55, 439–450, 2003.
41. Cattaneo, G., Flocchini, P., Mauri, G., Quaranta Vogliotti, C., and Santoro, N., Cellular automata in fuzzy backgrounds, *Physica D*, 105, 105–120, 1997.
42. Liu, Y. and Phinn, S.R., Modelling urban development with cellular automata incorporating fuzzy-set approaches, *Computers Environment and Urban Systems*, 27, 637–658, 2003.

Section VI

Appendices

Appendix 1

A Tutorial for Geodatabase and Modelshed Tools Operation

Contents

A1 Orientation

The Modelshed Tools are a collection of utilities designed to function with a Modelshed-extended ArcHydro geodatabase. They serve functions of database management, index assignment, raster cataloging, data ingestion, relationship building, and data conversion. The tools are integral with ArcGIS and use the functionality of the ArcObjects, especially the Spatial Analyst extension for geoprocessing. The ArcGIS software and its wide array of database utilities are compatible with this database model, and may be used in the same manner as with a normal geodatabase. Likewise, the ArcHydro Tools may be used with the Modelshed Tools to provide complementary functionality. The Modelshed Tools are a work in progress, and this documentation describes the basic functionality of the beta version of the software.

A basic level of proficiency with the ArcGIS software, geodatabases, and a familiarity with the ArcHydro and Modelshed geodata model structure is assumed in this tutorial. This walkthrough includes pictures and instructions to help guide the user through some basic operations. In addition to the ArcGIS software, the ArcHydro Tools, ESRI CASE Tools, and other ArcHydro tools are referenced. These toolsets are available free online from ESRI, Inc.

and from the Center for Research in Water Resources at the University of Texas at Austin, who publish and distribute the software.

A1.1 Constructing a Modelshed-Compatible Geodatabase

The steps to the construction of a geodatabase are not complicated, and they are covered in much more technical detail by ESRIs training and technical documentation. This section will serve as a short summary of that information, which is essential for the creation and population of a geodatabase with the Modelshed geodata model.

A1.1.1 Collecting Data

The first obvious step to building a geodatabase is to collect your data. This is also the hardest step. Because the modelshed framework will take many types of geospatial data, there is a lot of flexibility here. Decide what is needed, and how that data will be put together in a way that makes sense within the Modelshed geodata model. In other words, decide which data to use and what data model structures will accept each dataset.

The data does not need to be in a native ArcGIS format (i.e., GRID for rasters), but it does need to be in a format that the ArcGIS conversion tools, or other tools, can read into a geodatabase. Time series (nonspatial) data will often come in a tabular format, stored in a spreadsheet or a delimited text file. With this loose type of data it is wise to wait until after the database is populated, and use a combination of applications such as Microsoft Excel™ and Microsoft Access™ to manually import that data into a database table. It is better to use automated import tools, but that option is not always available.

There is a huge variety of data available, in many different formats, from many sources. Online sources are a great resource. State and federal resource agencies, particularly the USGS, NOAA, and NASA, have lots of online data available for download.

A1.1.2 Building a Geodatabase with ArcCatalog

ArcCatalog makes it easy to construct a new Personal Geodatabase (MSAccess-format DBMS), populate it with data, and apply a schema. Start by navigating in ArcCatalog to the folder where you want the database, right click → new → Personal Geodatabase, as shown in Figure A1.1.

In a similar fashion, create feature datasets within the geodatabase to hold you feature classes/spatial objects. You will want to name these feature datasets the same as they will appear in the data model (e.g., "Modelshed" or "Drainage"): this will make it easier to apply the schema. Now here is a tricky part. As you create the feature datasets, you will need to specify the spatial reference (projection and datum) and extents. This reference and extents will apply to all feature classes stored therein, so it is important to use the same reference and equal or greater extents than the extents of the most extensive feature class you plan to store there. If you have spatially referenced data already available in a format ArcGIS recognizes, you can import that spatial reference and extents during the wizard-driven dataset creation process. After naming your dataset, hit "Edit" next to the window displaying an unknown spatial reference. Use the "Import" button and browse to the spatial dataset that has the right reference and extent. You can examine the settings by flipping through the tabs in this

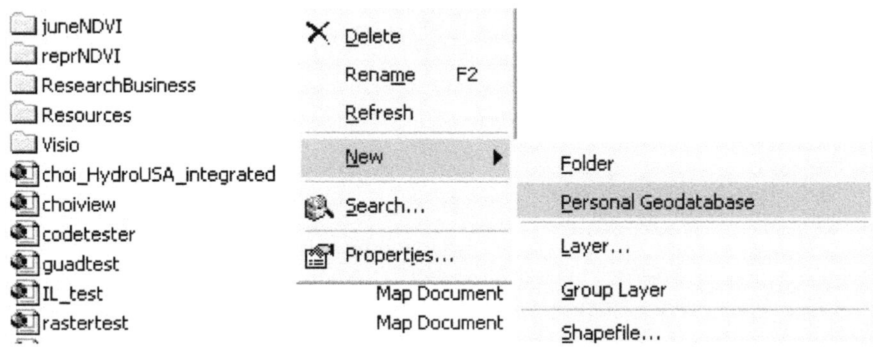

Figure A1.1
Creating a new Personal Geodatabase (MSAccess format).

form — make sure they are correct. It is hard to change the reference and extent after the feature dataset is created.

Next we will import our spatial data. ESRI handles most of this for us by providing a toolbox for conversion of many data types, and one-right-click import and export capabilities within the catalog, as shown below. Data import can wait until after the data model schema has been applied, but it is best done now. As before, name the imported feature classes what they should be called in the data model — in Figure A1.2, the "rcm_grid_polygon" polygon layer is being imported to the "modelshed" feature class. Once we apply the data model schema, this imported data will be converted, and so it has the proper attributes.

A1.1.3 Applying the Modelshed Schema with the CASE Tools

The CASE Tools (Computer Aided Software Engineering) are part of the Geodatabase Designer toolkit provided by ESRI. This toolkit may be downloaded and added to the catalog user interface using the Tools → Customize menu item. The schema creation tool is easy to miss — it looks like:

Select your geodatabase full of data and hit the schema creation button. You will see something like Figure A1.3. You can automatically apply the schema/data model in one database to this one, or you can get the schema from an XMI file. An XMI is an XML-formatted document that describes the structure of a database. In this example we are using an XMI schema document that was exported from Modelshed geodata model diagrams in Microsoft Visio™, using the ESRI semantics checker and ESRI XMI Export utility.

Next, a tree structure will appear displaying the layout of the schema that is about to be applied to the database. Objects in the database that are detected as already existing are highlighted in red. The feature classes and datasets that we added before and named properly should be detected in this way. Select one of those classes and hit the "Properties" button, then look through the options. It may be important to set some of these depending on your data. In particular, examine your feature classes to make sure they contain the proper fields. It may be necessary to manually edit the fields, or come back later with another software package such as MS Access™, to make sure the correct fields are all accounted for. Remember, if the feature class in your database is named something other than the name in the data model schema, the schema will not be applied to that feature class unless you set it up manually. Make sure that your database is not being accessed by any other software (i.e., ArcMap), then proceed. The schema is applied to your database, and now the database should be properly formed with the data model.

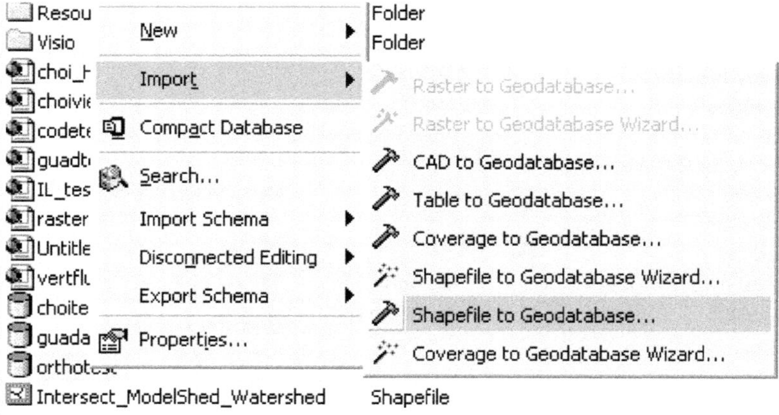

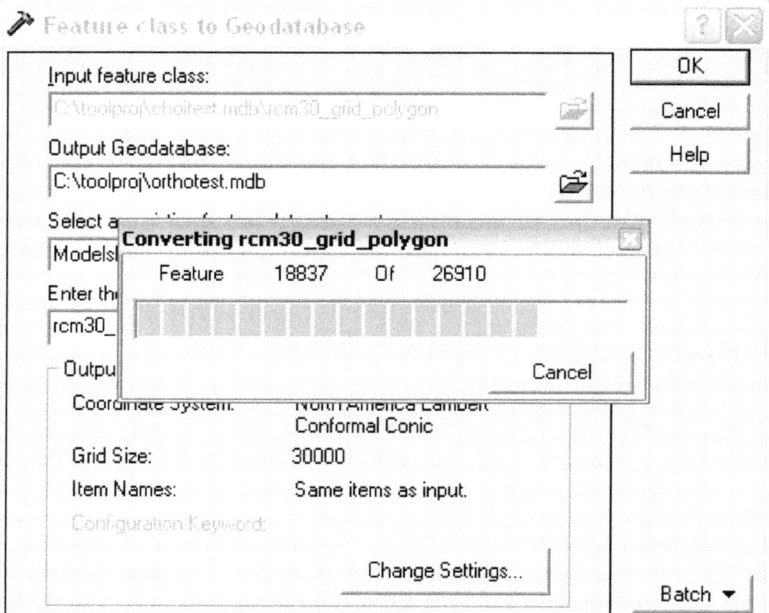

Figure A1.2
Importing a polygon shapefile to the geodatabase.

A1.1.4 Assigning HydroIDs with the ArcHydro Tools

Before you can work with feature classes in the database you must assign unique HydroIDs to the features in the database. This is true for ArcHydro databases as well as the Modelshed database model, which is based on ArcHydro. Fortunately, the ArcHydro Tools have a utility that will assign HydroIDs properly. Load ArcMap and add the ArcHydro Tools with the Customize menu dialog if you have not already done so. The ArcHydro Tools provide a lot of capabilities, and they look like Figure A1.4.

Add your feature classes to the ArcMap table of contents using the button on the toolbar, then go to Arc Hydro Tools → Attribute Tools → Assign HydroID. Select your classes in the form that appears, and run the tool. Now all the features in these classes

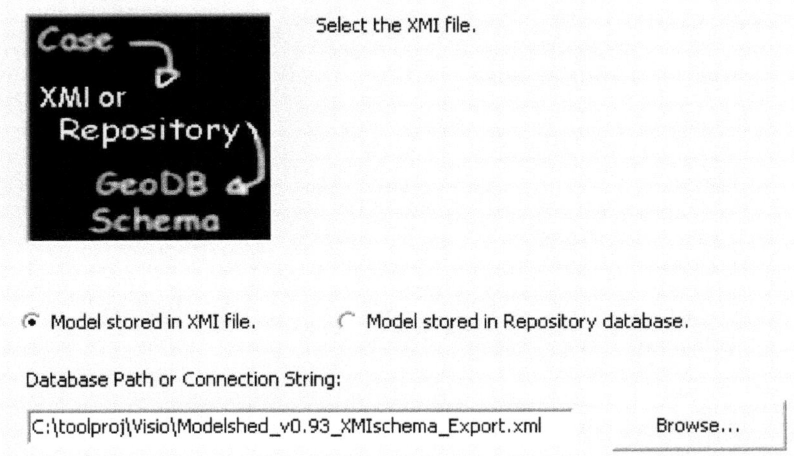

FIGURE A1.3
The ESRI CASE tools for applying a schema to a geodatabase.

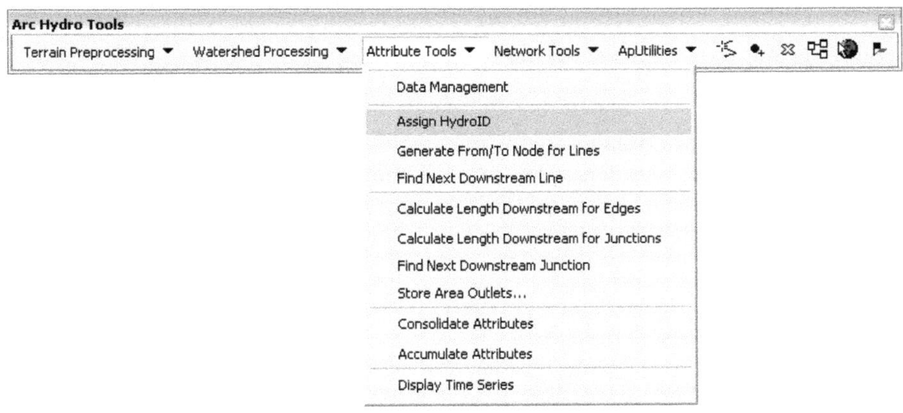

FIGURE A1.4
The AcrHydro Toolbox.

should have unique HydroIDs. Now the spatial feature classes are present and properly formed, and all the database structures specified in the data model are present (tables and relationships) although the tables are probably still empty. At this point the database is ready for more advanced operations using the ArcHydro Tools and Modelshed Tools, and for you to add time series data and other data.

A1.2 Applying the Modelshed Tools

To apply the Modelshed Tools, we will be working with an example geodatabase of the Illinois River Basin. This geodatabase is supplied with the tutorial, and is named ILRDB.mdb. The database was derived from the ArcHydroUSA database, which is a compilation of the watersheds, stream networks, and gauging stations in the continental

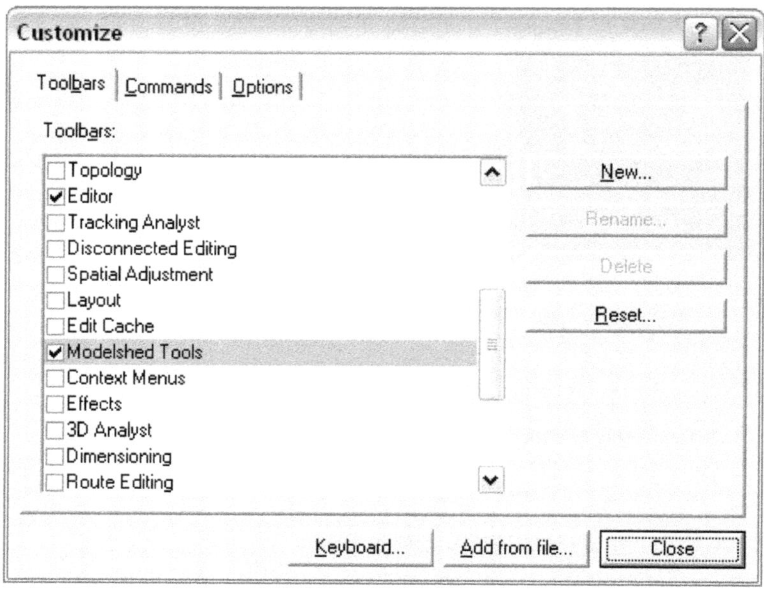

FIGURE A1.5
Adding a custom toolbar to the Graphical User Interface.

United States. Additional datasets including vegetation, elevation, land cover, and soils will be added to this database using the Modelshed Framework and Tools. This database is already populated with the basic watershed data and has the proper schema, so we are ready to install and use the Modelshed Tools to expand this database.

A1.2.1 Installing the Tools

Installing the tools is simple. The tools are compiled as a Dynamic Linked Library called something similar to "ModelshedTools_vXX.dll," where XX is the software version. The tools will function in any ArcGIS 8.x or ArcGIS 9.x software, but ArcMap is the most convenient and versatile host. In ArcMap Select Tools → Customize → Add from file... → locate the ModelshedTools_vXX.dll on the disk. Before they will appear, the checkbox next to the Modelshed Tools name must be selected as shown in Figure A1.5. Then the tools will appear as a drop-down menubar on the GUI.

Click on the tools and inspect the options. Most menu items will be unavailable; before the tools can be used, the operational geodatabase must be selected (see Figure A1.6). A file dialog opens — it will only allow the selection of personal geodatabases. When a geodatabase is selected, it is validated. In this case validation means that the geodatabase is checked to see that it has the necessary specific tables and fields for Modelshed Tools operations, and not that the database is perfectly formed based on the schema. Select the ILRDB.mdb personal geodatabase. Now other commands are available — they all operate within this selected personal geodatabase. Now that we have selected the geodatabase, let us add the data in the database to the ArcMap view. Hit the button on the standard toolbar and navigate to the ILRDB.mdb database. Add all of the tables and both feature datasets (ArcHydro and Modelshed) to the view. One important note here: the Modelshed tools do not use the layers in the ArcMap view for any processing, or access the database

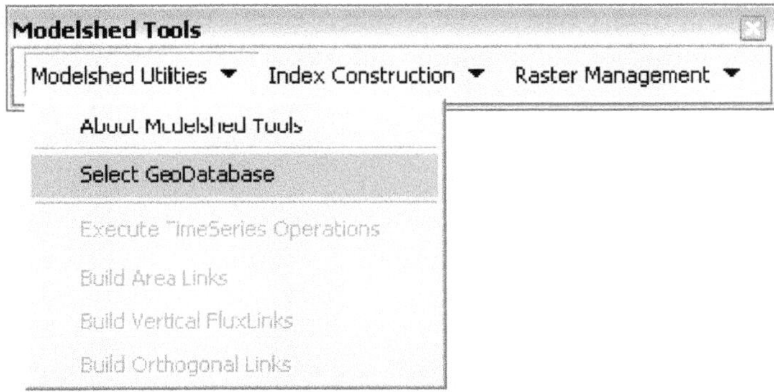

FIGURE A1.6
The Modelshed toolbox.

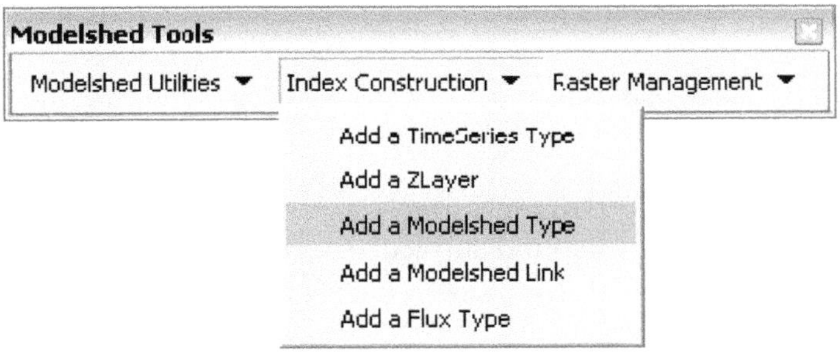

FIGURE A1.7
Adding a ModelshedType index using the Modelshed tools.

through the view at all. In fact, we do not need to view the layers at all for the Modelshed Tools to function; it is only for our visual reference that we need to look at them at this point.

A1.2.2 Specifying Database Indexes

There is nothing complicated about assigning database indexes. Each of the forms under "index construction" provides a list of text boxes and drop-down menus where metadata is entered. These tools exist to make it easier for a user to assign well-formed metadata. The same task may be accomplished by editing the database directly. This metadata describes the information in the database, and is used in automated database operations. Five indexes may be specified: see Figure A1.7. Start with the first items on the menu — Time series Type and Zlayer — and proceed to the later items. The Modelshed Type, Modelshed Link, and Flux Type tools assign indexes that cross-reference other indexes, so those cross-referenced indexes must be specified first. The database we are working with already is populated with geospatial data in feature classes, but it lacks the indexes that will allow relational time series and raster-aggregated statistical data to be associated with those classes. Let us put in some indexes.

First let us add some Time Series Types. Before we do so, however, we should include streamflow, NDVI, and CONUS-SOIL soils data in our database, so we need three types of indexes to describe this data. Go to Modelshed Tools → Index Construction → Add a Time Series Type. Use the form to add these three indexes:

TSTypeID	1	2	3	4
Variable	Streamflow	NDVI	CONUS-SOIL Sand	Elevation
Units	cfs	Fractional	Fractional	m
IsRegular	True	True	True	True
TSInterval	1Day	1Month	Other	Other
DataType	Average	Average	Instantaneous	Average
Origin	Recorded	Recorded	Recorded	Recorded

Now add ZLayer indexes in a similar manner. We need a ground surface ZLayer and two CONUS-SOIL ZLayers, as follows:

ZLayerID	1	2	3
Altitude Units	m	cm	cm
Altitude Datum	Ground surface	Ground surface	Ground surface
Description	Ground surface	CONUS-SOIL 1	CONUS-SOIL 2
Layer Bottom Altitude	0	−5	−10
Layer Top Altitude	0	0	−5
ZLayer Above ID		1	2
ZLayer Below ID	2	3	

Now it is necessary to add some Modelshed Types. This index is very important to the database; it describes the data stored in a feature class, in association with its relationship to other data in the database. Groups of modelsheds belong to a Modelshed Type, and data records are associated with a Modelshed Type. Add six Modelshed Types to represent the six base datasets from which we will be building our database:

ModelshedTypeID	ModelshedClassName	Description
1	Watershed	USGS Level II Watershed
2	IL_state	Illinois State Boundary
3	IL_counties	Illinois County Boundaries
4	rcm30_grid_polygon	30-km land-atmosphere grid
5	Ecoregion	Eco-Region
6	Basin	Illinois River Basin

That takes care of the essential indexes. There are two more, which we can set if we want the added functionality of Modelshed Links and Flux Links. We will not build any Flux Links in this tutorial, so just add a Modelshed Link between the watershed and basin types as shown in Figure A1.8. This link establishes the hierarchical relationships between the Illinois River Basin and its watersheds.

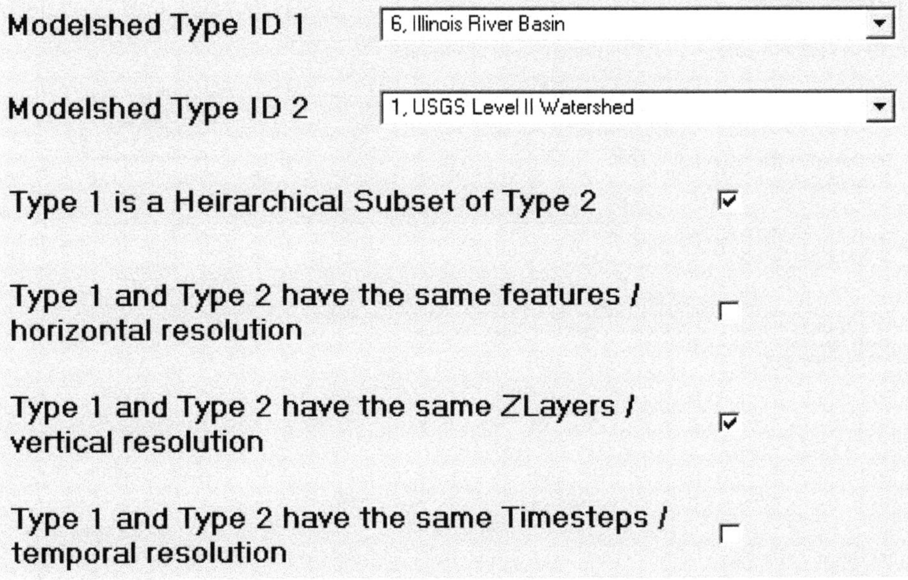

FIGURE A1.8
The dialog box used for building a ModelshedLink.

A1.2.3 Building Area Links

Area Links are constructed to allow the cross-referential querying of polygon features when overlapping, schematically related features from another Modelshed Type are accessed. In this example, we want to associate data stored with the 30 km climate grid Modelshed Type with data linked to the Watershed Modelshed Type. Run the Modelshed Tools → Modelshed Utilities → Build AreaLinks form to build the Area Links. The fractional area limit will exclude links that which fail to overlap with more than *X* fractional overlap. This parameter defaults to zero, including all links. An intersect operation is performed by the Modelshed Tools, and a shapefile with the intersected areas is stored on the hard drive in the folder specified on the form. If this path is incorrect or unavailable, the tool will fail. The shapefile may be deleted after running the tool, if desired, because it will not be used again. Repeat this step to build AreaLinks between the Eco-Region feature class and the Watershed class, as in Figure A1.9.

A1.2.4 Cataloging Raster Datasets

Now it is time to use those indexes we have established to catalog some raster files. These rasters will be organized in the RasterTSIndex table in association with a ZLayer and a time stamp. We are using three raster datasets: a 1 km DEM, a collection of CONUS-SOIL percentage rasters, and a time series of NDVI images spanning four years' vegetation in July. Use the form at Modelshed Tools → Raster Management → Add Rasters to the Index to add these rasters automatically, along with the proper ZLayer and TSType indexes describing the vertical reference of the raster and the data type stored in it (see Figure A1.10). The downside of this tool is that it does not automatically assign time stamps to the rasters. Most raster images do not have time stamp metadata, so it is not possible to do this using a generic tool. After setting the indexes for the rasters select a group of rasters on disk that

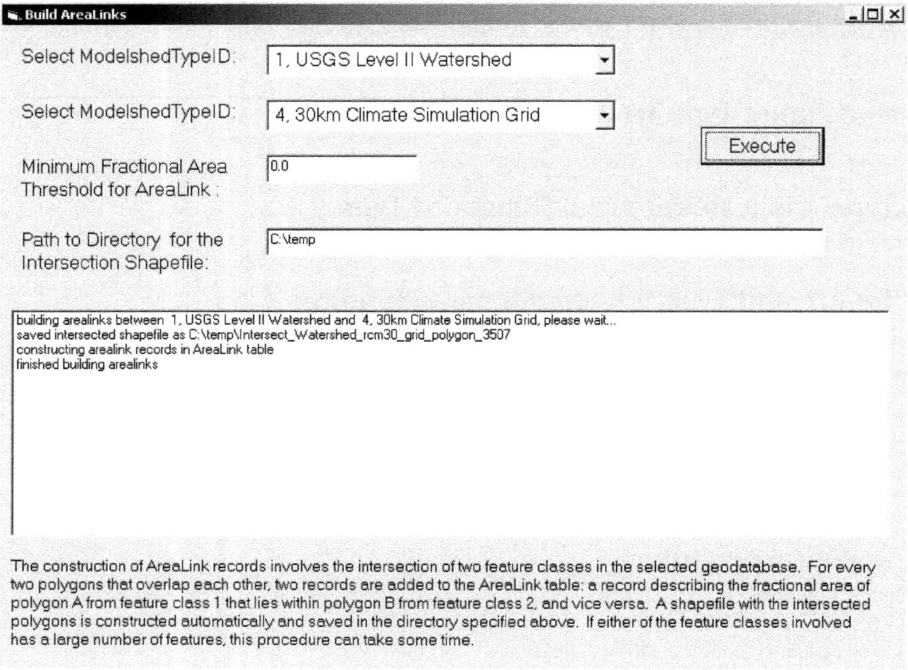

FIGURE A1.9
The dialog box used for building AreaLinks.

match those indexes and hit OK. The rasters are added to the index. Repeat this process for the DEM, CONUS-SOIL, and NDVI raster sets. Then, open an edit session in ArcMap (Editor toolbar → Editor → Start Editing), and manually set the correct time stamps for the rasters. Note that there is a default value; this does not need to be changed except for the NDVI rasters, because it is the only dataset that varies with time. When you have finished editing, hit the "save edits" button under the same editor menu bar.

A1.2.5 Ingesting Raster Data

Automatically ingesting raster data into the Modelshed framework is what the Modelshed Tools are most useful for. With the previous steps complete, we are ready to go ahead with this process. Load the Execute Time series Operations form, as shown in Figure A1.11. For starters, let us get the second CONUS-SOIL layer into the database. Select the CONUS-SOIL Sand Fractions data type, the CONUS-SOIL 2 ZLayer, and the Watershed Modelshed Type. Hit "Execute." The tool will find all the rasters in the RasterTSIndex catalog with matching data types and ZLayers, then statistically summarize the data in the raster cells by overlaying the Modelshed layer (Watershed in this case). Those statistics are stored in the StatisticalTS table, associated with the TSType data type, the ZLayer, and the Feature ID/Hydro ID of the feature (Watershed here) that the statistics are based on. Repeat this process for the other rasters we added — the layer 1 CONUS-SOIL, NDVI, and Elevation rasters. Do not forget to select the correct ZLayers that go with each dataset: elevation datum for elevation, ground surface for NDVI, and CONUS-SOIL 1 for the first CONUS-SOIL layer. We have just converted 4D raster data into vector data using the Modelshed framework and tools.

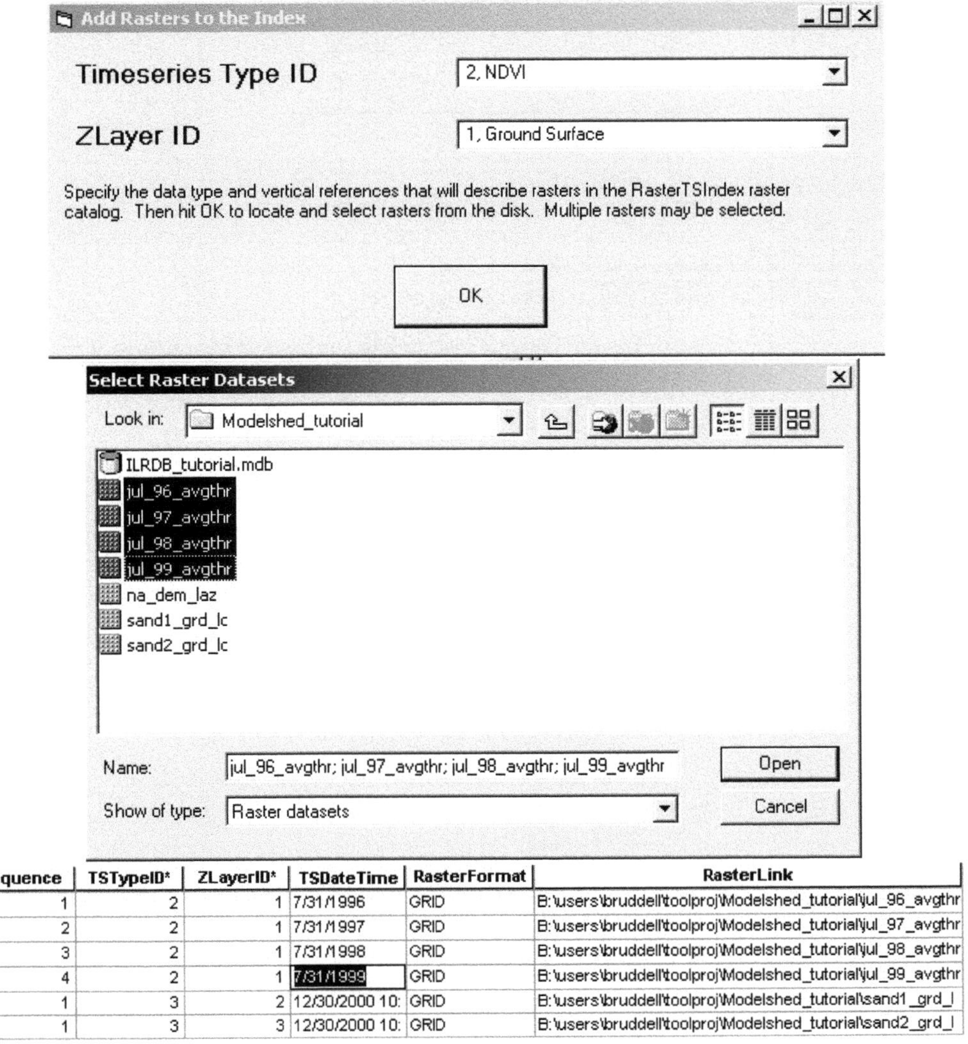

Sequence	TSTypeID*	ZLayerID*	TSDateTime	RasterFormat	RasterLink
1	2	1	7/31/1996	GRID	B:\users\bruddell\toolproj\Modelshed_tutorial\jul_96_avgthr
2	2	1	7/31/1997	GRID	B:\users\bruddell\toolproj\Modelshed_tutorial\jul_97_avgthr
3	2	1	7/31/1998	GRID	B:\users\bruddell\toolproj\Modelshed_tutorial\jul_98_avgthr
4	2	1	7/31/1999	GRID	B:\users\bruddell\toolproj\Modelshed_tutorial\jul_99_avgthr
1	3	2	12/30/2000 10:	GRID	B:\users\bruddell\toolproj\Modelshed_tutorial\sand1_grd_l
1	3	3	12/30/2000 10:	GRID	B:\users\bruddell\toolproj\Modelshed_tutorial\sand2_grd_l

FIGURE A1.10
Adding raster datasets to the RasterIndex.

An important technical point is that the ESRI Spatial Analyst for ArcGIS is used during the raster statistics generation. This means that unlike the other utilities in the Modelshed Tools, the raster processor requires a valid Spatial Analyst Extension license and installation. Also, since the Spatial Analyst uses some behind-the-scenes analysis settings and environments, it may be possible that the results could vary from computer to computer if the analysis environment defaults are misconfigured. Since the Spatial Analyst converts all rasters and polygons into ESRI GRID format before geoprocessing, this will take some time and it is better to have your datasets in GRID format before beginning the process. The Spatial Analyst will use an on-the-fly 16-point reprojection of raster datasets to match the polygon dataset; this means that minor inaccuracies may occur for datasets with very large extent, covering more than a few degrees of arc on the globe, or datasets projected at high latitudes on the globe. If a more accurate projection is needed, reproject your rasters to match the polygon before using the tool.

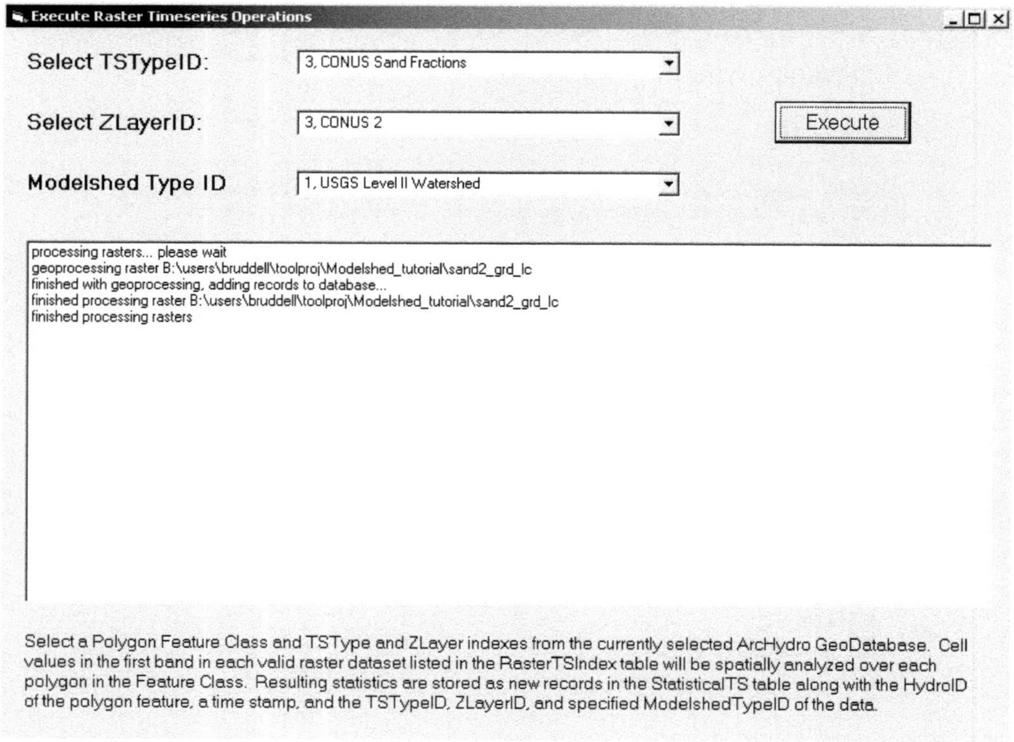

Execute Raster Timeseries Operations

Select TSTypeID: 3, CONUS Sand Fractions

Select ZLayerID: 3, CONUS 2

Modelshed Type ID 1, USGS Level II Watershed

Execute

processing rasters... please wait
geoprocessing raster B:\users\bruddell\toolproj\Modelshed_tutorial\sand2_grd_lc
finished with geoprocessing, adding records to database...
finished processing raster B:\users\bruddell\toolproj\Modelshed_tutorial\sand2_grd_lc
finished processing rasters

Select a Polygon Feature Class and TSType and ZLayer indexes from the currently selected ArcHydro GeoDatabase. Cell values in the first band in each valid raster dataset listed in the RasterTSIndex table will be spatially analyzed over each polygon in the Feature Class. Resulting statistics are stored as new records in the StatisticalTS table along with the HydroID of the polygon feature, a time stamp, and the TSTypeID, ZLayerID, and specified ModelshedTypeID of the data.

FIGURE A1.11
The raster time series processing tool.

A1.3 Data Analysis Using the Modelshed Geodata Model

The unique advantage of data analysis within a relational object-oriented database is the ability to query related objects and attributes, and examine these associations from different perspectives. This sort of analysis may take the form of data mining, which is currently used mostly in business marketing applications to uncover new sorts of correlations among the data, or directed query, in which case a user will select related objects in the database and join them to look at the effect of one object on another. We will examine a simple example of directed query-based analysis and visualization of database features according to associated data attributes. Then we will look at the similar task of selecting related features.

A1.3.1 Query-Based Visualization and Extraction in ArcGIS

Make sure that all the feature classes and tables from the ILRDB_tutorial database are added to the ArcMap view. We will perform some joins on the StatisticalTS table and the Watershed feature class, for the purpose of identifying the watersheds with better flooding characteristics. Let us assume that a watershed with more vegetation and more permeable soils will be generally less susceptible to flooding because rainwater will infiltrate into the soils and runoff will be slowed. We have in our database a vegetation index, NDVI, which measures the relative amount of vegetation in an area using a fraction from zero to one, with

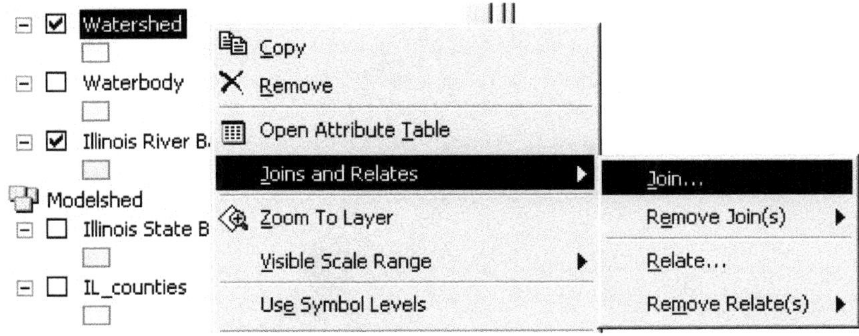

Figure A1.12
Joining layers in ArcGIS.

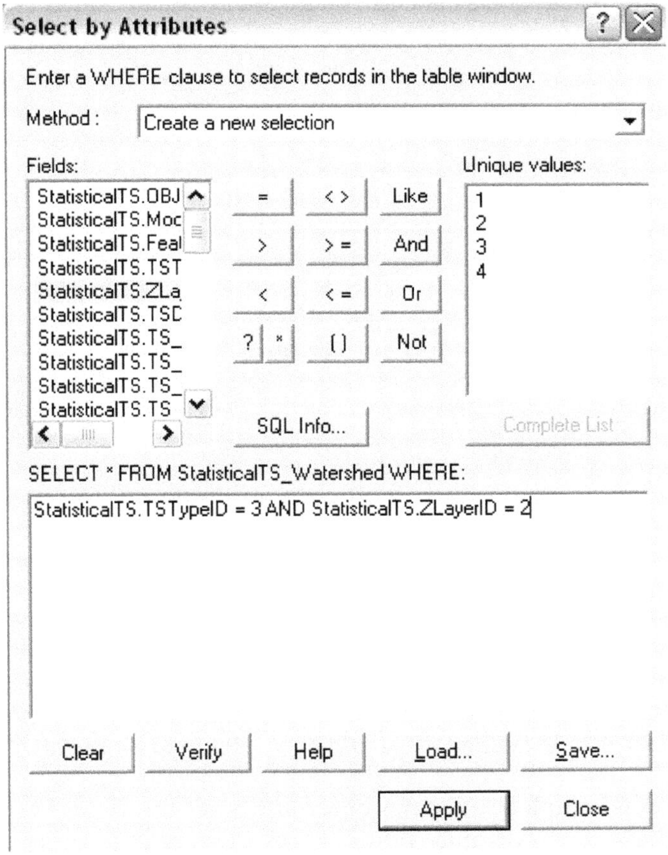

Select by Attributes
Enter a WHERE clause to select records in the table window.
Method : Create a new selection
Fields:
StatisticalTS.OBJ
StatisticalTS.Moc
StatisticalTS.Feal
StatisticalTS.TST
StatisticalTS.ZLa
StatisticalTS.TSC
StatisticalTS.TS_
StatisticalTS.TS_
StatisticalTS.TS_
StatisticalTS.TS
= <> Like
> >= And
< <= Or
? * () Not
SQL Info...
Unique values:
1
2
3
4
Complete List
SELECT * FROM StatisticalTS_Watershed WHERE:
StatisticalTS.TSTypeID = 3 AND StatisticalTS.ZLayerID = 2
Clear Verify Help Load... Save...
Apply Close

Figure A1.13
Querying a table in ArcGIS for desirable attributes.

one being more vegetation. We also have a soil sand percentage dataset, which measures the percentage of sand in soil in a location (soils also include clay and silt). Sandier soils are generally more permeable and well-drained. We want to look for the watersheds in the Illinois River Basin that have the most troublesome flooding characteristics in the month of July 1997.

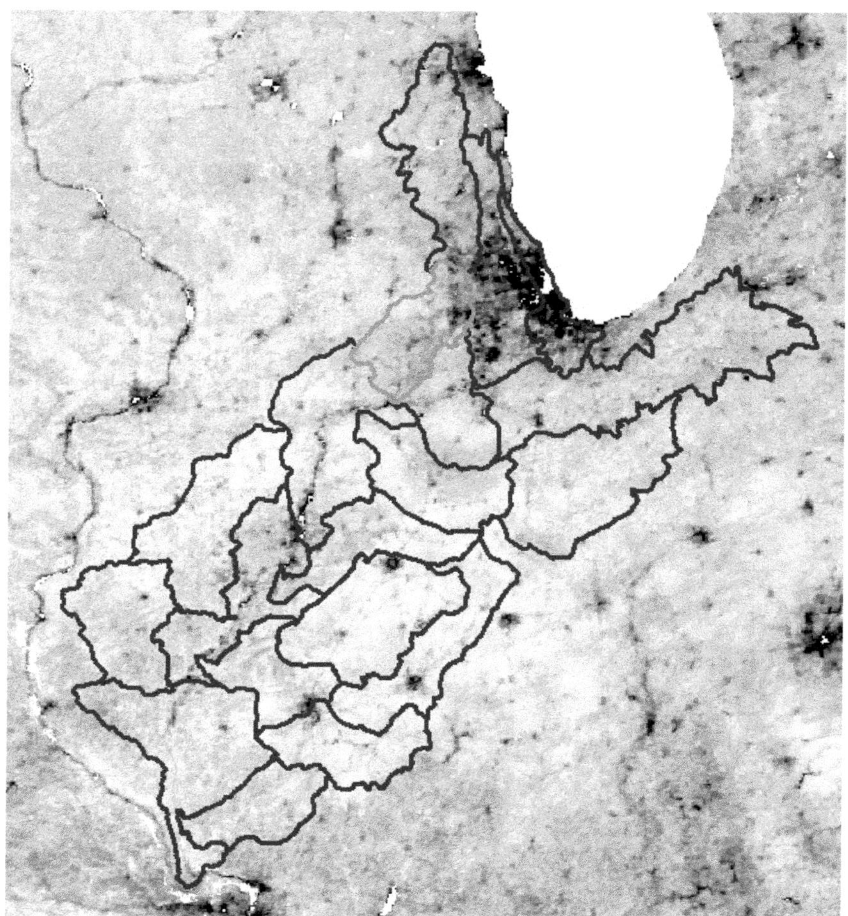

FIGURE A1.14
The Illinois River Basin and sub-basins overlaid on an NDVI dataset.

Join the Watershed and StatisticalTS attribute tables in ArcMap. This is accomplished by right-clicking on the "StatisticalTS" feature class in the Table of Contents, then selecting Joins and Relates → Join ..., and then filling out the form (see Figure A1.12). We can join tables based on an attribute field, a predefined relationship class, or spatial location. Join based on attributes, using the FeatureID field from StatisticalTS as the primary joining attribute, and select the Watershed table and HydroID field as the attribute on which to base the join.

Once this process is complete, open the joined attribute table by right clicking on StatisticalTS and selecting "Open Attribute Table." Many more records are now in the table; all the records from both Watershed and StatisticalTS, connected based on the HydroIDs of the watersheds. This information must be filtered based on attributes to find what we need — we are seeking watersheds with low NDVI and low sand fraction values. In the attribute table, hit the Options button and choose "Select by Attributes" (see Figure A1.13). Use the form to pick out only the records with at TSTypeID of 3 and ZLayer of 2 as shown below; these are the CONUS-SOIL layer 1 sand fraction records.

Push "Selected" in the table view window to show only the selected records matching our criteria. Right-click on the StatisticalTS.TS_MEAN column heading and select "sort ascending." Note that the least sandy soils are found in the Vermillion, South

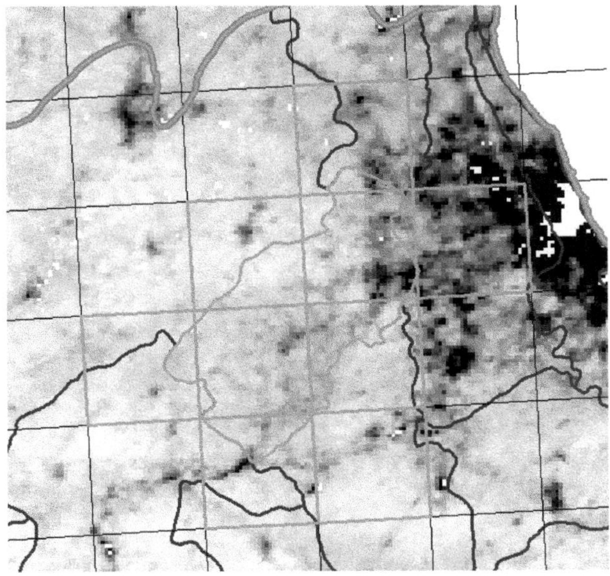

FIGURE A1.15
Querying AreaLink relationships to select overlapping areas in ArcGIS.

Fork Sangamon, and Lower Fox watersheds, with CONUS-SOIL layer 1 sand percentages of 14.7, 15.6, and 16.8, respectively. Roughly half of the watersheds have sand percentages less than 20%, so we will select this as our query criteria. Run the "Select by Attributes" form once more. Select all records from July 1997, then sort those records by StatisticalTS.TS_MEAN. The Chicago, Des Plaines, Upper Fox and Lower Fox are the four least vegetated watersheds in July 1997, with NDVI values of 0.44, 0.49, 0.59, and 0.59, respectively. Our query threshold here might be roughly 0.60. The only watershed in the Illinois River Basin matching our dual query criteria of 20% or less soil fraction and 0.60 or less NDVI for July 1997 is the Lower Fox River watershed. It looks like the Lower Fox is more susceptible to flooding than the rest of the basin, according to these criteria. Figure A1.14 shows the watersheds of the Illinois River Basin, with the Lower Fox selected via query on its assocated attributes, and the 1997 NDVI image overlaid. In the NDVI image, darker areas are less vegetated.

A1.3.2 Accessing Database Features by Traversing Relationships

A simple and similar problem involves selecting related features, rather than related data records. The principles involved are the same. Using joins and relates, one feature class may be selected or loaded into a program via its relational link to another feature class. One good example of this is in ArcHydro, where each Watershed is related via the data model to a watershed outlet HydroJunction. If we select a watershed, we can also select its network outlet because the outlet's HydroID is stored as an attribute of the Watershed. This example uses the Modelshed geodata model's AreaLink table to associate the Lower Fox River Watershed with the overlapping modelsheds of the 30 km climate grid and the Eco-Region zones. Because we built AreaLinks earlier in the tutorial, we can use them in this way. Open the AreaLink table in ArcMap and query using "Select By Attributes" all records with the Area1FeatureID field equal to 27020 (27020 is the Lower Fox River's HydroID).

For each queried record, the Area2FeatureID is the HydroID of an overlapping polygon feature. We know which feature class to find these in by looking at the ModelshedTypeID2 field of each record and then checking the ModelshedType table for that ModelshedTypeID. Add all the features overlapping the Lower Fox River Watershed to the selection in ArcMap (see Figure A1.15). This whole process may be automated using the ArcObjects and Visual Basic scripts.

Appendix 2

XSL Transformation File Example

Contents

A2.1 Original Schema HydroProject.xsd

```
<!-- edited with XMLSPY v2004 rel. 2 U (http://www.xmlspy.com) by Francina Dominguez
(University of Illinois) -->
<xs:schema xmlns:xs="http://www.w3.org/2001/XMLSchema" elementFormDefault="qualified"
attributeFormDefault="unqualified">
    <xs:element name="HydroProject">
        <xs:annotation>
            <xs:documentation>Contains data from the GLERL described with a logic similar to their original
design. </xs:documentation>
        </xs:annotation>
        <xs:complexType>
            <xs:sequence>
                <xs:element name="Description" type="xs:string"/>
                <xs:element name="HydroData">
                    <xs:complexType>
                        <xs:sequence>
                            <xs:element name="Geographic-Location" maxOccurs="unbounded">
                                <xs:complexType>
                                    <xs:sequence>
                                        <xs:element ref="basin-name"/>
                                        <xs:element name="area-description" maxOccurs="unbounded">
                                            <xs:complexType>
                                                <xs:sequence>
                                                    <xs:element name="area-type" type="xs:string"/>
                                                    <xs:element name="area" type="xs:decimal"/>
                                                </xs:sequence>
                                            </xs:complexType>
                                        </xs:element>
                                        <xs:element name="shoreline-delineation">
                                            <xs:complexType>
                                                <xs:sequence>
                                                    <xs:element ref="external-file"/>
                                                </xs:sequence>
                                            </xs:complexType>
                                        </xs:element>
                                        <xs:element name="bounding-coordinates">
                                            <xs:complexType>
                                                <xs:sequence>
```

```
                                    <xs:element name="max-lat" type="lattype"/>
                                    <xs:element name="min-lat" type="lattype"/>
                                    <xs:element name="max-lon" type="longtype"/>
                                    <xs:element name="min-lon" type="longtype"/>
                                </xs:sequence>
                            </xs:complexType>
                        </xs:element>
                    </xs:sequence>
                </xs:complexType>
            </xs:element>
            <xs:element name="Gridded-Data" maxOccurs="unbounded">
                <xs:complexType>
                    <xs:sequence>
                        <xs:element ref="basin-name"/>
                        <xs:element name="hydro-data-type" type="xs:string"/>
                        <xs:element name="grid-data-description" type="xs:string"/>
                        <xs:element name="base-lat" type="lattype"/>
                        <xs:element name="base-lon" type="longtype"/>
                        <xs:element name="gridXdim" type="xs:string"/>
                        <xs:element name="gridYdim" type="xs:string"/>
                        <xs:element name="numXgrids" type="xs:nonNegativeInteger"/>
                        <xs:element name="numYgrids" type="xs:nonNegativeInteger"/>
                        <xs:element ref="external-file"/>
                    </xs:sequence>
                </xs:complexType>
            </xs:element>
            <xs:element name="Timeseries-Data" maxOccurs="unbounded">
                <xs:complexType>
                    <xs:sequence>
                        <xs:element ref="basin-name"/>
                        <xs:element name="hydro-data-type" type="xs:string"/>
                        <xs:element ref="data-description"/>
                        <xs:element name="timeseries-data" type="timeseries"/>
                        <xs:element name="statistics">
                            <xs:complexType>
                                <xs:sequence>
                                    <xs:element name="mean" type="xs:decimal"/>
                                    <xs:element name="stdev" type="xs:decimal"/>
                                </xs:sequence>
                            </xs:complexType>
                        </xs:element>
                    </xs:sequence>
                </xs:complexType>
            </xs:element>
            <xs:element name="Forecasting-Data" maxOccurs="unbounded">
                <xs:complexType>
                    <xs:sequence>
                        <xs:element ref="basin-name"/>
                        <xs:element name="quantile" type="xs:decimal"/>
                        <xs:element name="hydro-data-type" type="xs:string"/>
                        <xs:element ref="data-description"/>
                        <xs:element name="forecasting-data" type="timeseries"/>
                    </xs:sequence>
                </xs:complexType>
            </xs:element>
        </xs:sequence>
    </xs:complexType>
</xs:element>
</xs:sequence>
</xs:complexType>
</xs:element>
<xs:simpleType name="timeseries">
    <xs:list itemType="xs:decimal"/>
</xs:simpleType>
<xs:simpleType name="longtype">
    <xs:restriction base="xs:decimal">
        <xs:minInclusive value="-180"/>
        <xs:maxInclusive value="180"/>
    </xs:restriction>
</xs:simpleType>
<xs:simpleType name="lattype">
    <xs:restriction base="xs:decimal">
        <xs:minInclusive value="-90"/>
        <xs:maxInclusive value="90"/>
    </xs:restriction>
</xs:simpleType>
```

```
    <xs:element name="basin-name">
        <xs:complexType>
            <xs:simpleContent>
                <xs:extension base="xs:anySimpleType">
                    <xs:attribute name="basin-code"/>
                </xs:extension>
            </xs:simpleContent>
        </xs:complexType>
    </xs:element>
    <xs:element name="external-file">
        <xs:complexType>
            <xs:sequence>
                <xs:element name="ext-file-desc" type="xs:string"/>
                <xs:element name="units" type="xs:string"/>
                <xs:element name="filepath" type="xs:string"/>
            </xs:sequence>
        </xs:complexType>
    </xs:element>
    <xs:element name="data-description">
        <xs:complexType>
            <xs:sequence>
                <xs:element name="information" type="xs:string"/>
                <xs:element name="units" type="xs:string"/>
                <xs:element name="temporal-scale" type="xs:string"/>
                <xs:element name="start-date" type="xs:date"/>
                <xs:element name="end-date" type="xs:date"/>
                <xs:element name="duration" type="xs:duration"/>
                <xs:element name="model" type="xs:string" minOccurs="0"/>
            </xs:sequence>
        </xs:complexType>
    </xs:element>
</xs:schema>
```

A2.2 Transformed Schema HydroProjectTransformed.xsd

```
<?xml version="1.0" encoding="UTF-8"?>
<!- edited with XMLSPY v2004 rel. 2 U (http://www.xmlspy.com) by Francina Dominguez
(University of Illinois) -->
<xs:schema xmlns:xs="http://www.w3.org/2001/XMLSchema" elementFormDefault="qualified"
attributeFormDefault="unqualified">
    <xs:element name="Basin-Data">
        <xs:annotation>
            <xs:documentation>Comment describing your root element</xs:documentation>
        </xs:annotation>
        <xs:complexType mixed="true">
            <xs:sequence>
                <xs:element name="introduction" type="xs:string"/>
                <xs:element name="Basin-Characterization">
                    <xs:complexType>
                        <xs:sequence>
                            <xs:element name="location">
                                <xs:complexType>
                                    <xs:sequence>
                                        <xs:element name="max-lat" type="lattype"/>
                                        <xs:element name="max-lon" type="longtype"/>
                                        <xs:element name="min-lat" type="lattype"/>
                                        <xs:element name="min-lon" type="longtype"/>
                                    </xs:sequence>
                                </xs:complexType>
                            </xs:element>
                            <xs:element name="area" maxOccurs="2">
                                <xs:complexType>
                                    <xs:simpleContent>
                                        <xs:extension base="xs:decimal">
                                            <xs:attribute name="area-type" type="xs:string"/>
                                        </xs:extension>
                                    </xs:simpleContent>
                                </xs:complexType>
                            </xs:element>
                        </xs:sequence>
                    </xs:complexType>
```

```
            </xs:element>
            <xs:element name="Physical-Data">
                <xs:complexType>
                    <xs:sequence>
                        <xs:element name="Gridded-Data" maxOccurs="unbounded">
                            <xs:complexType>
                                <xs:sequence>
                                    <xs:element name="data-information" type="xs:string"/>
                                    <xs:element name="units" type="xs:string"/>
                                    <xs:element name="grid-description">
                                        <xs:complexType>
                                            <xs:sequence>
                                                <xs:element name="base-lat" type="lattype"/>
                                                <xs:element name="base-lon" type="longtype"/>
                                                <xs:element name="gridXdim" type="xs:string"/>
                                                <xs:element name="gridYdim" type="xs:string"/>
                                                <xs:element name="gridXnum" type="xs:nonNegativeInteger"/>
                                                <xs:element name="gridYnum" type="xs:nonNegativeInteger"/>
                                            </xs:sequence>
                                        </xs:complexType>
                                    </xs:element>
                                    <xs:element name="file-description" type="xs:string"/>
                                    <xs:element name="filepath" type="xs:string"/>
                                </xs:sequence>
                                <xs:attribute name="basin" type="xs:string"/>
                                <xs:attribute name="hydro-data-type" type="xs:string"/>
                            </xs:complexType>
                        </xs:element>
                    </xs:sequence>
                </xs:complexType>
            </xs:element>
            <xs:element name="Hydrologic-Data">
                <xs:complexType>
                    <xs:sequence>
                        <xs:element name="Historical">
                            <xs:complexType>
                                <xs:sequence>
                                    <xs:element name="timeseries-data" maxOccurs="unbounded">
                                        <xs:complexType>
                                            <xs:sequence>
                                                <xs:element ref="data-description"/>
                                                <xs:element name="historical-data" type="data"/>
                                            </xs:sequence>
                                            <xs:attribute name="basin" type="xs:string"/>
                                            <xs:attribute name="hydro-data-type" type="xs:string"/>
                                        </xs:complexType>
                                    </xs:element>
                                </xs:sequence>
                            </xs:complexType>
                        </xs:element>
                        <xs:element name="Forecast">
                            <xs:complexType>
                                <xs:sequence>
                                    <xs:element name="forecasted-data" maxOccurs="unbounded">
                                        <xs:complexType>
                                            <xs:sequence>
                                                <xs:element ref="data-description"/>
                                                <xs:element name="forecast-data" type="data"/>
                                            </xs:sequence>
                                            <xs:attribute name="basin" type="xs:string"/>
                                            <xs:attribute name="hydro-data-type" type="xs:string"/>
                                            <xs:attribute name="quantile" type="xs:decimal"/>
                                        </xs:complexType>
                                    </xs:element>
                                </xs:sequence>
                            </xs:complexType>
                        </xs:element>
                    </xs:sequence>
                </xs:complexType>
            </xs:element>
        </xs:sequence>
    </xs:complexType>
</xs:element>
<xs:simpleType name="data">
    <xs:list itemType="xs:decimal"/>
</xs:simpleType>
```

```
    <xs:simpleType name="longtype">
        <xs:restriction base="xs:decimal">
            <xs:minInclusive value="-180"/>
            <xs:maxInclusive value="180"/>
        </xs:restriction>
    </xs:simpleType>
    <xs:simpleType name="lattype">
        <xs:restriction base="xs:decimal">
            <xs:minInclusive value="-90"/>
            <xs:maxInclusive value="90"/>
        </xs:restriction>
    </xs:simpleType>
    <xs:element name="data-description">
        <xs:complexType>
            <xs:sequence>
                <xs:element name="data-information" type="xs:string"/>
                <xs:element name="units" type="xs:string"/>
                <xs:element name="temporal-scale" type="xs:string"/>
                <xs:element name="start-date" type="xs:date"/>
                <xs:element name="end-date" type="xs:date"/>
                <xs:element name="duration" type="xs:duration"/>
            </xs:sequence>
        </xs:complexType>
    </xs:element>
</xs:schema>
```

A2.3 Transformation.xsl

```
<?xml version="1.0" encoding="utf-8"?>
<xsl:stylesheet version="2.0" xmlns:xsl="http://www.w3.org/1999/XSL/Transform">
    <!-- This transform creates an xml file with different hydrologic variables for -->
    <!-- each basin. It groups the information in a form more suitable for hydrologic modeling. -->
    <!-- Level one. -->
    <xsl:template match="HydroProject" xml:space="preserve">
        <Basin-Data>
            <introduction>This transformed xml file selects data from the xml file HydroProject.xml that
applies specifically to Lake Superior. This is useful when
            modeling only one basin. The data is organized in a different way from the original data.
            </introduction>
            <xsl:apply-templates/>
        </Basin-Data>
    </xsl:template>
    <!-- Level two. -->
    <xsl:template match="HydroData" xml:space="preserve">
        <Basin-Characterization>
            <xsl:apply-templates select="Geographic-Location">
            </xsl:apply-templates>
        </Basin-Characterization>
        <Physical-Data>
            <xsl:apply-templates select="Gridded-Data">
            </xsl:apply-templates>
        </Physical-Data>
        <Hydrologic-Data>
            <Historical>
                <xsl:apply-templates select="Timeseries-Data">
                </xsl:apply-templates>
            </Historical>
            <Forecast>
                <xsl:apply-templates select="Forecasting-Data">
                </xsl:apply-templates>
            </Forecast>
        </Hydrologic-Data>
    </xsl:template>
    <!-- Level three. -->
    <xsl:template match="Geographic-Location" xml:space="preserve">
        <xsl:if test="basin-name[@basin-code = 1]">
            <location>
                <max-lat>
                    <xsl:value-of select="bounding-coordinates/max-lat"/>
                </max-lat>
                <max-lon>
```

```
                <xsl:value-of select="bounding-coordinates/max-lon"/>
            </max-lon>
            <min-lat>
                <xsl:value-of select="bounding-coordinates/min-lat"/>
            </min-lat>
            <min-lon>
                <xsl:value-of select="bounding-coordinates/min-lon"/>
            </min-lon>
        </location>
        <xsl:apply-templates select="area-description">
        </xsl:apply-templates>
    </xsl:if>
</xsl:template>
<xsl:template match="Gridded-Data" xml:space="preserve">
    <xsl:if test="basin-name[@basin-code = 1]">
        <Gridded-Data><xsl:attribute name="basin"><xsl:value-of select="basin-name"/></xsl:attribute><xsl:
attribute name="hydro-data-type"><xsl:value-of select="hydro-data-type"/></xsl:attribute>
            <data-information>
                <xsl:value-of select="grid-data-description"/>
            </data-information>
            <units>
                <xsl:value-of select="hydro-data-type"/>
            </units>
            <grid-description>
                <base-lat>
                    <xsl:value-of select="base-lat"/>
                </base-lat>
                <base-lon>
                    <xsl:value-of select="base-lon"/>
                </base-lon>
                <gridXdim>
                    <xsl:value-of select="gridXdim"/>
                </gridXdim>
                <gridYdim>
                    <xsl:value-of select="gridYdim"/>
                </gridYdim>
                <gridXnum>
                    <xsl:value-of select="numXgrids"/>
                </gridXnum>
                <gridYnum>
                    <xsl:value-of select="numYgrids"/>
                </gridYnum>
            </grid-description>
            <file-description>
                <xsl:value-of select="external-file/ext-file-desc"/>
            </file-description>
            <filepath>
                <xsl:value-of select="external-file/filepath"/>
            </filepath>
        </Gridded-Data>
    </xsl:if>
</xsl:template>
<xsl:template match="Timeseries-Data" xml:space="preserve">
    <xsl:if test="basin-name[@basin-code = 1]">
        <timeseries-data><xsl:attribute name="basin"><xsl:value-of select="basin-name"/>
</xsl:attribute><xsl:attribute name="hydro-data-type"><xsl:value-of select="hydro-data-type"/></xsl:attribute>
            <data-description>
                <data-information>
                    <xsl:value-of select="data-description/information"/>
                </data-information>
                <units>
                    <xsl:value-of select="data-description/units"/>
                </units>
                <temporal-scale>
                    <xsl:value-of select="data-description/temporal-scale"/>
                </temporal-scale>
                <start-date>
                    <xsl:value-of select="data-description/start-date"/>
                </start-date>
                <end-date>
                    <xsl:value-of select="data-description/end-date"/>
                </end-date>
                <duration>
                    <xsl:value-of select="data-description/duration"/>
                </duration>
            </data-description>
```

```
                <historical-data>
                    <xsl:value-of select="timeseries-data"/>
                </historical-data>
            </timeseries-data>
        </xsl:if>
    </xsl:template>
    <xsl:template match="Forecasting-Data" xml:space="preserve">
        <xsl:if test="basin-name[@basin-code = 1]">
            <forecasted-data><xsl:attribute name="basin"><xsl:value-of select="basin-
name"/></xsl:attribute><xsl:attribute name="hydro-data-type"><xsl:value-of select="hydro-data-
type"/></xsl:attribute><xsl:attribute name="quantile"><xsl:value-of select="quantile"/></xsl:attribute>
                <data-description>
                    <data-information>
                        <xsl:value-of select="data-description/information"/>
                    </data-information>
                    <units>
                        <xsl:value-of select="data-description/units"/>
                    </units>
                    <temporal-scale>
                        <xsl:value-of select="data-description/temporal-scale"/>
                    </temporal-scale>
                    <start-date>
                        <xsl:value-of select="data-description/start-date"/>
                    </start-date>
                    <end-date>
                        <xsl:value-of select="data-description/end-date"/>
                    </end-date>
                    <duration>
                        <xsl:value-of select="data-description/duration"/>
                    </duration>
                </data-description>
                <forecast-data>
                    <xsl:value-of select="forecasting-data"/>
                </forecast-data>
            </forecasted-data>
        </xsl:if>
    </xsl:template>
    <!-- Level four. -->
    <xsl:template match="area-description" xml:space="preserve">
        <area><xsl:attribute name="area-type"><xsl:value-of select="area-type"/></xsl:attribute><xsl:value-of
select="area"/>
        </area>
    </xsl:template>
</xsl:stylesheet>
```

A2.4 Original XML: HydroProject.xml

```
  <?xml version="1.0" encoding="UTF-8" ?>
- <HydroProject xmlns:xsi="http://www.w3.org/2001/XMLSchema-instance"
    xsi:noNamespaceSchemaLocation="C:\ogre\examples\HydroProject.xsd">
  <Description>Hydrologic data for the Great Lakes obtained from the Great Lakes
    Environmental Research Laboratory through their Web site
    http://www.glerl.noaa.gov/data/. It contains physical data, as well as historic and
    forecasted hydrologic variables. In order to reduce the length of this file, only the data
    from Lake Superior and Lake Michigan are described. Each basin is assigned a code Lake
    Superior=1, Lake Michigan=2.</Description>
- <HydroData>
- <Geographic-Location>
  <basin-name basin-code="1">Lake Superior</basin-name>
- <area-description>
  <area-type>Lake Area</area-type>
  <area>82100</area>
  </area-description>
- <area-description>
  <area-type>Land Area</area-type>
  <area>128000</area>
  </area-description>
- <shoreline-delineation>
- <external-file>
  <ext-file-desc>The shoreline deliniation has been digitized from the National Ocean Survey
    1:400,000 charts. The average segment length is 1.24 km. Each line endpoint's coordinates
```

```
    are ordered in pairs and form a closed loop.</ext-file-desc>
  <units>latitude-longitude</units>
  <filepath>ftp://ftp.glerl.noaa.gov/bathy/shores.dat</filepath>
  </external-file>
  </shoreline-delineation>
- <bounding-coordinates>
  <max-lat>49.02</max-lat>
  <min-lat>46.07</min-lat>
  <max-lon>-84.35</max-lon>
  <min-lon>-89.58</min-lon>
  </bounding-coordinates>
  </Geographic-Location>
- <Gridded-Data>
  <basin-name basin-code="1">Lake Superior</basin-name>
  <hydro-data-type>Bathymetric Data</hydro-data-type>
  <grid-data-description>Bathymetric data for Lake Superior on a 2-km grid as described in
   Robertson and Jordan (unpublished) was obtained from the Canada Center for Inland
   Waters. They superimposed 2-km grids on standard bathymetric charts and averaged a
   mean depth in each grid square by eye. Depths are relative to the Great Lakes Datum of
   1955. All grids are aligned with the central meridian of the bathymetric chart. The datum is
   182.88 m.</grid-data-description>
  <base-lat>46.32</base-lat>
  <base-lon>-92.10</base-lon>
  <gridXdim>2km</gridXdim>
  <gridYdim>2km</gridYdim>
  <numXgrids>304</numXgrids>
  <numYgrids>147</numYgrids>
- <external-file>
  <ext-file-desc>The depths on bathymetric data fill the batymetric grid from west to east and
    then south to north. Lines 3-4 describe the transformation values from geographic to map
    coordinates : x = a Dlon + b Dlat + c Dlon Dlat + d Dlon**2 and y = e Dlon + f Dlat + g Dlon
    Dlat + h Dlon**2 lines 5-6 describe the transformation from map to geographic coordinates
    : Dlon = Ax + By + Cxy + Dx**2 and Dlat = Ex + Fy + Gxy + Hx**2.</ext-file-desc>
  <units>m</units>
  <filepath>ftp://ftp.glerl.noaa.gov/bathy/superior.dat</filepath>
  </external-file>
  </Gridded-Data>
- <Gridded-Data>
  <basin-name basin-code="2">Lake Michigan</basin-name>
  <hydro-data-type>Bathymetric Data</hydro-data-type>
  <grid-data-description>For Lake Michigan, a 2-km grid aligned with the 86.50° meridian was
    defined. A computer subroutine was written to interpolate the 2-min grid depths of Hughes
    et al. to arbitrary latitude and longitude by bilinear interpolation within each 2-min square.
    These data were used to calculate the average depth in each 2-km grid square by
    averaging the depth at the center of the square and the four depths at coordinates (0.5,0.5),
    (0.5,-0.5), (-0.5,-0.5), and (-0.5,0.5) km relative to the center of the square. Depths at some grid
    squares near the shoreline were adjusted by hand so that they better fit the actual lake
    shoreline. The data for Lake Michigan were then put into the same format as those for the
    rest of the lakes. The datum is 175.81 m.</grid-data-description>
  <base-lat>41.61</base-lat>
  <base-lon>-87.94</base-lon>
  <gridXdim>2km</gridXdim>
  <gridYdim>2km</gridYdim>
  <numXgrids>160</numXgrids>
  <numYgrids>250</numYgrids>
- <external-file>
  <ext-file-desc>The depths on bathymetric data fill the batymetric grid from west to east and
    then south to north. Lines 3-4 describe the transformation values from geographic to map
    coordinates : x = a Dlon + b Dlat + c Dlon Dlat + d Dlon**2 and y = e Dlon + f Dlat + g Dlon
    Dlat + h Dlon**2 lines 5-6 describe the transformation from map to geographic coordinates
    : Dlon = Ax + By + Cxy + Dx**2 and Dlat = Ex + Fy + Gxy + Hx**2.</ext-file-desc>
  <units>m</units>
  <filepath>ftp://ftp.glerl.noaa.gov/bathy/michigan.dat</filepath>
  </external-file>
  </Gridded-Data>
- <Timeseries-Data>
  <basin-name basin-code="1">Lake Superior</basin-name>
  <hydro-data-type>Lake Precipitation</hydro-data-type>
- <data-description>
  <information>We computed 1948-1999 monthly precipitation from all available daily data from
    stations in the basin or within approximately 0-30 km of the basin, depending upon
    station density near the edge of the basin. The distance was chosen to assure that we
    obtain the same nonzero Thiessen weights as if no stations were eliminated. Station data
    for the United States were obtained from the National Climatic Data Center (2000), and station data
    for Canada were obtained from the Meteorological Service of Canada (1999). We then used
    Thiessen weighting (Croley and Hartmann, 1985) similar to the above, but defined for a
```

1-km grid and recomputed every day for all watersheds within the basin and the lake. The
basin watersheds were defined by the U.S. Geological Survey and the Water Survey of
Canada. We constructed daily over-land basin precipitation by areally weighting the
areally averaged daily precipitation for each watershed and summing over the days in each
month. We also constructed total monthly over-lake precipitation by summing the areally
averaged daily values for the lake surface.</information>
<units>mm</units>
<temporal-scale>total monthly</temporal-scale>
<start-date>1990-01-01</start-date>
<end-date>1999-01-01</end-date>
<duration>P10Y</duration>
</data-description>
<timeseries-data>44.8 32.3 37.9 56.1 59.9 114.8 66.7 44.9 99.6 108.0 45.3 46.4 55.2 23.2
60.7 69.1 84.4 75.5 105.8 39.6 113.1 98.8 107.9 46.3 46.1 35.9 22.1 55.7 61.9 55.9 118.3
86.4 115.3 57.9 67.3 80.9 53.7 14.8 18.2 72.6 104.3 78.1 108.1 68.3 77.3 73.0 51.9 38.3
58.5 15.4 38.5 69.6 75.2 77.0 74.4 102.6 76.5 55.7 59.3 14.5 41.3 49.5 38.4 49.4 88.7 26.7
113.1 86.9 115.9 137.6 67.4 74.3 91.4 54.1 36.7 70.2 45.6 107.8 116.1 80.8 86.3 97.1 69.7
77.3 99.8 20.2 64.6 26.9 69.0 75.3 72.0 53.5 60.6 75.2 58.4 28.6 46.0 18.2 68.5 25.6 52.0
92.0 56.6 64.9 78.0 81.6 82.6 62.1 75.9 51.7 22.3 39.1 137.0 77.4 129.1 84.4 96.8 90.4
41.6 39.3</timeseries-data>
- <statistics>
<mean>67.1</mean>
<stdev>28.43</stdev>
</statistics>
</Timeseries-Data>
- <Timeseries-Data>
<basin-name basin-code="2">Lake Michigan</basin-name>
<hydro-data-type>Lake Precipitation</hydro-data-type>
- <data-description>
<information>We computed 1948-1999 monthly precipitation from all available daily data from
stations in the basin or within approximately 0-30 km of the basin, depending upon
station density near the edge of the basin. The distance was chosen to assure that we
obtain the same nonzero Thiessen weights as if no stations were eliminated. Station data
for the United States were obtained from the National Climatic Data Center (2000), and station data
for Canada were obtained from the Meteorological Service of Canada (1999). We then used
Thiessen weighting (Croley and Hartmann, 1985) similar to the above, but defined for a
1-km grid and recomputed every day for all watersheds within the basin and the lake. The
basin watersheds were defined by the U.S. Geological Survey and the Water Survey of
Canada. We constructed daily over-land basin precipitation by areally weighting the
areally averaged daily precipitation for each watershed and summing over the days in each
month. We also constructed total monthly over-lake precipitation by summing the areally
averaged daily values for the lake surface.</information>
<units>mm</units>
<temporal-scale>total monthly</temporal-scale>
<start-date>1990-01-01</start-date>
<end-date>1999-01-01</end-date>
<duration>P10Y</duration>
</data-description>
<timeseries-data>57.4 37.6 68.4 46.9 125.5 150.0 63.4 95.4 99.1 98.0 91.7 53.0 29.5 16.2
85.1 93.7 87.7 45.0 113.2 50.5 87.7 158.0 82.5 49.2 34.2 34.9 53.4 71.6 27.8 53.8 96.4
67.5 121.8 54.2 137.9 64.9 59.9 19.5 33.0 131.3 72.4 150.3 95.0 100.3 107.1 53.5 58.0
29.3 62.7 57.5 28.3 75.3 38.6 81.1 103.3 105.6 79.7 45.9 89.8 20.2 61.5 22.2 51.2 94.7
75.9 37.8 72.0 108.2 52.7 116.6 82.3 41.7 57.9 33.2 34.7 83.8 83.8 155.3 87.7 45.6 78.5
91.8 37.9 62.1 76.4 65.3 39.8 30.5 89.7 91.9 68.4 105.5 63.6 49.3 45.5 33.1 72.3 28.5
103.1 76.7 62.2 95.7 43.7 107.4 69.8 79.3 68.1 37.5 92.4 32.5 14.9 92.2 89.6 99.4 117.1
62.7 66.2 34.6 23.7 62.5</timeseries-data>
- <statistics>
<mean>70.5</mean>
<stdev>31.93</stdev>
</statistics>
</Timeseries-Data>
- <Timeseries-Data>
<basin-name basin-code="1">Lake Superior</basin-name>
<hydro-data-type>Land Precipitation</hydro-data-type>
- <data-description>
<information>We computed 1948-1999 monthly precipitation from all available daily data from
stations in the basin or within approximately 0-30 km of the basin, depending upon
station density near the edge of the basin. The distance was chosen to assure that we
obtain the same nonzero Thiessen weights as if no stations were eliminated. Station data
for the United States were obtained from the National Climatic Data Center (2000), and station data
for Canada were obtained from the Meteorological Service of Canada (1999). We then used
Thiessen weighting (Croley and Hartmann, 1985) similar to the above, but defined for a
1-km grid and recomputed every day for all watersheds within the basin and the lake. The
basin watersheds were defined by the U.S. Geological Survey and the Water Survey of
Canada. We constructed daily over-land basin precipitation by areally weighting the
areally averaged daily precipitation for each watershed and summing over the days in each

```
    month. We also constructed total monthly over-lake precipitation by summing the areally
    averaged daily values for the lake surface.</information>
  <units>mm</units>
  <temporal-scale>total monthly</temporal-scale>
  <start-date>1990-01-01</start-date>
  <end-date>1999-01-01</end-date>
  <duration>P10Y</duration>
  </data-description>
  <timeseries-data>57.0 36.7 39.3 61.0 60.1 130.6 80.1 63.7 93.4 105.5 48.0 52.3 49.4 20.3
    54.1 57.3 89.2 80.1 106.9 47.4 124.8 94.7 95.3 45.8 36.7 34.0 21.2 58.7 69.0 64.9 119.6
    107.7 135.4 50.9 60.1 76.9 44.0 10.0 19.1 76.0 103.3 87.7 140.4 90.4 101.1 74.8 51.5 34.4
    49.0 17.6 33.4 73.5 69.3 89.5 115.4 100.0 79.6 61.6 66.4 20.3 45.3 48.9 32.3 45.6 84.9
    44.6 112.1 81.7 131.1 145.6 69.3 61.1 101.2 62.4 29.3 68.3 52.5 85.8 139.2 79.4 91.1
    113.4 70.1 76.5 84.5 21.3 56.9 31.1 65.3 91.5 80.2 54.2 66.6 81.4 65.7 29.6 44.6 18.6 59.6
    30.9 53.6 87 72.9 90.8 91.4 108.6 70.1 50.8 50.1 54.7 28.9 38.3 115.3 89.8 150.1 93.6
    119.4 98.5 35.2 38.3</timeseries-data>
- <statistics>
  <mean>70.2</mean>
  <stdev>31.62</stdev>
  </statistics>
  </Timeseries-Data>
- <Timeseries-Data>
  <basin-name basin-code="2">Lake Michigan</basin-name>
  <hydro-data-type>Land Precipitation</hydro-data-type>
- <data-description>
  <information>We computed 1948-1999 monthly precipitation from all available daily data from
    stations in the basin or within approximately 0-30 km of the basin, depending upon
    station density near the edge of the basin. The distance was chosen to assure that we
    obtain the same nonzero Thiessen weights as if no stations were eliminated. Station data
    for the United States were obtained from the National Climatic Data Center (2000), and station data
    for Canada were obtained from the Meteorological Service of Canada (1999). We then used
    Thiessen weighting (Croley and Hartmann, 1985) similar to the above, but defined for a
    1-km grid and recomputed every day for all watersheds within the basin and the lake. The
    basin watersheds were defined by the U.S. Geological Survey and the Water Survey of
    Canada. We constructed daily over-land basin precipitation by areally weighting the
    areally averaged daily precipitation for each watershed and summing over the days in each
    month. We also constructed total monthly over-lake precipitation by summing the areally
    averaged daily values for the lake surface.</information>
  <units>mm</units>
  <temporal-scale>total monthly</temporal-scale>
  <start-date>1990-01-01</start-date>
  <end-date>1999-01-01</end-date>
  <duration>P10Y</duration>
  </data-description>
  <timeseries-data>44.8 43.1 59.5 49.5 120.5 138.0 74.4 97.3 103.7 101.6 80.9 57.8 28.2 18.8
    79.9 100.8 97.2 60.0 121.6 63.1 84.8 148.4 93.7 46.6 32.6 31.3 55.3 73.4 41.7 50.7 114.4
    69.7 137.5 51.7 122.9 64.3 58.5 19.0 28.4 111.9 85.5 154.5 97.4 98.5 109.7 65.2 49.3 24.6
    46.7 36.0 30.3 81.4 44.9 93.0 128.6 114.6 81.4 53.5 84.4 21.3 47.7 21.8 47.0 78.3 80.0
    58.0 78.8 133.5 52.6 113.2 81.7 37.9 63.3 33.9 34.7 87.3 71.0 151.1 91.3 57.3 71.0 85.9
    47.6 66.2 76.8 57.4 52.6 27.9 93.3 90.5 79.8 108.6 76.7 54.8 35.2 28.9 72.5 33.4 94.8 65.3
    57.9 97.8 45.6 92.5 66.3 73.8 56.1 33.2 84.2 37.0 20.8 93.1 94.5 105.5 147.2 66.7 58.2
    42.4 29.0 48.3</timeseries-data>
- <statistics>
  <mean>71.4</mean>
  <stdev>32.45</stdev>
  </statistics>
  </Timeseries-Data>
- <Timeseries-Data>
  <basin-name basin-code="1">Lake Superior</basin-name>
  <hydro-data-type>Lake Levels</hydro-data-type>
- <data-description>
  <information>Beginning of month lake levels. The beginning-of-month level for a gage is
    defined as the level at 12:00:00 a.m. (midnight) on the first day of that month. Since this
    value is generally unknown for practical reasons, it is standard practice to compute the
    value by averaging the daily mean level for the last day of the previous month with the
    daily mean level for the first day of the current month, when both are available. When one
    or both of the daily mean values are unavailable, several alternative methods for filling in
    the missing data are available.</information>
  <units>m</units>
  <temporal-scale>average annual</temporal-scale>
  <start-date>1990-01-01</start-date>
  <end-date>1999-01-01</end-date>
  <duration>P10Y</duration>
  </data-description>
  <timeseries-data>183.25 183.34 183.39 183.46 183.45 183.36 183.54 183.60 183.33
    183.26</timeseries-data>
```

```
- <statistics>
  <mean>183.4</mean>
  <stdev>0.11</stdev>
  </statistics>
  </Timeseries-Data>
- <Timeseries-Data>
  <basin-name basin-code="2">Lake Michigan</basin-name>
  <hydro-data-type>Lake Levels</hydro-data-type>
- <data-description>
  <information>Beginning of month lake levels. The beginning-of-month level for a gage is
    defined as the level at 12:00:00 a.m. (midnight) on the first day of that month. Since this
    value is generally unknown for practical reasons, it is standard practice to compute the
    value by averaging the daily mean level for the last day of the previous month with the
    daily mean level for the first day of the current month, when both are available. When one
    or both of the daily mean values are unavailable, several alternative methods for filling in
    the missing data are available.</information>
  <units>m</units>
  <temporal-scale>average annual</temporal-scale>
  <start-date>1990-01-01</start-date>
  <end-date>1999-01-01</end-date>
  <duration>P10Y</duration>
  </data-description>
  <timeseries-data>176.34 176.46 176.47 176.68 176.66 176.53 176.65 176.97 176.73
    176.24</timeseries-data>
- <statistics>
  <mean>176.59</mean>
  <stdev>0.22</stdev>
  </statistics>
  </Timeseries-Data>
- <Timeseries-Data>
  <basin-name basin-code="1">Lake Superior</basin-name>
  <hydro-data-type>Evaporation</hydro-data-type>
- <data-description>
  <information>Monthly evaporation estimates (spreadsheet entitled evaporation.xls) were
    derived from daily evaporation estimates generated by the Great Lakes Evaporation Model
    (Croley, 1989a,b, 1992; Croley and Assel, 1994). This is a lumped-parameter surface flux
    and heat-storage model. It uses areal-average daily air temperature, windspeed, humidity,
    precipitation, and cloudcover. These data are sufficiently available since 1948 (1953 for
    Georgian Bay), and 2 years are used for model initialization. Over-land data are adjusted
    for over-water or over-ice conditions. Surface flux processes are represented for short-
    wave radiation and reflection, net long-wave radiation exchange, and advection.
    Atmospheric stability effects on the bulk transfer coefficients are formulated and used with
    the aerodynamic equation for sensible and latent heat surface fluxes.</information>
  <units>mm</units>
  <temporal-scale>total monthly</temporal-scale>
  <start-date>1990-01-01</start-date>
  <end-date>1999-01-01</end-date>
  <duration>P10Y</duration>
  </data-description>
  <timeseries-data>74.09 73.43 50.38 26.31 2.75 -4.81 -0.98 16.63 53.63 74.76 90.51 124.96
    110.21 56.74 50.02 13.5 0.49 -3.8 6.92 20.85 80.28 74.8 109.48 108 93.09 64.59 61.1
    26.94 3.72 3.08 8.27 29.83 55.61 69.22 91.26 113.68 103 80.63 49.26 23.26 0.67 -3.87
    -1.5 9.34 73.27 86.19 105.95 98.58 126.17 46.37 26.71 19.16 0.94 -4.71 -4.2 19.29 29.58
    57.65 100.67 86.04 111.16 98.29 45.65 31.9 -0.07 -4.74 -0.78 12.72 59.96 65.07 127.01
    124.19 131.82 47.29 37 22.14 5.17 -3.99 -5.22 -1.51 33.57 69.84 112.91 110.23 91.73
    34.59 32.99 19.87 4.28 -3.84 -3.36 8.21 35.18 67.32 86.53 80.05 83.96 29.92 59.61 13.83
    2.09 4.46 20.13 30.96 60.81 73.88 91.02 111.11 113.04 64.28 51.26 13.35 7.32 12.16
    10.48 50.76 64.38 91.08 84.89 106.8</timeseries-data>
- <statistics>
  <mean>49.11</mean>
  <stdev>40.86</stdev>
  </statistics>
  </Timeseries-Data>
- <Timeseries-Data>
  <basin-name basin-code="2">Lake Michigan</basin-name>
  <hydro-data-type>Evaporation</hydro-data-type>
- <data-description>
  <information>Monthly evaporation estimates (spreadsheet entitled evaporation.xls) were
    derived from daily evaporation estimates generated by the Great Lakes Evaporation Model
    (Croley, 1989a,b, 1992; Croley and Assel, 1994). This is a lumped-parameter surface flux
    and heat-storage model. It uses areal-average daily air temperature, windspeed, humidity,
    precipitation, and cloudcover. These data are sufficiently available since 1948 (1953 for
    Georgian Bay), and 2 years are used for model initialization. Over-land data are adjusted
    for over-water or over-ice conditions. Surface flux processes are represented for short-
    wave radiation and reflection, net long-wave radiation exchange, and advection.
    Atmospheric stability effects on the bulk transfer coefficients are formulated and used with
```

```
    the aerodynamic equation for sensible and latent heat surface fluxes.</information>
  <units>mm</units>
  <temporal-scale>total monthly</temporal-scale>
  <start-date>1990-01-01</start-date>
  <end-date>1999-01-01</end-date>
  <duration>P10Y</duration>
  </data-description>
  <timeseries-data>54.15 56.65 37.61 18.06 4.28 1.79 15.18 37.44 83.08 95.97 91.99 115.33
    101.27 47.45 35.31 8.77 0.44 7.99 33.32 45.74 111.50 86.13 113.52 99.87 82.10 52.50
    50.12 18.85 8.10 17.58 23.23 53.92 72.74 90.32 91.07 105.63 88.27 69.86 37.28 13.29
    1.69 1.31 13.49 31.20 103.09 104.06 100.14 96.30 108.10 40.85 18.88 10.46 1.04 -0.22
    8.70 44.30 57.94 80.50 105.91 81.89 95.51 83.08 31.80 23.73 1.66 2.03 16.61 41.13 99.43
    88.05 133.75 119.22 96.00 43.02 38.89 14.72 3.06 -2.94 9.79 22.75 70.23 89.25 113.81
    91.23 295.24 39.73 33.19 17.99 4.78 1.63 9.36 41.23 59.35 88.40 98.97 88.93 278.78
    29.57 47.21 14.55 9.17 19.99 45.70 56.93 78.90 95.16 110.38 117.22 309.26 59.18 49.07
    15.71 15.01 22.12 27.81 77.91 90.19 107.03 101.26 115.02</timeseries-data>
- <statistics>
  <mean>59.88</mean>
  <stdev>53.79</stdev>
  </statistics>
  </Timeseries-Data>
- <Forecasting-Data>
  <basin-name basin-code="1">Lake Superior</basin-name>
  <quantile>95</quantile>
  <hydro-data-type>Lake Precipitation</hydro-data-type>
- <data-description>
  <information>Three NOAA agencies support estimates of present hydrology and long-term
    hydrological probabilities for each of the Great Lake Basins:NOAAs Great Lakes
    Environmental Research Laboratory (GLERL) , NOAAs Climate Prediction Center (CPC) and
    The Midwestern Climate Center (MCC).</information>
  <units>mm</units>
  <temporal-scale>monthly</temporal-scale>
  <start-date>2003-08-01</start-date>
  <end-date>2004-08-01</end-date>
  <duration>P1Y1M</duration>
  </data-description>
  <forecasting-data>23.94 48.00 70.28 64.80 46.86 34.70 39.40 67.15 111.77 135.78 76.69
    58.50 49.47</forecasting-data>
  </Forecasting-Data>
- <Forecasting-Data>
  <basin-name basin-code="1">Lake Superior</basin-name>
  <quantile>95</quantile>
  <hydro-data-type>Evaporation</hydro-data-type>
- <data-description>
  <information>Three NOAA agencies support estimates of present hydrology and long-term
    hydrological probabilities for each of the Great Lake Basins:NOAAs Great Lakes
    Environmental Research Laboratory (GLERL) , NOAAs Climate Prediction Center (CPC) and
    The Midwestern Climate Center (MCC).</information>
  <units>mm</units>
  <temporal-scale>monthly</temporal-scale>
  <start-date>2003-08-01</start-date>
  <end-date>2004-08-01</end-date>
  <duration>P1Y1M</duration>
  </data-description>
  <forecasting-data>6.56 56.52 83.73 113.47 129.41 122.06 80.94 63.40 25.90 7.93 0.12 5.33
    33.37</forecasting-data>
  </Forecasting-Data>
- <Forecasting-Data>
  <basin-name basin-code="1">Lake Superior</basin-name>
  <quantile>5</quantile>
  <hydro-data-type>Lake Precipitation</hydro-data-type>
- <data-description>
  <information>Three NOAA agencies support estimates of present hydrology and long-term
    hydrological probabilities for each of the Great Lake Basins:NOAAs Great Lakes
    Environmental Research Laboratory (GLERL) , NOAAs Climate Prediction Center (CPC) and
    The Midwestern Climate Center (MCC).</information>
  <units>mm</units>
  <temporal-scale>monthly</temporal-scale>
  <start-date>2003-08-01</start-date>
  <end-date>2004-08-01</end-date>
  <duration>P1Y1M</duration>
  </data-description>
  <forecasting-data>41.65 39.14 29.01 33.33 27.52 27.84 17.49 19.14 19.83 26.39 45.63 48.83
    47.68</forecasting-data>
  </Forecasting-Data>
- <Forecasting-Data>
```

```
  <basin-name basin-code="1">Lake Superior</basin-name>
  <quantile>5</quantile>
  <hydro-data-type>Evaporation</hydro-data-type>
- <data-description>
  <information>Three NOAA agencies support estimates of present hydrology and long-term
    hydrological probabilities for each of the Great Lake Basins:NOAAs Great Lakes
    Environmental Research Laboratory (GLERL), NOAAs Climate Prediction Center (CPC) and
    The Midwestern Climate Center (MCC).</information>
  <units>mm</units>
  <temporal-scale>monthly</temporal-scale>
  <start-date>2003-08-01</start-date>
  <end-date>2004-08-01</end-date>
  <duration>P1Y1M</duration>
  </data-description>
  <forecasting-data>-1.63 25.45 37.6 67.47 80.85 73.53 21.52 14.36 3.17 -3.11 -6.74 -6.11
    1.01</forecasting-data>
  </Forecasting-Data>
- <Forecasting-Data>
  <basin-name basin-code="2">Lake Michigan</basin-name>
  <quantile>95</quantile>
  <hydro-data-type>Lake Precipitation</hydro-data-type>
- <data-description>
  <information>Three NOAA agencies support estimates of present hydrology and long-term
    hydrological probabilities for each of the Great Lake Basins:NOAAs Great Lakes
    Environmental Research Laboratory (GLERL), NOAAs Climate Prediction Center (CPC) and
    The Midwestern Climate Center (MCC).</information>
  <units>mm</units>
  <temporal-scale>monthly</temporal-scale>
  <start-date>2003-08-01</start-date>
  <end-date>2004-08-01</end-date>
  <duration>P1Y1M</duration>
  </data-description>
  <forecasting-data>93.46 138.25 112.09 98.37 78.08 73.62 59.91 102.16 108.34 127.74 151.15
    132.26 141.09</forecasting-data>
  </Forecasting-Data>
- <Forecasting-Data>
  <basin-name basin-code="2">Lake Michigan</basin-name>
  <quantile>95</quantile>
  <hydro-data-type>Evaporation</hydro-data-type>
- <data-description>
  <information>Three NOAA agencies support estimates of present hydrology and long-term
    hydrological probabilities for each of the Great Lake Basins:NOAAs Great Lakes
    Environmental Research Laboratory (GLERL), NOAAs Climate Prediction Center (CPC) and
    The Midwestern Climate Center (MCC).</information>
  <units>mm</units>
  <temporal-scale>monthly</temporal-scale>
  <start-date>2003-08-01</start-date>
  <end-date>2004-08-01</end-date>
  <duration>P1Y1M</duration>
  </data-description>
  <forecasting-data>59.72 128.73 122.34 125.81 128.6 118.33 81.01 47.84 19.13 5.99 18.48
    34.42 87.89</forecasting-data>
  </Forecasting-Data>
  </HydroData>
  </HydroProject>
```


A2.5 Transformed XML: BasinData.xml


```
  <?xml version="1.0" encoding="UTF-8" ?>
- <Basin-Data>
  <introduction>This transformed xml file selects data from the xml file HydroProject.xml that
     applies specifically to Lake Superior. This is useful when modeling only one basin. The data
     is organized in a different way from the original data.</introduction>
   Hydrologic data for the Great Lakes obtained from the Great Lakes Environmental Research
     Laboratory through their Web site http://www.glerl.noaa.gov/data/. It contains physical
     data, as well as historic and forecasted hydrologic variables. In order to reduce the length
     of this file, only the data from Lake Superior and Lake Michigan are described. Each basin is
     assigned a code Lake Superior=1, Lake Michigan=2.
- <Basin-Characterization>
- <location>
  <max-lat>49.02</max-lat>
```

```
<max-lon>-84.35</max-lon>
<min-lat>46.07</min-lat>
<min-lon>-89.58</min-lon>
  </location>
<area area-type="Lake Area">82100</area>
<area area-type="Land Area">128000</area>
  </Basin-Characterization>
- <Physical-Data>
- <Gridded-Data basin="Lake Superior" hydro-data-type="Bathymetric Data">
  <data-information>Bathymetric data for Lake Superior on a 2-km grid as described in Robertson
     and Jordan (unpublished) was obtained from the Canada Center for Inland Waters. They
     superimposed 2-km grids on standard bathymetric charts and averaged a mean depth in
     each grid square by eye. Depths are relative to the Great Lakes Datum of 1955. All grids are
     aligned with the central meridian of the bathymetric chart. The datum is 182.88 m.</data-
     information>
  <units>Bathymetric Data</units>
- <grid-description>
  <base-lat>46.32</base-lat>
  <base-lon>-92.10</base-lon>
  <gridXdim>2km</gridXdim>
  <gridYdim>2km</gridYdim>
  <gridXnum>304</gridXnum>
  <gridYnum>147</gridYnum>
  </grid-description>
  <file-description>The depths on bathymetric data fill the batymetric grid from west to east and
      then south to north. Lines 3-4 describe the transformation values from geographic to map
      coordinates : x = a Dlon + b Dlat + c Dlon Dlat + d Dlon**2 and y = e Dlon + f Dlat + g Dlon
      Dlat + h Dlon**2 lines 5-6 describe the transformation from map to geographic coordinates
      : Dlon = Ax + By + Cxy + Dx**2 and Dlat = Ex + Fy + Gxy + Hx**2.</file-description>
  <filepath>ftp://ftp.glerl.noaa.gov/bathy/superior.dat</filepath>
    </Gridded-Data>
    </Physical-Data>
- <Hydrologic-Data>
- <Historical>
- <timeseries-data basin="Lake Superior" hydro-data-type="Lake Precipitation">
- <data-description>
  <data-information>We computed 1948-1999 monthly precipitation from all available daily data
     from stations in the basin or within approximately 0-30 km of the basin, depending upon
     station density near the edge of the basin. The distance was chosen to assure that we
     obtain the same nonzero Thiessen weights as if no stations were eliminated. Station data
     for the United States were obtained from the National Climatic Data Center (2000), and station data
     for Canada were obtained from the Meteorological Service of Canada (1999). We then used
     Thiessen weighting (Croley and Hartmann, 1985) similar to the above, but defined for a
     1-km grid and recomputed every day for all watersheds within the basin and the lake. The
     basin watersheds were defined by the U.S. Geological Survey and the Water Survey of
     Canada. We constructed daily over-land basin precipitation by areally weighting the
     areally averaged daily precipitation for each watershed and summing over the days in each
     month. We also constructed total monthly over-lake precipitation by summing the areally
     averaged daily values for the lake surface.</data-information>
  <units>mm</units>
  <temporal-scale>total monthly</temporal-scale>
  <start-date>1990-01-01</start-date>
  <end-date>1999-01-01</end-date>
  <duration>P10Y</duration>
  </data-description>
  <historical-data>44.8 32.3 37.9 56.1 59.9 114.8 66.7 44.9 99.6 108.0 45.3 46.4 55.2 23.2 60.7
      69.1 84.4 75.5 105.8 39.6 113.1 98.8 107.9 46.3 46.1 35.9 22.1 55.7 61.9 55.9 118.3 86.4
      115.3 57.9 67.3 80.9 53.7 14.8 18.2 72.6 104.3 78.1 108.1 68.3 77.3 73.0 51.9 38.3 58.5
      15.4 38.5 69.6 75.2 77.0 74.4 102.6 76.5 55.7 59.3 14.5 41.3 49.5 38.4 49.4 88.7 26.7
      113.1 86.9 115.9 137.6 67.4 74.3 91.4 54.1 36.7 70.2 45.6 107.8 116.1 80.8 86.3 97.1 69.7
      77.3 99.8 20.2 64.6 26.9 69.0 75.3 72.0 53.5 60.6 75.2 58.4 28.6 46.0 18.2 68.5 25.6 52.0
      92.0 56.6 64.9 78.0 81.6 82.6 62.1 75.9 51.7 22.3 39.1 137.0 77.4 129.1 84.4 96.8 90.4
      41.6 39.3</historical-data>
    </timeseries-data>
- <timeseries-data basin="Lake Superior" hydro-data-type="Land Precipitation">
- <data-description>
  <data-information>We computed 1948-1999 monthly precipitation from all available daily data
     from stations in the basin or within approximately 0--30 km of the basin, depending upon
     station density near the edge of the basin. The distance was chosen to assure that we
     obtain the same nonzero Thiessen weights as if no stations were eliminated. Station data
     for the United States were obtained from the National Climatic Data Center (2000), and station data
     for Canada were obtained from the Meteorological Service of Canada (1999). We then used
     Thiessen weighting (Croley and Hartmann, 1985) similar to the above, but defined for a
     1-km grid and recomputed every day for all watersheds within the basin and the lake. The
     basin watersheds were defined by the U.S. Geological Survey and the Water Survey of
     Canada. We constructed daily over-land basin precipitation by areally weighting the
```

areally averaged daily precipitation for each watershed and summing over the days in each
month. We also constructed total monthly over-lake precipitation by summing the areally
averaged daily values for the lake surface.</data-information>
<units>mm</units>
<temporal-scale>total monthly</temporal-scale>
<start-date>1990-01-01</start-date>
<end-date>1999-01-01</end-date>
<duration>P10Y</duration>
</data-description>
<historical-data>57.0 36.7 39.3 61.0 60.1 130.6 80.1 63.7 93.4 105.5 48.0 52.3 49.4 20.3 54.1
57.3 89.2 80.1 106.9 47.4 124.8 94.7 95.3 45.8 36.7 34.0 21.2 58.7 69.0 64.9 119.6 107.7
135.4 50.9 60.1 76.9 44.0 10.0 19.1 76.0 103.3 87.7 140.4 90.4 101.1 74.8 51.5 34.4 49.0
17.6 33.4 73.5 69.3 89.5 115.4 100.0 79.6 61.6 66.4 20.3 45.3 48.9 32.3 45.6 84.9 44.6
112.1 81.7 131.1 145.6 69.3 61.1 101.2 62.4 29.3 68.3 52.5 85.8 139.2 79.4 91.1 113.4
70.1 76.5 84.5 21.3 56.9 31.1 65.3 91.5 80.2 54.2 66.6 81.4 65.7 29.6 44.6 18.6 59.6 30.9
53.6 87 72.9 90.8 91.4 108.6 70.1 50.8 50.1 54.7 28.9 38.3 115.3 89.8 150.1 93.6 119.4
98.5 35.2 38.3</historical-data>
</timeseries-data>
- <timeseries-data basin="Lake Superior" hydro-data-type="Lake Levels">
- <data-description>
<data-information>Beginning of month lake levels. The beginning-of-month level for a gage is
defined as the level at 12:00:00 a.m. (midnight) on the first day of that month. Since this
value is generally unknown for practical reasons, it is standard practice to compute the
value by averaging the daily mean level for the last day of the previous month with the
daily mean level for the first day of the current month, when both are available. When one
or both of the daily mean values are unavailable, several alternative methods for filling in
the missing data are available.</data-information>
<units>m</units>
<temporal-scale>average annual</temporal-scale>
<start-date>1990-01-01</start-date>
<end-date>1999-01-01</end-date>
<duration>P10Y</duration>
</data-description>
<historical-data>183.25 183.34 183.39 183.46 183.45 183.36 183.54 183.60 183.33
183.26</historical-data>
</timeseries-data>
- <timeseries-data basin="Lake Superior" hydro-data-type="Evaporation">
- <data-description>
<data-information>Monthly evaporation estimates (spreadsheet entitled evaporation.xls) were
derived from daily evaporation estimates generated by the Great Lakes Evaporation Model
(Croley, 1989a,b, 1992; Croley and Assel, 1994). This is a lumped-parameter surface flux
and heat-storage model. It uses areal-average daily air temperature, windspeed, humidity,
precipitation, and cloudcover. These data are sufficiently available since 1948 (1953 for
Georgian Bay), and 2 years are used for model initialization. Over-land data are adjusted
for over-water or over-ice conditions. Surface flux processes are represented for short-
wave radiation and reflection, net long-wave radiation exchange, and advection.
Atmospheric stability effects on the bulk transfer coefficients are formulated and used with
the aerodynamic equation for sensible and latent heat surface fluxes.</data-information>
<units>mm</units>
<temporal-scale>total monthly</temporal-scale>
<start-date>1990-01-01</start-date>
<end-date>1999-01-01</end-date>
<duration>P10Y</duration>
</data-description>
<historical-data>74.09 73.43 50.38 26.31 2.75 -4.81 -0.98 16.63 53.63 74.76 90.51 124.96
110.21 56.74 50.02 13.5 0.49 -3.8 6.92 20.85 80.28 74.8 109.48 108 93.09 64.59 61.1
26.94 3.72 3.08 8.27 29.83 55.61 69.22 91.26 113.68 103 80.63 49.26 23.26 0.67 -3.87
-1.5 9.34 73.27 86.19 105.95 98.58 126.17 46.37 26.71 19.16 0.94 -4.71 -4.2 19.29 29.58
57.65 100.67 86.04 111.16 98.29 45.65 31.9 -0.07 -4.74 -0.78 12.72 59.96 65.07 127.01
124.19 131.82 47.29 37 22.14 5.17 -3.99 -5.22 -1.51 33.57 69.84 112.91 110.23 91.73
34.59 32.99 19.87 4.28 -3.84 -3.36 8.21 35.18 67.32 86.53 80.05 83.96 29.92 59.61 13.83
2.09 4.46 20.13 30.96 60.81 73.88 91.02 111.11 113.04 64.28 51.26 13.35 7.32 12.16
10.48 50.76 64.38 91.08 84.89 106.8</historical-data>
</timeseries-data>
</Historical>
- <Forecast>
- <forecasted-data basin="Lake Superior" hydro-data-type="Lake Precipitation" quantile="95">
- <data-description>
<data-information>Three NOAA agencies support estimates of present hydrology and long-term
hydrological probabilities for each of the Great Lake Basins:NOAAs Great Lakes
Environmental Research Laboratory (GLERL), NOAAs Climate Prediction Center (CPC) and
The Midwestern Climate Center (MCC).</data-information>
<units>mm</units>
<temporal-scale>monthly</temporal-scale>
<start-date>2003-08-01</start-date>
<end-date>2004-08-01</end-date>

```
<duration>P1Y1M</duration>
  </data-description>
<forecast-data>23.94 48.00 70.28 64.80 46.86 34.70 39.40 67.15 111.77 135.78 76.69 58.50
    49.47</forecast-data>
   </forecasted-data>
- <forecasted-data basin="Lake Superior" hydro-data-type="Evaporation" quantile="95">
- <data-description>
<data-information>Three NOAA agencies support estimates of present hydrology and long-term
   hydrological probabilities for each of the Great Lake Basins:NOAAs Great Lakes
   Environmental Research Laboratory (GLERL), NOAAs Climate Prediction Center (CPC) and
   The Midwestern Climate Center (MCC).</data-information>
<units>mm</units>
<temporal-scale>monthly</temporal-scale>
<start-date>2003-08-01</start-date>
<end-date>2004-08-01</end-date>
<duration>P1Y1M</duration>
   </data-description>
<forecast-data>6.56 56.52 83.73 113.47 129.41 122.06 80.94 63.40 25.90 7.93 0.12 5.33
    33.37</forecast-data>
   </forecasted-data>
- <forecasted-data basin="Lake Superior" hydro-data-type="Lake Precipitation" quantile="5">
- <data-description>
<data-information>Three NOAA agencies support estimates of present hydrology and long-term
   hydrological probabilities for each of the Great Lake Basins:NOAAs Great Lakes
   Environmental Research Laboratory (GLERL), NOAAs Climate Prediction Center (CPC) and
   The Midwestern Climate Center (MCC).</data-information>
<units>mm</units>
<temporal-scale>monthly</temporal-scale>
<start-date>2003-08-01</start-date>
<end-date>2004-08-01</end-date>
<duration>P1Y1M</duration>
   </data-description>
<forecast-data>41.65 39.14 29.01 33.33 27.52 27.84 17.49 19.14 19.83 26.39 45.63 48.83
    47.68</forecast-data>
   </forecasted-data>
- <forecasted-data basin="Lake Superior" hydro-data-type="Evaporation" quantile="5">
- <data-description>
<data-information>Three NOAA agencies support estimates of present hydrology and long-term
   hydrological probabilities for each of the Great Lake Basins:NOAAs Great Lakes
   Environmental Research Laboratory (GLERL), NOAAs Climate Prediction Center (CPC) and
   The Midwestern Climate Center (MCC).</data-information>
<units>mm</units>
<temporal-scale>monthly</temporal-scale>
<start-date>2003-08-01</start-date>
<end-date>2004-08-01</end-date>
<duration>P1Y1M</duration>
   </data-description>
<forecast-data>-1.63 25.45 37.6 67.47 80.85 73.53 21.52 14.36 3.17 -3.11 -6.74 -6.11
    1.01</forecast-data>
</forecasted-data>
</Forecast>
</Hydrologic-Data>
</Basin-Data>
```

Appendix 3

The UTM Northern Hemisphere Projection

Contents

The purpose of this appendix is to give a synopsis of a projection that is commonly used for current maps of regions within the continental United States.

A3.1 Transverse Mercator Projection

In the Transverse Mercator Projection, a cylinder passes through the Earth model, slicing through the ellipsoid and then emerging on the other side. Any line of longitude can be chosen to be in the center of the piece of the ellipsoid that sits on the outside of the cylinder. The ellipsoid surface is then projected onto the cylinder and the unfolded cylinder then becomes our map.

The Transverse Mercator Projection is shown graphically in Figure A3.1. Notice in figure that the ends of the cylinder are quite far from the ellipsoid surface, causing a significant amount of distortion. The most accurate (and useful) part of our map is the exposed piece in the center. Its distance from the cylinder is small so the distortion is minimized at these points. This "slicing" of our ellipsoid leads naturally to the concept of UTM zones, discussed in the next section of Appendix 1.3.2.

A3.2 UTM Zones

The Transverse Mercator projection can provide a highly accurate map for only a relatively small span of longitude at any one time. To accurately cover the entire planet with this projection, one could make a projection for a certain "slice" of the ellipsoid, rotate the ellipsoid, make another projection, rotate again, and so on. The only piece of each projection used would be the center slice, but after making many slices, one would get the entire Earth. To that end, the UTM Zone system was developed to standardize this slicing.

In the UTM Zone system, the Earth is split up into 60 zones, each six degrees of longitude wide. The center longitude of each zone serves as the projection center for a map of any

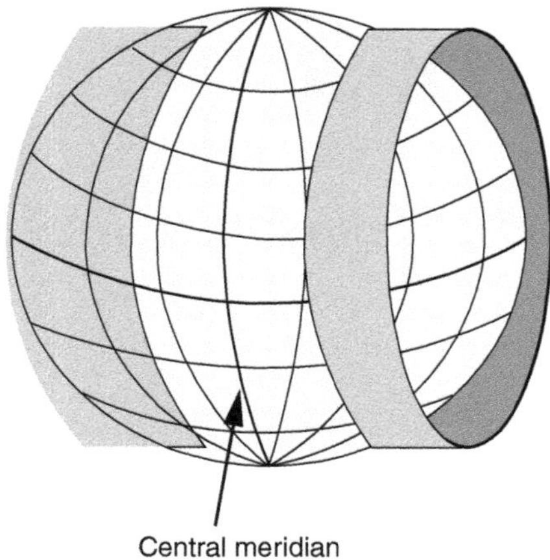

Figure A3.1
Transverse mercator projection.

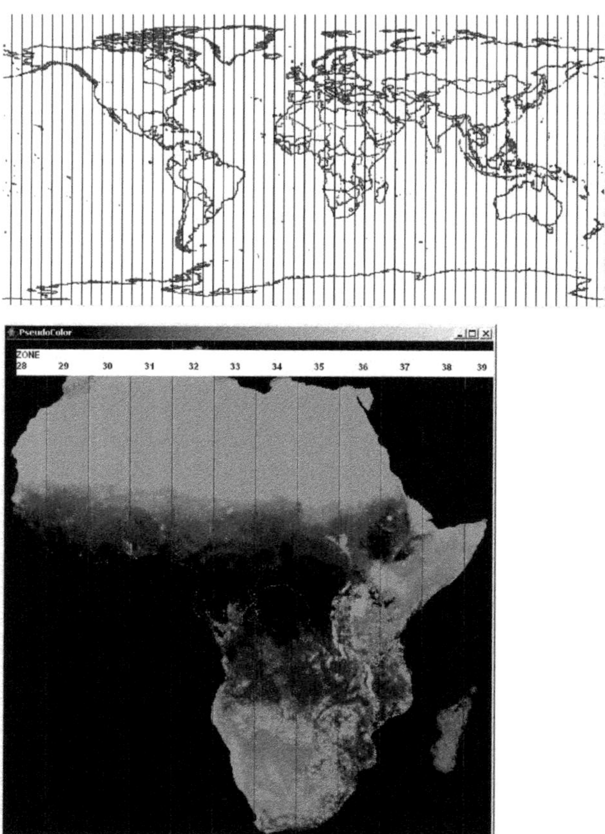

Figure A3.2
The UTM zones for the world (top) and for Africa (bottom).

region inside that zone. Any map using the UTM Northern Hemisphere projection would then only show a region within one particular zone. As demonstrated by Figure A3.2, it is often impossible to show an entire state and remain completely within a single UTM zone. When this occurs, the tie point defined in Chapter 18 determines which UTM zone will be used for the given image.

Appendix 4

Molodensky Equations

Contents

In this appendix we present both the *forward* and *reverse* Molodensky equations used for geographical transformations 1 and 2 described in Chapter 18. We make heavy use of Reference 1 in this appendix, adding additional comments for clarity. The forward Molodensky equations take a latitude/longitude value to its northing/easting pair, and the reverse equations take a northing/easting pair to its latitude/longitude value.

A4.1 Forward Equations

The forward Molodensky equations can be thought of as a function, F, which takes a latitude and longitude, along with all of the parameters of the UTM projection, and returns the northing and easting. F is specified as:

$$F\colon (\phi, \lambda, \phi_0, \lambda_0, k_0, a, e^2, FE, FN) \rightarrow (N, E).$$

The parameters for F are: ϕ — the latitude, λ — the longitude, ϕ_0 — the projection center latitude, λ_0 — the projection center longitude, k_0 — the scale factor of the projection, a — the semi-major axis of the ellipsoid, e^2 — the square of the eccentricity of the ellipsoid, FE — the false easting of the projection, FN — the false northing of the projection, N — the northing, E — the easting. The function F makes use of two equations that compute easting and northing values:

$$E = FE + k_0\nu\left[A + (1 - T + C)\frac{A^3}{6} + (5 - 18T + T^2 + 72C - 58e'^2)\frac{A^5}{120}\right],$$

$$N = FN + k_0\left\{M - M_0 + \nu \tan\varphi\left[\frac{A^2}{2} + (5 - T + 9C + 4C^2)\frac{A^4}{24}\right.\right.$$
$$\left.\left. + (61 - 58T + T^2 + 600C - 330e'^2)\frac{A^6}{720}\right]\right\}.$$

The equation parameters T, C, and A are specified as follows:

$$T = \tan^2\varphi,$$
$$C = \frac{e^2}{1 - e^2}\cos^2\varphi = e'^2\cos^2\varphi,$$
$$A = (\lambda - \lambda_0)\cos\varphi.$$

The parameter e'^2 is defined as:

$$e'^2 = e^2/[1 - e^2].$$

M is defined here, with M_0 calculated in the same manner substituting ϕ_0 for ϕ:

$$M = a\left[\left(1 - \frac{e^2}{4} - \frac{3e^4}{64} - \frac{5e^6}{256} - \cdots\right)\varphi - \left(\frac{3e^2}{8} + \frac{3e^4}{32} + \frac{45e^6}{1024} + \cdots\right)\sin 2\varphi\right.$$
$$\left. + \left(\frac{15e^4}{256} + \frac{45e^6}{1024} + \cdots\right)\sin 4\varphi - \left(\frac{35e^6}{3072} + \cdots\right)\sin 6\varphi + \cdots\right].$$

The ν term in the easting formula is defined by the ν_1 formula presented in the following section on the Reverse Equations, using ϕ instead of ϕ_1.

A4.2 Reverse Equations

The reverse Molodensky equations require the same projection parameters as the forward equations, but take northing/easting and return a latitude and longitude. They can be thought of as a function R specified as:

$$R\colon (N, E, \phi_0, \lambda_0, k_0, a, e^2, FE, FN) \rightarrow (\phi, \lambda).$$

The parameters in this R reverse function are the same as those described earlier for the F forward function.

The function R makes use of two equations that compute the latitude and longitude:

$$\varphi = \varphi_1 - \frac{\nu_1 \tan\varphi_1}{\rho_1}\left[\frac{D^2}{2} - (5 + 3T_1 + 10C_1 - 4C_1^2 - 9e'^2)\frac{D^4}{24}\right.$$
$$\left. + (61 + 90T_1 + 298C_1 + 45T_1^2 - 252e'^2 - 3C_1^2)\frac{D^6}{720}\right],$$

$$\lambda = \lambda_0 + \frac{D - (1 + 2T_1 + C_1)(D^3/6) + (5 - 2C_1 + 28T_1 - 3C_1^2 + 8e'^2 + 24T_1^2)(D^5/120)}{\cos\varphi_1}.$$

The equation parameters ν_1 and p_1 are defined as:

$$\nu_1 = \frac{a}{\sqrt{1 - e^2 \sin^2 \varphi_1}},$$

$$p_1 = \frac{a(1 - e^2)}{(1 - e^2 \sin^2 \varphi_1)3/2}.$$

The equation parameters D, T_1, and C_1 are defined as:

$$D = \frac{E - FE}{\nu_1 k_0},$$

$$T_1 = \tan^2 \varphi_1,$$

$$C_1 = e'^2 \cos^2 \varphi.$$

The parameter ϕ_1 is defined by the sum:

$$\varphi_1 = \mu_1 + (3e_1/2 - 27e_1^3/32 + \cdots) \sin^2 \mu_1 + (21e_1^2/16 - 55e_1^4/32 + \cdots) \sin^4 \mu_1 + (151e_1^3/96 + \cdots) \sin^6 \mu_1 + (1097e_1^4/512 - \cdots) \sin^8 \mu_1 + \cdots .$$

The parameters e_1 and μ_1 are defined as:

$$e_1 = \frac{1 - (1 - e^2)^{1/2}}{1 + (1 - e^2)^{1/2}},$$

$$\mu_1 = \frac{M_1}{a(1 - e^2/4 - 3e^4/64 - 5e^6/256 - \cdots)}.$$

Finally, M_1 is defined in terms of M_0 from above:

$$M_1 = M_0 + (N - FN)/k_0.$$

Reference

1. Molodensky equations reference http://www.posc.org/Epicentre.2_2/DataModel/ExamplesofUsage/eu_cs34h.html.

Appendix 5

Section IV Review Questions

Chapter 14:

- What would be the main components of future Hydroinformatics Systems based on current NSF projects?
- Why would one consider data storage, data communication and visualization as parts of data processing and analysis?
- Can you identify scenarios similar to the motivation example.

Chapter 15:

- How would one model a visible spectrum measurement (a pixel value) that came from a patch covered 50% with soil and 50% with concrete?
- List several ways how to classify data sources.

Chapter 16:

- What are the tradeoffs when representing vector data and raster data?
- What representation would you recommend for a user-driven real-time computation of spatial statistics over a set of boundaries?

Chapter 17:

- What are the differences between georegistration and registration?
- Describe registration steps.
- What is the major difference between linear and nonlinear registration transformation?
- How would you widen the peaks in a search space by data preprocessing?

Chapter 18:

- What is Earth's flattening?
- Is it possible to georeference two files in WGS84 and NAD27 datums (geographic coordinate systems? Describe how.
- Is it possible to reproject maps in different projections, for example, a map in equal area projection into a map in cylindrical projection?
- Can the private TIFF tags provide conflicting values with values defined in geokeys of the same TIFF file?

Chapter 19:

- What types of data heterogeneity one encounters in multiple raster files?
- What types of data heterogeneity one encounters in multiple vector files?
- How would you propose to evaluate reprojection spatial accuracy for categorical variables (raster data)?
- How would you fuse multiple vector files?

Chapter 20:

- Why is feature extraction important for databases?
- How would you extract arc features from raster data?
- What is the difference between segmentation and clustering of raster data?
- What features would you recommend for discriminating dashed curves from solid curves?
- How would you compute a boundary orientation based on the order of boundary points?

Chapter 21:

- Why would redundant features degrade prediction model accuracy? Give at least two reasons.
- Which has a higher entropy measure: A feature with four equally likely values or a feature with one very likely value, and three equally unlikely values?
- Why might the continuous prediction graphs have a more definite "U" shape than the categorical predictions?
- Can you name at least one example of hydrology related application that would need feature analysis and decision support?
- What does one have to decide about when preparing data for decision making?
- How would you compare aggregation results obtained from different boundary definitions, for example, zip codes and census tracks?
- How would you weigh the relevance of features for territorial aggregation?

Appendix 6

Section IV Project Assignment

Contents

A6.1 Project Assignment

Given input data, such as, Digital Elevation Map (Raster Data) stored in the USGS DEM file format (File Name: sampDEMCD1.hdr and sampDEMCD1) and accurate point elevation measurements (Tabular Data) stored in the XML file format (File Name: TabularInfo.xml), design and implement algorithms for updating raster elevation maps with accurate point elevation measurements. The assignment should be completed by saving out (1) updated digital elevation map based on point elevation measurements (Raster Data) in the HDF5 file format (File Name: rasterDEMCD1_result.hdf), and (2) updated DEM points (Tabular Data) in the XML file format (File Name: pointDEMCD1_result.xml), as well as, by summarizing the data processing steps and challenges. The input datasets are shown in Figure A6.1.

A6.2 Prerequisites and Recommended Approach

The following software was provided for the assignment. A sample C/C++ code that loads a DEM file, loads tab-delimited file with accurate elevation measurements and converts points from latitude, longitude coordinate system to row, column coordinate system. The C/C++ code is using Java native interface (JNI) and java based I2K library with DEM loader, tab-delimited loader, and GIS coordinate conversion functions. Furthermore, a Makefile for Windows (Visual C++ .NET compiler) and linux (gnu C++ compiler) was

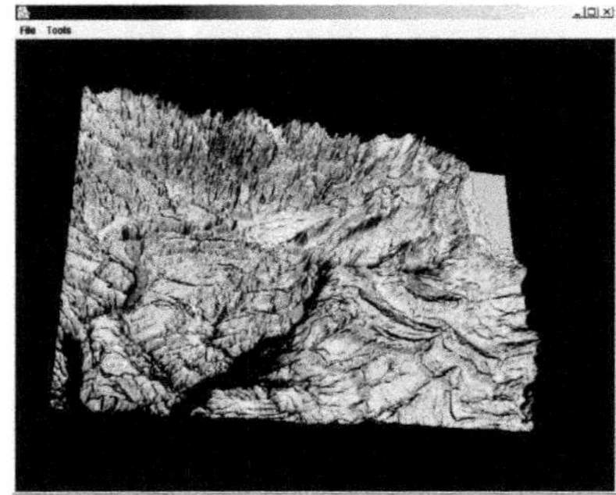

1	lat	lng	value
2	41.32110	-89.5531	194.100
3	41.07110	-89.4697	197.830
4	40.94611	-89.4697	138.479
5	40.94611	-91.2197	217.650
6	42.42111	-91.2197	279.200
7	41.07111	-91.2197	228.070
8	41.07111	-90.9697	161.900

FIGURE A6.1
Input raster data (left) and tabular data (right) with latitude, longitude, and accurate elevation values.

provided, as well as, application programming interface (API) for visualization and least square fit functions in I2K library.

The following data processing steps were recommended. First, load DEM file (use JNI + I2K Library). Second, load XML file (write C/C++ code or use XML library). Third, visualize DEM data (use JNI + I2K Library). Fourth, convert point information (Lat/Long) to DEM coordinates (row/col) (use JNI+I2K Library). Fifth, extract elevation values of point neighbors (write C/C++ Code). Sixth, compute new DEM values (write C/C++ Code). Seventh, compute local spatial error after updating DEM data (write C/C++ Code). Eight, update DEM raster data and save updated DEM file in HDF file format (use HDF library). Ninth, save only updated DEM pixel values in XML file format (write C/C++ code or use XML library).

As for the sixth step (compute new DEM values), it was recommended to implement the least square fit and one more method of choice. Furthermore, propose one more approach that would be more accurate than the least square fit but it would take a large effort to implement the method.

A6.3 Example Solution

The DEM is obtained from the United States Geological Survey (USGS) and covers northern Illinois, along with small portions of Iowa, Wisconsin, Indiana, and Lake Michigan. The georeferenced accurate point elevation measurements are shown in Figure A6.2 and the point conversions are provided in Table A6.1.

A6.4 Find Nearest Neighbors

The four nearest neighbors to the point measurements are determined based on the coordinates of each point [row, column] as follows: neighbor 1 = [row, column]; neighbor 2 = [row, column-1]; neighbor 3 = [row-1, column]; and neighbor 4 = [row-1, column-1].

Figure A6.3 illustrates the four nearest neighbors to a point measurement. Table A6.2 is a summary of the data associated with each point measurement and its four nearest neighbors.

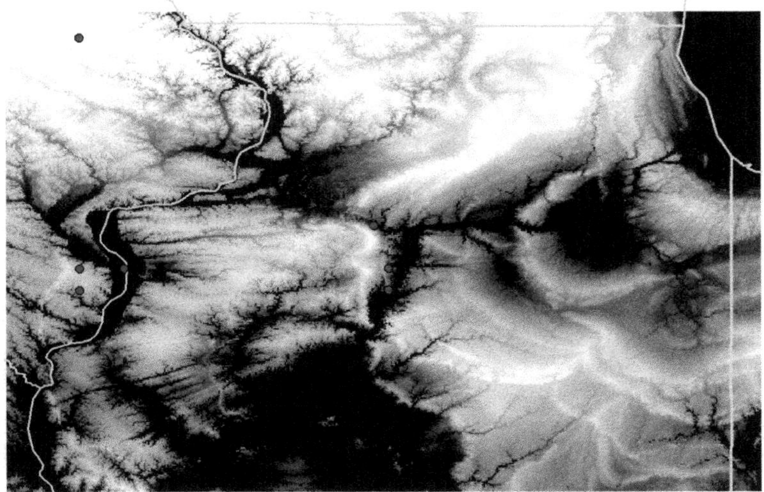

FIGURE A6.2
USGS DEM overlaid with the accurate point elevation measurements and the state boundaries.

TABLE A6.1
Latitude, longitude, elevation, row, and column of the accurate point measurements. The row/column values are the output after converting from latitude/longitude.

Latitude	Longitude	Elevation	Row	Column
41.3211	−89.5531	194.100	150.001	249.995
41.0711	−89.4697	197.830	180.001	260.003
40.9461	−89.4697	138.479	195.000	260.003
40.9461	−91.2197	217.650	195.000	50.003
42.4211	−91.2197	279.200	18.000	50.003
41.0711	−91.2197	228.070	180.000	50.003
41.0711	−90.9697	161.900	180.000	80.003

A6.5 Compute New Elevations

Two methods are implemented to determine the new elevations of the four nearest neighbors to an accurate point measurement: least squares and nearest neighbors. Both are methods used to determine missing or inaccurate values in a grid of continuous data.

Another method that could be used to determine the new elevations of the four nearest neighbors is punctual kriging. In order of accuracy, punctual kriging would be the most accurate, followed by least squares, and finally nearest neighbors. Punctual kriging uses a semivariogram of the known values to determine the unknown or inaccurate values. The semivariogram is used to determine weights, which estimate the unknown value as follows:

$$Y_p = \sum_{i=1}^{N} W_i Y_i, \tag{A6.1}$$

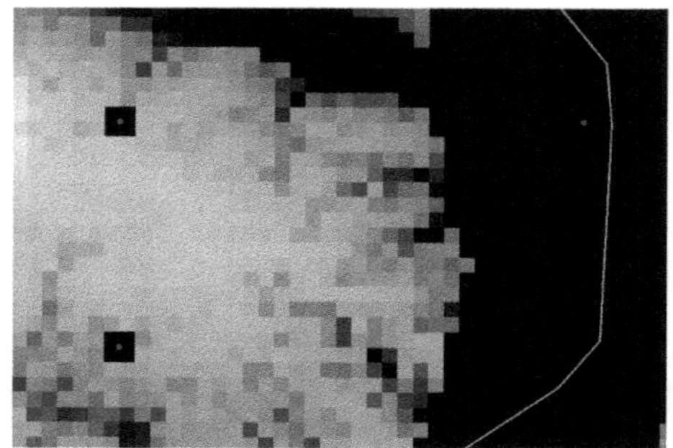

FIGURE A6.3
Illustration of four nearest neighbors to a point. Red dots are point measurements along the Illinois–Iowa border and the blackened pixels behind them are the nearest neighbors.

where: Y_p is the unknown value at point p, W_i is the weight at point i, Y_i is the known value at point i, and N is the number of known points.

The least squares method is arranged for each point measurement using an over-determined system of five equations and three unknowns $Ax = b$ where

$$A = \begin{bmatrix} x_1 & y_1 & z_1 \\ x_2 & y_2 & z_2 \\ x_3 & y_3 & z_3 \\ x_4 & y_4 & z_4 \\ x_5 & y_5 & z_5 \end{bmatrix} \qquad x = \begin{bmatrix} a \\ b \\ c \end{bmatrix} \qquad b = \begin{bmatrix} -z_5 \\ -z_5 \\ -z_5 \\ -z_5 \\ -z_5 \end{bmatrix}. \tag{A6.2}$$

x_i, y_i, and z_i represent the row, column, and elevation of neighbor i, respectively. The fifth index ($i = 5$) signifies the row, column, and elevation associated with the accurate point measurement. The form of each equation is: $ax_i + by_i + cz_i + d = 0$, where the value for d is set to the elevation of the accurate point measurement such that the least squares fit goes through that elevation, as it is the most accurate of the five points. The x vector contains the coefficients a, b, and c of the five equations, and the new elevation for each of the four nearest neighbors is computed as:

$$z_{i_{\text{new}}} = \frac{-(ax_i + by_i + d)}{c}. \tag{A6.3}$$

In the nearest neighbors method, the point elevation is assigned to each of the four nearest neighbors. This is a simple way to adjust continuous raster data for missing or inaccurate values, however it is still used with much success. Table A6.3 presents the results of both the least squares and nearest neighbors methods and Table A6.4 presents the coefficients a, b, c, and d used to compute the new values for the least squares method.

A6.6 Compute Error

The local error spatial error, or residual, is computed for each of the four nearest neighbors to each point measurement as follows:

$$\text{Residual} = z_{i_{\text{new}}} - z_{i_{\text{old}}}, \tag{A6.4}$$

Table A6.2

The latitude, longitude, elevation, row, and column associated with each accurate point measurement and its four nearest neighbors. The numbers in bold represent the point measurement data and the following four entries represent the data associated with the point.

Latitude	Longitude	Elevation	Row	Column
Point 1				
41.3211	**−89.5531**	**194.100**	**150**	**250**
41.3253	−89.5572	198.580	149.5	249.5
41.3253	−89.5489	194.438	149.5	250.5
41.3169	−89.5572	191.704	150.5	249.5
41.3169	−89.5489	187.562	150.5	250.5
Point 2				
41.0711	**−89.4697**	**197.830**	**180**	**260**
41.0753	−89.4739	204.033	179.5	259.5
41.0753	−89.4656	197.452	179.5	260.5
41.0669	−89.4739	203.430	180.5	259.5
41.0669	−89.4656	196.850	180.5	260.5
Point 3				
40.9461	**−89.4697**	**138.479**	**195**	**260**
40.9503	−89.4739	145.500	194.5	259.5
40.9503	−89.4656	139.884	194.5	260.5
40.9419	−89.4739	140.937	195.5	259.5
40.9419	−89.4656	135.322	195.5	260.5
Point 4				
40.9461	**−91.2197**	**217.650**	**195**	**50**
40.9503	−91.2239	246.657	194.5	49.5
40.9503	−91.2156	238.821	194.5	50.5
40.9419	−91.2239	194.869	195.5	49.5
40.9419	−91.2156	187.033	195.5	50.5
Point 5				
42.4211	**−91.2197**	**279.200**	**18**	**50**
42.4253	−91.2239	287.181	17.5	49.5
42.4253	−91.2156	262.896	17.5	50.5
42.4169	−91.2239	301.099	18.5	49.5
42.4169	−91.2156	276.814	18.5	50.5
Point 6				
41.0711	**−91.2197**	**228.070**	**180**	**50**
41.0753	−91.2239	222.536	179.5	49.5
41.0753	−91.2156	216.183	179.5	50.5
41.0669	−91.2239	235.763	180.5	49.5
41.0669	−91.2156	229.410	180.5	50.5
Point 7				
41.0711	**−90.9697**	**161.900**	**180**	**80**
41.0753	−90.9739	161.792	179.5	79.5
41.0753	−90.9656	161.677	179.5	80.5
41.0669	−90.9739	160.577	180.5	79.5
41.0669	−90.9656	160.463	180.5	80.5

Table A6.3

The original (old) elevation, new elevation, and residual error of the four nearest neighbors to each point for both the least squares and nearest neighbors methods.

Least squares			Nearest neighbors		
Old elevation	**New elevation**	**Residual**	**Old elevation**	**New elevation**	**Residual**
Point 1					
196.697	198.580	1.88295	196.697	194.100	−2.59681
193.261	194.438	1.17694	193.261	194.100	0.83891
188.824	191.704	2.88022	188.824	194.100	5.27615
192.469	187.562	−4.90712	192.469	194.100	1.63053
Point 2					
203.516	204.033	0.51612	203.516	197.830	−5.68640
198.174	197.452	−0.72194	198.174	197.830	−0.34424
203.667	203.430	−0.23663	203.667	197.830	−5.83676
198.994	196.850	−2.14440	198.994	197.830	−1.16431
Point 3					
145.919	145.500	−0.41922	145.919	138.479	−7.44001
139.919	139.884	−0.03459	139.919	138.479	−1.44001
140.799	140.937	0.13782	140.799	138.479	−2.32006
136.919	135.322	−1.59750	136.919	138.479	1.55999
Point 4					
220.559	246.657	26.09800	220.559	217.650	−2.90855
214.945	238.821	23.87620	214.945	217.650	2.70527
207.751	194.869	−12.88240	207.751	217.650	9.89893
223.038	187.033	−36.00540	223.038	217.650	−5.38838
Point 5					
276.207	287.181	10.97400	276.207	279.200	2.99251
271.938	262.896	−9.04256	271.938	279.200	7.26168
304.430	301.099	−3.33041	304.430	279.200	−25.2297
277.971	276.814	−1.15730	277.971	279.200	1.22914
Point 6					
225.024	222.536	−2.48818	225.024	228.070	3.04603
214.747	216.183	1.43635	214.747	228.070	13.3233
232.814	235.763	2.94897	232.814	228.070	−4.74395
229.207	229.410	0.20338	229.207	228.070	−1.13677
Point 7					
160.726	161.792	1.06585	160.726	161.900	1.17428
161.475	161.677	0.20197	161.475	161.900	0.42504
160.865	160.577	−0.28820	160.865	161.900	1.03464
160.664	160.463	−0.20132	160.664	161.900	1.23617

where $z_{i_{\text{new}}}$ is the new elevation and $z_{i_{\text{old}}}$ is the original (old) elevation. Table A6.3 contains the residuals associated with both the least squares and nearest neighbors methods.

A6.7 Summary

This project provides an opportunity to exercise different aspects of data processing and analysis, as well as make use of the previously attained abilities in XML and HDF data

Table A6.4

Coefficients $a, b, c,$ and d for the least squares method.

a	*b*	*c*	*d*
Point 1			
−0.590556	−0.355735	−0.085890	194.100
Point 2			
−0.059004	−0.644524	−0.097949	197.830
Point 3			
−0.253744	−0.312272	−0.055610	138.479
Point 4			
−1.052710	−0.159278	−0.020327	217.650
Point 5			
3.119250	−5.442910	−0.224120	279.200
Point 6			
−1.641950	0.788601	0.124135	228.070
Point 7			
−0.505572	−0.047728	−0.416310	161.900

format processing. Two different data sources and representations are used: surveying information represented as vector point data and remotely sensed elevation values represented as raster data. Both datasets are georeferenced, or referenced to coordinates on the earth (latitude/longitude). Each dataset is then converted into row/column coordinates in order to overlay/fuse them and extract the nearest pixels in the raster dataset to the points in the vector dataset. Different schemes for filling in missing data or replacing erroneous data are used to create a more locally accurate digital elevation model. Least squares and nearest neighbors schemes were used and, according to the errors associated with each, the least squares is the most accurate overall, although the nearest neighbors method was a close second. Punctual kriging would presumably be the most accurate, although more difficult to implement. Thus, the method selection would become a battle of accuracy vs. time. Finally, the results of the project were exported into two data formats, XML and HDF, which exercised the skills of students in handling these popular data formats.

Index

D

E

F

G

H

I

Milton Keynes UK
Ingram Content Group UK Ltd.
UKHW052026071024
449327UK00027B/2450